The Blood C

FactsBook

THE BLOOD GROUP ANTIGEN
FactsBook

Third Edition

Marion E. Reid, PhD, FIBMS, DSc (Hon.)
Head, Laboratory of Immunochemistry
New York Blood Center, New York

Christine Lomas-Francis, MSc, FIBMS
Technical Director, Laboratory of Immunohematology and Genomics
New York Blood Center, New York

Martin L. Olsson, MD, PhD
Deputy Head, Department of Laboratory Medicine
Lund University, Lund, Sweden

AMSTERDAM • BOSTON • HEIDELBERG • LONDON
NEW YORK • OXFORD • PARIS • SAN DIEGO
SAN FRANCISCO • SINGAPORE • SYDNEY • TOKYO

Academic Press is an imprint of Elsevier

ELSEVIER

Academic Press is an imprint of Elsevier
32 Jamestown Road, London NW1 7BY, UK
225 Wyman Street, Waltham, MA 02451, USA
525 B Street, Suite 1800, San Diego, CA 92101-4495, USA

First edition 1997
Second edition 2004
Third Edition 2012

Notice
No responsibility is assumed by the publisher for any injury and/or damage to persons
or property as a matter of products liability, negligence or otherwise, or from any use
or operation of any methods, products, instructions or ideas contained in the material
herein.

Because of rapid advances in the medical sciences, in particular, independent verifica-
tion of diagnoses and drug dosages should be made

British Library Cataloguing-in-Publication Data
A catalogue record for this book is available from the British Library

Library of Congress Cataloging-in-Publication Data
A catalog record for this book is available from the Library of Congress

ISBN: 978-0-12-415849-8

For information on all Academic Press publications
visit our website at elsevierdirect.com

Typeset by MPS Limited, Chennai, India
www.adi-mps.com

Contents

Section II
The Blood Group Systems and Antigens

Preface

We are particularly indebted to Christine Halter Hipsky, Kim Hue-Roye, and Jill Storry for information gathering and proofreading. We also thank Patricia Arndt, Olga Blumenfeld, Gail Coghlan, Laura Cooling, George Garratty, Gregory Halverson, Åsa Hellberg, Annika Hult, Peter Issitt, Cyril Levene, Joann Moulds, John Moulds, Marilyn Moulds, Joyce Poole, Axel Seltsam, Maryse St. Louis, Jannica Samuelsson Johannesson, Britt Thuresson, Connie Westhoff, Julia Westman, Lung-Chih Yu and Teresa Zelinski for providing information and/or proofreading and helpful discussion. We thank Mary Preap and Denise Penrose from Elsevier for their help in shepherding us through the multiple phases of publishing the book, also Caroline Johnson (Production Editor) and Diane Chandler (Copyeditor). Last but not least, we are indebted to Robert Ratner for his skill in various aspects of preparing the manuscript and figures.

In compiling the entries for this text we were again impressed by the rapid pace with which new information became available. We have done our best to be consistent, accurate, and up-to-date (as of going to press). We encourage comments from readers on any errors, omissions, and improvements. Please email the authors:

mereid2010@gmail.com

clomas-francis@nybloodcenter.org

Martin_L.Olsson@med.lu.se

AChE	Acetylcholinesterase
AET	2-aminoethylisothiouronium bromide
AIDS	Acquired immune deficiency syndrome
AIHA	Autoimmune hemolytic anemia
ART	ADP-ribosyltransferase
BCAM	Basal cell adhesion molecule
BGM	Blood group modified
bp	Base pair
CAT	Column agglutination technology
CCP	Complement control protein
CD	Cluster of differentiation
CDA	Congenital dyserythropoietic anemia
CDG	Congenital disorder of glycosylation
cDNA	complementary DNA
Cer	Ceramide
CHAD	Cold hemagglutinin disease
CHIP	Channel-forming integral protein
CHO	Carbohydrate moiety
CGD	Chronic granulomatous disease
COOH	Carboxyl terminus
CR1	Complement receptor 1
CSF	Cerebrospinal fluid
CTH	Ceramide trihexoside
DAF	Decay accelerating factor
DAT	Direct antiglobulin test
DNA	Deoxyribonucleic acid
DTT	Dithiothreitol
ep	epitope
ER-Golgi	Endoplasmic reticulum-Golgi apparatus
Fuc	L-fucose
Gal	D-galactose
GalNAc	N-acetyl-D-galactosamine
Glc	Glucose
GlcNAc	N-acetyl-D-glucosamine
GP	Glycophorin
GPI	Glycosylphosphatidylinositol
GYP	Glycophorin gene
HA	Hemolytic anemia
HDFN	Hemolytic disease of the fetus and newborn
HLA	Human leukocyte antigen
HEMPAS	Hereditary erythoblastic multinuclearity with positive acidified serum test
HUS	Hemolytic uremic syndrome
IAT	Indirect antiglobulin test

Ig	Immunoglobulin
ISBT	International Society of Blood Transfusion
ITP	Immune thrombocytopenia
kbp	Kilo base pair
LAD	Leukocyte adhesion deficiency
LISS	Low-ionic strength solution
MAb	Monoclonal antibody
MAIEA	Monoclonal antibody immobilization of erythrocyte antigens
2-ME	2-Mercaptoethanol
MDS	Myelodysplastic syndromes
M_r	Apparent relative molecular mass
NeuNAc	N-acetyl neuraminic acid
NH_2	Amino terminus
NSAID	Nonsteroidal anti-inflammatory drug
NT/nt	Nucleotide
ORF	Open reading frame
PCH	Paroxysmal cold hemoglobinuria
PEG	Polyethylene glycol
PNH	Paroxysmal nocturnal hemoglobinuria
R	Remainder of carbohydrate chain
RBC	Red blood cell
RT	Room temperature
SCD	Sickle cell disease
SCR	Short consensus repeat
SDS-PAGE	Sodium dodecyl sulfate-polyacrylamide gel electrophoresis
Se/se	Secretor/non-secretor
SGP	Sialoglycoprotein
SLE	Systemic lupus erythematosus
SNP	Single nucleotide polymorphism
SSEA	Stage-specific embryonic antigen

Useful Websites

The easiest and quickest way is to access via a search engine.
dbRBC (for information about all blood group systems)
nybloodcenter.org (for information about *RHCE* alleles)
Rhesus Base (for information about *RHD* alleles)
www.genenames.org/ (for HUGO gene names)
www.isbt-web.org (for information about terminology and all blood group systems)
www.emina.med.uni-muenchen.de or http://www.euro-hd.net/html/na/registry
(for information about McLeod syndrome).

Section I

The Introductory Chapters

Introduction

Aims of this FactsBook

The purpose of this FactsBook is to provide key information relating to the erythrocyte membrane components carrying blood group antigens, the genes encoding them, the molecular basis of the antigens and phenotypes, characteristics, and the clinical significance of blood group antibodies. Only key references are given to allow the interested reader to obtain more details. The book is designed to be a convenient, easy-to-use reference for those involved in the field of transfusion medicine, as well as medical technologists, students, physicians, and researchers interested in erythrocyte blood group antigens.

This FactsBook contains information about the blood group antigens that have been numbered by the International Society of Blood Transfusion Working Party on Red Cell Immunogenetics and Blood Group Terminology[1–4]. The blood group systems and the antigens within each system are listed by their traditional name, and are arranged in the same order as described by the ISBT Working Party for Red Cell Immunogenetics and Blood Group Terminology. See Table 1.1 for an overview of the blood group systems. Those antigens not in a blood group system are accommodated in Collections (200 series), in the 700 Series of Low-Incidence Antigens or in the 901 Series of High-Incidence Antigens (for latest ISBT terminology, see www.isbt-web.org).

Selection of entries

Blood group antigens are surface markers on the outside of the red blood cell (RBC) membrane. They are proteins and carbohydrates attached to lipid or protein. A model for the types of membrane components carrying blood group antigens is shown in Figure 1.1. A blood group antigen is defined serologically by antibodies made by a human, and in order to be assigned a number by the ISBT Working Party the antigen must be shown to be inherited. Historically, antigens associated with forms of polyagglutination have not been numbered by the ISBT; however, in Section III we have included a table summarizing the characteristics of T, Tn, Tk, and Cad.

The Blood Group Antigen (3/e). DOI: http://dx.doi.oxg/10.1016/B978-0-12-415849-8.00001-6

TABLE 1.1 Blood group systems with gene name and chromosome location

System name	ISBT symbol^	ISBT number	Number of antigens	Gene names	Chromosome location	CD number
ABO	ABO	001	4	*ABO*	9q34.2	
MNS	MNS	002	46	*GYPA, GYPB, GYPE*	4q31.21	CD235
P1PK	P1PK	003	3	*A4GALT*	22q13.2	CD77 (P^k)
Rh	RH	004	52	*RHD, RHCE*	1p36.11	CD240
Lutheran	LU	005	20	*LU, BCAM*	19q13.32	CD239
Kell	KEL	006	34	*KEL*	7q34	CD238
Lewis	LE	007	6	*FUT3*	19p13.3	
Duffy	FY	008	5	*FY, DARC*	1q23.2	CD234
Kidd	JK	009	3	*JK, SLC14A1*	18q12.3	
Diego	DI	010	22	*DI, SLC4A1*	17q21.31	CD233
Yt	YT	011	2	*YT, ACHE*	7q22.1	
Xg	XG	012	2	*XG*	Xp22.33	CD99
Scianna	SC	013	7	*SC, ERMAP*	1p34.2	
Dombrock	DO	014	8	*DO, ART4*	12p12.3	CD297
Colton	CO	015	4	*CO, AQP1*	7p14.3	
Landsteiner-Wiener	LW	016	3	*LW, ICAM4*	19p13.2	CD242
Chido/Rodgers	CH/RG	017	9	*CH/RG, C4A, C4B*	6p21.32	
H	H	018	1	*FUT1*	19q13.33	CD173

System name	ISBT symbol^	ISBT number	Number of antigens	Gene names	Chromosome location	CD number
Kx	XK	019	1	XK	Xp21.1	
Gerbich	GE	020	11	GE, GYPC	2q14.3	CD236
Cromer	CROM	021	17	CROM, CD55	1q32.2	CD55
Knops	KN	022	9	KN, CR1	1q32.2	CD35
Indian	IN	023	4	IN, CD44	11p13	CD44
Ok	OK	024	3	OK, BSG	19p13.3	CD147
Raph	RAPH	025	1	RAPH, CD151	11p15.5	CD151
John Milton Hagen	JMH	026	6	JMH, SEMA7A	15q24.1	CD108
I	I	027	1	GCNT2	6p24.2	
Globoside	GLOB	028	1	B3GALNT1	3q26.1	
Gill	GIL	029	1	GIL, AQP3	9p13.3	
Rh-associated glycoprotein	RHAG	030	4	RHAG	6p12.3	CD241
Forssman	FORS	031	1	GBGT1	9q34.2	
JR	JR	032	1	JR, ABCG2	4q22.1	CDw338
LAN	LAN	033	1	LAN, ABCB6	2q36	

^For up to date HUGO gene names see http://www.genenames.org/

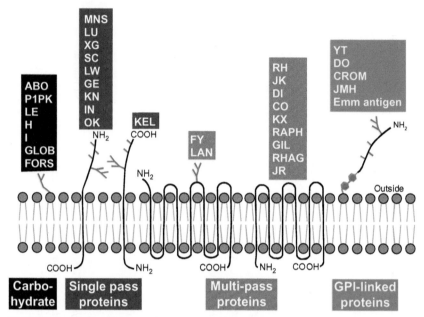

FIGURE 1.1 Model of RBC membrane components that carry blood group antigens. Not shown is the Ch/Rg blood group system. Ch/Rg antigens are carried on C4d, which is adsorbed from plasma onto RBC membrane components.

Terminology

The nomenclature used for erythrocyte blood group antigens is inconsistent. While several antigens were named after the proband whose RBCs carried the antigen or who made the first known antibody, others were assigned an alphabetical or a numerical notation. Even within the same blood group system, antigens have been named using different schemes, and this has resulted in a cumbersome terminology for describing phenotypes. The ISBT Working Party established a system of upper case letters and numbers to represent blood group systems and blood group antigens in a format that will allow both infinite expansion and computer-based storage. These symbols and numbers are designed for use in computer databases (no lower case letters) and are short (for column headings). A comprehensive review of terminology and its recommended usage can be found in Garratty, et al.[5].

Throughout this book, the systems and antigens are named by the traditional name, but we also give the ISBT symbol, the ISBT number, and obsolete names that have been used in the literature. We have included a brief description of how the blood group systems and antigens were named.

The following are examples of how to write antigens, antibodies, phenotypes, and genotypes.

List of antigens: M, N, P1, K, Kp^b, K11, Fy^a, Fy^b, Lu3
List of antibodies: Anti-M, anti-K, anti-Fy^a, anti-Jk3 or Anti-M, -K,
                    -Fy^a, -Jk3 or antibodies directed against M, K, Fy^a,
                    and Jk3 antigens
Phenotype: D+C–E–c+e+; M+N–S–s+,Vw+; K+k–K11–;
                    Fy(a+b+); Jk(a+b–) or RH:1,–2,–3,4,5;
                    MNS:1,–2,–3,4,9; KEL:1,–2,–11; FY:1,2; JK:1,–2

	Traditional	ISBT
Antigen	Fy^a	FY1, 008001 or 8.1
Phenotype	Fy(a+b–)	FY:1,–2
Gene	Fy^a	FY*01 or FY*A
	FY	FY*N, FY*01N or FY*02N
Genotype	Fy^{a}Fy^a	FY*01/FY*01 or FY*A/FY*A
	Fy^{a}Fy	FY*01/FY*N or FY*A/FY*N

In addition to the ISBT terminology for antigens and traditional allele names, the ISBT Working Party recently agreed on a proposed terminology for blood group alleles[4]. Since the introduction of a new naming system for thousands of alleles is a complicated and cumbersome procedure, it is anticipated that certain changes to the new terminology are unavoidable. Despite this, we have used the currently agreed nomenclature to familiarize the reader with the new allele names. However, we encourage the use of the official and constantly updated lists of allele names found at *www.isbt-web.org*. These allele names are restricted to those with a phenotypic effect and intended for use in Transfusion Medicine. Their use does not require sequencing of the entire allele. The rules for naming alleles and for obtaining a number for a new allele may be found at *www.isbt-web.org*.

References

1 Daniels, G.L., et al., 1995. Blood group terminology 1995. ISBT working party on terminology for red cell surface antigens. Vox Sang 69, 265–279.
2 Daniels, G.L., et al., 2004. Blood group terminology 2004. Vox Sang 87, 316.
3 Daniels, G., et al., 2007. International Society of Blood Transfusion committee on terminology for red cell surface antigens: Cape Town report. Vox Sang 92, 250–253.
4 Storry, J.R., et al., 2011. International Society of Blood Transfusion working party on red cell immunogenetics and blood group terminology: Berlin report. Vox Sang 101, 77–82.
5 Garratty, G., et al., 2000. Terminology for blood group antigens and genes: Historical origins and guidelines in the new millennium. Transfusion 40, 477–489.

Organization of the data

Section II is organized into four main parts: (i) the blood group systems; (ii) the blood group collections; (iii) the 700 Series of Low-Incidence Antigens; and (iv) the 901 Series of High-Incidence Antigens. For the systems, facts pertaining to the gene, carrier molecule, and antigens, and the clinical relevance of antibodies are given. For the collections, series of low-incidence antigens, and series of high-incidence antigens, facts pertaining to the antigens and antibodies are given. All are listed in ISBT numerical order. The format for the facts sheets displaying the data about the systems and the antigens is explained below; that for the collections and high-incidence antigens is similar. The 700 Series of Low-Incidence Antigens are given in a table. Section III consists of facts in lists and tables where the information encompasses more than one antigen or blood group system. Other related information is also included. We will use "prevalence" throughout this book, because it is the word used to describe a permanent/inherited characteristic on the phenotypic level.

ISBT Blood Group Systems

Both the ISBT system symbol and traditional name are used in the section headings. Information is provided under the following headings.

Number of Antigens

The total number of antigens in the system is indicated. The antigens are listed under the headings: "polymorphic" (in blue font); high prevalence (in red font); and low prevalence (in green font). Table 2.1A lists the blood group systems and antigens currently recognized by the ISBT Working Party for Red Cell Immunogenetics and Blood Group Terminology. If a number assigned to an antigen becomes inappropriate, the number becomes obsolete and is not reused; such antigens are noted with three small dots (...).

Terminology

Both the traditional name and the ISBT symbol for the blood group system are given as headers. Under this subheading, the ISBT symbol, ISBT number (parenthetically)[1-4], other, obsolete, names that have been associated with the

The Blood Group Antigen (3/e). DOI: http://dx.doi.oxg/10.1016/B978-0-12-415849-8.00002-8

TABLE 2.1A Blood group antigens assigned to each system

System		Antigen number																
	001	002	003	004	005	006	007	008	009	010	011	012	013	014	015	016	017	018
001 ABO	A	B	A,B	A1	...													
002 MNS	M	N	S	s	U	He	Mi^a	M^c	Vw	Mur	M^g	Vr	M^e	Mt^a	St^a	Ri^a	Cl^a	Ny^a
003 P1PK	P1	...	P^k	NOR														
004 RH	D	C	E	c	e	f	Ce	C^w	C^x	V	E^w	G					Hr_0	Hr
005 LU	Lu^a	Lu^b	Lu3	Lu4	Lu5	Lu6	Lu7	Lu8	Lu9	...	Lu11	Lu12	Lu13	Lu14	...	Lu16	Lu17	Au^a
006 KEL	K	k	Kp^a	Kp^b	Ku	Js^a	Js^b	...	...	UJ^a	K11	K12	K13	K14	...	K16	K17	K18
007 LE	Le^a	Le^b	Le^{ab}	Le^{bH}	ALe^b	BLe^b												
008 FY	Fy^a	Fy^b	Fy3	...	Fy5	Fy6												
009 JK	Jk^a	Jk^b	Jk3															
010 DI	Di^a	Di^b	Wr^a	Wr^b	Wd^a	Rb^a	WARR	ELO	Wu	Bp^a	Mo^a	Hg^a	Vg^a	Sw^a	BOW	NFLD	Jn^a	KREP
011 YT	Yt^a	Yt^b																
012 XG	Xg^a	CD99																
013 SC	Sc1	Sc2	Sc3	Rd	STAR	SCER	SCAN											
014 DO	Do^a	Do^b	Gy^a	Hy	Jo^a	DOYA	DOMR	DOLG										
015 CO	Co^a	Co^b	Co3	Co4														
016 LW	...	...	...	...	LW^a	LW^{ab}	LW^b											
017 CH/RG	Ch1	Ch2	Ch3	Ch4	Ch5	Ch6	WH				Rg1	Rg2						

Antigen number

System	001	002	003	004	005	006	007	008	009	010	011	012	013	014	015	016	017	018
018	H																	
019	Kx																	
020	...	Ge2	Ge3	Ge4	Wb	Ls^a	An^a	Dh^a	GEIS	GEPL	GEAT	GETI						
021	Cr^a	Tc^a	Tc^b	Tc^c	Dr^a	Es^a	IFC	WES^a	WES^b	UMC	GUTI	SERF	ZENA	CROV	CRAM	CROZ		
022	Kn^a	Kn^b	McC^a	Sl1	Yk^a	McC^b	Sl2	Sl3	KCAM									
023	In^a	In^b	INFI	INJA														
024	Ok^a	OKGV	OKVM															
025	MER2																	
026	JMH	JMHK	JMHL	JMHG	JMHM	JMHQ												
027	I																	
028	P																	
029	GIL																	
030	Duclos	Ol^a	DSLK	RHAG4														
031	FORS																	
032	Jr^a																	
033	Lan																	

system, the CD number (if any), and a brief description of how the system was named are given.

Expression

This section relates to the component on which the antigens in a blood group system are carried. Several blood group antigens occur naturally in soluble form in body fluids (e.g., saliva, urine, plasma). If a naturally occurring soluble form of the carbohydrate antigen or carrier protein is available, it will be indicated in this section. Soluble forms of an antigen can be used for inhibition tests to confirm or eliminate antibody activity. If the soluble form of the antigen is to be used for inhibition studies, the substance must be obtained from a person who inherited the antigen of interest and be made isotonic. An ideal negative dilution control for this test would be the particular fluid from a person who did not inherit the antigen of interest[5]. A soluble form of many antigens can be produced through recombinant technology.

Some of the components carrying blood group antigens have been detected on other blood cells and tissues by use of various methods, including testing with polyclonal and monoclonal antibodies or by Northern blot analysis. However, it cannot be assumed that detection of the carrier molecule equates with the expression of the RBC antigen(s).

Under the subheading "Other blood cells," we refer to cells in the peripheral blood excluding RBCs because it goes without saying that RBCs express all the components carrying blood groups (except in null phenotypes).

Gene

The chromosomal location for the genes encoding proteins associated with blood group systems is taken from original papers and reviews of the chromosome assignments[6] and how blood groups were cloned[7]. The chromosome number, arm [short p (upper on diagrams); long q (lower on diagrams)], and band number are given. The name of the gene used is that recommended by the ISBT, with alternative names in parenthesis. If the presence of a blood group antigen is detected by serological means, the gene is named by the corresponding ISBT system symbol, an asterisk followed by the antigen number, in italics to indicate the specific allele, e.g., *FY*01* for the allele encoding Fy^a. The organization of the gene in terms of number of exons, kilobase pairs of gDNA, and an overview map are provided. The gene product name (and alternatives) is given.

Database accession numbers

Key GenBank accession numbers and the Entrez gene number are given. The Blood Group Antigen Gene Mutation Database (http://www.ncbi. nlm.nih.gov/gv/rbc/xslcgi.fcgi?cmd=bgmut/home or enter "dbRBC" in

TABLE 2.1B Blood group antigens assigned to each system: extension 1

System		Antigen number																		
		019	020	021	022	023	024	025	026	027	028	029	030	031	032	033	034	035	036	037	038
002	MNS	Hut	Hil	M^v	Far	s^D	Mit	Dantu	Hop	Nob	En^a	ENKT	"N"	Or	DANE	TSEN	MINY	MUT	SAT	ERIK	Os^a
004	RH	hr^s	VS	C^G	CE	D^W	...	...	c-like	cE	hr^H	Rh29	Go^a	hr^B	Rh32	Rh33	Hr^B	Rh35	Be^a	Evans	...
005	LU	Au^b	Lu20	Lu21	LURC																
006	KEL	K19	Km	Kp^c	K22	K23	K24	VLAN	TOU	RAZ	VONG	KALT	KTIM	KYO	KUCI	KANT	KASH	KELP	KETI	KHUL	
010	DI	Tr^a	Fr^a	SW1	DISK																

TABLE 2.1C Blood group antigens assigned to each system: extension 2

System		Antigen number																			
		039	040	041	042	043	044	045	046	047	048	049	050	051	052	053	054	055	056	057	058	059
002	MNS	ENEP	ENEH	HAG	ENAV	MARS	ENDA	ENEV	MNTD													
004	RH	Rh39	Tar	Rh41	Rh42	Crawford	Nou	Riv	Sec	Dav	JAL	STEM	FPTT	MAR	BARC	JAHK	DAK	LOCR	CENR	CEST	CELO	CEAG

TABLE 2.2 Blood group antigens in the collections

Symbol	Name	Number	001	002	003	004	005	006
COST	Cost	205	Cs^a	Cs^b				
I	Ii	207	...	i				
ER	Er	208	Er^a	Er^b	Er3			
GLOB	Globoside	209	...	...	LKE	PX2		
	Unnamed	210	Le^c	Le^d				
VEL	Vel	212	Vel	ABTI				
MN CHO^		213	Hu	M$_1$	Tm	Can	Sext	Sj

Obsolete Collections: 201 (GE), 202 (CROMER), 203 (IN), 204 (AU), 206 (GY), and 211 (WR). For obsolete antigens see www.isbt-web.org.
^ = M and N antigens associated with different sialic acid-carrying oligosaccharides on GPA.

TABLE 2.3 Series of low-prevalence and high-prevalence antigens

Series	002	003	005	006	008	009	012	014	016	017	018	019	021	028	039	040	044	045	047	049	050	052	054
Low-Incidence	700 By	Chr^a	Bi	Bx^a	...	...	...	...	...	To^a	Pt^a	Re^a	Je^a	Li^a	Milne	RASM	JFV	Kg	JONES	HJK	HOFM	SARA	REIT
High-Incidence	901 ...	At^a	...	...	Emm	AnWj	Sd^a	PEL	MAM														

For obsolete antigens see www.isbt-web.org.

a search engine) is a useful website which gives information about alleles relevant to blood groups and hyperlinks to other websites and original articles.

Molecular bases of antigens and phenotypes

Tables listing alleles encoding blood group variants, together with relevant information, are given. The alleles and their names are those assigned by the ISBT Working Party as of December 2011. For updates, go to the ISBT website (www.isbt-web.org). The accession number for the reference allele is indicated, and key nucleotide differences are noted for each allele. Cases where the nucleotide substitution introduces or ablates a restriction enzyme site (indicated, respectively, by "+" or "–") are documented. A blank indicates that, despite our best efforts, the information could not be ascertained and not, necessarily, that there is no change. The approximate prevalence of the allele in different ethnicities are noted using the following general terms: many = 13 or more examples; several = 6 to 12 examples; few = 2 to 5; and rare = 1. Amino acid changes (using the three-letter code; see Table 2.2) are also given for each variant. In line with the ISBT allele nomenclature, silent nucleotide changes are not given unless there is an effect on antigen expression. For each protein-based blood group system, the reference allele encodes all high prevalence antigens in that system. The protein encoded by a variant allele will differ by the unique expression, or absence, of an antigen. The variant protein will express all high-prevalence antigens of that blood group system, except for the specific variant antigen, e.g., the protein encoded by *KEL*02.06* lacks Js^b, expresses the antithetical antigen Js^a, and also all other Kell high-prevalence antigens, k, Kp^b, Ku, K11, etc.

The molecular bases for each antigen are also given on the individual antigen pages. Other tables list alleles encoding so-called mod (greatly reduced antigen expression that may require adsorption/elution for detection) or null phenotypes. If the allelic backbone of the altered allele is known, this is noted, e.g., *CO*01N.03* and *CO*01N.04* means that the silencing nucleotide change(s) is on a Co(a+) background. If the background was not reported, the allele is written, e.g., *CO*N.01* and *CO*N.02*. "Compound heterozygosity" means two different alleles encoding a weak phenotype, one allele encoding a weak phenotype and one allele encoding a null phenotype or two different null alleles.

For all alleles, the numbering for nucleotides and amino acids follows the ISBT system, i.e., nucleotides are counted as #1 being the "A" of the initiation "AUG" and amino acids are counted as #1 from the initiation methionine. This ISBT consistency policy means that the numbers may differ from those published; for example, nucleotide numbers in Kell decrease by 120 and in Knops decrease by 27, and amino acids numbers in MNS increase by 19, in LW they increase by 30 (see Section III for the complete list).

For references to original reports of the alleles, go to dbRBC and use the hyperlinks.

Amino acid sequence

The amino acid sequences in this section are shown in the single-letter code (Table 2.4). The predicted transmembrane sequence for a single-pass

TABLE 2.4 Amino acids and their three-letter and single-letter codes[8]

Amino acid	Three-letter code	Single-letter code	Properties	Molecular weight (Daltons)
Alanine	Ala	A	Nonpolar	89
Arginine	Arg	R	Polar, positively charged	174
Asparagine	Asn	N	Polar, uncharged	132
Aspartic acid	Asp	D	Polar, negatively charged	133
Cysteine	Cys	C	Polar, uncharged	121
Glutamine	Gln	Q	Polar, uncharged	146
Glutamic acid	Glu	E	Polar, negatively charged	147
Glycine	Gly	G	Polar, uncharged	75
Histidine	His	H	Polar, positively charged	155
Isoleucine	Ile	I	Nonpolar	131
Leucine	Leu	L	Nonpolar	131
Lysine	Lys	K	Polar, positively charged	146
Methionine	Met	M	Nonpolar	149
Phenylalanine	Phe	F	Nonpolar	165
Proline	Pro	P	Nonpolar	115
Serine	Ser	S	Polar, uncharged	105
Threonine	Thr	T	Polar, uncharged	119
Tryptophan	Trp	W	Nonpolar	204
Tyrosine	Tyr	Y	Polar, uncharged	181
Valine	Val	V	Nonpolar	117

membrane protein is underlined. Amino acids are counted as number 1 being the initiation methionine. The number of amino acids that are believed to be cleaved, and thus, not present in the mature membrane-bound protein is stated.

Carrier molecule

Molecules carrying blood group antigens are glycoconjugates (the carbohydrate portions of glycolipids or glycoproteins), single-pass membrane proteins, multi-pass membrane proteins or glycosylphosphatidylinositol (GPI)-linked proteins. Single-pass proteins can be oriented with the N-terminus outside (type I) or inside (type II) the membrane (Figure 1.1).

Carbohydrate antigens are depicted by the critical immunodominant sugars and linkages. Proteins carrying blood group antigens are depicted by a stick diagram within a gray band that represents the RBC membrane lipid bilayer. The inside (cytoplasmic surface) of the membrane is always to the bottom of the page and the outside (exofacial surface) is to the top of the page. The predicted topology of the protein in the membrane is shown in the models. The predicted orientation of the N-terminus and the C-terminus is indicated, as are the total number of amino acids. O-glycans are depicted by an open circle (○) and N-glycans by a closed circle on a line (lollipop). Glycosylphosphatidylinositol linkage will be depicted by the zigzag symbol. On the RBC membrane, the presence of a third fatty acid chain on GPI-linked proteins makes the protein harder to cleave by phospholipases. For background reading about membrane proteins the interested reader is referred to Alberts, et al.[8].

The locations of single amino acid substitutions for antigens within the blood group system are shown on a diagram. While one protein molecule can carry numerous high-prevalence antigens, with few exceptions, it is unlikely to carry more than one antigen of low prevalence. Certain characteristics will be given:

M_r (SDS-PAGE): The relative molecular mass (M_r) of a protein as determined by SDS-PAGE. The M_r of a protein (and in particular a glycoprotein) usually differs from the actual molecular weight and the molecular weight calculated from the amino acid sequence deduced from the nucleotide sequence.

Glycosylation: Potential N-linked glycosylation sites (Asn-X-Ser/Thr where X is any amino acid except Pro) are indicated. O-linked glycosylation occurs at Ser and Thr residues. Not all Ser and Thr residues are glycosylated.

Cysteine residues: The total number present is indicated.

Copies per RBC: The number of copies in the RBC membrane of the protein carrying a blood group antigen is indicated. Human polyclonal antibodies to a specific antigen and monoclonal antibodies to the protein have been used as intact immunoglobulin molecules and as Fab fragments to ascertain copy number. Depending on the technology used, these numbers can vary dramatically in different publications. This is particularly true of carbohydrate antigens. The figures given are only a guide, and the interested investigator is encouraged to perform a thorough literature search.

Function

Function of the carrier protein, or the predicted function, based on homology with other proteins of known function, is given.

Disease association

This entry includes diseases caused by an absence of the protein or carbohydrate carrying the blood group antigens and disease susceptibilities associated with an absence, an altered form or a reduced number of copies/RBC of the protein. We also include diseases associated with altered carbohydrate structures.

Phenotypes

The prevalence of phenotypes associated with the blood group system, the null phenotype, and any unusual phenotypes (unusual in expression, not in prevalence) are given. In general, the figures given are for Caucasian populations (northern European), because that is the best studied group by hemagglutination but where possible and where particularly relevant due to differences between populations, other ethnic groups have also been included. Information was obtained from original publications, *Blood Groups in Man*[9], and the AABB *Technical Manual*[5]. The numbers are an average estimate.

In addition, tables with useful facts pertaining to antigens or phenotypes in one system are given here.

Comments

Any fact or interesting information relevant to the blood group system and/or the carrier protein or carbohydrate that does not fit elsewhere is placed here.

References

It is incompatible with the format of this book to provide a comprehensive list of references. However, when appropriate, key or recent references and reviews have been included. For references for antigens and/or alleles, go to GenBank, dbRBC, Rhesus Base (*RHD*), or nybloodcenter.org (*RHCE*) for hyperlinks to the original publications. Certain reference books have been used throughout and rather than list them all on each set of sheets they are listed below. Some references for specific antigens are not given on the system pages, but are given in the antigen pages. These texts are also a good source of references[5,9–17].

ISBT Blood Group Antigens

Terminology

The traditional name for the blood group antigen is given at the top of the page. Also given are the ISBT symbol, ISBT number (parenthetically), other, obsolete, names that have been associated with the antigen, and a brief history about the antigen, including how it was named.

As use of the ISBT number without sinistral zeros was originally stated to be an acceptable terminology, we have given this as well as the more commonly used six-digit number[18].

Occurrence

Antigen and phenotype prevalence are often obtained by averaging several series of tests, and are given as a guide. The prevalence of an antigen is given for Caucasians; notable ethnic differences also are given. Where no ethnicity is given, the figures refer to all populations tested. In general, the information was obtained from *Blood Groups in Man*[9], *Distribution of the Human Blood Groups and Other Polymorphisms*[19], the *AABB Technical Manual*[5], and original articles.

Antithetical antigen

If an antigen is polymorphic, the antithetical partner is indicated.

Expression

Entries here relate to serologically detectable antigens; however, in most instances the information also will apply to the carrier molecule.

Molecular basis associated with antigen

The name and position of specific amino acid(s) (using the three-letter code) associated with the antigen and the position of the base pair (bp) change(s) are both indicated. For the amino acid associated with the allele encoding a blood group antigen, it will be necessary to refer to the pages for the antithetical antigen. For those antigens that do not have defined antithetical partners, the amino acid associated with the wild-type protein or with the absence of a high-prevalence antigen, are given. Where the nucleotide or amino acid number has changed due to counting #1 from the initiation codon, we give this information.

For antigens on hybrid molecules or that are more complex than a single amino acid change, a stick diagram is given.

Effect of enzymes and chemicals on intact RBCs

The options for entries are: resistant; sensitive; weakened; variable. In those instances when reactions between antibody and antigen are markedly enhanced, this information is noted; enhancement may not have been studied for all antigens in a particular system. If no information is available, we have taken the liberty of using "presumed," and extrapolated our interpretation based on the behavior of other antigens in the same system. The information given is to be used only as a guide, because with all chemical treatment of RBCs, the effect varies depending on the exact conditions of treatment, purity of reagents, and the age (condition) of the RBCs. It should be noted that the effect of enzymes on an isolated protein may not be the same when the protein is within the milieu of the RBC membrane.

We provide information regarding the effect of pronase treatment of RBCs only when it is known to differ from the effect of papain or ficin treatment. Namely, this applies to antigens in Lutheran, Diego (3rd extracellular loop), Dombrock, Landsteiner-Wiener, and Cromer blood group systems and Ge3. Similarly, we only give the effect of sialidase treatment of RBCs when the antigen is sensitive to such treatment.

It is important to remember that antibodies to enzyme-treated RBCs exist naturally in some plasma, which can cause false-positive results. This is particularly true for pronase.

If an antigen is sensitive to treatment of RBCs with 200 mM dithiothreitol (DTT), an additional entry will be made for the effect of 50 mM DTT. Other thiol-containing reagents, which include 2-mercaptoethanol (2-ME) and 2-aminoethylisothiouronium bromide (AET), would be expected to give similar results to those indicated for DTT treatment. The commonly used reagents, WARM™ and ZZAP[20], are a combination of DTT and papain.

The effect of acid treatment on antigens is included for those who wish to type RBCs after *in vivo* bound immunoglobulin has been removed by EDTA/glycine/acid treatment[21]. RBCs treated in this way do not express antigens in the Kell blood group system, the Er collection or Bg antigens.

Most information for this section was obtained from original papers and from references[11,20,22,23].

In vitro characteristics of alloantibody

An alloantibody can be made by a person who lacks the corresponding antigen. The immunoglobulin class of a blood group antibody is usually IgG and/or IgM. Blood bank techniques do not routinely include methods to detect IgA antibodies. IgD and IgE have not been described as blood group specific antibodies. In general, naturally-occurring antibodies are IgM and react best by direct agglutination tests, while immune antibodies are IgG and react best by the indirect antiglobulin test (IAT). Readers who are interested in information about the IgG subclass of blood group antibodies are referred to Petz and Garratty[16].

The optimal technique for detection of an antibody to a given antigen is listed as room temperature (RT) or IAT. "RT" means incubation at ambient temperature followed by centrifugation and examination for hemagglutination. "IAT" represents the indirect antiglobulin test, regardless of which enhancement medium (e.g., LISS, albumin, PEG) was used. "Enzymes" means that the antibody agglutinates protease-treated (usually ficin or papain) RBCs, usually after incubation at 37°C. Enzyme-treated RBCs also may be used by the IAT. If column agglutination technology is being used, then "RT" indicates use of the neutral (buffer) cassette, and "IAT" indicates use of the antiglobulin cassette.

Complement binding is used to convey whether the alloantibody is known to bind complement during the *in vitro* interaction with its antigen. It is not intended to indicate the potential of an alloantibody to cause *in vivo* hemolysis of transfused antigen-positive blood.

Clinical significance of alloantibody

This section summarizes the type and degree of transfusion reaction(s), and the degree of clinically significant hemolytic disease of the fetus and newborn (HDFN) that have been associated with the alloantibody in question. Many factors influence the clinical significance of a blood group antibody, and the interested reader is referred to the following references[14,24,25].

Under "Transfusion reaction" the entries are: No/+DAT/mild/moderate/ severe; immediate/delayed/hemolytic, and no data. "Severe" usually means an immediate transfusion reaction as indicated by symptoms that may include but are not limited to lower back pain, change in blood pressure, shortness of breath, feeling of "impending doom," nausea, and/or vomiting, restlessness, flush, passing of red or dark urine. Hemoglobinemia, hemoglobinuria, rapid drop of haptoglobin, reduced RBC count or hematocrit can often be registered. This type of reaction may be fatal. "Delayed" transfusion reaction means a reduced RBC survival associated with a positive DAT, absence of the expected rise in hemoglobin levels, and/or shortened transfusion interval (for chronically transfused patients). This type of reaction is typically not dramatic and often asymptomatic, although jaundice and fatigue may develop. Hemoglobinemia and hemoglobinuria may be recorded but is not typical, whilst a reduced RBC count or hematocrit is expected. The options for HDFN are No/+DAT but no clinical HDFN/mild/moderate/severe and rare.

Autoantibody

If autoantibodies directed to the antigen in question have been described, they are indicated here[11,16,26].

Comments

Any fact or interesting information relevant to the antigen and that does not fit elsewhere is placed here.

References

It is incompatible with the format of this book to provide a comprehensive list of references. However, appropriate key references for reviews or recent papers have been selected as a source of further relevant references. References given on the system page will not necessarily be repeated on each antigen page. Where no reference is given, refer to the system page. For references for alleles, go to GenBank or dbRBC for hyperlinks to the original publications. Certain textbooks have been used throughout, and rather than list them on each antigen page they are listed below. These textbooks are a good source of references[5,9–17].

ISBT Blood Group Collections

Antigens within each collection have a serological, biochemical or genetic relationship, but do not fulfill the criteria for system status. Information relating to each collection of blood group antigens [COST, I, ER, GLOB, Unnamed, VEL, and MN CHO (MNS carbohydrate antigens)] are given on separate pages.

ISBT 700 Series of Low-Incidence Antigens

Antigens in this section occur in less than 1% of most populations studied, and are not known to belong to a blood group system.

ISBT 901 Series of High-Incidence Antigens

Antigens in this section occur in more than 90% of the population, and are not known to belong to a blood group system. Originally, this Series had number 900 but was renumbered 901 in 1988 after many of the antigens were relocated to Systems or Collections.

References

[1] Daniels, G., et al., 2007. International society of blood transfusion committee on terminology for red cell surface antigens: Cape Town report. Vox Sang 92, 250–253.

[2] Daniels, G.L., et al., 2004. Blood group terminology 2004. Vox Sang 87, 316.

[3] Daniels, G.L., et al., 1996. Terminology for red cell surface antigens – Makuhari report. Vox Sang 71, 246–248.

[4] Storry J.R., et al., 2011. International society of blood transfusion working party on red cell immunogenetics and blood group terminology: Berlin report. Vox Sang 101, 77–82.

[5] Roback, J.D. (Ed.), 2011. Technical Manual, seventeenth ed. American Association of Blood Banks, Bethesda, MD.

[6] Reid, M.E., et al., 1998. Chromosome location of genes encoding human blood groups. Transfus Med Rev 12, 151–161.

[7] Lögdberg, L., et al., 2011. Human blood group genes 2010: Chromosomal locations and cloning strategies revisited. Transfus Med Rev 25, 36–46.

[8] Alberts, B., et al., 2007. Molecular Biology of the Cell, fifth ed. Garland Science, New York, NY.

[9] Race, R.R., Sanger, R., 1975. Blood Groups in Man, sixth ed. Blackwell Scientific Publications, Oxford, UK.

[10] Cartron, J.P., Rouger, P., 1995. Molecular Basis of Human Blood Group Antigens. Plenum Press, New York.

[11] Daniels, G., 2002. Human Blood Groups, second ed. Blackwell Science Ltd., Oxford, UK.

[12] Garratty, G. (Ed.), 1994. Immunobiology of Transfusion Medicine. Marcel Dekker, Inc., New York, NY.

[13] Issitt, P.D., Anstee, D.J., 1998. Applied Blood Group Serology, fourth ed. Montgomery Scientific Publications, Durham, NC.

[14] Klein, H.G., Anstee, D.J., 2006. Mollison's Blood Transfusion in Clinical Medicine, eleventh ed. Wiley-Blackwell, Oxford, UK.

[15] Mollison, P.L., et al., 1997. Blood Transfusion in Clinical Medicine, tenth ed. Blackwell Science, Oxford, UK.

[16] Petz, L.D., Garratty, G., 2004. Immune Hemolytic Anemias, second ed. Churchill Livingstone, Philadelphia, PA.

[17] Schenkel-Brunner, H., 2000. Human Blood Groups: Chemical and Biochemical Basis of Antigen Specificity, second ed. Springer-Verlag Wien, New York, NY.

[18] Lewis, M., et al., 1990. Blood group terminology 1990. ISBT working party on terminology for red cell surface antigens. Vox Sang 58, 152–169.

[19] Mourant, A.E., et al., 1976. Distribution of the Human Blood Groups and Other Polymorphisms, second ed. Oxford University Press, London, UK.

[20] Branch, D.R., Petz, L.D., 1982. A new reagent (ZZAP) having multiple applications in immunohematology. Am J Clin Pathol 78, 161–167.

[21] Byrne, P.C., 1991. Use of a modified acid/EDTA elution technique. Immunohematology 7, 46–47.

[22] Committee ARCNRLMM, 1993. Immunohematology Methods. American Red Cross National Reference Laboratory, Rockville, MD.

[23] Judd, W.J., et al., 2008. Judd's Methods in Immunohematology, third ed. AABB Press, Bethesda, MD.

[24] Daniels, G., et al., 2002. The clinical significance of blood group antibodies. Transfusion Med 12, 287–295.

[25] Reid, M.E., et al., 2000. Summary of the clinical significance of blood group alloantibodies. Semin Hematol 37, 197–216.

[26] Garratty, G., 1989. Factors affecting the pathogenicity of red cell auto- and alloantibodies. In: Nance, S.J. (Ed.), Immune Destruction of Red Blood Cells. American Association of Blood Banks, Arlington, VA, pp. 109–169.

The Blood Group Systems and Antigens

ABO Blood Group System

Number of antigens 4

Polymorphic | A, B, A,B, and A1

Terminology

ISBT symbol (number) ABO (001)

History In 1900, Landsteiner mixed sera and RBCs from his colleagues and observed agglutination. On the basis of the agglutination pattern, he named the first two blood groups A and B, using the first letters of the alphabet. RBCs not agglutinated by either sera were first called C, but became known as "ohne A" and "ohne B" (*ohne* is German for "without"), and finally O. In 1907, Jansky proposed using Roman numerals I, II, III, IV for O, A, B, and AB respectively, and in 1910, Moss proposed using I, II, III, and IV for AB, A, B, and O, respectively. These numerical terminologies were used respectively in Europe and America until 1927 when Landsteiner suggested, in order to avoid confusion, to use throughout the world the symbols A, B, O, and AB. When the ISBT nomenclature was first described in 1982, there were five ABO antigens, but ABO5 is now obsolete after the H antigen was removed to form the H system in 1990.

Expression

Soluble form Saliva and all body fluids except CSF (in secretors)

Other blood cells Lymphocytes, but also other leucocytes express A/B antigens, most prominently in secretors (antigens adsorbed from plasma), platelets (very strongly in ~5% of individuals)

The Blood Group Antigen (3/e). DOI: http://dx.doi.oxg/10.1016/B978-0-12-415849-8.00003-X

| Tissues | On most epithelial cells (particularly glandular epithelia), and on endothelial cells. Broad tissue distribution (often termed "histo-blood group" antigens) |

Gene[1,2]

Chromosome	9q34.1–q34.2
Name	*ABO*
Organization	7 exons distributed over 19.5 kbp of gDNA
Product	3-α-*N*-acetylgalactosaminyltransferase for A
	3-α-galactosyltransferase for B

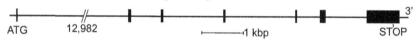

ATG   12,982   ⊢————1 kbp   STOP   3'

Database accession numbers

| GenBank | NG_006669.1 (gDNA reference used by the NCBI website, http://www.ncbi.nlm.nih.gov/sites/varvu?gene=28); AF134412–AF134416 (represent different *A*, *B* and *O*-mRNA) |
| Entrez Gene ID | 28 |

Amino acid sequence

Residues differing between the highly homologous A and B transferases are shown in bold; the two residues most important for donor sugar specificity are underlined.

A transferase (encoded by *ABO*A1.01*, based on AF134412)^

```
MAEVLRTLAG  KPKCHALRPM  ILFLIMLVLV  LFGYGVLSPR  SLMPGSLERG   50
FCMAVREPDH  LQRVSLPRMV  YPQPKVLTPC  RKDVLVVTPW  LAPIVWEGTF  100
NIDILNEQFR  LQNTTIGLTV  FAIKKYVAFL  KLFLETAEKH  FMVGHRVHYY  150
VFTDQPAAVP  RVTLGTGRQL  SVLEVRAYKR  WQDVSMRRME  MISDFCERRF  200
LSEVDYLVCV  DVDMEFRDHV  GVEILTPLFG  TLHPGFYGSS  REAFTYERRP  250
QSQAYIPKDE  GDFYYLGGFF  GGSVQEVQRL  TRACHQAMMV  DQANGIEAVW  300
HDESHLNKYL  LRHKPTKVLS  PEYLWDQQLL  GWPAVLRKLR  FTAVPKNHQA  350
VRNP                                                       354
```

^B transferase (encoded by *ABO*B.01*, based on AF134414) has G, S, M, and A instead of R, G, L, and G at the four positions that differ between A and B, in bold above.

ABO

Carrier molecule description[3]

A and B antigens are not the primary gene products.

Antigens are defined by immunodominant terminal sugars (α3GalNAc for A; α3Gal for B) attached to one of several different types of acceptor molecules (see below), which are oligosaccharide chains carried on either glycoproteins (~90%) or glycosphingolipids (~10%). On glycoproteins, ABO antigens are expressed mainly on N-glycans containing polylactosaminyl units predominantly on band 3 (DI), the glucose transporter, RhAG, and CHIP-1 (CO). Some precursor types (3 and 4) may only be expressed on glycolipids and not on glycoproteins. The precursor of A and B antigens is the H antigen (H1).

Peripheral core	Structure^	Predominantly found in
Type 1	Galβ(1–3)GlcNAcβ(1–3)-R	Secretions, plasma, endodermal tissues (small amounts adsorbed onto RBCs)
Type 2	Galβ(1–4)GlcNAcβ(1–3)-R	Ecto- and mesodermal tissues (main structure on RBCs)
Type 3	Galβ(1–3)GalNAcα(1–3)-R	O-linked mucin type, repetitive A (also on RBCs)
Type 4	Galβ(1–3)GalNAcβ(1–3)-R	Glycolipids in kidney (also on RBCs)

^Shown without the H-specific Fuc α2-linked to Gal.
R = Inner core structure or linkage (towards protein or lipid anchor).

Molecular basis associated with the various ABO phenotypes[4-6]

Close to 200 alleles at the *ABO* locus have been described (see dbRBC), and a selection of them is listed below. Some are associated with the common A_1, A_2, and B phenotypes, but most convey weak (A subgroups and B subgroups) or null (group O) phenotypes. Some alleles encode glycosyltransferases that have the ability to synthesize both A and B antigens and give rise to the cisAB or B(A) phenotypes. Even the normal A and B glycosyltransferases may be able to synthesize trace amounts of the "wrong" antigen (respectively, B and A), but the *cisAB* and *B(A)* alleles make more than normal amounts of the "wrong" antigen, to a degree that it may be detected serologically by routine reagents.

The *ABO* allele nomenclature listed here is under consideration by the ISBT, and must be considered provisional. Since various terminologies have been in use for several years[4], previously used allele names are given in parallel with the provisional ISBT allele names. The new allele names follow the ISBT naming format (see www.isbt-web.org) and nucleotide differences from the reference allele, *ABO*A1.01* (AF134412), and amino acids affected are given for all alleles except the *B* and *B*^*weak* alleles for which *ABO*B.01* (AF134414) is used as reference. Alleles that differ from consensus or other alleles only by silent mutations are not listed here due to space restrictions. For more alleles, finally approved names, and details including the original reference for each allele, see dbRBC and the ISBT website.

Molecular bases associated with the A₁ and A₂ phenotypes

Pheno-type	Provisional ISBT allele name	dbRBC allele name	Other allele names§	Exon	Nucleotide	Amino acid change	Ethnicity (prevalence)
A₁	*ABO*A1.01*	A101	A¹, A¹–1, A(Pro)	–	–	–	All (Common)
A₁	*ABO*A1.02*	A102	A1ᵛ, A¹–2, A(Leu)	7	467C>T	Pro156Leu	Mainly Asians (Common)
A₂	*ABO*A2.01*	A201	A², A²–1	7	467C>T	Pro156Leu	Non-Asians
				7	1061delC	354fs^	(Common)
A₂	*ABO*A2.02*	A202	A106	7	1054C>T	Arg352Trp	Japanese (Few)
A₂	*ABO*A2.03*	A203	A107	7	1054C>G	Arg352Gly	Japanese (Few)
A₂	*ABO*A2.04*#	A204	R101	6	297A>G	–	Japanese
				7	526C>G	Arg176Gly	(Rare)
				7	657C>T	–	
				7	703G>A	Gly235Ser	
				7	771C>T	–	
				7	829G>A	Val277Met	
A₂	*ABO*A2.05*	A205	A111	7	467C>T	Pro156Leu	Japanese
				7	1009G>A	Arg337Gly	(Few)
A₂	*ABO*A2.06*	A206	A¹–A², A²⁽⁴⁶⁷C⁾	7	1061delC	354fs^	All (Rare)
A₂	*ABO*A2.07*	A207	–	7	539G>C	Arg180Pro	Taiwanese (Rare)
A₂	*ABO*A2.08*	A208	–	7	467C>T	Pro156Leu	Chinese
				7	539G>C	Arg180Pro	(Rare)
A₂	*ABO*A2.09*	A209	Avar, A207	7	467C>T	Pro156Leu	Taiwanese,
				7	527G>A	Arg176His	Kuwaitis
				7	1061delC	354fs^	(Rare)
A₂	*ABO*A2.10*	A210	–	6	268T>C	Trp90Arg	Chinese
				7	467C>T	Pro156Leu	(Rare)

(Continued)

(Continued)

Pheno-type	Provisional ISBT allele name	dbRBC allele name	Other allele names[§]	Exon	Nucleotide	Amino acid change	Ethnicity (prevalence)
A_2	ABO*A2.11	A211	–	6	266C>T	Pro89Leu	Chinese
				7	467C>T	Pro156Leu	(Rare)
A_2	ABO*A2.12	A212	–	4	190G>A	Val64Ile	American
				7	527G>A	Arg176His	(Rare)
				7	1061delC	354fs^	
A_2	ABO*A2.13	A213	–	7	467C>T	Pro156Leu	Chinese
				7	742C>T	Arg248Cys	(Rare)
A_2	ABO*A2.16	A216	A2.16.01.1	3	106G>T	Val36Phe	Austrian
				4	188G>A	Arg63His	(Rare)
				4	189C>T	–	
				7	467C>T	Pro156Leu	
				7	1061delC	354fs^	
A_2	ABO*A2.17	A217	–	7	407C>T	Thr136Met	Chinese
				7	467C>T	Pro156Leu	(Rare)
A_2	ABO*A2.18	A218	–	7	467C>T	Pro156Leu	Chinese
				7	722G>A	Arg241Gln	(Rare)
A_2	ABO*A2.19	A219	–	7	467C>T	Pro156Leu	Chinese
				7	778G>A	Glu260Lys	(Rare)
A_2	ABO*A2.20	A220	–	7	467C>T	Pro156Leu	Chinese
				7	829G>A	Val277Met	(Rare)

[§]For instance, those introduced by the authors of the original publications.
^A frame-shift extends the glycosyltransferase by 21 amino acids.
#May be a hybrid between a *B* and an *O* allele.

Comments

The *A103–A107* alleles in dbRBC do not give rise to an altered amino acid sequence compared to other alleles, and so are not included here. *A108* and *A109* are listed as unpublished, and had no phenotype registered in dbRBC. *A214* and *A215* represent the same coding sequence as *ABO*A2.01*, but have been registered under other names due to intron polymorphisms. Also, their phenotypes are not given in dbRBC. Some alleles listed above are unpublished, but have been submitted to GenBank/dbRBC.

It is also notable that many of the alleles registered as associated with the rare A_2 phenotype in Asia (e.g., *A2.08*, *A2.13*, *A2.17*, *A2.18*, and *A2.20*) cause amino acid substitutions that have been associated with weaker A subgroups in other studies. In the case of *A2.18* and *A2.19*, the phenotype was given as A, not A_2.

Molecular bases associated with weak A subgroup phenotypes (most of them are rare)

Pheno-type	Provisional ISBT allele name	dbRBC allele name	Other allele names§	Exon	Nucleotide	Amino acid change	Ethnicity
A₃	ABO*A3.01	A301	A3, A³–1	7	871G>A	Asp291Asn	French, American
A₃	ABO*A3.02	A302	A₃	7	829G>A	Val277Met	Brazilian
				7	1061delC	354fs^	
A₃	ABO*A3.03	A303	A³	7	838C>T	Leu280Phe	Taiwanese
A₃	ABO*A3.04	A304	A³, A²(539G>A)	7	539G>A♯	Arg180His	Swedish
A₃	ABO*A3.05	A305	–	7	820G>A	Val274Met	Taiwanese
A₃	ABO*A3.06	A306	–	7	467C>T	Pro156Leu	Taiwanese
				7	820G>A	Val274Met	
A₃	ABO*A3.07	A307	–	7	467C>T	Pro156Leu	Taiwanese
				7	745C>T	Arg249Trp	
A_weak	ABO*AW.01	Aw01	Aᵂ–1	7	407C>T♯	Thr136Met	English
A_weak	ABO*AW.02	Aw02	Aᵂ–2	6	350G>C♯	Gly117Ala	Caucasian
A_weak	ABO*AW.03	Aw03	Aᵂ–3	4	203G>C♯	Arg68Thr	Scandinavian
A_weak	ABO*AW.04	Aw04	Aᵂ–4	7	721C>T	Arg241Trp	German
A_weak	ABO*AW.05	Aw05	Aᵂ–5	7	965A>G	Glu322Gly	Finnish
A_weak	ABO*AW.06	Aw06	–	7	502C>G	Arg168Gly	Caucasian
A_weak	ABO*AW.07	Aw07	–	7	592C>T♯	Arg198Trp	German
A_weak	ABO*AW.08	Aw08	O²–4	5	220C>T	Pro74Ser	Caucasian
				6	297A>G	–	
				7	488C>T	Thr163Met	
				7	526C>G	Arg176Gly	
				7	802G>A	Gly268Arg	
A_weak	ABO*AW.09	Aw09	Oᵀᵛ–A² hybrid	2	46G>A♯	Ala16Thr	African
				3	106G>T	Val36Phe	
				4	188G>A	Arg63His	
				5	220C>T	Pro74Ser	
A_weak	ABO*AW.10	Aw10	Avar	7	784G>A	Asp262Asn	Korean
A_weak	ABO*AW.11	Aw11	–	7	523G>A	Val175Met	German
				7	721C>T	Arg241Trp	
A_weak	ABO*AW.12	Aw12	–	7	467C>T	Pro156Leu	Chinese
				7	556A>G	Met186Val	

(Continued)

ABO

(Continued)

Pheno-type	Provisional ISBT allele name	dbRBC allele name	Other allele names§	Exon	Nucleotide	Amino acid change	Ethnicity
A_weak	ABO*AW.13	Aw13	–	1	2T>C	Start codon disrupted	Caucasian
A_weak	ABO*AW.14	Aw14	–	7 7	467C>T 699C>A	Pro156Leu His233Gln	Korean
A_weak	ABO*AW.15	Aw15	–	6–7	IVS6:+4a>t	Altered splicing	Turkish
A_weak	ABO*AW.16	Aw16	–	1	1A>G#	Start codon disrupted	Swiss
A_weak	ABO*AW.17	Aw17	–	5	236C>T#	Pro79Leu	Caucasian
A_weak	ABO*AW.18	Aw18	–	6	347T>C#	Ile116Thr	Swedish
A_weak	ABO*AW.19	Aw19	–	7	434A>G#	His145Arg	Swiss
A_weak	ABO*AW.20	Aw20	–	7	607G>A#	Glu203Lys	Canadian
A_weak	ABO*AW.21	Aw21	–	7	607G>C#	Glu203Gln	Portuguese
A_weak	ABO*AW.22	Aw22	–	7	634G>A#	Val212Met	Swiss
A_weak	ABO*AW.23	Aw23	–	7	722G>A#	Arg241Gln	French
A_weak	ABO*AW.24	Aw24	–	7	742C>T#	Arg248Cys	American
A_weak	ABO*AW.25	Aw25	–	7	829G>A#	Val277Met	African
A_weak	ABO*AW.26	Aw26	O^1–A^2 hybrid	7	527G>A#	Arg176His	Turkish
A_weak	ABO*AW.27	Aw27	–	7 7	527G>A 1061delC	Arg176His 354fs^	Syrian
A_weak	ABO*AW.28	Aw28	–	1–2	IVS2:+2t>c	Altered splicing	German
A_weak	ABO*AW.29	Aw29	–	6	311T>A	Ile104Asn	Caucasian
A_x/A_weak	ABO*AW.30.01	Ax01	A^x, A^{x-1}, A108	7	646T>A	Phe216Ile	Caucasian
A_x/A_weak	ABO*AW.30.02	Ax04	–	7 7	646T>A 681G>A	Phe216Ile –	Japanese
A_x/A_weak	ABO*AW.31.01	Ax02	A^{x-3}	6 7 7 7 7	297A>G 646T>A 681G>A 771C>T 829G>A	– Phe216Ile – – Val277Met	Swedish
A_x/A_weak	ABO*AW.31.02	Ax03	A^{x-2}	7 7 7 7	646T>A 681G>A 771C>T 829G>A	Phe216Ile – – Val277Met	Swedish†

(Continued)

ABO

(Continued)

Phenotype	Provisional ISBT allele name	dbRBC allele name	Other allele names§	Exon	Nucleotide	Amino acid change	Ethnicity
A$_x$/A$_{weak}$	ABO*AW.31.03	Ax05	A^x–4	7	646T>A	Phe216Ile	Polish[†]
				7	681G>A	–	
				7	771C>T	–	
				7	829G>A	Val277Met	
A$_x$/A$_{weak}$	ABO*AW.31.04	Ax06	A^x–5	7	646T>A	Phe216Ile	American[†]
				7	681G>A	–	
				7	771C>T	–	
				7	829G>A	Val277Met	
A$_x$/A$_{weak}$	ABO*AW.31.05	Ax08	–	7	646T>A	Phe216Ile	German[†]
				7	681G>A	–	
				7	771C>T	–	
				7	829G>A	Val277Met	
A$_x$/A$_{weak}$	ABO*AW.32	Ax07	A^x–6	7	996G>A	Trp332Stop	New Zealand
A$_x$/A$_{weak}$	ABO*AW.33	Ax09	A^x–4	7	467C>T	Pro156Leu	Chinese
				7	543G>T	Trp181Cys	
A$_x$/A$_{weak}$	ABO*AW.34	Ax10	–	7	467C>T	Pro156Leu	Chinese
				7	829G>A	Val277Met	
				7	1009A>G	Arg337Gly	
A$_x$/A$_{weak}$	ABO*AW.35	Ax11	–	7	467C>T	Pro156Leu	Taiwanese
				7	860C>T	Ala287Val	
A$_x$/A$_{weak}$	ABO*AW.36	Ax12	–	7	607G>A	Glu203Lys	Chinese
A$_x$/A$_{weak}$	ABO*AW.37	Ax13	–	7	940A>G	Lys314Glu	Chinese
A$_x$/A$_{weak}$	ABO*AW.38	Ax14	–	7	426G>C	Met142Ile	Chinese
A$_x$/A$_{weak}$	ABO*AW.39	Ax15	–	7	385T>C	Phe129Leu	Chinese
A$_x$/A$_{weak}$	ABO*AW.40	Ax16	–	7	499G>T	Gly167Cys	Chinese
A$_x$/A$_{weak}$	ABO*AW.41	Ax17	–	6	370A>G	Lys124Glu	Chinese
A$_x$/A$_{weak}$	ABO*AW.42	Ax18	–	7	467C>T	Pro156Leu	Chinese
				7	905A>G	Asp302Gly	
A$_x$/A$_{weak}$	ABO*AW.43	Ax19	–	7	467C>T	Pro156Leu	Chinese
				7	721C>T	Arg241Trp	
A$_{finn}$/A$_{weak}$	ABO*AW.44	–	A^{finn}	6–7	IVS6:+4a>g	Altered splicing	Finns

(Continued)

ABO

(Continued)

Pheno-type	Provisional ISBT allele name	dbRBC allele name	Other allele names§	Exon	Nucleotide	Amino acid change	Ethnicity
A$_{bantu}$/A$_{weak}$	ABO*AW.45	Abantu 01	A^{bantu}, O^{1bantu}–A^2	4–5	IVS4:+1delG$^\#$	Altered splicing	Bantu
A$_m$	ABO*AM.01	Am.01	A112	7	467C>T	Pro156Leu	Japanese
					761C>T	Ala254Val	
A$_m$	ABO*AM.02	Am.02	–	7	664G>A	Val222Met	Taiwanese
A$_{el}$	ABO*AEL.01	Ael01	A^{el}–1, A109	7	804insG	269fs$^\square$	Caucasians
A$_{el}$	ABO*AEL.02	Ael02	A110	7	467C>T	Pro156Leu	Japanese
				7	646T>A	Phe216Ile	
				7	681G>A	–	
A$_{el}$	ABO*AEL.03	Ael03	Aelvar	7	804delG	269fs288Stop	African
A$_{el}$	ABO*AEL.04	Ael04		6–7	IVS6:+5g>a	Altered splicing	Taiwanese
A$_{el}$	ABO*AEL.05	Ael05		7	467C>T	Pro156Leu	Chinese
				7	767T>C	Ile256Thr	
A$_{el}$	ABO*AEL.06	Ael06		7	425T>C	Met142Thr	Chinese
				7	467C>T	Pro156Leu	
A$_{el}$	ABO*AEL.07	Ael07	A^{1v}–O^{1v} hybrid	7	467C>T	Pro156Leu	Taiwanese
				7	681G>A	–	
				7	771C>T	–	
				7	829G>A	Val277Met	
A$_{el}$	ABO*AEL.08	Ael08		7	467C>T	Pro156Leu	Chinese
				7	804insG	269fs$^\square$	

§For instance, those introduced by the authors of the original publications.
^A frame-shift extends the glycosyltransferase by 21 amino acids.
$^\#$This allele also carries the two *ABO*A2.01*-related SNPs 467C > T and 1061delC which result in Pro156Leu and 354fs and extension of the protein by 21 amino acids.
$^\square$This frame-shifting mutation theoretically extends the protein by 37 amino acids.
†These alleles differ in their intron 6 sequence, and are probable hybrid alleles with different crossing-over points between common *ABO* alleles.

Molecular bases associated with B phenotype^

(The seven B-associated polymorphisms are only shown for the first allele but are present in the others.)

Phenotype	Provisional ISBT allele name	dbRBC allele name	Other allele names§	Exon	Nucleotide	Amino acid change	Ethnicity
B	ABO*B.01	B101	B	6	297A>G	–	All
				7	526C>G	Arg176Gly	(Common)
				7	657C>T	–	
				7	703G>A	Gly235Ser	
				7	796C>A	Leu266Met	
				7	803G>C	Gly268Ala	
				7	930G>A	–	
B	ABO*B.02	B108	Bvar	7	892G>T	Ala298Ser	Taiwanese (Rare)
B	ABO*B.03	B112	–	7	559C>T	Arg187Cy	Chinese (Rare)

§For instance, by the authors of the original publication.

^Other variants of B alleles exist, but the ones listed in dbRBC are either based on: (1) lack of one of the silent A vs. B SNPs (e.g., B102 has 930G, B103 has 657C); (2) silent mutations (B109 has 498C > T); (3) intron SNPs (e.g., B107, B113, B114, B116); (4) a sequence identical to a proven B$_{weak}$ (B110); (5) unpublished (B113–B116).

Molecular bases associated with weak B subgroup phenotypes (all rare)

Differences compared to ABO*B.01 are given.

Phenotype	Provisional ISBT allele name	dbRBC allele name	Other allele names§	Exon	Nucleotide	Amino acid change	Ethnicity
B$_3$	ABO*B3.01	B301	B^3	7	1054C>T	Arg352Trp	Japanese
B$_3$	ABO*B3.02	B302	–	7	646T>A	Phe216Ile	Japanese
B$_3$	ABO*B3.03	B303	–	3–4	IVS3:+5g>a	Altered splicing	Asians
B$_3$	ABO*B3.04	B304	–	6	247G>T	Asp83Tyr	Taiwanese
B$_3$	ABO*B3.05	B305	–	7	425T>C	Met142Thr	Chinese
B$_3$	ABO*B3.06	B306	Bvar	7	547G>A	Asp183Asn	Koreans
B$_3$	ABO*B3.07	B307	–	7	410T>C	Ala137Val	Chinese

(Continued)

(Continued)

Pheno-type	Provisional ISBT allele name	dbRBC allele name	Other allele names§	Exon	Nucleotide	Amino acid change	Ethnicity
B₃	ABO*B3.08	B308	–	7	938A>C	His313Pro	Koreans
Bₓ/B_weak	ABO*BW.01	Bx01	B^w–1	7	871G>A	Asp291Asn	Japanese
B_weak	ABO*BW.02	Bw02	B^w–2	7	873G>C	Asp291Glu	French
B_weak	ABO*BW.03	Bw03	B^w–3	7	721C>T	Arg241Trp	Swedish
B_weak	ABO*BW.04	Bw04	B^w–4	7	548A>G	Asp183Gly	Swedish
B_weak	ABO*BW.05	Bw05	B^w–5	7	539G>A	Arg180His	American
B_weak	ABO*BW.06	Bw06	B^w–6	7	1036A>G	Lys346Glu	Finnish
B_weak	ABO*BW.07	Bw07	B^w–7	7	1055G>A	Arg352Gln	Indian, American
B_weak	ABO*BW.08	Bw08	B^w–8	7	863T>G	Met288Arg	Turkish
B_weak	ABO*BW.09	Bw09	Bw08	7	1037A>T	Lys346Glu	German
B_weak	ABO*BW.10	Bw10	–	7	556A>G	Met186Val	Brazilian
B_weak	ABO*BW.11	Bw11	B^w–11	7	695T>C	Leu232Pro	Chinese, Thai
B_weak	ABO*BW.12	Bw12	–	6	278C>T	Leu93Pro	Chinese
B_weak	ABO*BW.14	Bw14	B^w–14	7	523G>A	Val175Met	Caucasian, Chinese^
B_weak	ABO*BW.15	Bw15	B^w–15	7	565A>G	Met189Val	Turkish
B_weak	ABO*BW.16	Bw16	B^w–16	7	575T>C	Ile192Thr	Hindustani
B_weak	ABO*BW.17	Bw17	B^w–17	7	784G>A	Asp262Asn	Indian
B_weak	ABO*BW.18	Bw18	B^w–18	7	802G>A	Ala268Thr	French
B_weak	ABO*BW.19	Bw19	B^w–19	7 7	646T>A 681G>A	Phe216Ile –	Chinese
B_weak#	ABO*BW.20	Bw20	–	7	816insG	272fs392Stop	German
B_weak	ABO*BW.21	Bw21	–	7	688G>C	Gly230Arg	Turkish
B_weak	ABO*BW.22	Bw22	–	7	503G>T	Arg168Leu	Chinese
B_weak	ABO*BW.23	Bw23	–	7	743G>C	Arg248Pro	Chinese
B_weak	ABO*BW.24	Bw24	–	7	558G>T	Met186Ile	Chinese
B_weak	ABO*BW.25	Bw25	–	3 7	103G>A 619C>G	Gly35Arg Leu207Val	German
B_weak	ABO*BW.26	Bw26	O²–B hybrid¤	2	53G>T	Arg18Leu	Swiss
B_weak	ABO*BW.27	Bx02	Bel06	7	905A>G	Asp302Gly	Chinese
B_weak	ABO*BW.28	Bx03	–	7	541T>C	Trp181Arg	Chinese

(Continued)

(Continued)

Pheno-type	Provisional ISBT allele name	dbRBC allele name	Other allele names§	Exon	Nucleotide	Amino acid change	Ethnicity
B_weak	ABO*BW.29	Bx04	–	7	588C>G	Cys196Trp	Chinese
B_weak	ABO*BW.30	Bx05	–	7	976G>T	Asp326Tyr	Chinese
B_weak	ABO*BW.31	Bx06	–	7	900G>C	Trp300Cys	Chinese
B_weak	ABO*BW.32	Bx07	–	7	808T>A	Phe270Ile	Chinese
B_weak	ABO*BW.33	Bx08	–	7	550G>A	Val184Met	Chinese
B_weak	ABO*BW.34	Bx09	–	7	889G>A	Glu297Lys	Chinese
B_el	ABO*BEL.01	Bel01	B105	7	641T>G	Met214Arg	Japanese
B_el	ABO*BEL.02	Bel02	B106	7	669G>T	Glu223Asp	Japanese
B_el	ABO*BEL.03	Bel03	–	7	502C>T	Arg168Trp	Taiwanese
B_el	ABO*BEL.04	Bel04	–	7	†	†	Brazilian Japanese
B_el	ABO*BEL.05	Bel05	–	7	952G>A	Val318Met	Chinese

§For instance, by the authors of the original publication.
^Apparently normal B phenotype in two Chinese donors.
#According to the original publication, adsorption/elution was negative so allele designation questionable.
□A hybrid between an O allele (exons 1–4) and B (exons 5–7).
†467C > T;646T > A;681G > A;771C > T;796C > A ;803G > C;829G > A, resulting in Pro156Leu;Phe216Ile; Met266Leu;Gly268Ala;Met277Val, i.e., an unusual combination of allelic markers from ABO*A1.02, *B.01 and a common O allele. Possibly a hybrid allele.

Molecular bases associated with cisAB and B(A) phenotypes (all rare except ABO*cisAB.01, which is infrequent in Asians)

Differences compared to ABO*A1.01 are given.

Pheno-type	Provisional ISBT allele name	dbRBC allele name	Other allele names	Exon	Nucleotide	Amino acid change	Ethnicity
cisAB	ABO*cisAB.01	cis–AB01	cisAB–1	7	467C>T	Pro156Leu	Asian
				7	803G>C	Gly268Ala	
cisAB	ABO*cisAB.02	cis–AB02	cisAB–2	7	526C>G	Arg176Gly	Australian
				7	657C>T	–	
				7	703G>A	Gly235Ser	
				7	803G>C	Gly268Ala	

(Continued)

(Continued)

Pheno-type	Provisional ISBT allele name	dbRBC allele name	Other allele names	Exon	Nucleotide	Amino acid change	Ethnicity
cisAB	ABO*cisAB.03	cis–AB03	cis AB. tlse(*)01, cisAB–3	6	297A>G	–	French
				7	526C>G	Arg176Gly	
				7	657C>T	–	
				7	700C>T	Pro234Ser	
				7	703G>A	Gly235Ser	
				7	796C>A	Leu266Met	
				7	803G>C	Gly268Ala	
				7	930G>A	–	
cisAB	ABO*cisAB.04	cis–AB04	cisAB–4	7	467C>T	Pro156Leu	Chinese
				7	796C>A	Leu266Met	
cisAB	ABO*cisAB.05	cis–AB05	cisAB–5	6	297A>G	–	Chinese^
				7	526C>G	Arg176Gly	
				7	657C>T	–	
				7	703G>A	Gly235Ser	
				7	796C>A	Leu266Met	
				7	930G>A	–	
cisAB	ABO*cisAB.06	cis–AB06	–	6	297A>G	–	Chinese
				7	657C>T	–	
				7	703G>A	Gly235Ser	
				7	796C>A	Leu266Met	
				7	803G>C	Gly268Ala	
				7	930G>A	–	
B(A)	ABO*BA.01	B(A)01	B(A)–1	6	297A>G	–	
				7	526C>G	Arg176Gly	
				7	796C>A	Leu266Met	
				7	803G>C	Gly268Ala	
				7	930G>A	–	
B(A)	ABO*BA.02	B(A)02	B(A)–2	6	297A>G	–	Taiwanese
				7	526C>G	Arg176Gly	
				7	657C>T	–	
				7	700C>G	Pro234Ala	
				7	703G>A	Gly235Ser	
				7	796C>A	Leu266Met	
				7	803G>C	Gly268Ala	
				7	930G>A	–	
B(A)	ABO*BA.03	B(A)03	B(A)–3	6	297A>G	–	Caucasian
				7	526C>G	Arg176Gly	
				7	657C>T	–	
				7	796C>A	Leu266Met	
				7	803G>C	Gly268Ala	
				7	930G>A	–	
B(A)	ABO*BA.04	B(A)04	B(A)–4	6	297A>G	–	Chinese
				7	526C>G	Arg176Gly	
				7	640A>G	Met214Val	
				7	657C>T	–	
				7	703G>A	Gly235Ser	

(Continued)

ABO

ABO

(Continued)

Pheno-type	Provisional ISBT allele name	dbRBC allele name	Other allele names	Exon	Nucleotide	Amino acid change	Ethnicity
				7	796C>A	Leu266Met	
				7	803G>C	Gly268Ala	
				7	930G>A	–	
B(A)	ABO*BA.05	B(A)05	B(A)–5	6	297A>G	–	Chinese
				7	526C>G	Arg176Gly	
				7	641T>C	Met214Thr	
				7	657C>T	–	
				7	703G>A	Gly235Ser	
				7	796C>A	Leu266Met	
				7	803G>C	Gly268Ala	
				7	930G>A	–	
B(A)	ABO*BA.06	B(A)06	B(A)–6, B(A)ᵛ	6	297A>G	–	Chinese
				7	526C>G	Arg176Gly	
				7	657C>T	–	
				7	703G>A	Gly235Ser	
				7	796C>A	Leu266Met	
				7	930G>A	–	

^This allele may be identical to *ABO*BA.06*. The polymorphisms and amino acids reported for both alleles are the same according to dbRBC, and the *cis-AB05* entry has been accepted in dbRBC without publication or GenBank submission so difficult to compare.

Molecular bases associated with O (null) phenotype^

Differences compared to *ABO*A1.01* are given, and the genetic alteration inducing the null phenotype is noted in bold (if known).

Pheno-type	Provisional ISBT allele name	dbRBC allele name	Other allele names§	Exon	Nucleotide	Amino acid change	Ethnicity
O	ABO*O.01.01	O01	O¹	6	**261delG**	88fs118Stop	All (Common)
O	ABO*O.01.02	O02	O¹ᵛ	3	106G>T	Val36Phe	All (Common)
				4	188G>A	Arg63His	
				4	189C>T	–	
				5	220C>T	Pro74Ser	
				6	**261delG**	88fs118Stop	
				6	297A>G	–	
				7	646T>A	–	
				7	681G>A	–	
				7	771C>T	–	
				7	829G>A	–	

(Continued)

(Continued)

Pheno-type	Provisional ISBT allele name	dbRBC allele name	Other allele names§	Exon	Nucleotide	Amino acid change	Ethnicity
O?	ABO*O.02.01	O03	O²–1	2	53G>T	Arg18Leu	All but
				5	220C>T	Pro74Ser	Asians
				6	297A>G	–	(Many)
				7	526C>G	Arg176Gly	
				7	**802G>A**	Gly268Arg	
O?	ABO*O.02.02	O48	O²–2	2	53G>T	Arg18Leu	Israeli
				5	220C>T	Pro74Ser	(Rare)
				6	297A>G	–	
				7	526C>G	Arg176Gly	
				7	649C>T	Arg217Cys	
				7	689G>A	Gly229Asp	
				7	**802G>A**	Gly268Arg	
O?	ABO*O.02.03	O49	O²–3	2	53G>T	Arg18Leu	Caucasians
				5	220C>T	Pro74Ser	(Many)
				6	297A>G	–	
				7	526C>G	Arg176Gly	
				7	689G>A	Gly229Asp	
				7	**802G>A**	Gly268Arg	
O?	ABO*O.02.04	O50	O²–4	2	53G>T	Arg18Leu	Caucasians
				5	220C>T	Pro74Ser	(Several)
				6	297A>G	–	
				7	488C>T	Thr163Met	
				7	526C>G	Arg176Gly	
				7	**802G>A**	Gly268Arg	
O?	ABO*O.03	O08	O³	7	467C>T	Pro156Leu	Caucasians
				7	**804insG**	269fs#	(Rare)
				7	**1061delC**	354fs#	
O?	ABO*O.04	O51	O⁴, O41	2	**87_88insG**	29fs56Stop	Caucasians (Rare)
O?	ABO*O.05	O52	O⁵	6	**322C>T**	Gln108Stop	Caucasians (Rare)
O?	ABO*O.06	O53	O⁶	7	**542G>A**	Trp181Stop	Caucasians (Rare)
O	ABO*O.07	O14	O301	7	467C>T	Pro156Leu	Japanese
				7	**893C>T**	Ala298Val	(Rare)
O	ABO*O.08	O15	O302	7	**927C>A**	Tyr309Stop	Japanese (Rare)
O?	ABO*O.09.01	O19	R102	7	646T>A	Phe216Ile	Japanese□
				7	681G>A	–	(Rare)
				7	771C>T	–	
				7	829G>A	Val277Met	

(Continued)

ABO

(Continued)

Pheno-type	Provisional ISBT allele name	dbRBC allele name	Other allele names§	Exon	Nucleotide	Amino acid change	Ethnicity
O?	ABO*O.09.02	O20	R103	6	297A>G	–	Japanese□
				7	646T>A	Phe216Ile	(Rare)
				7	681G>A	–	
				7	771C>T	–	
				7	829G>A	Val277Met	
O?	ABO*O10	O72	–	2	**67insG**	23fs	

§For instance, by the authors of the original publication.
‡These two frame-shifts neutralize each other and the protein product has normal size (354 amino acids).
□These hybrid alleles involve exon 7 of ABO*O.01.02, and are similar or identical to so-called A^x hybrids (ABO*AW.31.01 to ABO*AW.31.05). The allele *in trans* may determine the resulting phenotype.

^Multiple variants of these alleles exist with numerous silent SNPs, but >95% of all *O* alleles depend on 261delG for their inactivation. The major alleles inactivated by principally different mechanisms are listed.

Comments

Hybrid *ABO* alleles are not uncommon. Null alleles including exon 6 from one of the common *O* alleles containing 261delG but with exon 7 elements from *A²* or *B* may interfere with *ABO* genotyping and cause risk for erroneous reporting of *A²* and *B* alleles, respectively. Since the products of these *O* alleles are truncated in the same way as *ABO*O.01.01* and *ABO*O.01.02*, they have not been included in the above table, as is also the case for other allelic variants featuring 261delG. If only position 261 has been tested to determine *O* allele status, *ABO*O.01* can be used for reporting.

Most of the *O* alleles lacking 261delG have been involved in serological ABO typing discrepancies. Despite the presence of premature stop codons in various exons, the phenotype produced varies from group O to weak expression of A antigen, but weakening of anti-A and anti-A1 is often observed. It can therefore be debated if they should be classified as *O* alleles or not. The mechanisms behind these phenomena have not yet been clarified.

Function

The A and B glycosyltransferase use, respectively, UDP-GalNAc and UDP-Gal as their donor substrates, and the various forms of H precursor structures as their acceptor substrates. There have been many speculative arguments as to the general function of ABO structures but it is clear that human "knock-outs," i.e., individuals with the null phenotype (group O), are not seriously affected. Instead, this phenotype is actually beneficial in many situations, perhaps best exemplified by its protective role in decreasing rosetting in severe malaria[7]. In an evolutionary perspective, glycan diversity (including ABO differences) is thought to have been a key survival factor, serving as a primitive immune system via the herd immunity concept[8]. Accordingly, it was important that members of a certain population had different carbohydrate epitopes serving as involuntary receptors for pathogen lectins, and also different naturally-occurring antibodies neutralizing pathogens with ABO-mimicking sugar coats on their surfaces. Only in this way would the population be likely to survive over time despite lethal pandemics. In addition to this fundamental function, other potential functions of ABO antigens have been proposed, including roles in embryogenesis, cell–cell interaction in carcinogenesis[9], and modulation of sialic acid recognition[10].

Disease association[5,7,11]

Expression of A and B antigens may be weakened in normal states such as during pregnancy, at young or old age, as well as due to disease. Weakening or even disappearance of A and B antigens can be the result of chromosome 9 translocations, development of (pre-) malignant hematological disease such as leukemia (especially in the acute and chronic myelogenous types) and myelodysplastic syndrome, and any disease inducing stress hematopoiesis, e.g., thalassemia and Diamond-Blackfan anemia. Stress hematopoiesis induces reduced branching of carbohydrate chains, and thus fewer A, B, H, and I antigens. Changes in sugar chains are often observed during carcinogenesis, and therefore altered ABH antigens can be considered tumor markers. Loss of A and B antigens have been described in various solid tumors. The acquired B antigen is a consequence of microbial infection in which the terminal GalNAc of the A antigen is deacetylated by bacterial enzymes, and thereby is made more similar to Gal. ABO phenotypes are associated with susceptibility to numerous diseases including cancer, thrombosis, and bleeding. There is also a strong correlation between susceptibility to certain infections and ABO status, most prominently severe forms of *Plasmodium falciparum* malaria.

Phenotypes (% occurrence)

	Caucasians^	Blacks^	Asian^	Mexican
A_1	33	19	27	22
A_2	10	8	rare	6
B	9	20	25	13
O	44	49	43	55
A_1B	3	3	5	4
A_2B	1	1	rare	rare
Null	O is the amorph; O_h (the Bombay phenotype) depends on the *FUT1/FUT2* loci, see H system [**018**]			
Unusual	Many subgroups of A and B (see phenotype tables in major text books)			

^Major variation occurs between ethnic subgroups within these main populations.

Comments

Aberrant ABO results created by modern medical practices (in addition to the mixed-field reactions seen after transfusion of ABO-non-identical RBCs) include: ABO-non-identical stem cell transplantation; ABO-incompatible solid organ transplantation; *in vitro* fertilization; artificial insemination; surrogate motherhood.

There is a vast literature on clinical, serological, microbiological, biochemical, enzymatic, structural, and molecular genetic aspects of the ABO system. The reader is recommended to search the databases for the numerous reviews written on various aspects of these different ABO-related topics.

References

[1] Yamamoto, F., et al., 1990. Cloning and characterization of DNA complementary to human UDP-GalNAc: Fuca1—>2Gala1—>3GalNAc transferase (histo-blood group A transferase) mRNA. J Biol Chem 265, 1146–1151.

[2] Yamamoto, F., et al., 1990. Molecular genetic basis of the histo-blood group ABO system. Nature 345, 229–235.

[3] Clausen, H., Hakomori, S., 1989. ABH and related histo-blood group antigens; immunochemical differences in carrier isotypes and their distribution. Vox Sang 56, 1–20.

[4] Chester, M.A., Olsson, M.L., 2001. The ABO blood group gene: a locus of considerable genetic diversity. Transfus Med Rev 15, 177–200.

[5] Storry, J.R., Olsson, M.L., 2009. The ABO blood group system revisited: a review and update. Immunohematology 25, 48–59.

[6] Yamamoto, F., 2004. Review: ABO blood group system – ABH oligosaccharide antigens, anti-A and anti-B, A and B glycosyltransferases and ABO genes. Immunohematology 20, 3–22.

[7] Cserti, C.M., Dzik, W.H., 2007. The ABO blood group system and *Plasmodium falciparum* malaria. Blood 110, 2250–2258.

[8] Gagneux, P., Varki, A., 1999. Evolutionary considerations in relating oligosaccharide diversity to biological function. Glycobiology 9, 747–755.

[9] Ichikawa, D., et al., 1997. Histo-blood group A/B versus H status of human carcinoma cells as correlated with haptotactic cell motility: approach with A and B gene transfection. Cancer Res 57, 3092–3096.

[10] Cohen, M., et al., 2009. ABO blood group glycans modulate sialic acid recognition on erythrocytes. Blood 114, 3668–3676.

[11] Anstee, D.J., 2010. The relationship between blood groups and disease. Blood 115, 4635–4643.

A Antigen

Terminology

ISBT symbol (number)   ABO1 (001001 or 1.1)
Other names and history   See System page

Occurrence

Caucasians	43%
Blacks	27%

Asians	28%
Mexicans	28%
South American Indians	0%

These numbers do not include group AB, which would increase the numbers (all except South American Indians) by approximately 4%.

Expression

Cord RBCs	Weak
Altered	Weak in A subgroup and other variants; some diseases

Molecular basis associated with A antigen

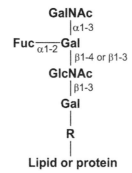

See ABO Blood Group System page for genetic basis of A subgroups.

Effect of enzymes and chemicals on A antigen on intact RBCs

Ficin/Papain	Resistant (markedly enhanced)
Trypsin	Resistant (markedly enhanced)
α-Chymotrypsin	Resistant (markedly enhanced)
DTT 200 mM	Resistant
Acid	Resistant

In vitro characteristics of alloanti-A

Immunoglobulin class	IgM; IgG
Optimal technique	RT or below; IAT for IgG component[^]
Neutralization	Saliva from A secretors
Complement binding	Yes; some hemolytic

[^]May be particularly relevant when testing or titrating the sera of platelet donors, pregnant women, and patients waiting for or having undergone ABO-incompatible organ transplantation or following ABO-non-identical stem cell transplantation complicated by pure red cell aplasia due to anti-A.

Clinical significance of alloanti-A

Transfusion reaction	None to severe; immediate/delayed; intravascular/extravascular
HDFN	No to moderate (rarely severe)

Autoanti-A

Rare

Comments

Serum from group A individuals contains naturally-occurring anti-B (see B antigen section).

B Antigen

Terminology

ISBT symbol (number)	ABO2 (001002 or 1.2)
Other names and history	See System page

Occurrence

Caucasians	9%
Blacks	20%
Asians	27%
Mexicans	13%
South American Indians	0%

These numbers do not include group AB, which would increase the numbers (all except South American Indians) by approximately 4%.

Expression

Cord RBCs	Weak
Altered	Weak in B subgroup and other variants; some diseases

Molecular basis associated with B antigen

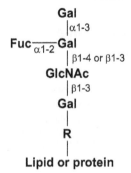

See ABO system page for genetic basis of B subgroups.

Effect of enzymes and chemicals on B antigen on intact RBCs

Ficin/Papain	Resistant (markedly enhanced)
Trypsin	Resistant (markedly enhanced)
α-Chymotrypsin	Resistant (markedly enhanced)
DTT 200 mM	Resistant
Acid	Resistant

In vitro characteristics of alloanti-B

Immunoglobulin class	IgM; IgG
Optimal technique	RT or below; IAT for IgG component[^]
Neutralization	Saliva from B secretors
Complement binding	Yes; some hemolytic

[^]May be particularly relevant when testing or titrating the sera of platelet donors, pregnant women, and patients waiting for or having undergone ABO-incompatible organ transplantation or following ABO-non-identical stem cell transplantation complicated by pure red cell aplasia due to anti-B.

Clinical significance of alloanti-B

Transfusion reaction	No to severe; immediate/delayed; intravascular/extravascular
HDFN	No to moderate (rarely severe)

Autoanti-B

Rare

Comments

Serum from group B individuals contains naturally-occurring anti-A and anti-A1 (see A and A1 antigen sections).

A,B Antigen

Terminology

ISBT symbol (number)	ABO3 (001003 or 1.3)
History	Discussed since the 1950s and many hypotheses about the molecular background of this antigen have been presented. At one point referred to as the C antigen of the ABO system. Acknowledged by the ISBT in the first workshop report in 1982.

Occurrence

Found in all individuals expressing A and/or B antigens (e.g., 56% of Caucasians according to the phenotype table in the ABO system section above).

Expression

Cord RBCs	Weak
Altered	Weakened when A and B antigens are weak

Molecular basis associated with A,B antigen[1]

Once the molecular differences between A and B antigens were revealed, it was hypothesized that the structures common to the oligosaccharide terminals ending with α3GalNAc and α3Gal, respectively, would make up the epitope recognized by anti-A,B. The only difference between A and B antigens is situated at carbon position 2 of the terminal residue, i.e., A has a NHAc group, whereas B has an OH group. The A,B epitope was shown to depend on the surface common to the terminal sugars in A and B. Monoclonal anti-A,B and polyclonal anti-A,B from group O sera react with this epitope.

Effect of enzymes and chemicals on A,B antigen on intact RBCs

Ficin/Papain	Resistant (markedly enhanced)
Trypsin	Resistant (markedly enhanced)
α-Chymotrypsin	Resistant (markedly enhanced)
DTT 200 mM	Resistant
Acid	Resistant

In vitro characteristics of alloanti-A,B

Immunoglobulin class	IgM; IgG
Optimal technique	RT or below; IAT for IgG component[^]
Neutralization	Saliva from A, B or AB secretors
Complement binding	Rare

[^]May be particularly relevant when testing or titrating the sera of platelet donors, pregnant women, and patients waiting for or having undergone ABO-incompatible organ transplantation or following ABO-non-identical stem cell transplantation complicated by pure red cell aplasia due to anti-A and anti-B.

Clinical significance of anti-A,B

Transfusion reaction	None to mild/delayed
HDFN	No to severe[^]

[^]No data to differentiate anti-A or anti-B from anti-A,B as the implicated antibody specificity in affected fetuses/infants to group O mothers in whom ABO-related HDFN is more common. However, group O individuals (who are the only anti-A,B makers) produce more IgG antibodies against ABO antigens compared to A and B individuals. Anti-A,B crosses the placenta more frequently than do anti-A and anti-B.

Autoantibody

Rare

Comments

Anti-A,B is an antibody specificity selectively found in the plasma of group O individuals and that cannot be separated into anti-A and anti-B.

Polyclonal anti-A,B from group O individuals reacts strongly with A_x subgroup RBCs, while anti-A may not react or react only weakly.

Reference

[1] Korchagina, E.Y., et al., 2005. Design of the blood group AB glycotope. Glycoconj J 22, 127–133.

A1 Antigen

Terminology

ISBT symbol (number)	ABO4 (001004 or 1.4)
History	The two major A subgroups, A_1 and A_2, were discovered already in 1911 by von Dungern and Hirszfeld, but remained unnamed until 1930 when the blood group A phenotype was subdivided by the presence or absence of (what later became known as) A1 antigen, recognized by antibodies in the serum of some individuals with the A_2 phenotype

Occurrence

Caucasians	34% (approximately 80% of group A)
Blacks	19%
Asians	27%

These numbers do not include group AB.

Expression

Cord RBCs	Weak
Altered	A_{int} (a subgroup with phenotype characteristics in between A_1 and A_2)

Acquired B syndrome (almost exclusively found in A1-positive individuals following modification of the terminal GalNAc of the A antigen on their RBCs by a bacterial deacetylase during infection to cause crossreactivity with some anti-B reagents).

Molecular basis associated with A1 antigen[1-6]

Even if the difference between the A1-positive A_1 phenotype and the A1-negative A_2 phenotype can easily be demonstrated by simple hemagglutination assays using lectin reagents diluted correctly (e.g., *Dolichos biflorus* positive with A_1 RBCs and negative with A_2, whilst *Ulex europaeus* is negative with A_1 and positive with A_2), the molecular identity of the A1 antigen is still under debate. It is unequivocal that a quantitative difference exists so that A_1 RBCs possess approximately 4–5 times more A antigen than A_2 RBCs. In addition, the common A^2 allele has distinct features (mainly the 1061delC polymorphism) compared to A^1 alleles, and encodes a glycosyltransferase that is qualitatively different (e.g., regarding its pH optimum, enzymic activity, and molecular size) from the A_1 transferase.

A_1 RBCs have higher amounts of A type 3 glycolipid (the repetitive A epitope) and A type 4 (globo-A). A_2 RBCs may have A type 3 at levels that are difficult to detect serologically on whole RBCs compared to A_1, and may lack A type 4.

Effect of enzymes and chemicals on A1 antigen on intact RBCs

Ficin/Papain	Resistant (markedly enhanced)
Trypsin	Resistant (markedly enhanced)
α-Chymotrypsin	Resistant (markedly enhanced)
DTT 200 mM	Resistant
Acid	Resistant

In vitro characteristics of alloanti-A1

Immunoglobulin class	IgM more common than IgG
Optimal technique	RT or below
Neutralization	Saliva from A (A_1) secretors
Complement binding	Rare

Clinical significance of alloanti-A1

Transfusion reaction	None to mild/delayed
HDFN	No

Autoantibody

Rare

Comments

Anti-A1 is found in serum from 1–2% of A_2 and 25% of A_2B individuals, and is a component of anti-A from group O and B people.

References

[1] Clausen, H., et al., 1984. Blood group A glycolipid (Ax) with globo-series structure which is specific for blood group A1 erythrocytes: one of the chemical bases for A1 and A2 distinction. Biochem Biophys Res Commun 124, 523–529.

[2] Clausen, H., et al., 1985. Repetitive A epitope (type 3 chain A) defined by blood group A1-specific monoclonal antibody TH-1: chemical basis of qualitative A1 and A2 distinction. Proc Natl Acad Sci USA 82, 1199–1203.

[3] Clausen, H., et al., 1986. Novel blood group H glycolipid antigens exclusively expressed in blood group A and AB erythrocytes (type 3 chain H). II. Differential conversion of different H substrates by A1 and A2 enzymes, and type 3 chain H expression in relation to secretor status. J Biol Chem 261, 1388–1392.

[4] Schachter, H., et al., 1973. Qualitative differences in the N-acetyl-D-galactosaminyltransferases produced by human A1 and A2 genes. Proc Natl Acad Sci USA 70, 220–224.

[5] Svensson, L., et al., 2009. Blood group A(1) and A(2) revisited: an immunochemical analysis. Vox Sang 96, 56–61.

[6] Yamamoto, F., et al., 1992. Human histo-blood group A2 transferase coded by A2 allele, one of the A subtypes, is characterized by a single base deletion in the coding sequence, which results in an additional domain at the carboxyl terminal. Biochem Biophys Res Commun 187, 366–374.

ABO

MNS Blood Group System

Number of antigens 46

Polymorphic	M, N, S, s
Low prevalence	He, Mi^a, M^c, Vw, Mur, M^g, Vr, M^e, Mt^a, St^a, Ri^a, Cl^a, Ny^a, Hut, Hil, M^v, Far, s^D, Mit, Dantu, Hop, Nob, Or, DANE, TSEN, MINY, MUT, SAT, ERIK, Os^a, HAG, MARS, MNTD
High prevalence	U, En^a, ENKT, "N", ENEP, ENEH, ENAV, ENDA, ENEV

Terminology

ISBT symbol (number)	MNS (002)
CD number	CD235A (GPA); CD235B (GPB)
Obsolete name	MNSs
History	Discovered in 1927 by Landsteiner and Levine; named after the first three antigens identified in this system: M, N, and S.

Expression

Tissues	Renal endothelium and epithelium (the GPA may not be fully glycosylated, as only sialic acid independent anti-M and -N react)

Gene

Chromosome	4q31.21
Name	*GYPA, GYPB*
Organization	*GYPA*: 7 exons distributed over 60 kbp *GYPB*: 5 exons (and 1 pseudoexon) distributed over 58 kbp
Product	Glycophorin A (GPA; MN sialoglycoprotein; SGPα) Glycophorin B (GPB; Ss sialoglycoprotein; SGPδ)

The Blood Group Antigen (3/e). DOI: http://dx.doi.org/10.1016/B978-0-12-415849-8.00004-1

A third gene (*GYPE*), which is adjacent to *GYPB*, may not encode an RBC
membrane component, but participates in gene rearrangements resulting in
variant alleles.

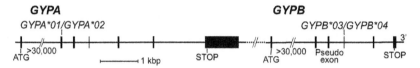

GenBank accession numbers

Database	*GYPA*	*GYPB*	*GYPE*
GenBank	NM_002099 and L31860	NM_002100 and J02982	NM_002102
Entrez Gene ID	2993	2994	2996

Exon numbering accounts for the presence of the pseudoexon(s) in *GYPB* and
GYPE. Thus, *GYPB* pseudoexon 3 corresponds to the *GYPA* exon 3 sequence.
This *GYPB* pseudoexon is involved in many gene rearrangements encoding
hybrid glycophorins in this blood group system. Similarly, *GYPE* pseudoexons
3 and 4 correspond to *GYPA* exon 3 and 4 sequences. These *GYPE* pseudoexons
are involved in gene rearrangements encoding hybrids.

Molecular basis of MNS Phenotypes

MNS alleles with single nucleotide changes in GYPA that generate blood group antigens

Reference allele *GYPA*01* or *GYPA*M* (Accession number L31860) encodes
M (**MNS1**), MNS28, MNS29, MNS39, MNS40, MNS42, MNS44, and
MNS45. Nucleotide differences, and amino acids affected, are given. Note: In
most cases, the nucleotide changes can also occur on a *GYPA*N* allele.

Allele encodes	Allele name	Exon	Nucleotide	Amino acid[‡]	Ethnicity (prevalence)
N+ or MNS:2	*GYPA*02* or *GYPA*N*	2	59C>T; 71G>A; 72T>G	Ser20Leu, Gly24Glu	(Common)
M^c+ or MNS:1,–2,8[†]	*GYPA*08* or *GYPA*Mc*	2	71G>A; 72T>G	Gly24Glu	(Several)

(Continued)

(Continued)

Allele encodes	Allele name	Exon	Nucleotide	Amino acid[‡]	Ethnicity (prevalence)
Vw+ or GP.Vw MNS:7,9,−40	GYPA*09 or GYPA*Vw	3	140C>T	Thr47Met	Europeans, especially Swiss (Many)
Mᵍ+ or MNS:−1, −2,11,32	GYPA*11 or GYPA*Mg	2	68C>A	Thr23Asn	Europeans, especially Swiss & Sicilians (Many)
Vr+ or MNS:12	GYPA*12 or GYPA*Vr	3	197C>A	Ser66Tyr	Dutch (Few)
Mt(a+) or MNS:14	GYPA*14 or GYPA*Mta	3	230C>T	Thr77Ile	(Many)
Ri(a+) or MNS:16	GYPA*16 or GYPA*Ria	3	226G>A	Glu76Lys	(Rare)
Ny(a+) or MNS:18	GYPA*18 or GYPA*Nya	3	138T>A	Asp46Glu	Norwegians (Many)
Hut+ or MNS:7,19,−40	GYPA*19 or GYPA*Hut	3	140C>A	Thr47Lys	(Many)
Or+ or MNS:31	GYPA*31 or GYPA*Or	3	148C>T	Arg50Trp	(Few)
ERIK+ or MNS:37	GYPA*37 or GYPA*ERIK^	4	232G>A	Gly78Arg	(Few)
Os(a+) or MNS:38	GYPA*38 or GYPA*Osa	3	217C>T	Pro73Ser	Japanese (Rare)
HAG+ or MNS:−39,41	GYPA*41 or GYPA*HAG	4	250G>C	Ala84Pro	Israeli (Rare)
MARS+ or MNS:−42,43	GYPA*43 or GYPA*MARS	4	244C>A	Gln82Lys	Choctaw Indians (Several)
ENEV− or MNS:−45	GYPA*−45	4	242T>G	Val81Gly	(Rare)
MNTD+or MNS:46	GYPA*46 or GYPA*MNTD	3	107C>G	Thr36Arg	(Rare)

[†] = Most anti-M but only few anti-N react with Mᶜ+ RBCs.
^ = Transcript 1; see also GYP*EBH in hybrid table for transcript 2.
[‡] = Amino acid #1 is the initiation methionine, which is +19 from the number given in earlier reports.

MNS alleles with single nucleotide changes in GYPB that generate blood group antigens

Reference allele *GYPB*04* or *GYPB*s* (Accession number J02982) encodes "N" (**MNS30**), s (**MNS4**). Nucleotide differences from this reference allele, and amino acids affected, are given. Expression of the U antigen involves GPB and another protein, probably RhAG.

Allele encodes	Allele name	Exon (intron)	Nucleotide	Amino acid[‡]	Ethnicity (prevalence)
S+ or MNS:3	GYPB*03 or GYPB*S	4	143C>T	Thr48Met	(Common)
s+He+ or MNS:4,6	GYPB*06.01	2	59T>G	Leu20Trp	Africans (Many)
		2	60A>G		
		2	67A>T	Thr23Ser	
		2	71A>G	Glu24Gly	
		2	72G>T		
S+He+ or MNS:3,6	GYPB*06.02	2	59T>G	Leu20Trp	Africans (Many)
		2	60A>G		
		2	67A>T	Thr23Ser	
		2	71A>G	Glu24Gly	
		2	72G>T		
		4	143C>T	Thr48Met	
Mᵛ+ or MNS:21	GYPB*21 or GYPB*Mv	2	65C>G	Thr22Ser	(Several)
sᴰ+ or MNS:23	GYPB*23 or GYPB*sD	4	173C>G	Pro58Arg	South African (Rare)
Mit+ or MNS:24	GYPB*24 or GYPB*Mit	4	161G>A	Arg54His	(Many)
S–U+ʷ or MNS:–3,w5	GYPB*03N.01 or GYPB*NY	4	143C>T	Thr48Met	Africans (Many)
		5	208G>T	Silent	
		5	230C>T	Silent	
		5	251C>G	Ser84Thr	
		(Intron 5)	+5g>t		
S–U+ʷ or MNS:–3,w5	GYPB*03N.02 or GYP*He(NY)	2	59T>G	Leu20Trp	Africans (Many)
		2	60A>G		
		2	67A>T	Thr23Ser	
		2	71A>G	Glu24Gly	
		2	72G>T		
		4	143C>T	Thr48Met	
		5	208G>T	Silent	
		5	230C>T	Silent	
		5	251C>G	Ser84Thr	
		(Intron 5)	+5g>t		

(Continued)

(Continued)

Allele encodes	Allele name	Exon (intron)	Nucleotide	Amino acid[‡]	Ethnicity (prevalence)
S–U+^{w} or MNS:–3,w5	GYPB*03N.03 or GYPB*P2	4 (Intron 5)	143C>T +5g>t	Thr48Met	Africans (Many)
S–U+^{w} or MNS:–3,w5	GYPB*03N.04 or GYP*He(P2)	2 2 2 2 2 4 (Intron 5)	59T>G 60A>G 67A>T 71A>G 72G>T 143C>T +5g>t	Leu20Trp Thr23Ser Glu24Gly Thr48Met	Africans (Many)

[‡] = Amino acid #1 is the initiation methionine, which is +19 from the number given in earlier reports.

MNS alleles created by gene rearrangement events within the *GYP* gene family

Parent allele *GYPA*

Allele encodes	Allele name	Nucleotide	Protein	Comments
GYP(A-A) hybrid series				
MNS:15 or St(a+); GP.Zan	GYP*101.01 or GYP*Zan	GYPA del exon 3	GPA del46-77	GYP(A1–2–BΨ3–A4–7) Africans (Many)
MNS:15, or St(a+); GP.EBH	GYP*101.02 or GYP*EBH^	GYPA 232G> A del exon 3	GPA del46-77	Nucleotide change at 232 destabilizes normal splicing. St^a is encoded by a GYPA transcript that lacks exon 3. Full-length transcript encodes ERIK (MNS37).
MNS:15 or St(a+); GP.Mar	GYP*101.03 or GYP*Mar	GYPA del exon 3	GPA del46-77	GYP(A1–2–EΨ3–A4–7)
MNS:6,15 or He+ St(a+); GP.Cal	GYP*101.04 or GYP*Cal	GYPA 58G>T, 67A>T	GPA Ser20Trp, Thr23Ser GPA del 46-77	GYP(A1–2–BΨ3–A4–7)

(Continued)

MNS

(Continued)

Allele encodes	Allele name	Nucleotide	Protein	Comments
GYP(A-B) hybrid series				
MNS:-3,4,20,34 or S- s+ Hil+ MINY+;GP.Hil	GYP*201.01 or GYP*Hil	GYP(A1–232– B233–312)	GP(A1–77– B78–104)	
MNS:3,-4,32, 33,34 or S+s- TSEN+ MINY+;GP.JL	GYP*202.01 or GYP*JL	GYP(A1–232– B233–312) 239C>T	GP(A1–77– B78–104) Thr80Met	
MNS:-1,2,-3,-4, -5,36 or M- N+ S-s-U- SAT+; GP.SAT	GYP*203.01 or GYP*SAT	GYP(A1–271– B272–369) 59C>T; 71G>A; 72T>G	GP(A1–90– B91–123) Ser20Leu, Gly24Glu	Previously GYP*TK
GYP(A-B-A) hybrid series				
MNS:10,32 or Mur+ Dane+; GP.Dane	GYP*301.01 or GYP*Dane	GYP(A1–159– BΨ160–177– A178–450); 191T>A	GP(A1–52– B53–58– A59–149) Ile65Asn	
MNS:10,32,-44 or Mur+ Dane+ ENDA-	GYP*301.02	GYP(A1–159– BΨ160–177– A178–450)	GP(A1–52– B53–58– A59–149)	Does not express ENDA
MNS:26,27 or Hop+ Nob+; GP.Joh	GYP*302.01 or GYP*Joh	GYP(A1–202– BΨ203–A204– 450)	GP(A1–67– B68–A69–150) Arg68Thr	Gene conversion in exon 3 replaces GYPA nucleotide 203 with the corresponding nucleotide from GYPBΨ3
MNS:-26,27,-29 or Hop–Nob+ ENKT–; GP.Nob	GYP*302.02 or GYP*Nob	GYP(A1–202– BΨ203–212– A213–450) 203G>C 212A>C	GP(A1–67– B68–72– A73–150) Arg68Thr; Tyr71Ser	Gene conversion in exon 3 replaces GYPA nucleotides (203–212) with corresponding nucleotides from GYPBΨ3
MNS:20,-34 or Hil+ MINY–; GP.KI	GYP*303 or GYP*KI	GYP(A1–238– B239–242– A243–450) 239G>C 242T>G	GP(A1–79– B80–81– A82–150) Arg80Thr Val81Gly	Gene conversion in exon 4 replaces GYPA nucleotides (239–242) with corresponding nucleotides from GYPB

^ = Transcript 2; see also GYP*ERIK in table above for transcript 1.

MNS alleles created by gene rearrangement events within the GYP gene family

Parent allele GYPB

Allele encodes	Allele name	Nucleotide	Amino acid	Comments
GYP(B-A) hybrid series				
MNS:15 or St(a+); GP.Sch	GYP*401 or GYP*Sch	GYP(B1–136–A137–354)	GP(B1–46–A47–118)	Reciprocal product is GYP*Hil
MNS:-3,4,25 or Dantu+; GP.Dantu	GYP*402 or GYP*Dantu	GYP(B1–175–A176–354)	GPB(1–58)-A(59–118)	Reciprocal product is GYP*Tk
GYP(B-A-B) hybrid series				
MNS:-3,4,7, 10,20,34,35 or S– s+ Mi(a+) Mur+ Hil+ MINY+ MUT+; GP.Mur	GYP*501 or GYP*Mur	GYP(B1–136– Bψ137–204– A205–229– B230–366)	GP(B1–69–A70– 77–B78–122) GPB^s ins 46–77 DKHKRDTYPAH TANEVSEISVRTV YPPEEET	Novel sequence derived from composite exon; GYPB 5' pseudoexon 3 + GYPA 3' exon 3
MNS:3,–4,7,10, 26,33,34,35 or S+s– Mi(a+) Mur+ Hop+ TSEN+ MINY+ MUT+; GP.Hop	GYP*502 or GYP*Hop	GYP(B1–136– Bψ137–204– A205–229– B230–366)	GP(B1–69–A70– 77–B78–122) GPB^S ins 46–77 DKHKRDTYPAH TANEVSEISVRTV YPPEEET	Novel sequence derived from composite exon; GYPB 5' pseudoexon 3 + GYPA 3' exon 3
MNS:-3,4,7,10, 20,26,34,35 or S– s+Mi(a+) Mur+ Hil+ MINY+ MUT+; GP.Bun	GYP*503 or GYP*Bun	GYP(B1–136– Bψ137–210– A211–229– B230–366)	GP(B1–71–A72– 77–B78–122) GPB^S ins 46–77 DKHKRDTYPAH TANEVSEISVRTVY PPEEET	Novel sequence derived from composite exon; GYPB 5' pseudoexon 3 + GYPA 3' exon 3
MNS:-3,4,7,20,34,35 or S– s+ Mi(a+) Hil+ MINY+ MUT+; GP.HF	GYP*504 or GYP*HF	GYP(B1–136– Bψ137–159– A160–232– B233–369)	GP(B1–53–A54– 78–B79–123) in GPB^s DKHKRDTYAAT PRAHEVSEISVRT VYPPEEET46– 77ins	Novel sequence derived from composite exon; GYPB 5' pseudoexon 3 + GYPA 3' exon 3
MNS:-3,–4,–5,6 or S– s– U– He+; GP.GL	GYP*505 or GYP*He(GL)	GYP(B1–12– A13–78– B79–168)	GP(B1–4–A5–26– B27–59)	Blacks (Rare)

For the molecular bases of other MNS antigens see dbRBC.

Molecular basis of silencing *GYPA* or *GYPB* or *GYPA* and *GYPB*

Phenotype	Allele name	Nucleotide	Amino acid	Ethnicity (number)
MNS:–1,–2,–28 or M– N– En(a–)	*GYPA*01N*	Del *GYPA* exons 2–7; *GYPB* exon 1	GPA absent	(Rare)
MNS:–3,–4,–5 or S– s– U–	*GYPB*01N*	Del *GYPB* exons 2–5; *GYPE* exon 1	GPB absent	Blacks (Many)
MNS:–1,–2,–3,– 4,–5 or M– N– S– s– or M^kM^k	*GYP*01N*	Del *GYPA* exons 2–7; *GYPB* exons 1–5 ; GYPE exon 1	GPA and GPB absent	(Rare)

Amino acid sequence

Glycophorin A^M:

```
MYGKIIFVLL  LSAIVSISAS  STTGVAMHTS  TSSSVTKSYI  SSQTNDTHKR  50
DTYAATPRAH  EVSEISVRTV  YPPEEETGER  VQLAHHFSEP  EITLIIFGVM  100
AGVIGTILLI  SYGIRRLIKK  SPSDVKPLPS  PDTDVPLSSV  EIENPETSDQ  150
```

Glycophorin B^s:

```
MYGKIIFVLL  LSEIVSISAL  STTEVAMHTS  TSSSVTKSYI  SSQTNGETGQ  50
LVHRFTVPAP  VVIILIILCV  MAGIIGTILL  ISYSIRRLIK  A           91
```

Both *GYPA* and *GYPB* encode a signal peptide of 19 amino acids, which are cleaved from the membrane bound protein.
Amino acid numbers are now +19 from previous reports.

Carrier molecule[1,2]

GPA and GPB are single-pass membrane sialoglycoproteins (type I).

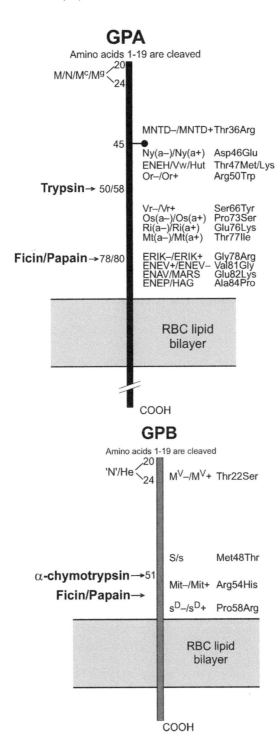

GPA is cleaved by trypsin at residues 50 and 58 on intact RBCs.
GPB is cleaved by α-chymotrypsin at residue 51 on intact RBCs.

	GPA	GPB
M_r (SDS-PAGE)	43,000	25,000
CHO: N-glycan	1 site	none
CHO: O-glycan	15 sites	11 sites
Copies per RBC	800,000	200,000

Function

Chaperone for band 3 transport to RBC membrane. Major component contributing to the negatively-charged RBC glycocalyx.

Disease association

Receptor for complement, bacteria, and viruses[3,4]. Involved in *Plasmodium falciparum* invasion of RBCs[5-7].

Phenotypes (% occurrence)

Phenotype	Caucasians	Blacks
M+N–S+s–	6	2
M+N–S+s+	14	7
M+N–S–s+	8	16
M+N+S+s–	4	2
M+N+S+s+	24	13
M+N+S–s+	22	33
M–N+S+s–	1	2
M–N+S+s+	6	5
M–N+S–s+	15	19
M+N–S–s–	0	0.4

(Continued)

(Continued)

Phenotype	Caucasians	Blacks
M+N+S–s–	0	0.4
M–N+S–s–	0	0.7
Null:	M^kM^k. An absence of GPA=En(a–) and an absence of GPB=U–	
Unusual:	Various hybrids, GPB is decreased in Rh_{null} and Rh_{mod} RBCs	

Glycophorin phenotypes and associated antigens [previously the Miltenberger (Mi) subsystem]

	MNS antigen name and ISBT number (MNS#)										
Phenotype (Previous Name)	Mi^a (7)	Vw (9)	Mur (10)	Hut (19)	Hil (20)	Hop (26)	Nob (27)	DANE (32)	TSEN (33)	MINY (34)	MUT (35)
GP.Vw (Mi.I)	+	+	0	0	0	0	0	0	0	0	0
GP.Hut (Mi.II)	+	0	0	+	0	0	0	0	0	0	+
GP.Mur (Mi.III)	+	0	+	0	+	0	0	0	0	+	+
GP.Hop (Mi.IV)	+	0	+	0	0	+	0	0	+	+	+
GP.Hil (Mi.V)	0	0	0	0	+	0	0	0	0	+	0
GP.Bun (Mi.VI)	+	0	+	0	+	+	0	0	0	+	+
GP.Nob (Mi.VII)	0	0	0	0	0	0	+	0	0	0	0
GP.Joh (Mi.VIII)	0	0	0	0	0	+	+	0	NT	0	0
GP.Dane (Mi.IX)	0	0	+	0	0	0	0	+	0	0	0
GP.HF (Mi.X)	+	0	0	0	+	0	0	0	0	+	+
GP.JL (Mi.XI)	0	0	0	0	0	0	0	0	+	+	0

Hybrid glycophorin molecules, phenotypes, and associated low incidence antigens

Hybrid allele	Glycophorin	Phenotype symbol	Associated novel antigens
GYP(A-B)	GP(A-B)	GP.Hil (Mi.V) GP.JL (Mi.XI) GP.TK	Hil, MINY TSEN, MINY SAT
GYP(B-A)	GP(B-A)	GP.Sch (M^r) GP.Dantu	St^a Dantu
GYP(A-B-A)	GP(A-B-A)	GP.Mg GP.KI	M^g, DANE Hil
GYP(B-A-B)	GP(B-A-B)	GP.Mur (Mi.III) GP.Bun (Mi.VI) GP.HF (Mi.X) GP.Hop (Mi.IV)	Mi^a, Mur, MUT, Hil, MINY Mi^a, Mur, MUT, Hop, Hil, MINY Mi^a, MUT, Hil, MINY Mi^a, Mur, MUT, Hop, TSEN, MINY
	GP(A-B)	GP.He; (P2, GL)	He
GYP(B-A-ψB-A)	GP(A-A)	GP.Cal	He, St^a
GYP(A-ψB-A)	GP(A-B-A)	GP.Vw (Mi.I) GP.Hut (Mi.II) GP.Nob (Mi.VII) GP.Joh (Mi.VIII) GP.Dane (Mi.IX)	Mi^a, Vw Mi^a, Hut, MUT Nob Nob, Hop Mur, DANE
	GP(A-A)	GP.Zan (M^z)	St^a
GYPA 179G>A	GPA GP(A-A)	GP.EBH GP.EBH	ERIK (from transcript 1) St^a (from transcript 2)
GYP(A-ψE-A)	GPA-A	GP.Mar	St^a

Comments

Linkage disequilibrium exists with M/N and S/s antigens.

MNS antigens associated with atypical glycosylation have been placed into Collection 213 (MN CHO) and include: Hu, M_1, Tm, Can, Sext, and Sj.

GPA and GPB are the major sialic acid-containing structures of the RBC membrane. The majority of the sialic acid is on the O-glycans.

GPA-deficient RBCs have a weak expression of Ch and Rg antigens.

References

[1] Huang, C.-H., Blumenfeld, O.O., 1995. MNSs blood groups and major glycophorins: molecular basis for allelic variation. In: Cartron, J.-P., Rouger, P. (Eds.), Molecular Basis of Human Blood Group Antigens. Plenum Press, New York, NY, pp. 153–188.

[2] Reid, M.E., 1994. Some concepts relating to the molecular genetic basis of certain MNS blood group antigens. Transf Med 4, 99–111.

[3] Daniels, G., 1999. Functional aspects of red cell antigens. Blood Rev 13, 14–35.

[4] Moulds, J.M., et al., 1996. Human blood groups: incidental receptors for viruses and bacteria. Transfusion 36, 362–374.

[5] Hadley, T.J., et al., 1991. Recognition of red cells by malaria parasites: the role of erythrocyte-binding proteins. Transfusion Med Rev 5, 108–113.

[6] Ko, W.-Y., et al., 2011. Effects of natural selection and gene conversion on the evolution of human glycophorins coding for MNS blood polymorphisms in malaria-endemic African populations. Am J Hum Genet 88, 741–754.

[7] Miller, L.H., 1994. Impact of malaria on genetic polymorphism and genetic diseases in Africans and African Americans. Proc Natl Acad Sci USA 91, 2415–2419.

M Antigen

Terminology

ISBT symbol (number)	MNS1 (002001 or 2.1)
History	M, identified in 1927, was the first antigen of the MNS system. It was named after "immune," because anti-M was the result of immunizing rabbits with human RBCs.

Occurrence

Caucasians	78%
Blacks	74%

Antithetical antigen

N (**MNS2**)

Expression

Cord RBCs	Expressed
Altered	On some hybrid glycophorin molecules

Molecular basis associated with M antigen[1]

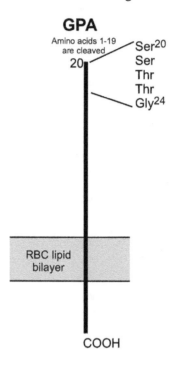

GPA

Amino acids 1-19 are cleaved

20

Ser20
Ser
Thr
Thr
Gly24

RBC lipid bilayer

COOH

Nucleotide	C at bp 59, G at bp 71, and T at bp 72 in exon 2 of *GYPA*

Recognition of antigen by anti-M is usually dependent on O-glycans attached to amino acid residues 21, 22, and 23 (previously 2, 3, and 4).

Effect of enzymes and chemicals on M antigen on intact RBCs

Ficin/Papain	Sensitive
Trypsin	Sensitive
α-Chymotrypsin	Resistant
Sialidase	Variable
DTT 200 mM	Resistant
Acid	Resistant

In vitro characteristics of alloanti-M

Immunoglobulin class	IgG (cold reactive; many direct agglutinins) and IgM
Optimal technique	4°C; RT; rarely also reactive by IAT

Clinical significance of alloanti-M

Transfusion reactions	No (except in extremely rare cases)
HDFN	No (except in extremely rare cases)

Autoanti-M

Rare; reactive at low temperatures.

Comments

Many examples of anti-M are naturally-occurring.

Acidification of serum enhances the reactivity of some anti-M. Anti-M often react more strongly with M+N– RBCs than with M+N+ RBCs (i.e., they show dosage).

Anti-M is more common in children than adults, and in patients with bacterial infections. It is not uncommon for pregnant M– women to produce anti-M, but to give birth to an M– baby.

Reference

[1] Dahr, W., et al., 1977. Different N-terminal amino acids in the MN-glycoprotein from MM and NN erythrocytes. Hum Genet 35, 335–343.

N Antigen

Terminology

ISBT symbol (number)	MNS2 (002002 or 2.2)
History	Identified shortly after the M antigen in 1927. It was named as the next letter after M, and for the fifth letter of "immune" because anti-N was the result of immunizing rabbits with human RBCs.

Occurrence

Caucasians	72%
Blacks	75%

Antithetical antigen

M (**MNS1**)

Expression

Cord RBCs	Expressed
Altered	On some hybrid glycophorin molecules

Molecular basis associated with N antigen[1]

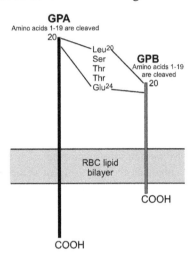

Nucleotide		T at bp 59, A at bp 71, and G at bp 72 in exon 2 of *GYPA*

Recognition of antigen by anti-N is often also dependent on O-glycans attached to amino acid residues 21, 22, and 23 (previously 2, 3, and 4).

Effect of enzymes and chemicals on N antigen on intact RBCs

	N on GPA	"N" on GPB
Ficin/Papain	Sensitive	Sensitive
Trypsin	Sensitive	Resistant
α-Chymotrypsin	Resistant	Sensitive
Sialidase	Variable	Variable
DTT 200 mM	Resistant	Resistant
Acid	Resistant	Resistant

In vitro characteristics of alloanti-N

Immunoglobulin class	IgM; IgG (some direct agglutinins)
Optimal technique	4°C; RT; rarely also reactive by IAT

Clinical significance of alloanti-N

Transfusion reaction	No
HDFN	No

Rare N–S–s–U– individuals make an antibody that reacts with N on GPA and GPB, and may be clinically significant.

Autoanti-N

Rare; found in patients on dialysis when equipment was sterilized with formal-dehyde (anti-Nf).

Comments

Most examples of anti-N are naturally-occurring.

The N antigen on GPB is denoted as "N" (**MNS30**) to distinguish it from N on GPA.

Anti-N typing reagents are formulated to detect N antigen on GPA but not on GPB.

Monoclonal anti-N are more specific for the N on GPA at alkaline pH.

Reference

[1] Dahr, W., et al., 1977. Different N-terminal amino acids in the MN-glycoprotein from MM and NN erythrocytes. Hum Genet 35, 335–343.

S Antigen

Terminology

ISBT symbol (number)	MNS3 (002003 or 2.3)
History	S was named after the city of Sydney (Australia), where the first example of anti-S was identified in 1947.

Occurrence

Caucasians	55%
Blacks	31%

Antithetical antigen

s (**MNS4**)

Expression

Cord RBCs	Expressed
Altered	On Rh_{null}, M^v+, Mit+ and TSEN+ RBCs

Molecular basis associated with S antigen[1]

Amino acid   Met48 (previously 29) of GPB
Nucleotide   T at bp 143 in exon 4 of *GYPB*

In addition to Met48, some anti-S also require Thr44 (previously 25) and/or the GalNAc attached to this residue, Glu47, His53, and Arg54 (previously 28, 34, and 35 respectively)[2].

Effect of enzymes and chemicals on S antigen on intact RBCs

Ficin/Papain   Variable
Trypsin   Resistant
α-Chymotrypsin   Sensitive
Sialidase   Variable
DTT 200 mM   Resistant
Acid   Resistant

In vitro characteristics of alloanti-S

Immunoglobulin class   IgM less common than IgG
Optimal technique   RT; IAT
Complement binding   Some

Clinical significance of alloanti-S

Transfusion reaction   No to moderate (rare)
HDFN   No to severe (rare)

Autoanti-S

Rare

Comments

Anti-S can be naturally-occurring.
There are approximately 1.5 times more copies of GPB in S+s− than in S−s+ RBCs. S+s+ RBCs have an intermediate amount of GPB[2].
S antigen is sensitive to trace amounts of chlorine[3,4].
Sera containing anti-S frequently contain antibodies to low prevalence antigens.

References

[1] Dahr, W., et al., 1980. Structure of the Ss blood group antigens, II: a methionine/threonine poly-morphism within the N-terminal sequence of the Ss glycoprotein. Hoppe-Seylers Z Physiol Chem 361, 895–906.

[2] Dahr, W., 1986. Immunochemistry of sialoglycoproteins in human red blood cell membranes. In: Vengelen-Tyler, V., Judd, W.J. (Eds.), Recent Advances in Blood Group Biochemistry. American Association of Blood Banks, Arlington, VA, pp. 23–65.

[3] Long, A., et al., 2002. Nondetection of the S antigen due to the presence of sodium hypochlorite. Immunohematology 18, 120–122.

[4] Rygiel, S.A., et al., 1985. Destruction of the S antigen by sodium hypochlorite. Transfusion, 274–277.

s Antigen

Terminology

ISBT symbol (number)	MNS4 (002004 or 2.4)
History	Anti-s was identified in 1951; it reacted with an antigen antithetical to S.

Occurrence

Caucasians	89%
Blacks	93%

Antithetical antigen

S (**MNS3**)

Expression

Cord RBCs	Expressed
Altered	Dantu+, Mit+, M^v+, s^D+, GP.Mur, GP.Hil, and some Rh_{null} RBCs

Molecular basis associated with s antigen[1]

Amino acid	Thr48 (previously 29) of GPB
Nucleotide	C at bp 143 in exon 4 in *GYPB*

In addition to Thr48, some anti-s also require Thr44 (previously 25) and/or GalNAc attached to this residue, Glu47, His53, and Arg54 (previously 28, 34, and 35, respectively)[2].

Effect of enzymes and chemicals on s antigen on intact RBCs

Ficin/Papain	Variable
Trypsin	Resistant
α-Chymotrypsin	Sensitive
Sialidase	Variable
DTT 200 mM	Resistant
Acid	Resistant

MNS

In vitro characteristics of alloanti-s

Immunoglobulin class	IgG; IgM
Optimal technique	IAT (often after incubation at RT or 4°C)
Complement binding	Rare

Clinical significance of alloanti-s

Transfusion reaction	No to mild (rare)
HDFN	No to severe (rare)

Comments

A pH of 6.0 enhances the reactivity of some anti-s.

References

[1] Dahr, W., et al., 1980. Structure of the Ss blood group antigens, II: a methionine/threonine polymorphism within the N-terminal sequence of the Ss glycoprotein. Hoppe-Seylers Z Physiol Chem 361, 895–906.

[2] Dahr, W., 1986. Immunochemistry of sialoglycoproteins in human red blood cell membranes. In: Vengelen-Tyler, V., Judd, W.J. (Eds.), Recent Advances in Blood Group Biochemistry. American Association of Blood Banks, Arlington, VA, pp. 23–65.

U Antigen

Terminology

ISBT symbol (number)	MNS5 (002005 or 2.5)
History	Described in 1953 and named "U" from "the almost universal distribution" of the antigen.

Occurrence

Caucasians	99.9%
Blacks	99%

Expression

Cord RBCs	Expressed
Altered	GPB variants and on regulator type of Rh_{null}, and on Rh_{mod} RBCs

Molecular basis associated with U antigen[1]
GPB

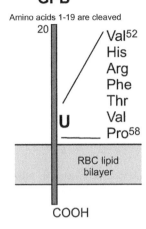

Amino acids 1-19 are cleaved

20

Val[52]
His
Arg
Phe
Thr
Val
Pro[58]

U

RBC lipid bilayer

COOH

Expression of U may require an interaction with another membrane protein, possibly the Rh associated glycoprotein (RhAG)[2,3].

The U-negative phenotype is associated with an absence of GPB or with altered forms of GPB [see He (MNS 6)][4].

Effect of enzymes and chemicals on U antigen on intact RBCs

Ficin/Papain	Resistant
Trypsin	Resistant
α-Chymotrypsin	Resistant
DTT 200 mM	Resistant
Acid	Resistant

In vitro characteristics of alloanti-U

Immunoglobulin class	IgG
Optimal technique	IAT

Clinical significance of alloanti-U

Transfusion reaction	Mild to severe
HDFN	Mild to severe (one fetus required an intrauterine transfusion)[5]

Autoanti-U

Yes

Comments

U– RBCs (except Dantu+ and some Rh_{null}/Rh_{mod} RBCs) are S–s–. Of S–s– RBCs, approximately 16% are U+, albeit weakly (U+^{var}), and are encoded by a hybrid glycophorin gene, of these approximately 23% are He+[4,6]). Antibodies that detect the altered protein should be more correctly called anti-U/GPB[6].

References

[1] Dahr, W., Moulds, J.J., 1987. High-frequency antigens of human erythrocyte membrane sialoglycoproteins, IV. Molecular properties of the U antigen. Biol Chem Hoppe-Seyler 368, 659–667.

[2] Ballas, S.K., et al., 1986. The blood group U antigen is not located on glycophorin B. Biochim Biophys Acta 884, 337–343.

[3] Mallinson, G., et al., 1990. Murine monoclonal antibody MB-2D10 recognizes Rh-related glycoproteins in the human red cell membrane. Transfusion 30, 222–225.

[4] Reid, M.E., et al., 1996. Expression and quantitative variation of the low incidence blood group antigen He on some S-s- RBCs. Transfusion 36, 719–724.

[5] Win, N., et al., 1996. Severe haemolytic disease of the newborn due to anti-U requiring intrauterine transfusion [abstract]. Transfusion Medicine 6, (suupl 2):39.

[6] Storry, JR, et al., 2003. Mutations in *GYPB* exon 5 drive the S-s-U+^{var} phenotype in persons of African descent: implications for transfusion. Transfusion 43: 1738–177.

He Antigen

Terminology

ISBT symbol (number)	MNS6 (002006 or 2.6)
Obsolete name	Henshaw
History	Named for the first He+proband, Mr. Henshaw; the original anti-He, present in a rabbit anti-M serum, was identified in 1951.

Occurrence

African Americans	3%
Blacks in Natal	Up to 7%
Caucasians	Not found

Antithetical antigen

"N" (MNS30)

Expression

Cord RBCs	Presumed expressed
Altered	On S–s– GPB variants[1]

Molecular basis associated with He antigen[1,2]

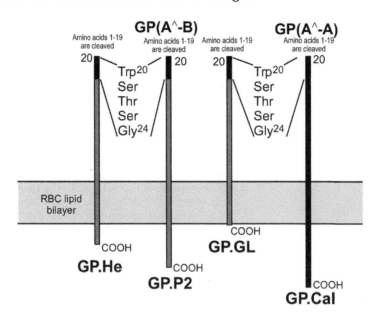

Variant glycophorin

GP.He	GPB(20–23^)-GPA^M^(24–45)-GPB(46–91)
GP.He(P2)	GPB(20–23^)-GPA^M^(24–45)-B^S^(46–58)-GPB(59–100^)
GP.He(GL)	GPB(20–23^)-GPA^M^(24–45)-B(47–78)
GP.He(Cal)	GPB(20–23^)-GPA^M^(24–118)

^An altered sequence.

Contribution by parent glycophorin

GP.He	GPB(20–23^)-GPA^M^(24–45)-GPB(46–91)
GP.He(P2)	GPB(20–23^)-GPA^M^(24–45)-GPB^S^(46–58 then new sequence 59–100)
GP.He(GL)	GPB(20–23^)-GPA^M^(24–45)-GPB(59–91)
GP.He(Cal)	GPB(20–23^)-GPA^M^(24–45)-GPA(78–150)

^An altered sequence.

Gene arrangement

GP.He	GYP(B-A-B)	
GP.He(P2)	GYP(B-A-B)	The G>T change in intron 5 causes altered splicing and chain elongation with a novel transmembrane amino acid sequence. GP.He(P2) is hard to detect in the RBC membrane; the S–s– RBCs are He+^{W} due to expression of low levels of GP.He.
GP.He(GL)	GYP(B-A-B)	There are 4 transcripts: t1 is GP.He; t2 has a T>G change in the acceptor splice site (Intron 3 at nt –6) leading to skipping of exon 4 [GP.He(GL)]; t3 has a partial deletion of exon 5 due to a C>G change in exon 5, a frame-shift, and a premature stop codon; t4 has the T>G in intron 3 and deletion of exon 4, and the C>G in exon 5, which results in a partial deletion of exon 5. Products of t3 and t4 have not been demonstrated in the RBC membrane[3].
GP.Cal	GYP(B-A-ψB-A)	The GYPA recombination site is in exon 2 so the mature protein, after cleavage of the leader peptide, is GP(A-A). The GYPB also contributes the pseudo exon, which is out-spliced. There are 2 He-specific transcripts: t1 has a junction of exon 2 to exon 4 and generates the amino acid sequence associated with the St^a antigen [GP.He(Cal)]; t2 has a junction of exon 2 to exon 5, which is unlikely to be translated[4].
GP.He (NY)	GYP(B-A-B)	Partial deletion of exon 5 alters the open reading frame, predicted to encode a protein of 43 amino acids, which has not been demonstrated in the RBC membrane. The S–s– RBCs are He+^{W} due to expression of low levels of GP.He[5].

Phenotype with antigen strength

Glycophorin	Antigens expressed		
	He	S/s	U
GP.He	Strong	S or s	Strong
GP.He(P2)	Weak/moderate^	No	Weak/moderate
GP.He(GL)	Strong^^	No	No
GP.He(Cal)	Weak/moderate	No	No
GP.He(NY)	Weak/moderate^	No	Weak/moderate

^The RBCs express He due to low levels of GP.He
^^The RBCs also express GP.He

Effect of enzymes and chemicals on He antigen on intact RBCs

Ficin/Papain	Sensitive
Trypsin	Resistant
α-Chymotrypsin	Variable
Sialidase	Variable
DTT 200 mM	Resistant

In vitro characteristics of alloanti-He

Immunoglobulin class	IgM; IgG
Optimal technique	RT; IAT

Clinical significance of alloanti-He

No data are available because human anti-He is rare.

Comments

Approximately 23% of S–s– RBCs express the He antigen[6], and approximately half have an altered *GYPB* [see U (MNS 5)]. GPB carrying He does not express "N."

References

[1] Dahr, W., et al., 1984. Structural analysis of the Ss sialoglycoprotein specific for Henshaw blood group from human erythrocyte membranes. Eur J Biochem 141, 51–55.

[2] Huang, C.-H., et al., 1994. Remodeling of the transmembrane segment in human glycophorin by aberrant RNA splicing. J Biol Chem 269, 10804–10812.

[3] Huang, C.-H., et al., 1997. Alternative splicing of a novel glycophorin allele GPHe(GL) generates two protein isoforms in the human erythrocyte membrane. Blood 90, 391–397.

[4] Huang, C.-H., et al., 1994. Glycophorin He(Stª) of the human red blood cell membrane is encoded by a complex hybrid gene resulting from two recombinational events. Blood 83, 3369–3376.

[5] Storry, J.R., et al., 2003. Mutations in *GYPB* exon 5 drive the S–s–U+ᵛᵃʳ phenotype in persons of African descent: Implications for transfusion. Transfusion 43, 1738–1747.

[6] Reid, M.E., et al., 1996. Expression and quantitative variation of the low incidence blood group antigen He on some S–s– RBCs. Transfusion 36, 719–724.

Miª Antigen

Terminology

ISBT symbol (number)	MNS7 (002007 or 2.7)
Obsolete name	Miltenberger
History	In 1951, the serum of Mrs. Miltenberger revealed a "new" low-prevalence antigen, named Miª. When other related antigens and antisera were found they formed a subsystem named Miltenberger. These antigens are in the MNS blood group system and a terminology based on glycophorin (e.g., GP.Vw, GP.Hop, etc.) is now commonly used[1].

Occurrence

Most populations	<0.01%
Chinese and SE Asians	Up to 15%

Expression

Cord RBCs	Expressed

Molecular basis associated with Miᵃ antigen[2]

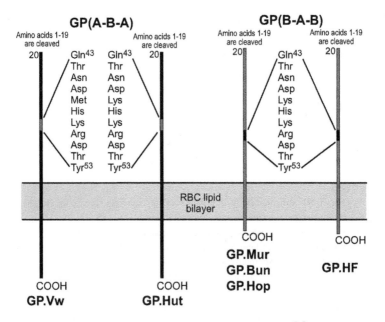

Anti-Miᵃ recognizes the amino acid sequence QTND$\overset{M}{K}$HKRDTY^{53}, but does not require residues C-terminal of the tyrosine at position 53 (previously 34)[2].

Effect of enzymes and chemicals on Miᵃ antigen on intact RBCs

	GP.Vw and GP.Hut	GP.Mur and GP.Hop	GP.Bun and GP.HF
Ficin/Papain	Sensitive	Sensitive/weakened	Sensitive/weakened
Trypsin	Sensitive	Resistant	Resistant
α-Chymotrypsin	Resistant	Sensitive	Sensitive/weakened

(Continued)

(Continued)

	GP.Vw and GP.Hut	GP.Mur and GP.Hop	GP.Bun and GP.HF
Sialidase	Sensitive	Variable	Variable
DTT 200 mM	Resistant	Resistant	Resistant
Acid	Resistant	Resistant	Resistant

In vitro characteristics and clinical significance of alloanti-Mi^a

Transfusion reaction	Rare (because antigen is rare in most populations)
HDFN	Mild to severe

Comments

Anti-Mi^a is often present in serum containing anti-Vw.

Production of monoclonal anti-Mi^a (GAMA 210 and CBC-172) showed that anti-Mi^a exists as a single specificity, and that Mi^a is a discrete antigen[3].

Due to the relatively high prevalence of some Mi(a+) phenotypes [(particularly GP.Mur (Mi.III), up to 15% in parts of Taiwan with a prevalence of 88% in some indigenous Taiwanese)][4,5] in Chinese and South East Asian populations, it is recommended by some to include GP.Mur phenotype RBCs in antibody investigations in these populations.

References

[1] Tippett, P., et al., 1992. The Miltenberger subsystem: Is it obsolescent? Transfus Med Rev 6, 170–182.

[2] Dahr, W., 1992. Miltenberger subsystem of the MNSs blood group system. Review and outlook. Vox Sang 62, 129–135.

[3] Chen, V., et al., 2001. Direct evidence for the existence of Miltenberger^a antigen. Vox Sang 80, 230–233.

[4] Broadberry, R.E., Lin, M., 1994. The incidence and significance of anti-"Mia" in Taiwan. Transfusion 34, 349–352.

[5] Mak, K.H., et al., 1994. A survey of the incidence of Miltenberger antibodies among Hong Kong Chinese blood donors. Transfusion 34, 238–241.

Mc Antigen

Terminology

ISBT symbol (number)	MNS8 (002008 or 2.8)
History	M^c was described in 1953. The antigen appeared to be intermediate between M and N and, as such, was analogous to the situation described in 1948 for the Rh antigen c^v; hence the name M^c was used.

Occurrence

Less than 0.01%; all probands are of European origin.

Molecular basis associated with M^c antigen[1]

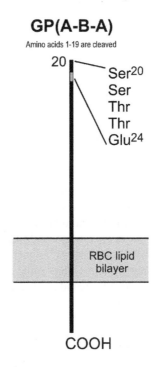

GP(A-B-A)

Amino acids 1-19 are cleaved

O-glycosylation of residues 21, 22, and 23 (previously 2, 3, and 4) is normal.

Variant glycophorin   GPA^M(20–23)-GPB(24)-GPA(25–150)
Gene arrangement   *GYP(A-B-A)*

Comments

No alloanti-M^c has been described. M^c is defined by the reaction of certain anti-M and anti-N: a majority of anti-M and a minority of anti-N react with M^c+ RBCs.

Reference

[1] Huang, C.-H., Blumenfeld, O.O., 1995. MNSs blood groups and major glycophorins: molecular basis for allelic variation. In: Cartron, J.-P., Rouger, P. (Eds.), Molecular Basis of Human Blood Group Antigens. Plenum Press, New York, NY, pp. 153–188.

Vw Antigen

Terminology

ISBT symbol (number)	MNS9 (002009 or 2.9)
Obsolete names	Gr; Verweyst; Mi.I
History	Identified in 1954; named for Mr. Verweyst; anti-Vw caused positive DAT on the RBCs in one of his children.

Occurrence

Caucasians	0.06%
South East Swiss	Up to 1.4%

Antithetical antigens

Hut (**MNS19**); ENEH (**MNS40**)

Expression

Cord RBCs	Expressed

Molecular basis associated with Vw antigen[1,2]

Amino acid	Met47 (previously 28) of GPA
Nucleotide	T at bp 140 in exon 3
Variant glycophorin	GPA(20–46)-GPB(47)-GPA(48–150)
Gene arrangement	$GYP(A\text{-}\psi B\text{-}A)$

The N-glycosylation consensus sequence is changed so that Asn45 (previously 26) is not N-glycosylated, which results in a decreased M_r of about 3,000.

Effect of enzymes and chemicals on Vw antigen on intact RBCs

Ficin/Papain	Sensitive
Trypsin	Sensitive
α-Chymotrypsin	Resistant
Sialidase	Resistant
DTT 200 mM	Resistant
Acid	Resistant

In vitro characteristics of alloanti-Vw

Immunoglobulin class	IgM; IgG
Optimal technique	RT; IAT

Clinical significance of alloanti-Vw

Transfusion reaction	No to severe
HDFN	Mild to severe

Comments

The altered GPA carrying Vw usually carries N (**MNS2**).

One *GYP*Vw* homozygote person has been described who made anti-En^{a}TS (anti-ENEH)[3].

Anti-Vw is found in 1% of sera and is a frequent component of multispecific sera.

Anti-Vw is commonly found in sera of patients with AIHA.

References

[1] Dahr, W., 1992. Miltenberger subsystem of the MNSs blood group system. Review and outlook. Vox Sang 62, 129–135.

[2] Huang, C.-H., Blumenfeld, O.O., 1995. MNSs blood groups and major glycophorins: molecular basis for allelic variation. In: Cartron, J.-P., Rouger, P. (Eds.), Molecular Basis of Human Blood Group Antigens. Plenum Press, New York, NY, pp. 153–188.

[3] Spruell, P., et al., 1993. An anti-En^{a}TS detected in the serum of an MiI homozygote. Transfusion 33, 848–851.

Mur Antigen

Terminology

ISBT symbol (number)	MNS10 (002010 or 2.10)
Obsolete names	Murrell; Mu
History	Identified in 1961 as the cause of HDFN in the Murrell family.

Occurrence

Most populations	<0.1%
Chinese	6%
Taiwanese	7% (up to 88% in some indigenous people)
Thai	9%

Expression

Cord RBCs	Expressed

Molecular basis associated with Mur antigen[1,2]

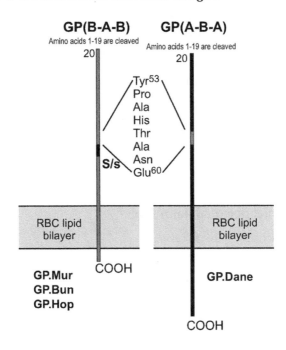

GP(B-A-B)

Amino acids 1-19 are cleaved

20

GP(A-B-A)

Amino acids 1-19 are cleaved

20

Tyr^{53}
Pro
Ala
His
Thr
Ala
S/s — Asn
Glu^{60}

RBC lipid bilayer

RBC lipid bilayer

COOH

GP.Mur
GP.Bun
GP.Hop

GP.Dane

COOH

Mur antigen is expressed when a sequence of amino acids (^{53}YPAHTANE^{60}) is encoded by the pseudoexon 3 of *GYPB*.

Variant glycophorins

GP.Mur (Mi.III)	GPB(20–45)-GPψB(46–67)-GPA(68–76)-GPB^s(77–122)
GP.Bun (Mi.VI)	GPB(20–45)-GPψB(46–69)-GPA(70–76)-GPB^s(77–122)
GP.Hop (Mi.IV)	GPB(20–45)-GPψB(46–69)-GPA(70–76)-GPB^S(77–122)
GP.Dane (Mi.IX)	GPA(20–53)-GPψB(54–59)-GPA(60–150)

Contribution by parent glycophorins

GP.Mur	GPB(20–45)-GPψB-GPA(68–76)-GPB(46–91)
GP.Bun, GP.Hop	GPB(20–45)-GPψB-GPA(70–76)-GPB(46–91)
GP.Dane	GPA(20–53)-GPψB-GPA(60–150)

Gene arrangement

GP.Mur, GP.Bun, GP.Hop	*GYP(B-A-B)*
GP.Dane	*GYP(A-ψB-A)*

Effect of enzymes and chemicals on Mur antigen on intact RBCs

Ficin/Papain	Sensitive
Trypsin	Resistant
α-Chymotrypsin	Sensitive (resistant on GP.Dane)
Sialidase	Presumed resistant
DTT 200 mM	Resistant
Acid	Resistant

In vitro characteristics of alloanti-Mur

Immunoglobulin class	IgM less common than IgG
Optimal technique	RT; IAT

Clinical significance of alloanti-Mur

Transfusion reaction	No to severe
HDFN	No to severe

Comments

Anti-Mur occurs as a single specificity and is often in sera containing anti-Mi^a. Sera with inseparable anti-Mur and anti-Hut are now considered to contain an additional specificity, anti-MUT (anti-MNS35).

Anti-Mur is common in South East Asian and Chinese populations (0.2%, 0.28%, and 0.06% of patients in Thailand, Taiwan, and Hong Kong, respectively).

M^g+(**MNS11**) RBCs reacted with serum from Mrs. Murrell, but not with other anti-Mur[3].

GP.Mur RBCs have increased levels (25 to 67%) of band 3 in the RBC membrane. This was correlated with functional changes including superior HCO_3^- transport, acid-base homeostasis, and osmotic resistance[4]. GP.Mur may confer resistance to invasion by *Plasmodium falciparum* or the higher levels of band 3 could help to alleviate major malarial symptoms such as acidosis[4].

References

[1] Huang, C.-H., Blumenfeld, O.O., 1995. MNSs blood groups and major glycophorins: molecular basis for allelic variation. In: Cartron, J.-P., Rouger, P. (Eds.), Molecular Basis of Human Blood Group Antigens. Plenum Press, New York, NY, pp. 153–188.

[2] Storry, J.R., et al., 2000. Identification of a novel hybrid glycophorin gene encoding GP.Hop. Transfusion 40, 560–565.

[3] Green, C., et al., 1994. Mg+ MNS blood group phenotype: further observations. Vox Sang 66, 237–241.

[4] Hsu, K., et al., 2009. Miltenberger blood group antigen type III (Mi.III) enhances the expression of band 3. Blood 114, 1919–1928.

M^g Antigen

Terminology

ISBT symbol (number)	MNS11 (002011 or 2.11)
Obsolete name	Gilfeather
History	Identified in 1958; RBCs of a patient, Mr. Gilfeather, reacted with the serum of a donor.

Occurrence

Most populations	<0.01%
Swiss and Sicilians	0.15%

One *M^{g}M^g* homozygote person has been described.

Expression

Cord RBCs	Expressed

Molecular basis associated with M^g antigen[1]

GP(A-B-A)

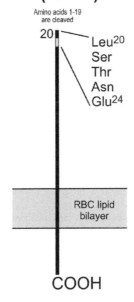

Amino acids 1-19
are cleaved

20

Leu^{20}
Ser
Thr
Asn
Glu^{24}

RBC lipid
bilayer

COOH

GP.Mg variant	GPA^N(20–23)-GPB(24)-GPA(25–150)
Gene arrangement	*GYP(A^N-B-A)*

The O-glycans attached to residues 21 and 22 are altered. There is no O-glycan attached to residue 23[2]. This causes a reduction in sialic acid content and decreased electrophoretic mobility.

Effect of enzymes and chemicals on M^g antigen on intact RBCs

Ficin/Papain	Sensitive
Trypsin	Sensitive
α-Chymotrypsin	Resistant
Sialidase	Resistant (usually)
DTT 200 mM	Resistant
Acid	Resistant

In vitro characteristics of alloanti-M^g

Immunoglobulin class	IgM more common than IgG
Optimal technique	RT; 37°C; IAT

Clinical significance of alloanti-M^g

No data available.

Comments

Human and rabbit anti-M and anti-N do not detect the M^g antigen. Some monoclonal anti-M react with $M- M^g+$ RBCs. Two of six anti-M^g reacted with a variant M^g+ RBC sample that had a higher level of glycosylation than other M^g+ samples.

M^g can travel with s and S, the latter combination may be indicative of a Sicilian background. M^g+ RBCs carry DANE (**MNS32**) and were agglutinated by anti-Mur from Mrs. Murrell, but not with other anti-Mur[3].

Anti-M^g is present in 1–2% of sera.

References

[1] Huang, C.-H., Blumenfeld, O.O., 1995. MNSs blood groups and major glycophorins: molecular basis for allelic variation. In: Cartron, J.-P., Rouger, P. (Eds.), Molecular Basis of Human Blood Group Antigens. Plenum Press, New York, NY, pp. 153–188.

[2] Dahr, W., et al., 1981. Amino acid sequence of the blood group M^g-specific major human erythrocyte membrane sialoglycoprotein. Hoppe-Seylers Z Physiol Chem 362, 81–85.

[3] Green, C., et al., 1994. Mg+ MNS blood group phenotype: further observations. Vox Sang 66, 237–241.

Vr Antigen

Terminology

ISBT symbol (number)	MNS12 (002012 or 2.12)
Obsolete name	Verdegaal
History	Identified in 1958; named for the family in which the antigen and antibody were found.

Occurrence

Only found in a few Dutch families.

Expression

Cord RBCs   Expressed

Molecular basis associated with Vr antigen[1]

Amino acid   Tyr66 (previously 47) of GPA
Nucleotide   A at bp 197 in exon 3 of *GYPA*
Vr– (wild type)   Ser66 and C at position 197

Effect of enzymes and chemicals on Vr antigen on intact RBCs[1]

Ficin/Papain   Sensitive
Trypsin   Resistant
α-Chymotrypsin   Sensitive
Sialidase   Resistant
DTT 200 mM   Resistant
Acid   Resistant

In vitro characteristics of alloanti-Vr

Immunoglobulin class   IgM and IgG
Optimal technique   RT; IAT

Clinical significance of alloanti-Vr

Transfusion reaction   No data
HDFN   The original maker of anti-Vr had three Vr+
                       children; none had HDFN

Comments

Inherited with Ms[2].

Anti-Vr has been found in multispecific sera.

The Ser66Tyr substitution in GPA is predicted to introduce a novel α-chymotrypsin cleavage site, which would explain the (unexpected) sensitivity of the Vr antigen to α-chymotrypsin. Vr is located to the carboxyl side of the major trypsin cleavage site on GPA, thus making it resistant to trypsin treatment.

References

[1] Storry, J.R., et al., 2000. The MNS blood group antigens, Vr (MNS 12) and Mta (MNS 14) each arise from an amino acid substitution on glycophorin A. Vox Sang 78, 52–56.

[2] van der Hart, M., et al., 1958. Vr, an antigen belonging to the MNSs blood group system. Vox Sang 3, 261–265.

Me Antigen

Terminology

ISBT symbol (number)	MNS13 (002013 or 2.13)
History	Anti-M^e was identified in 1961; named M^e because epitope expressed on M+ RBCs and on He+ RBCs, regardless of MN type.

Molecular basis associated with M^e antigen[1]

M^e antigen is expressed when glycine occupies residue 24 (previously 5) of either GPA (M antigen [**MNS1**]) or GPB (He antigen [**MNS6**]).

Comments

Some anti-M (anti-M^e) have a component that reacts with glycine at residue 24 of GPA^M or GPB^{He}. The characteristics of these antibodies are the same as for anti-M (see **MNS1**).

M^e on GPB^{He} is resistant to trypsin treatment and sensitive to α-chymotrypsin treatment.

Reference

[1] Dahr, W., 1986. Immunochemistry of sialoglycoproteins in human red blood cell membranes. In: Vengelen-Tyler, V., Judd, W.J. (Eds.), Recent Advances in Blood Group Biochemistry. American Association of Blood Banks, Arlington, VA, pp. 23–65.

Mt^a Antigen

Terminology

ISBT symbol (number)	MNS14 (002014 or 2.14)
Obsolete name	Martin
History	Reported in 1962; named for the first antigen-positive donor.

Occurrence

Thai	1%
Swiss	0.35%
White Americans	0.24%
Black Americans	0.1%

Expression

Cord RBCs	Expressed

Molecular basis associated with Mt^a antigen[1]

Amino acid	Ile77 (previously 58) of GPA
Nucleotide	T at bp 230 in exon 3 in *GYPA*
Mt(a−) (wild type)	Thr77 and C at bp 230

Effect of enzymes and chemicals on Mt^a antigen on intact RBCs

Ficin/Papain	Variable
Trypsin	Resistant
α-Chymotrypsin	Resistant
Pronase	Sensitive
Sialidase	Resistant
DTT 200 mM	Resistant
Acid	Resistant
Chloroquine	Sensitive

In vitro characteristics of alloanti-Mt^a

Immunoglobulin class	IgM; IgG
Optimal technique	RT; IAT

Clinical significance of alloanti-Mt^a

Transfusion reaction	No data
HDFN	No to severe[2]

Comments

Inherited with Ns[3,4].

The variable susceptibility of Mt^a to ficin and papain treatment may reflect slight differences in the epitope recognized by certain anti-Mt^a or may result from the proximity of residue 77 to Arg80, one of the two proteolytic sites on GPA.

Anti-Mt^a is found as a single specificity and occasionally in multispecific sera.

References

[1] Storry, J.R., et al., 2000. The MNS blood group antigens, Vr (MNS 12) and Mta (MNS 14) each arise from an amino acid substitution on glycophorin A. Vox Sang 78, 52–56.

[2] Cheung, C.C., et al., 2002. Anti-Mt^a associated with three cases of hemolytic disease of the newborn. Immunohematology 18, 37–39.

[3] Konugres, A.A., et al., 1965. Distribution and development of the blood factor Mt^a. Vox Sang 10, 206–207.

[4] Swanson, J., Matson, G.A., 1962. Mt^a, a "new" antigen in the MNSs system. Vox Sang 7, 585–590.

St^a Antigen

Terminology

ISBT symbol (number)   MNS15 (002015 or 2.15)
Obsolete name   Stones
History   Antigen named in 1962 after the first producer of the
                        antibody.

Occurrence

Caucasians   <0.1%
Asians   2%
Japanese   6%

Expression

Cord RBCs   Presumed expressed

Molecular basis associated with St^a antigen[1,2]

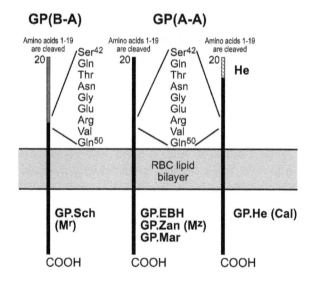

The St^a antigen arises when amino acid at residue 45 (previously 26) of GPB or GPA joins to GPA at residue 78 (previously 59).

Variant glycophorins

GP.Sch (M^r)   GPB(20–45)-GPA(46–118)
GP.Zan, GP.EBH t2   GPA(20–45)-GPA(46–118)
GP.Mar   GPA(20–45)-GPA(46–118)
GP.He (Cal)   GPA(20–24^)-GPA(25–118)
^Altered sequence.

Contribution by parent glycophorins

GP.Sch	GPB(20–45)-GPA(78–150)
GP.Zan, GP.EBH t2	GPA(20–45)-GPA(78–150)
GP.Mar	GPA(20–45)-GPA(78–150)
GP.He (Cal)	GPA(20–24^)-GPA(25–45)-GPA(78–150)

^Altered sequence.

Gene arrangement

GP.Sch	*GYP(B-A)*
GP.Zan	*GYP(A-ψB-A)*
GP.EBH	*GYPA*
GP.Mar	*GYP(A-ψE-A)*
GP.He (Cal)	*GYP(B-A-ψB-A)*

Effect of enzymes and chemicals on St^a antigen on intact RBCs

Ficin/Papain	Variable
Trypsin	Resistant
α-Chymotrypsin	Resistant
Pronase	Sensitive
Sialidase	Presumed resistant
DTT 200 mM	Resistant
Acid	Resistant

In vitro characteristics of alloanti-St^a

Immunoglobulin class	IgM; IgG
Optimal technique	RT; IAT

Clinical significance of alloanti-St^a

No data are available because anti-St^a is rare.

Comments

*GYP*Sch* is the reciprocal gene rearrangement product of *GYP*Hil* (see Hil antigen [**MNS20**]) and *GYP*JL* (see TSEN antigen [**MNS33**]).

The shortened product from transcript 2 (t2) of *GYP*EBH*, which lacks amino acids encoded by exon 3, expresses St^a but not ERIK antigen. The full length product (GP.EBH) expresses ERIK [**MNS37**] but not St^{a1} (see MNS system page).

GP.Zan and GP.Mar lack amino acids encoded by exon 3 and each has a trypsin-resistant M antigen.

One *GYP*Sta/GYP*Sta* homozygote person has been described. This person had decreased glycosylation on band 3.

Anti-St^a is a rare specificity, and often occurs in sera containing anti-S and in multispecific sera. Anti-St^a is notorious for deteriorating *in vitro*.

References

[1] Huang, C.-H., Blumenfeld, O.O., 1995. MNSs blood groups and major glycophorins: molecular basis for allelic variation. In: Cartron, J.-P., Rouger, P. (Eds.), Molecular Basis of Human Blood Group Antigens. Plenum Press, New York, NY, pp. 153–188.

[2] Huang, C.-H., et al., 1994. Gene conversion between glycophorins A and E results in St^a glycophorin in a family exhibiting the ERIK/St^a blood group phenotype [abstract]. Blood 84 (Suppl. 1), 238a.

Ri^a Antigen

Terminology

ISBT symbol (number)	MNS16 (002016 or 2.16)
Obsolete name	Ridley
History	Identified in 1962; named for the original Ri(a+) person.

Occurrence

Only found and studied in one large family[1].

Expression

Cord RBCs	Presumed expressed

Molecular basis associated with Ri^a antigen[2]

Amino acid	Lys76 (previously 57) of GPA
Nucleotide	A at bp 226 in exon 3 in *GYPA*
Ri(a–) (wild type)	Glu76 and G at bp 226

Effect of enzymes and chemicals on Ri^a antigen on intact RBCs

Ficin/Papain	Partially sensitive
Trypsin	Sensitive
α-Chymotrypsin	Resistant
Pronase	Resistant
DTT 200 mM	Resistant
Acid	Resistant

In vitro characteristics of alloanti-Ri^a

Immunoglobulin class	IgM (12 of 13 anti-Ri^a were IgM, one was IgG)
Optimal technique	RT; IAT
Complement binding	Some

Clinical significance of alloanti-Ri^a

No data are available because anti-Ri^a is rare.

Comments

The Glu76Lys substitution in GPA is predicted to introduce a novel trypsin cleavage site.

Anti-Ri[a], likely to be naturally-occurring, was found in sera containing multiple antibodies to low prevalence antigens[3]. Anti-S [see **MNS3**] often contain anti-Ri[a].

Inherited with MS[3].

References

[1] Cleghorn, T.E., 1962. Two human blood group antigens, St[a] (Stones) and Ri[a] (Ridley), closely related to the MNSs system. Nature 195, 297–298.

[2] Reid, M.E., Storry, J.R., 2001. Low-incidence MNS antigens associated with single amino acid changes and their susceptibility to enzyme treatment. Immunohematology 17, 76–81.

[3] Contreras, M., et al., 1984. The MNSs antigen Ridley (Ri[a]). Vox Sang 46, 360–365.

Cl[a] Antigen

Terminology

ISBT symbol (number)	MNS17 (002017 or 2.17)
Obsolete name	Caldwell
History	Identified in 1963; antibody found in an anti-B typing serum; named for the antigen-positive person.

Occurrence

Only found in one Scottish and one Irish family.

Expression

Cord RBCs	Presumed expressed

Effect of enzymes and chemicals on Cl[a] antigen on intact RBCs

Ficin/Papain	Sensitive
Trypsin	Sensitive
α-Chymotrypsin	Resistant
DTT 200 mM	Resistant

In vitro characteristics of alloanti-Cl[a]

Immunoglobulin class	IgM
Optimal technique	RT

Clinical significance of alloanti-Cl[a]

No data are available because antigen and antibody are rare.

Comments

Anti-Cl^a was found in serum samples from 24 of 5,000 British blood donors. Inherited with Ms[1].

Reference

[1] Wallace, J., Izatt, M.M., 1963. The Cl^a (Caldwell) antigen: a new and rare human blood group antigen related to the MNSs system. Nature 200, 689–690.

Ny^a Antigen

Terminology

ISBT symbol (number)	MNS18 (002018 or 2.18)
Obsolete name	Nyberg
History	Identified in 1964; named for Mr. Nyberg, the first Ny(a+) person.

Occurrence

Found in 0.2% of Norwegians, in one Swiss family and an American of non-Scandinavian descent.

Expression

Cord RBCs	Expressed

Molecular basis associated with Ny^a antigen[1]

Amino acid	Glu46 (previously 27) of GPA
Nucleotide	A at bp 138 in exon 3 in *GYPA*
Ny(a–) (wild type)	Asp46 and T at bp 138

Effect of enzymes and chemicals on Ny^a antigen on intact RBCs

Ficin/Papain	Sensitive
Trypsin	Sensitive
α-Chymotrypsin	Resistant
DTT 200 mM	Resistant
Acid	Resistant

In vitro characteristics of alloanti-Ny^a

Immunoglobulin class	IgM
Optimal technique	RT

Clinical significance of alloanti-Ny^a

No data are available because antigen and antibody are rare.

Comments

Inherited with Ns[2,3].

Anti-Ny[a] appears to be naturally-occurring, found in about 0.1% of sera studied, and has been produced in rabbits.

References

[1] Daniels, G.L., et al., 2000. The low-frequency MNS blood group antigens Ny[a] (MNS18) and Os[a] (MNS38) are associated with GPA amino acid substitutions. Transfusion 40, 555–559.

[2] Kornstad, L., et al., 1971. Further observations on the frequency of the Ny[a] blood-group antigen and its genetics. Am J Hum Genet 23, 612–613.

[3] Örjasaeter, H., et al., 1964. Studies on the Ny[a] blood group antigen and antibodies. Vox Sang 9, 673–683.

Hut Antigen

Terminology

ISBT symbol (number)	MNS19 (002019 or 2.19)
Obsolete name	Mi.II
History	Anti-Hut was reported in 1962 and redefined in 1982. It was first identified in 1958 as the cause of HDFN in the Hutchinson family. At the time, it was considered to be anti-Mi[a].

Occurrence

Most populations	<0.01%

Antithetical antigen

Vw (**MNS9**); ENEH (**MNS40**)

Expression

Cord RBCs	Expressed

Molecular basis associated with Hut antigen[1,2]

Amino acid	Lys47 (previously 28) of GPA
Nucleotide	A at bp140 in exon 3
Variant glycophorin	GPA(20–46)-GPB(47)-GPA(48–150)
Gene arrangement	*GYP(A-ψB-A)*

The N-glycosylation consensus sequence is changed so that Asn45 (previously 26) is not N-glycosylated, which results in a decreased M_r of about 3,000.

Effect of enzymes and chemicals on Hut antigen on intact RBCs

Ficin/Papain	Sensitive
Trypsin	Sensitive
α-Chymotrypsin	Resistant
Sialidase	Resistant
DTT 200 mM	Resistant
Acid	Resistant

In vitro characteristics of alloanti-Hut

Immunoglobulin class	IgM more common than IgG
Optimal technique	RT; IAT

Clinical significance of alloanti-Hut

Transfusion reaction	No data because antibody and antigen are rare
HDFN	No to moderate

Comments

Hut has been aligned with MS, Ns and Ms in decreasing order of frequency but not with NS.

The specificity originally called anti-Hut is now called anti-MUT (see MNS35) since Hut+, Mur+ RBCs are reactive. Anti-Hut reacts with Hut+ RBCs only.

References

[1] Dahr, W., 1992. Miltenberger subsystem of the MNSs blood group system. Review and outlook. Vox Sang 62, 129–135.

[2] Huang, C.-H., Blumenfeld, O.O., 1995. MNSs blood groups and major glycophorins: molecular basis for allelic variation. In: Cartron, J.-P., Rouger, P. (Eds.), Molecular Basis of Human Blood Group Antigens. Plenum Press, New York, NY, pp. 153–188.

Hil Antigen

Terminology

ISBT symbol (number)	MNS20 (002020 or 2.20)
Obsolete name	Hill
History	Antibody identified in 1963 as the cause of HDFN in the Hill family; in 1966 named Hil.

Occurrence

Most populations   <0.01%
Chinese   6%
One *GYP*Hil/GYP*Hil* homozygote has been described.

Expression

Cord RBCs   Expressed

Molecular basis associated with Hil antigen[1-4]

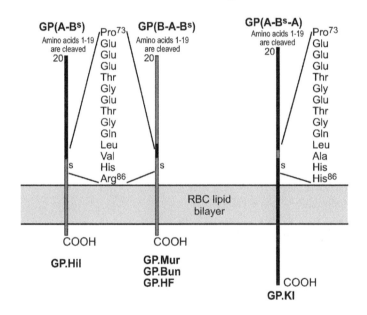

Variant glycophorins

GP.Hil (Mi.V)	GPA(20–77)-GPB^s(78–123)
GP.Mur (Mi.III)	GPB(20–45)-GPψB(46–67)-GPA(68–76)-GPB^s(77–122)
GP.Bun (Mi.VI)	GPB(20–45)-GPψB(46–69)-GPA(70–76)-GPB^s(77–122)
GP.HF (Mi.X)	GPB(20–45)-GPψB(46–53)-GPA(54–77)-GPB^s(78–123)
GP.KI	GPA(20–79)-GPB(80–81)-GPA(82–150)

Contribution by parent glycophorins

GP.Hil	GPA(20–77)-GPB(46–91)
GP.Mur	GPB(20–45)-GPψB–GPA(68–76)-GPB(46–91)
GP.Bun	GPB(20–45)-GPψB–GPA(70–76)-GPB(46–91)
GP.HF	GPB(20–45)-GPψB–GPA(54–77)-GPB(46–91)
GP.KI	GPA(20–79)-GPB(48–49)-GPA(82–150)

Gene arrangement

GP.Hil	*GYP(A-B)*
GP.Mur, GP.Bun, GP.HF	*GYP(B-A-B)*
GP.KI	*GYP(A-B-A)*

Effect of enzymes and chemicals on Hil antigen on intact RBCs

Ficin/Papain	Sensitive
Trypsin	Resistant
α-Chymotrypsin	Sensitive
Sialidase	Resistant
DTT 200 mM	Resistant
Acid	Resistant

In vitro characteristics of alloanti-Hil

Immunoglobulin class	IgM and IgG
Optimal technique	RT; IAT

Clinical significance of alloanti-Hil

Transfusion reaction	No data
HDFN	No to moderate

Comments

Reciprocal product to *GYP*Hil* is *GYP*Sch* (see St[a] antigen [**MNS15**]).
Hil+ RBCs are s+ and, except those carrying GP.KI, are also MINY+.

References

[1] Dahr, W., 1992. Miltenberger subsystem of the MNSs blood group system. Review and outlook. Vox Sang 62, 129–135.

[2] Huang, C.-H., Blumenfeld, O.O., 1995. MNSs blood groups and major glycophorins: molecular basis for allelic variation. In: Cartron, J.-P., Rouger, P. (Eds.), Molecular Basis of Human Blood Group Antigens. Plenum Press, New York, NY, pp. 153–188.

[3] Poole, J., et al., 1998. Novel molecular basis for the Hil (MNS20) antigen [abstract]. Transfusion 38 (Suppl), 103S.

[4] Poole, J., 2000. Red cell antigens on band 3 and glycophorin A. Blood Rev 4, 31–43.

M^v Antigen

Terminology

ISBT symbol (number) MNS21 (002021 or 2.21)
Obsolete name Armstrong
History Found in 1961 when a serum containing anti-N
agglutinated RBCs from 1 in 400 M+N–
Caucasians and described in detail in 1966. The "v"
is for "variant."

Occurrence

Most populations <0.01%

Expression

Cord RBCs Expressed

Molecular basis associated with M^v antigen[1]

Amino acid Ser22 (previously 3) of GPB
Nucleotide G at bp 65 in exon 2 in *GYPB*
M^v– (wild type) Thr22 and C at bp 65

Effect of enzymes and chemicals on M^v antigen on intact RBCs

Ficin/Papain Sensitive
Trypsin Resistant
α-Chymotrypsin Sensitive
Sialidase Sensitive
DTT 200 mM Resistant
Acid Resistant

In vitro characteristics of alloanti-M^v

Immunoglobulin class IgG and IgM
Optimal technique IAT

Clinical significance of alloanti-M^v

Transfusion reaction No data
HDFN No to moderate

Comments

M^v+ RBCs have a decreased level of GPB and a weak expression of s (**MNS4**), and may have a slight weakening of S when M^v is associated with $MS^{1,2}$.

Inherited with Ms in 14 families, and with MS in 2 families.

GPB carrying M^v does not express 'N'.

References

[1] Storry, J.R., et al., 2001. The low incidence MNS antigens, M^v, s^D, and Mit arise from single amino acid substitutions on glycophorin B. Transfusion 41, 269–275.

[2] Dahr, W., Longster, G., 1984. Studies of M^v red cells. II. Immunochemical investigations. Blut 49, 299–306.

Far Antigen

Terminology

ISBT symbol (number)	MNS22 (002022 or 2.22)
Obsolete names	Kam; Kamhuber
History	The Kam antigen reported in 1966 and the Far antigen, reported in 1968, were shown to be the same in 1977. The name Far was chosen. Anti-"Kam" caused a severe transfusion reaction in a multiply-transfused hemophiliac; probably immunized following transfusion with blood of the same donor 11 years previously!

Occurrence

Found in only two families.

Expression

Cord RBCs	Expressed

Effect of enzymes and chemicals on Far antigen on intact RBCs

Ficin/Papain	Resistant
Trypsin	Resistant
α-Chymotrypsin	Not known
DTT 200 mM	Resistant

In vitro characteristics of alloanti-Far

Immunoglobulin class	IgG
Optimal technique	IAT

Clinical significance of alloanti-Far

Transfusion reaction	Severe in one
HDFN	Severe in one

Comments

Travels with Ns[1] and MS[2].
Only two examples of anti-Far have been reported.

References

[1] Cregut, R., et al., 1974. A new rare blood group antigen, "FAR," probably linked to the MNSs system. Vox Sang 26, 194–198.

[2] Speiser, P., et al., 1966. "Kamhuber" a new human blood group antigen of familial occurrence, revealed by a severe transfusion reaction. Vox Sang 11, 113–115.

s^D Antigen

Terminology

ISBT symbol (number)	MNS23 (002023 or 2.23)
Obsolete name	Dreyer
History	Named in 1981; "s" was used because the s antigen is expressed weakly and "D" from the family name.

Occurrence

Found only in one white South African family.

Expression

Cord RBCs	Expressed

Molecular basis associated with s^D antigen[1]

Amino acid	Arg58 (previously 39) of GPB
Nucleotide	G at bp 173 in exon 4 in *GYPB*
s^D– (wild type)	Pro58 and C at position 173

Effect of enzymes and chemicals on s^D antigen on intact RBCs

Ficin/Papain	Partially sensitive
Trypsin	Resistant
α-Chymotrypsin	Resistant
DTT 200 mM	Presumed resistant

In vitro characteristics of alloanti-s^D

Immunoglobulin class	IgG
Optimal technique	IAT

Clinical significance of alloanti-s^D

Transfusion reaction	No data
HDFN	No to severe

Comments

S+s+s^D+ RBCs have a weakened expression of the s antigen (**MNS4**)[2]. Inherited with Ms[2].

The Pro58Arg introduces a novel papain cleavage site. However, the close proximity of the antigen to the lipid bilayer may make the site relatively inaccessible.

References

[1] Storry, J.R., et al., 2001. The low incidence MNS antigens, M^v, s^D, and Mit arise from single amino acid substitutions on glycophorin B. Transfusion 41, 269–275.

[2] Shapiro, M., Le Roux, M.E., 1981. Serology and genetics of a "new" red cell antigen: s^D (the Dreyer antigen) [abstract]. Transfusion 21, 614.

Mit Antigen

Terminology

ISBT symbol (number)	MNS24 (002024 or 2.24)
Obsolete name	Mitchell
History	Named in 1980 after the family where the antigen (father's RBCs) and antibody (mother's serum) were found.

Occurrence

Western Europeans	0.1%

Expression

Cord RBCs	Expressed

Molecular basis associated with Mit antigen[1]

Amino acid	His54 (previously 35) of GPB
Nucleotide	A at bp 161 in exon 4 in *GYPB*
Mit– (wild type)	Arg54 and G at bp 161

Effect of enzymes and chemicals on Mit antigen on intact RBCs

Ficin	Resistant
Papain	Partially sensitive
Trypsin	Resistant
α-Chymotrypsin	Resistant
Pronase	Weakened
Sialidase	Variable
DTT 200 mM	Resistant
Acid	Resistant

In vitro characteristics of alloanti-Mit

Immunoglobulin class	IgG
Optimal technique	IAT

Clinical significance of alloanti-Mit

Transfusion reaction	No data
HDFN	Positive DAT; no clinical HDFN

Comments

Mit+ RBCs have weakened expression of S antigen[2,3] or s antigen[1].
Mit is usually associated with MS, and rarely with NS or Ms.

References

[1] Storry, J.R., et al., 2001. The low incidence MNS antigens, M^v, s^D, and Mit arise from single amino acid substitutions on glycophorin B. Transfusion 41, 269–275.

[2] Eichhorn, M., et al., 1981. Suppression of the S antigen by the MIT antigen: Potential source of error in red cell typing [abstract]. Transfusion 21, 614.

[3] Skradski, K.J., et al., 1983. Further investigation of the effect of Mitchell (Mit) antigen on S antigen expression [abstract]. Transfusion 23, 409.

Dantu Antigen

Terminology

ISBT symbol (number)	MNS25 (002025 or 2.25)
History	Named in 1984 after the first proband.

Occurrence

Blacks	0.5%

Expression

Cord RBCs	Expressed

Molecular basis associated with Dantu antigen[1,2]

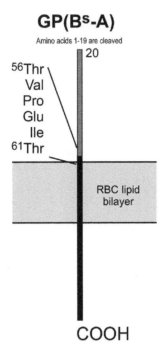

GP(B^s-A)

Amino acids 1-19 are cleaved

20

^{56}Thr
Val
Pro
Glu
Ile
^{61}Thr

RBC lipid bilayer

COOH

Variant glycophorin
GPB^s(20–58)-GPA(59–118)

Contribution by parent glycophorin
GPB(20–58)-GPA(89–150)

Gene arrangement
GYP(B-A)

Three variants expressing Dantu have been reported:

The MD type is associated with a chromosome carrying *GYPA*, *GYP(B-A)* and *GYPB* genes.
The NE type is associated with a chromosome carrying *GYPA*, *GYP(B-A)* and a duplicated *GYP(B-A)*.
The Ph type may be associated with a chromosome carrying *GYPA*, *GYP(B-A)*.

Effect of enzymes and chemicals on Dantu antigen on intact RBCs

Ficin/Papain	Resistant
Trypsin	Resistant

α-Chymotrypsin	Resistant
DTT 200 mM	Resistant
Acid	Resistant

In vitro characteristics of alloanti-Dantu

Immunoglobulin class	IgM and IgG
Optimal technique	RT; IAT

Clinical significance of alloanti-Dantu

Transfusion reaction	No data
HDFN	Positive DAT; no clinical HDFN

Comments

Dantu+ RBCs (NE type) have a weak expression of s and are U–; they have decreased glycosylation of band 3.

The reciprocal product of *GYP*Dantu* is *GYP*TK* (see SAT antigen **MNS36**).

References

[1] Blumenfeld, O.O., et al., 1987. Membrane glycophorins of Dantu blood group erythrocytes. J Biol Chem 262, 11864–11870.

[2] Huang, C.-H., Blumenfeld, O.O., 1995. MNSs blood groups and major glycophorins: molecular basis for allelic variation. In: Cartron, J.-P., Rouger, P. (Eds.), Molecular Basis of Human Blood Group Antigens. Plenum Press, New York, NY, pp. 153–188.

Hop Antigen

Terminology

ISBT symbol (number)	MNS26 (002026 or 2.26)
History	Reported in 1977 and named after the first donor whose RBCs expressed the antigen.

Occurrence

Most populations	<0.01%
Thai	0.68%

Expression

Cord RBCs	Presumed expressed

Molecular basis associated with Hop antigen[1–3]

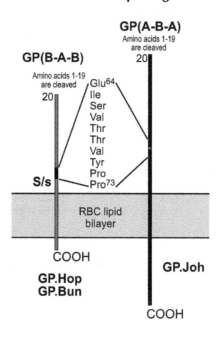

Variant glycophorins

GP.Hop (Mi.IV)	GPB(20–45)-GPψB(46–69)-GPA(70–76)-GPB^S(77–122)
GP.Bun (Mi.VI)	GPB(20–45)-GPψB(46–69)-GPA(70–76)-GPB^s(77–122)
GP.Joh (Mi.VIII)	GPA(20–67)-GPψB(68)-GPA(69–150)

Contribution by parent glycophorins

GP.Hop, GP.Bun	GPB(20–45)-GPψB-GPA(70–76)-GPB(46–91)
GP.Joh	GPA(20–67)-GPψB-GPA(69–150)

Gene arrangement

GP.Hop, GP.Bun	*GYP(B-A-B)*
GP.Joh	*GYP(A-B-A)*

Effect of enzymes and chemicals on Hop antigen on intact RBCs

	GP.Hop (Mi.IV)	GP.Bun (Mi.VI)	GP.Joh (Mi.VIII)
Ficin/Papain	Sensitive	Sensitive	Sensitive
Trypsin	Resistant	Resistant	Sensitive
α-Chymotrypsin	Sensitive	Variable	Resistant
Sialidase	Variable	Variable	Variable
DTT 200 mM	Resistant	Resistant	Resistant
Acid	Resistant	Resistant	Resistant

In vitro characteristics of alloanti-Hop

Immunoglobulin class   IgG
Optimal technique   IAT

Clinical significance of alloanti-Hop

No data are available because antibody is rare.

Comments

Antigen is defined by the Anek serum (predominantly anti-Hop, weak anti-Nob). Sera which contain anti-Hop may also contain anti-Nob (see **MNS27**).

References

[1] Dahr, W., 1992. Miltenberger subsystem of the MNSs blood group system. Review and outlook. Vox Sang 62, 129–135.

[2] Huang, C.-H., Blumenfeld, O.O., 1995. MNSs blood groups and major glycophorins: molecular basis for allelic variation. In: Cartron, J.-P., Rouger, P. (Eds.), Molecular Basis of Human Blood Group Antigens. Plenum Press, New York, NY, pp. 153–188.

[3] Storry, J.R., et al., 2000. Identification of a novel hybrid glycophorin gene encoding GP.Hop. Transfusion 40, 560–565.

Nob Antigen

Terminology

ISBT symbol (number)   MNS27 (002027 or 2.27)
History   The antigen is defined by the Lane serum and was named Nob after the person whose RBCs carried the antigen.

Occurrence

Most populations   <0.01%

Antithetical antigen

ENKT (**MNS29**)

Expression

Cord RBCs   Presumed expressed

Molecular basis associated with Nob antigen[1,2]

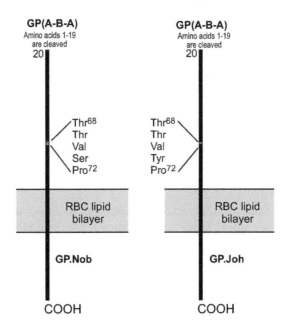

Variant glycophorins

GP.Nob (Mi.VII)	GPA(20–67)-GPψB(68–71)-GPA(72–150)
GP.Joh (Mi.VIII)	GPA(20–67)-GPψB(68)-GPA(69–150)

Contribution by parent glycophorins

GP.Nob	GPA(20–67)-GPψB-GPA(72–150)
GP.Joh	GPA(20–67)-GPψB-GPA(69–150)

Gene arrangement

GP.Nob, GP.Joh	*GYP(A-B-A)*

Effect of enzymes and chemicals on Nob antigen on intact RBCs

Ficin/Papain	Sensitive
Trypsin	Resistant
α-Chymotrypsin	Resistant
Sialidase	Variable
DTT 200 mM	Resistant
Acid	Resistant

In vitro characteristics of alloanti-Nob

Immunoglobulin class	IgM; IgG
Optimal technique	RT; IAT

Clinical significance of alloanti-Nob

Transfusion reaction	Mild in one case
HDFN	No data

Comments

The Raddon serum is predominantly anti-Nob with a weak anti-Hop. Sera which contain anti-Nob may also contain anti-Hop (see **MNS26**).

References

[1] Dahr, W., 1992. Miltenberger subsystem of the MNSs blood group system. Review and outlook. Vox Sang 62, 129–135.

[2] Huang, C.-H., Blumenfeld, O.O., 1995. MNSs blood groups and major glycophorins: molecular basis for allelic variation. In: Cartron, J.-P., Rouger, P. (Eds.), Molecular Basis of Human Blood Group Antigens. Plenum Press, New York, NY, pp. 153–188.

En^a Antigen

Terminology

ISBT symbol (number)	MNS28 (002028 or 2.28)
History	Named in 1965 when it was recognized that the antigen was carried on an important component of the envelope of the RBC. Joined the MNS system in 1985.

Occurrence

All populations	100%

Expression

Cord RBCs	Expressed

Molecular basis associated with En^a antigen[1]

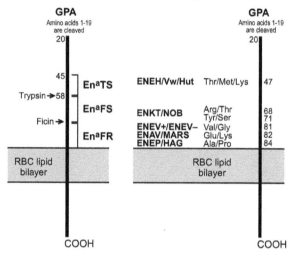

Legend: TS = trypsin sensitive; FS = ficin sensitive; FR = ficin resistant

Effect of enzymes and chemicals on En^a antigen on intact RBCs

Ficin/Papain	See figure
Trypsin	See figure
α-Chymotrypsin	Resistant
Pronase	Most are sensitive
Sialidase	Variable
DTT 200 mM	Resistant
Acid	Resistant

In vitro characteristics of alloanti-En^a

Immunoglobulin class	IgM and IgG
Optimal technique	RT; IAT
Complement binding	Rare

Clinical significance of alloanti-En^a

Transfusion reaction	No to severe
HDFN	No to severe

Autoantibody

Yes (Anti-En^{a}TS, anti-En^{a}FS, and anti-En^{a}FR).

Comments

RBCs that lack GPA lack all En^a antigens, type as Wr(b–), and have reduced levels of sialic acid (40% of normal).

Reference

[1] Issitt, P.D., et al., 1981. Proposed new terminology for En^a. Transfusion 21, 473–474.

ENKT Antigen

Terminology

ISBT symbol (number)	MNS29 (002029 or 2.29)
Obsolete names	En^{a}FS, En^{a}KT
History	Reported as En^{a}FS in 1985. In 1988, it was named "EN" because it is a high-prevalence antigen on GPA and "KT" for the initials of the first antigen-negative proband.

Occurrence

All populations	100%

Antithetical antigen

Nob (**MNS27**)

Expression

Cord RBCs	Presumed expressed

Molecular basis associated with ENKT antigen[1]

Amino acid	Arg68 (previously 49) and Tyr71 (previously 52) of GPA
Nucleotide	C at bp 203 and C at bp 212 of *GYPA*

Effect of enzymes and chemicals on ENKT antigen on intact RBCs

Ficin/Papain	Sensitive
Trypsin	Resistant
α-Chymotrypsin	Resistant
DTT 200 mM	Resistant
Acid	Resistant

In vitro characteristics of alloanti-ENKT

Immunoglobulin class	IgG
Optimal technique	IAT

Clinical significance of alloanti-ENKT

No data are available because anti-ENKT is rare.

MNS

Reference

[1] Dahr, W., 1992. Miltenberger subsystem of the MNSs blood group system. Review and outlook. Vox Sang 62, 129–135.

'N' Antigen

Terminology

ISBT symbol (number)	MNS30 (002030 or 2.30)
Other name	GPB^N
History	Named when it was realized that the N-terminal amino acid sequence of GPB was the same as GPA carrying the N antigen. Quotation marks were used to distinguish N on GPB from N on GPA. Assigned an MNS number in 1985 by the ISBT.

Occurrence

Present on all cells except those deficient in GPB or RBCs with GPB expressing He or M^v antigen.

Antithetical antigen

He (**MNS6**)

Expression

Cord RBCs	Expressed

Molecular basis associated with 'N' antigen[1]

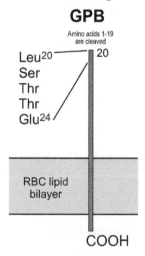

GPB

Amino acids 1-19 are cleaved

Leu^{20} ⎯⎯ 20
Ser
Thr
Thr
Glu^{24}

RBC lipid bilayer

COOH

Effect of enzymes and chemicals on 'N' antigen on intact RBCs

Ficin/Papain	Sensitive
Trypsin	Resistant
α-Chymotrypsin	Sensitive
Sialidase	Variable
DTT 200 mM	Resistant
Acid	Resistant

Comments

See N antigen (**MNS2**). Anti-'N' does not exist.

Reference

[1] Blanchard, D., et al., 1987. Glycophorins B and C from human erythrocyte membranes. Purification and sequence analysis. J Biol Chem 262, 5808–5811.

Or Antigen

Terminology

ISBT symbol (number)	MNS31 (002031 or 2.31)
Obsolete names	Orriss; Or[a]
History	Named in 1987 after the family in which the antigen was first found.

Occurrence

Found in two Japanese, one Australian, one African American, and one Jamaican.

Expression

Cord RBCs	Expressed

Molecular basis associated with Or antigen[1,2]

Amino acid	Trp50 (previously 31) of GPA
Nucleotide	T at bp 148 in exon 3 in *GYPA*
Or– (wild type)	Arg50 and C at bp 148

Effect of enzymes and chemicals on Or antigen on intact RBCs

Ficin/Papain	Sensitive
Trypsin	Variable
α-Chymotrypsin	Resistant
Sialidase	Sensitive
DTT 200 mM	Resistant
Acid	Resistant

MNS

In vitro characteristics of alloanti-Or

Immunoglobulin class	IgM more common than IgG
Optimal technique	RT

Clinical significance of alloanti-Or

Transfusion reactions	No data
HDFN	No to moderate[1]

Comments

The M (**MNS1**) antigen on Or+ RBCs is more resistant to trypsin treatment than normal M, presumably due to the close proximity of the amino acid change to the major trypsin cleavage site[3].

References

[1] Reid, M.E., et al., 2000. First example of hemolytic disease of the newborn caused by anti-Or and confirmation of the molecular basis of Or. Vox Sang 79, 180–182.

[2] Tsuneyama, H., et al., 1998. Molecular basis of Or in the MNS blood group system [abstract]. Vox Sang 74 (Suppl. 1), 1446.

[3] Bacon, J.M., et al., 1987. Evidence that the low frequency antigen Orriss is part of the MN blood group system. Vox Sang 52, 330–334.

DANE Antigen

Terminology

ISBT symbol (number)	MNS32 (002032 or 2.32)
History	Named in 1991 after it was found in four Danish families.

Occurrence

Most populations	<0.01%
Danes	0.43%

Antithetical antigen

ENDA (**MNS44**)

Expression

Cord RBCs	Presumed expressed

Molecular basis associated with DANE antigen[1]

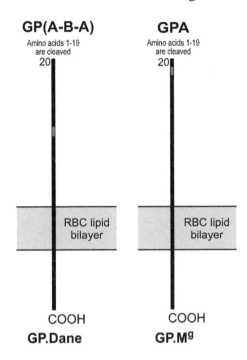

GP(A-B-A)
Amino acids 1-19
are cleaved
20

GPA
Amino acids 1-19
are cleaved
20

RBC lipid
bilayer

RBC lipid
bilayer

COOH
GP.Dane

COOH
GP.M^g

Variant glycophorin

GP.Dane (Mi.IX)   GPA(20–53)-GPψB(54–59)-GPA(60–150)

Contribution by parent glycophorin

GPA(20–53)-GPψB-GPA(60–150)

Gene arrangement

GYP(A-B-A). Found with an Ile65 (previously 46) change in GPA to Asn64 of
GP.Dane in one case[1], and without the Ile65 change in one case, which was
ENDA− (**MNS44**)[2].

Effect of enzymes and chemicals on DANE antigen on intact RBCs

Ficin/Papain	Sensitive
Trypsin	Sensitive
α-Chymotrypsin	Resistant
DTT 200 mM	Resistant
Acid	Resistant

In vitro characteristics of alloanti-DANE

Immunoglobulin class	IgG
Optimal technique	IAT

Clinical significance of alloanti-DANE

Unknown since only one example, in an untransfused male, has been described.

Comments

GP.Dane has trypsin resistant M (**MNS1**) and Mur (**MNS10**) antigens, but does not express other low prevalence MNS antigens[3]. Mg+ RBCs are DANE+, maybe due to the presence of Asn64 in the hybrid[4].
DANE is inherited with MS.

References

[1] Huang, C.-H., et al., 1992. Molecular analysis of human glycophorin MiIX gene shows a silent segment transfer and untemplated mutation resulting from gene conversion via sequence repeats. Blood 80, 2379–2387.

[2] Velliquette, R.W., et al., 2008. Novel GYP(A-B-A) hybrid gene in a DANE+ person who made an antibody to a high prevalence MNS antigen. Transfusion 48, 2618–2623.

[3] Skov, F., et al., 1991. Miltenberger class IX of the MNS blood group system. Vox Sang 61, 130–136.

[4] Green, C., et al., 1994. Mg+ MNS blood group phenotype: further observations. Vox Sang 66, 237–241.

TSEN Antigen

Terminology

ISBT symbol (number)	MNS33 (002033 or 2.33)
History	Named in 1992 after the last name of the first antibody producer.

Occurrence

Most populations	<0.01%

Expression

Cord RBCs	Expressed

Molecular basis associated with TSEN antigen[1,2]

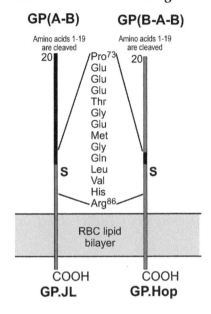

Variant glycophorins

GP.JL (Mi.XI)	GPA(20–77)-GPB^S(78–123)
GP.Hop (Mi.IV)	GPB(20–45)-GPψB(46–69)-GPA(70–76)-GPB^S(77–123)

Contribution by parent glycophorins

GP.JL	GPA(20–77)-GPB(46–91)
GP.Hop	GPB(20–45)-GPψB-GPA(70–76)-GPB(46–91)

Gene arrangement

GP.JL	*GYP(A-B)*
GP.Hop	*GYP(B-A-B)*

Effect of enzymes and chemicals on TSEN antigen on intact RBCs

Ficin/Papain	Sensitive
Trypsin	Resistant
α-Chymotrypsin	Sensitive
Sialidase	Resistant
DTT 200 mM	Resistant
Acid	Resistant

In vitro characteristics of alloanti-TSEN

Immunoglobulin class	IgM and IgG
Optimal technique	RT; IAT

Clinical significance of alloanti-TSEN

Transfusion reactions	No
HDFN	No

Comments

Reciprocal product of *GYP*JL* is *GYP*Sch* (see St^a antigen [**MNS15**]).
TSEN+ RBCs are also MINY+ (**MNS34**).
Several examples of anti-TSEN have been described[3].
Some anti-S do not agglutinate S+s+ TSEN+ RBCs.

References

[1] Huang, C.-H., Blumenfeld, O.O., 1995. MNSs blood groups and major glycophorins: molecular basis for allelic variation. In: Cartron, J.-P., Rouger, P. (Eds.), Molecular Basis of Human Blood Group Antigens. Plenum Press, New York, NY, pp. 153–188.

[2] Reid, M.E., et al., 1992. TSEN: a novel MNS-related blood group antigen. Vox Sang 63, 122–128.

[3] Storry, J.R., et al., 2000. Four examples of anti-TSEN and three of TSEN-positive erythrocytes. Vox Sang 79, 175–179.

MINY Antigen

Terminology

ISBT symbol (number)	MNS34 (002034 or 2.34)
History	Named in 1992 after the only producer of the antibody.

Occurrence

Most populations	<0.01%
Chinese	6%

Expression

Cord RBCs	Presumed expressed

Molecular basis associated with MINY antigen[1]

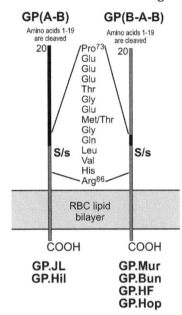

GP(A-B)

GP(B-A-B)

For details of variant glycophorin, contribution by parent glycophorin and gene arrangement, see Hil (**MNS20**) and TSEN (**MNS33**).

Effect of enzymes and chemicals on MINY antigen on intact RBCs

Ficin/Papain	Sensitive
Trypsin	Resistant
α-Chymotrypsin	Sensitive
Sialidase	Resistant
DTT 200 mM	Resistant
Acid	Resistant

In vitro characteristics of alloanti-MINY

Immunoglobulin class	IgM
Optimal technique	RT

Clinical significance of alloanti-MINY

No data because antibody is rare.

Comments

All Hil+ (**MNS20**) and TSEN+ (**MNS33**) RBCs are MINY+ except when the Hil antigen is carried on GP.KI[2].

References

[1] Reid, M.E., et al., 1992. MINY: A novel MNS-related blood group antigen. Vox Sang 63, 129–132.

[2] Poole, J., et al., 1998. Novel molecular basis for the Hil (MNS20) antigen [abstract]. Transfusion 38 (Suppl), 103S.

MUT Antigen

Terminology

ISBT symbol (number)	MNS35 (002035 or 2.35)
History	The specificity originally called anti-Hut was renamed anti-MUT in 1984, because both <u>Mu</u>r+ and <u>Hu</u>t+ RBCs are reactive.

Occurrence

Most populations	<0.01%
Chinese	6%

Expression

Cord RBCs	Presumed expressed

Molecular basis associated with MUT antigen[1]

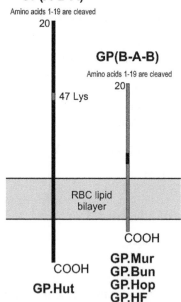

GP(A-B-A)

Amino acids 1-19 are cleaved

20

47 Lys

GP(B-A-B)

Amino acids 1-19 are cleaved

20

RBC lipid bilayer

COOH

COOH

GP.Hut

GP.Mur
GP.Bun
GP.Hop
GP.HF

For details of variant glycophorin, contributions by parent glycophorin, and gene arrangement, see Hut (**MNS19**), Hil (**MNS20**), and Hop (**MNS26**).

Effect of enzymes and chemicals on MUT antigen on intact RBCs

	GP(A-B-A)	GP(B-A-B)
Ficin/Papain	Sensitive	Sensitive
Trypsin	Sensitive	Resistant
α-Chymotrypsin	Resistant	Sensitive
DTT 200 mM	Resistant	Resistant
Acid	Resistant	Resistant

In vitro characteristics of alloanti-MUT

Immunoglobulin class	IgM and IgG
Optimal technique	RT; IAT

Clinical significance of alloanti-MUT

Transfusion reactions	No data are available because antibody is rare
HDFN	One case[2]

Comments

Anti-MUT is often in serum with (and is separable from) anti-Hut (see **MNS19**).

References

[1] Huang, C.-H., Blumenfeld, O.O., 1995. MNSs blood groups and major glycophorins: molecular basis for allelic variation. In: Cartron, J.-P., Rouger, P. (Eds.), Molecular Basis of Human Blood Group Antigens. Plenum Press, New York, NY, pp. 153–188.

[2] van den Bos, A.G., Steiner, K., 2004. Haemolytic disease of the newborn caused by anti-MUT (MNS 35). Vox Sang 87, 208–209.

SAT Antigen

Terminology

ISBT symbol (number)	MNS36 (002036 or 2.36)
History	Reported in 1991 and named after the first proband whose RBCs carried the antigen; joined the MNS system in 1994.

Occurrence

Most populations   <0.01%

Expression

Cord RBCs   Presumed expressed

Molecular basis associated with SAT antigen[1,2]

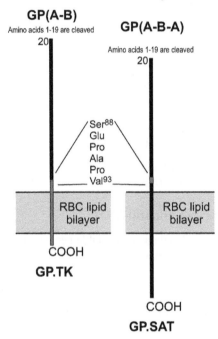

Variant glycophorins

GP.TK	GPA(20–90)-GPB(91–123)
GP.SAT	GPA(20–90)-GPB(91–93)-GPA(94–153)

Contribution by parent glycophorins

GP.TK	GPA(20–90)-GPB(59–91)
GP.SAT	GPA(20–90)-GPB(59–61)-GPA(91–150)

Gene arrangement

GP.TK	*GYP(A-B)*
GP.SAT	*GYP(A-B-A)*

Effect of enzymes and chemicals on SAT antigen on intact RBCs

Ficin/Papain	Sensitive
Trypsin	Sensitive (GP.TK); resistant (GP.SAT)
α-Chymotrypsin	Sensitive
Sialidase	GP.SAT resistant
DTT 200 mM	Resistant
Acid	Resistant

In vitro characteristics of alloanti-SAT

Immunoglobulin class	IgG
Optimal technique	IAT

Clinical significance of alloanti-SAT

No data are available because anti-SAT is rare.

Comments

The reciprocal product of *GYP*TK* is *GYP*Dantu* (see **MNS25**).

References

[1] Huang, C.-H., et al., 1995. Glycophorin SAT of the human erythrocyte membrane is specified by a hybrid gene reciprocal to glycophorin Dantu gene. Blood 85, 2222–2227.

[2] Uchikawa, M., et al., 1994. A novel amino acid sequence result in the expression of the MNS related private antigen, SAT [abstract]. Vox Sang 67 (S2), 116.

ERIK Antigen

Terminology

ISBT symbol (number)	MNS37 (002037 or 2.37)
History	Named in 1993 after the proband whose St(a+) RBCs expressed another low-prevalence antigen.

Occurrence

Most populations	<0.01%

Expression

Cord RBCs	Presumed expressed

Molecular basis associated with ERIK antigen[1,2]

Amino acid	Arg78 (previously 59) of GPA
Nucleotide	A at bp 232 in exon 4 in *GYPA*
ERIK– (wild type)	Gly78 and G at position 232

ERIK has been associated with a *GYP(A-E-A)*, which encodes a variant of GPA carrying St^a.

Effect of enzymes and chemicals on ERIK antigen on intact RBCs

Ficin/Papain	Variable/sensitive
Trypsin	Weakened
α-Chymotrypsin	Resistant
Pronase	Sensitive
Sialidase	Resistant
DTT 200 mM	Resistant
Acid	Resistant

In vitro characteristics of alloanti-ERIK

Immunoglobulin class	IgG
Optimal technique	IAT

Clinical significance of alloanti-ERIK

Transfusion reaction	No data because the antibody is rare
HDFN	Positive DAT

Comments

Alternative splicing of *GYP*EBH* gives rise to a variant glycophorin GP.EBH(t2) expressing the St^a antigen (see **MNS15**). Thus, in ERIK+ RBCs, ERIK and St^a antigens are carried on different glycophorin molecules (see table on MNS system page).

The Gly78Arg change introduces a trypsin cleavage site.

References

[1] Huang, C.-H., et al., 1993. Alteration of splice site selection by an exon mutation in the human glycophorin A gene. J Biol Chem 268, 25902–25908.

[2] Huang, C.-H., et al., 1994. Gene conversion between glycophorins A and E results in St^a glycophorin in a family exhibiting the ERIK/St^a blood group phenotype [abstract]. Blood 84 (Suppl. 1), 238a.

Os^a Antigen

Terminology

ISBT symbol (number)	MNS38 (002038 or 2.38)
Obsolete name	700033
History	Named in 1983 after Osaka, the town where the antibody and antigen were first found; joined the MNS system in 1994.

Occurrence

Only studied in one Japanese family[1].

Expression

Cord RBCs   Presumed expressed

Molecular basis associated with Os^a antigen[2]

Amino acid	Ser73 (previously 54) of GPA
Nucleotide	T at bp 217 in exon 3 in *GYPA*
Os(a–) (wild type)	Pro73 and C bp 217

Effect of enzymes and chemicals on Os^a antigen on intact RBCs

Ficin/Papain	Sensitive
Trypsin	Resistant
α-Chymotrypsin	Resistant
Sialidase	Resistant
DTT 200 mM	Resistant
Acid	Resistant

In vitro characteristics of alloanti-Os^a

Immunoglobulin class	IgG
Optimal technique	IAT

Clinical significance of alloanti-Os^a

No data are available.

Comments

Anti-Os^a found in several sera containing antibodies to multiple low-prevalence antigens.

References

[1] Seno, T., et al., 1983. OS^a, a "new" low-frequency red cell antigen. Vox Sang 45, 60–61.

[2] Daniels, G.L., et al., 2000. The low-frequency MNS blood group antigens Ny^a (MNS18) and Os^a (MNS38) are associated with GPA amino acid substitutions. Transfusion 40, 555–559.

ENEP Antigen

Terminology

ISBT symbol (number)	MNS39 (002039 or 2.39)
History	Reported in 1995 and named "EN" because it is a high-prevalence antigen on GPA, and "EP" for the name of the first antigen-negative proband.

MNS

Occurrence

All populations   100%

Antithetical antigen

HAG (**MNS41**)

Expression

Cord RBCs   Presumed expressed

Molecular basis associated with ENEP antigen[1]

Amino acid   Ala84 (previously 65) of GPA
Nucleotide   G at bp 250 in exon 4 in *GYPA*

Effect of enzymes and chemicals on ENEP antigen on intact RBCs

Ficin/Papain   Ficin resistant/Papain sensitive
Trypsin   Resistant
α-Chymotrypsin   Resistant
DTT 200 mM   Presumed resistant

In vitro characteristics of alloanti-ENEP

Immunoglobulin class   IgG
Optimal technique   IAT

Clinical significance of alloanti-ENEP

No data are available because antibody is rare.

Comments

Anti-ENEP (anti-En^{a}FR) was made by a person homozygous for *GYP*HAG*.
RBCs lacking ENEP (HAG+; **MNS41**) have an altered expression of Wr^b
(**DI4**) antigen[1].

Reference

[1] Poole, J., et al., 1999. Glycophorin A mutation Ala65 --> Pro gives rise to a novel pair of MNS
alleles ENEP (MNS39) and HAG (MNS41) and altered Wr^b expression: direct evidence for
GPA/band 3 interaction necessary for normal Wr^b expression. Transfus Med 9, 167–174.

ENEH Antigen

Terminology

ISBT symbol (number)   MNS40 (002040 or 2.40)
History   Named "EN" because it is a high-prevalence antigen
                         on GPA, and "EH" from the initials of the first
                         antigen-negative proband.

Occurrence

All populations   100%

Antithetical antigen

Vw (**MNS9**); Hut (**MNS19**)

Expression

Cord RBCs   Expressed

Molecular basis associated with ENEH antigen[1]

Amino acid   Thr47 (previously 28) of GPA
Nucleotide   C at bp 140 in exon 3 in *GYPA*

Effect of enzymes and chemicals on ENEH antigen on intact RBCs

Ficin/Papain   Sensitive
Trypsin   Sensitive
α-Chymotrypsin   Resistant
Sialidase   Resistant
DTT 200 mM   Presumed resistant
Acid   Resistant

In vitro characteristics of alloanti-ENEH

Immunoglobulin class   IgM and IgG is the only example of anti-ENEH
                               described[2]
Optimal technique   RT; IAT

Clinical significance of alloanti-ENEH

Transfusion reactions   No data are available
HDFN   The anti-ENEH (anti-En^aTS) did not cause HDFN[2]

Comment

ENEH– RBCs have an altered expression of Wr^b (**DI4**).

References

[1] Huang, C.-H., et al., 1992. Molecular basis for the human erythrocyte glycophorin specifying the Miltenberger class I (MiI) phenotype. Blood 80, 257–263.

[2] Spruell, P., et al., 1993. An anti-En^aTS detected in the serum of an MiI homozygote. Transfusion 33, 848–851.

HAG Antigen

Terminology

ISBT symbol (number)   MNS41 (002041 or 2.41)
History   Reported in 1995 and named after the transfused
                       man whose serum contained an antibody to a high-
                       prevalence antigen (ENEP), and who's RBCs had a
                       double dose of this low-prevalence antigen.

Occurrence

Two probands, both Israeli.

Antithetical antigen

ENEP (**MNS39**)

Expression

Cord RBCs   Presumed expressed

Molecular basis associated with HAG antigen[1]

Amino acid   Pro84 (previously 65) of GPA
Nucleotide   C at bp 250 in exon 4 in *GYPA*

Effect of enzymes and chemicals on HAG antigen on intact RBCs

Ficin/Papain   Resistant
Trypsin   Resistant
α-Chymotrypsin   Resistant
DTT 200 mM   Presumed resistant
Acid   Resistant

In vitro characteristics of alloanti-HAG

Immunoglobulin class   IgG
Optimal technique   IAT

Clinical significance of alloanti-HAG

No data are available because antibody is rare.

Comments

RBCs with a double dose expression of HAG (ENEP–; [**MNS39**]) have an
altered expression of Wr^b (**DI4**)[1].

Reference

[1] Poole, J., et al., 1999. Glycophorin A mutation Ala65 --> Pro gives rise to a novel pair of MNS alleles ENEP (MNS39) and HAG (MNS41) and altered Wr^b expression: direct evidence for GPA/band 3 interaction necessary for normal Wr^b expression. Transfus Med 9, 167–174.

ENAV Antigen

Terminology

ISBT symbol (number)	MNS42 (002042 or 2.42)
Obsolete names	Avis
History	Reported in 1996; named "EN" because it is a high-prevalence antigen on GPA, and "AV" after the name of the proband whose serum contained the antibody.

Occurrence

All populations	100%

Antithetical antigen

MARS (**MNS43**)

Expression

Cord RBCs	Expressed

Molecular basis associated with ENAV antigen[1]

Amino acid	Gln82 (previously 63) of GPA
Nucleotide	C at bp 244 in exon 4 in *GYPA*

Effect of enzymes and chemicals on ENAV antigen on intact RBCs

Ficin/Papain	Resistant
Trypsin	Resistant
α-Chymotrypsin	Resistant
Sialidase	Resistant
DTT 200 mM	Resistant
Acid	Resistant

In vitro characteristics of alloanti-ENAV

Immunoglobulin class	IgG
Optimal technique	IAT

Clinical significance of alloanti-ENAV

No data are available because antibody is rare.

Comments

ENAV– RBCs have a weak expression of Wr^b (see **DI4**).

Reference

[1] Jarolim, P., et al., 1997. Molecular basis of the MARS and AVIS blood group antigens [abstract]. Transfusion 37 (Suppl), 90S.

MARS Antigen

Terminology

ISBT symbol (number)	MNS43 (002043 or 2.43)
History	Reported in 1996 and named after the Native American proband (Marsden) whose serum contained antibodies to several low-prevalence antigens and reacted with ENAV– RBCs (**MNS42**).

Occurrence

Most populations	None found
Choctaw Indians	15%

Antithetical antigen

ENAV (**MNS42**)

Expression

Cord RBCs	Presumed expressed

Molecular basis associated with MARS antigen[1]

Amino acid	Lys82 (previously 63) of GPA
Nucleotide	A at bp 244 in exon 4 in *GYPA*

Effect of enzymes and chemicals on MARS antigen on intact RBCs

Ficin/Papain	Resistant
Trypsin	Resistant
α-Chymotrypsin	Resistant
DTT 200 mM	Resistant
Acid	Resistant

In vitro characteristics of alloanti-MARS

Immunoglobulin class	IgG
Optimal technique	IAT

Clinical significance of alloanti-MARS

No data are available because antibody is rare.

Comments

RBCs with a double dose of the MARS antigen (ENAV–) have a weak expression of Wr^b (see **DI4**).

Reference

[1] Jarolim, P., et al., 1997. Molecular basis of the MARS and AVIS blood group antigens [abstract]. Transfusion 37 (Suppl), 90S.

ENDA Antigen

Terminology

ISBT symbol (number)	MNS44 (002044 or 2.44)
History	Reported in 2008; named "EN" because it is a high-prevalence antigen on GPA, and "DA" for the association with DANE.

Occurrence

Only one ENDA– proband and her ENDA– brother, have been reported.

Antithetical antigen

DANE (**MNS32**)

Expression

Cord RBCs	Presumed expressed

Molecular basis associated with ENDA antigen[1]

Variant glycophorin
GP.Dane: GPA(20–53)-GPψB(54–59)-GPA(60–150)

Contribution by parent glycophorin
GPA(20–53)-GPψB-GPA(60–150)

Gene arrangement
GYP(A-B-A)

Effect of enzymes and chemicals on ENDA antigen on intact RBCs

Ficin/Papain	Sensitive
Trypsin	Sensitive
α-Chymotrypsin	Resistant
DTT 200 mM	Resistant

In vitro characteristics of alloanti-ENDA

Immunoglobulin class   IgM; IgG (see Comments)
Optimal technique   RT; 37°C; IAT

Clinical significance of alloanti-ENDA

No data are available because antibody is rare.

Comments

The only known example of anti-ENDA was found in the serum of an untransfused woman during her first pregnancy. The antibody was IgM, but after the birth of the proband's baby a trace of IgG was found. In this proband, the novel $GYP(A\text{-}B\text{-}A)$ was in trans to M^k.

Reference

[1] Velliquette, R.W., et al., 2008. Novel GYP(A-B-A) hybrid gene in a DANE+ person who made an antibody to a high prevalence MNS antigen. Transfusion 48, 2618–2623.

ENEV Antigen

Terminology

ISBT symbol (number)   MNS45 (002045 or 2.45)
History   Reported in 2010; named "EN" because it is a high-prevalence antigen on GPA, and "EV" from the ENEV– proband's name.

Occurrence

One ENEV– proband has been reported.

Expression

Cord RBCs   Presumed expressed

Molecular basis associated with ENEV antigen[1]

Amino acid   Val81 (previously 62) of GPA
Nucleotide   T at bp 242 in exon 4 in *GYPA*
ENEV–   Gly81 and G at bp 242

Effect of enzymes and chemicals on ENEV antigen on intact RBCs

Ficin/Papain   Resistant
Trypsin   Resistant
α-Chymotrypsin   Resistant
DTT 200 mM   Resistant

In vitro characteristics of alloanti-ENEV

Immunoglobulin class	IgM; IgG
Optimal technique	IAT; RT; 37°C

Clinical significance of alloanti-ENEV

Only one example of anti-ENEV has been reported. The proband, who had a history of three pregnancies, received four units of crossmatch-compatible RBCs. Ten days after transfusion she presented with a drop in hemoglobin. Testing of her RBCs suggested that no transfused RBCs remained in her circulation.

Comments

ENEV– RBCs have an altered expression of the Wr^b (**DI4**) antigen.
Anti-ENEV gave marginally weaker reactions with ENEP– and ENAV– RBCs.

Reference

[1] Velliquette, R.W., et al., 2010. Novel single nucleotide change in *GYP*A* in a person who made an alloantibody to a new high prevalence MNS antigen called ENEV. Transfusion 50, 856–860.

MNTD Antigen

Terminology

ISBT symbol (number)	MNS46 (002046 or 2.46)
History	Reported in 2006; named "MN" for the system, and "TD" from the name of the antigen-positive index case.

Occurrence

Most populations	None found
Japanese	0.02%

Expression

Cord RBCs	Presumed expressed

Molecular basis associated with MNTD antigen[1]

Amino acid	Arg36 (previously 17) on GPA
Nucleotide	G at bp 107 in exon 3 of *GYPA*
MNTD– (wild type)	Thr36 and C at bp107

Effect of enzymes and chemicals on MNTD antigen on intact RBCs

Ficin/Papain	Sensitive
Trypsin	Sensitive
α-Chymotrypsin	Sensitive
DTT 200 mM	Presumed resistant

In vitro characteristics of alloanti-MNTD

Immunoglobulin class	IgG
Optimal technique	IAT

Clinical significance of alloanti-MNTD

No data are available because antibody and antigen are rare.

Reference

[1] Uchikawa, M, et al., 2006. Molecular basis for a novel low frequency antigen in the MNS blood group system, Td [abstract]. Vox Sang 91 (Suppl. 3), 133.

P1PK Blood Group System

Number of antigens 3

Polymorphic	P1
High prevalence	P^k (albeit weakly expressed on most RBC phenotypes)
Low prevalence	NOR

Terminology

ISBT symbol (number)	P1PK (003)
History	The P1 antigen (originally named P) was discovered by Landsteiner and Levine in 1927. The antigen was named P because this was the first letter after the already assigned M, N, and O. In addition to P1, the P system previously also contained P, P^k, and LKE antigens. However, uncertainty about the genetic loci and biochemical pathways underlying these antigens arose, so in 1994 they were moved to the Globoside Collection. In 2010, the P^k antigen was replaced into the P system, which was renamed the P1PK system. In 2011, the NOR antigen was provisionally assigned to the same system.

Expression

Soluble form	P1: Pigeon egg white, hydatid cyst fluid, *Echinococcus* cyst fluid
Other blood cells	P1/P^k: Weakly on lymphocytes, granulocytes, monocytes, platelets

Gene

Chromosome	22q13.2
Name	*A4GALT*

The Blood Group Antigen (3/e). DOI: http://dx.doi.org/10.1016/B978-0-12-415849-8.00005-3

Organization	Four exons distributed over ~26.6 kbp
Product	4-α-galactosyltransferase

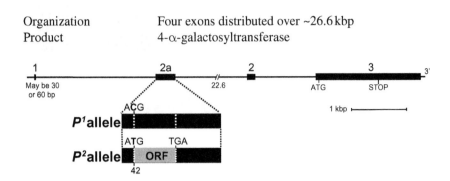

Database accession numbers

GenBank	NG_007495 (gDNA); AJ245581 (cDNA; partial exon 3); GU902278 (cDNA, exon 1 and 2a)
Entrez Gene ID	53947

Molecular bases of the P_1 and P_2 phenotypes due to changes in exon 2a of *A4GALT*[1]

The open reading frame in exon 3 (accession number AJ245581) encodes 4-α-galactosyltransferase, the enzyme that synthesizes P1 and P^k. Changes in exon 2a distinguish P_1 and P_2 phenotypes. Differences from reference allele *A4GALT*P1.01* (accession number GU902278) are given below. A nucleotide change of C > T (ACG > ATG) in exon 2a of *A4GALT* introduces an open reading frame in transcripts that include exons 1 and 2a. This is associated with fewer enzyme-encoding transcripts (comprising exons 1, 2, and 3) in the presence of the P^2 allele, but it is unknown how this transcriptional regulation occurs.

Phenotype	Allele name	Exon	Nucleotide	Amino acid change^	Ethnicity (prevalence)
P_1 phenotype					
P1+P^k+(P_1)	A4GALT*P1.01				
P_2 phenotype					
P1–P^k+(P_2)	A4GALT*P2.01	2a	42C>T	Start codon introduced	(Common)
P1–P^k+(P_2)	A4GALT*P2.02	2a	42C>T; 122T>G	Start codon introduced; Gly28Trp	(Several)

^ = Investigations are in progress to determine if the open reading frame in exon 2a is translated.

Molecular bases of P1+/− P^k+, NOR+, and p (PP1P^k−, PP1P$^k_{null}$) phenotypes due to changes in exon 3 of *A4GALT*[2,3]

Nucleotide differences from reference allele *A4GALT*01* (Accession number AJ245581), and amino acids affected, are given. This reference allele (comprising the open reading frame in exon 3) encodes 4-α-galactosyltransferase, which adds α-galactose to paragloboside (lacto-*N*-neotetraosylceramide) to form the P1 antigen. It also adds α-galactose to lactosylceramide (CDH) to form the P^k antigen (CTH). As the CTH is the precursor for the P antigen (see **GLOB** System and Section III), changes in *A4GALT* that prevent addition of galactose to CDH (the precursor of P^k antigen) also prevent addition of 3-β-*N*-acetylgalactosamine to form P, and therefore give rise to the p [P1PP^k−, previously known as the Tj(a−)] phenotype. An altered form of the transferase can also add α-galactose to globoside (Gb4) to form the NOR antigen, in addition to P1 and P^k.

Phenotype	Allele name	Exon	Nucleotide	Amino acid	Ethnicity (prevalence)
P1+/− P^k+	*A4GALT*01*^	3			(Common)
P1+/− P^k+	*A4GALT*02*^	3	109A>G	Met37Val	(Common)
NOR+, P1+, P^k+	*A4GALT*04*#	3	631G>C	Gln211Glu	(Rare)
Null phenotypes					
p	*A4GALT*01N.01.01*	3	241_243delTTC	Phe81del	Japanese, English (Few)
p	*A4GALT*01N.01.02*	3	241_243delTTC (with 903G>C)	Phe81del	Italian (Rare)
p	*A4GALT*01N.02*	3	287G>A	Cys96Tyr	Italian (Rare)
p	*A4GALT*01N.03.01*	3	299C>T	Ser100Leu	Amish (Several)
p	*A4GALT*01N.03.02*	3	299C>T (with 903G>C)	Ser100Leu	Pakistani (Rare)
p	*A4GALT*01N.04*	3	301delG	Ala101fs 113Stop	Chinese (Rare)
p	*A4GALT*01N.05*	3	418_428delins	Gln140fs 218Stop	Asian (Rare)
p	*A4GALT*01N.06*	3	470_496delins	Asp157fs 276Stop	English (Rare)
p	*A4GALT*01N.07*	3	473G>A	Trp158Stop	Brazilian (Rare)

(Continued)

P1PK

(Continued)

Phenotype	Allele name	Exon	Nucleotide	Amino acid	Ethnicity (prevalence)
p	A4GALT*01N.08	3	502_504insC (with 914C>T)	Tyr169fs282 Stop	Maghreb (Few)
p	A4GALT*01N.09.01	3	548T>A	Met183Lys	Swedish (Several)
p	A4GALT*01N.09.02	3	548T>A (with 987G>A)	Met183Lys	Swedish (Rare)
p	A4GALT*01N.10	3	559G>C	Gly187Arg	Thai (Rare)
p	A4GALT*01N.11	3	560G>A	Gly187Asp	Swedish (Rare)
p	A4GALT*01N.12	3	656C>T	Ala219Val	French (Rare)
p	A4GALT*01N.13	3	657delG	Ala219fs349 Stop	Israel (Few)
p	A4GALT*01N.14	3	731_732insG	Ile245fs281 Stop	Norwegian (Rare)
p	A4GALT*01N.15	3	751C>T	Pro251Ser	(Rare)
p	A4GALT*01N.16	3	752C>T	Pro251Leu	Japanese (Rare)
p	A4GALT*01N.17	3	769delG	Val257fs349 Stop	Polish (Rare)
p	A4GALT*01N.18	3	783G>A	Trp261Stop	Japanese (Rare)
p	A4GALT*01N.19	3	972_997del	Thr324fs436 Stop	USA (Rare)
p	A4GALT*01N.20	3	1026_1029 insC	Thr344fs446 Stop	Japanese (Few)
p	A4GALT*01N.21	3	196_201insC	Thr68fs282 Stop	Thai (Rare)
p	A4GALT*02N.01	3	68_69insT	Leu23fs53Stop	Israel (Rare)
p	A4GALT*02N.02	3	290C>T	Ser97Leu	Polish (Rare)
p	A4GALT*02N.03	3	752C>T	Pro251Leu	Japanese (Rare)
p	A4GALT*02N.04	3	902delC	Pro301fs349 Stop	(Rare)
p	A4GALT*02N.05	3	972_997del	Thr324fs436 Stop	French (Rare)

For changes in *B3GALNT1* giving rise to P_1^k and P_2^k phenotypes, see **GLOB** System.
^Can be *in cis* to P^1 or P^2 polymorphism in exon 2a, i.e., can travel with either P_1 or P_2 phenotype.
‡*In cis* to P^1 allele, i.e., travels with the P_1 phenotype.

Amino acid sequence of α4GalT (protein accession #AAH55286)

```
MSKPPDLLLR  LLRGAPRQRV  CTLFIIGFKF  TFFVSIVIYW  HVVGEPKEKG  50
QLYNLPAEIP  CPTLTPPTPP  SHGPTPGNIF  FLETSDRTNP  NFLFMCSVES  100
AARTHPESHV  LVLMKGLPGG  NASLPRHLGI  SLLSCFPNVQ  MLPLDLRELF  150
RDTPLADWYA  AVQGRWEPYL  LPVLSDASRI  ALMWKFGGIY  LDTDFIVLKN  200
LRNLTNVLGT  QSRYVLNGAF  LAFERRHEFM  ALCMRDFVDH  YNGWIWGHQG  250
PQLLTRVFKK  WCSIRSLAES  RACRGVTTLP  PEAFYPIPWQ  DWKKYFEDIN  300
PEELPRLLSA  TYAVHVWNKK  SQGTRFEATS  RALLAQLHAR  YCPTTHEAMK  350
MYL                                                         353
```

Carrier molecule[4,5]

The P1, P^k, and NOR antigens are not primary gene products; they are located on glycolipids. The terminal linkage of each antigen is synthesized by the primary gene product (4-α-galactosyltransferase).

All antigens in the P1PK system are based on lactosylceramide, which is also the immediate precursor for P^k antigen. Paragloboside is the precursor for P1 antigen, and globoside (P antigen) is the precursor for NOR antigen (see "Antigens with lactosylceramide as the precursor" and "Biosynthetic Pathways" in Section III).

Copies per RBC   Highly variable and depending on the P_1/P_2 phenotype, P^1 allele zygosity, and status of the B3GALNT1 gene (see **GLOB** System)

Function

The enzyme transfers Gal to the terminal sugar of paragloboside (for P1) or lactosylceramide (for P^k). It is unclear how acceptor preference is determined, but the higher levels of A4GALT transcripts in the P_1 phenotype indicate that it may be at least partially a quantitative question.

Disease association

See individual antigens under Comments.

Phenotypes (% occurrence)

Phenotype	Caucasians	Blacks	Cambodians & Vietnamese
P_1	79	94	20
P_2	21	6	80
Null: p (very rare)			
See GLOB System (028).			

P1PK

Comments

RBCs with either the P_1 or the P_2 phenotype express P^k (weakly), P, and LKE antigens, whilst RBCs with the p phenotype lack all these antigens and P1.

References

[1] Thuresson, B., et al., 2011. Identification of a novel *A4GALT* exon reveals the genetic basis of the P_1/P_2 histo-blood groups. Blood 117, 678–687.

[2] Steffensen, R., et al., 2000. Cloning and expression of the histo-blood group Pk UDP-galactose: Gal-beta1-4Glc-beta1-Cer alpha1,4-galactosyltransferase. Molecular genetic basis of the p phenotype. J Biol Chem 275, 16723–16729.

[3] Furukawa, K., et al., 2000. Molecular basis for the p phenotype. Identification of distinct and multiple mutations in the alpha1,4-galactosyltransferase gene in Swedish and Japanese individuals. J Biol Chem 275, 37752–37756.

[4] Bailly, P., Bouhours, J.F., 1995. P blood group and related antigens. In: Cartron, J.-P., Rouger, P. (Eds.), Molecular Basis of Human Blood Group Antigens. Plenum Press, New York, NY, pp. 300–329.

[5] Spitalnik, P.F., Spitalnik, S.L., 1995. The P blood group system: biochemical, serological, and clinical aspects. Transfusion Med Rev 9, 110–122.

P1 Antigen

Terminology

ISBT symbol (number)	P1PK1 (003001 or 3.1)
Obsolete names	P; P_1
History	Discovered in 1927; named P antigen because the letters M, N, and O had been used; renamed P_1 and then P1.

Occurrence

Caucasians	79%
Blacks	94%
Cambodian/Vietnamese	20%

Expression

Cord RBCs	Weaker than on RBCs from adults
Altered	There is considerable variation in the strength of P1 expression on RBCs. This variation is inherited, and at least partially dependent on the zygosity of P^l alleles[1]. P1 expression is also weakened in the In(Lu) phenotype.

Molecular basis associated with P1 antigen[2,3]

P1 antigen is derived by the addition of an α-galactosyl residue to paragloboside.

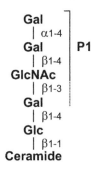

Effect of enzymes and chemicals on P1 antigen on intact RBCs

Ficin/Papain	Resistant (markedly enhanced)
Trypsin	Resistant (markedly enhanced)
α-Chymotrypsin	Resistant (markedly enhanced)
Pronase	Resistant (markedly enhanced)
Sialidase	Resistant
DTT 200 mM	Resistant
Acid	Resistant

In vitro characteristics of alloanti-P1

Immunoglobulin class	IgM (IgG rare)
Optimal technique	RT (or lower)
Neutralization	Hydatid cyst fluid, pigeon egg white, *Echinococcus* cyst fluid
Complement binding	Rare

Clinical significance of alloanti-P1

Transfusion reaction	No to moderate/delayed (rare)
HDFN	No

Comments

The P1 determinant is widely distributed throughout nature. It has been detected in, for example, liver flukes and pigeon egg white. The determinant is a receptor for a variety of microorganisms, including P-fimbriated strains of *E. coli* with the PapG adhesion, and *Streptococcus suis*[4].

Anti-P1 is a naturally-occurring antibody in many P1– individuals. Anti-P1 is frequently present in serum from patients with hydatid disease, liver fluke disease, and acute hepatic fascioliasis.

References

[1] Thuresson, B., et al., 2011. Identification of a novel *A4GALT* exon reveals the genetic basis of the P_1/P_2 histo-blood groups. Blood 117, 678–687.

[2] Bailly, P., Bouhours, J.F., 1995. P blood group and related antigens. In: Cartron, J.-P., Rouger, P. (Eds.), Molecular Basis of Human Blood Group Antigens. Plenum Press, New York, N.Y., pp. 300–329.

[3] Spitalnik, P.F., Spitalnik, S.L., 1995. The P blood group system: biochemical, serological, and clinical aspects. Transfusion Med Rev 9, 110–122.

[4] Moulds, J.M., et al., 1996. Human blood groups: Incidental receptors for viruses and bacteria. Transfusion 36, 362–374.

P^k Antigen

Terminology

ISBT symbol (number)	P1PK3 (003003 or 3.3)
Obsolete names	Trihexosylceramide; Ceramide trihexose (CTH); Globotriaosylceramide (Gb$_3$Cer); Gb3
CD number	CD77
History	Named in 1959 when the relationship to P was recognized; the "k" comes from the last name of the first proband to produce anti-P^k.

Occurrence

Strongly expressed on RBCs from <0.01% of the population, i.e., individuals with the P_1^k and P_2^k phenotypes. RBCs from all other individuals, except those with the p phenotype, have varying and often small amounts of P^k depending on the genotype ($P^1/P^1 > P^1/P^2 > P^2/P^2$).

Expression

Cord RBCs	Expressed

Molecular basis associated with P^k antigen[1,2]

P^k antigen is derived by the addition of an α-galactosyl residue to lactosylceramide.

$$
\begin{array}{l}
\textbf{Gal} \\
\quad | \; \alpha 1\text{-}4 \quad \textbf{pk} \\
\textbf{Gal} \\
\quad | \; \beta 1\text{-}4 \\
\textbf{Glc} \\
\quad | \; \beta 1\text{-}1 \\
\textbf{Ceramide}
\end{array}
$$

Effect of enzymes and chemicals on P^k antigen on intact RBCs

Ficin/Papain	Resistant (markedly enhanced)
Trypsin	Resistant (markedly enhanced)
α-Chymotrypsin	Resistant (markedly enhanced)
Pronase	Resistant (markedly enhanced)
Sialidase	Resistant (markedly enhanced)
DTT 200 mM	Resistant
Acid	Resistant

In vitro characteristics of alloanti-PP1P^k

Immunoglobulin class	IgM; IgG
Optimal technique	RT; 37°C; IAT
Neutralization	Hydatid cyst fluid, *Echinococcus* cyst fluid or pigeon egg white
Complement binding	Yes; some hemolytic

Clinical significance of alloanti-PP1P^k

Transfusion reaction	No to severe (rare) because anti-PP1P^k is rare (cross-match would be incompatible)
HDFN	No to severe
Spontaneous abortions	Cytotoxic IgM and IgG3 antibodies directed against P and/or P^k antigens are associated with a higher than normal rate of spontaneous abortion in women with the rare p [Tj(a–)], P_1^k, and P_2^k phenotypes

Autoanti-P^k

Yes

Comments

P^k was thought to be expressed only by P_1^k/P_2^k phenotype RBCs until it was realized that most RBCs express P^k antigen, albeit weakly. P^k antigen is more strongly expressed on LKE– RBCs than on LKE+ RBCs (see **GLOB Collection**).

Neuraminidase treatment of RBCs exposes neutral glycosphingolipids (e.g., P^k and P antigens) and gangliosides. Anti-P^k can be separated from some anti-PP1P^k by absorption with P1 RBCs.

Siblings of patients with anti-PP1P^k should be tested for compatibility, and the patient urged to donate blood for cryogenic storage when his/her clinical state permits.

Gb3 (P^k) is the physiologic receptor for shiga toxin from Shigella (Stx) or certain *E. coli* strains (Stx1 and Stx2) on renal epithelium, platelets, and endothelium.

P^k is a receptor for P-fimbriated pyelonephritogenic *E. coli* with the PapG adhesin and *Streptococcus suis*. The p phenotype confers resistance to urinary tract infection with P-fimbriated *E. coli* due to lack of P, P1, and P^k receptors on urothelium. Transcriptional up-regulation of P^k by inflammatory mediators (IFN, IL1) and increased P^k levels contribute to susceptibility to Stx toxicity in renal and vascular tissue in the development of *E. coli*-associated HUS. P^k is involved in signal modulation of α-interferon receptor and CXCR4 (an HIV co-receptor). Some of these effects may be mediated through lipid rafts. P^k may provide protection against HIV-1 infection[3].

References

[1] Bailly, P., Bouhours, J.F., 1995. P blood group and related antigens. In: Cartron, J.-P., Rouger, P. (Eds.), Molecular Basis of Human Blood Group Antigens. Plenum Press, New York, N.Y., pp. 300–329.

[2] Spitalnik, P.F., Spitalnik, S.L., 1995. The P blood group system: biochemical, serological, and clinical aspects. Transfusion Med Rev 9, 110–122.

[3] Lund, N., et al., 2009. The human P(k) histo-blood group antigen provides protection against HIV-1 infection. Blood 113, 4980–4991.

NOR Antigen

Terminology

ISBT symbol (number)	P1PK4 (003.004 or 3.4)
History	Reported in 1982, and named after the city where the original propositus resided (Norton, VA, USA)[1]. Assigned to the P1PK system in 2012.

Occurrence

Only found in two families so far, American and Polish.

Expression

Altered	Considerable variation in the strength of NOR expression.

Molecular basis associated with NOR antigen[2–4]

NOR antigen (α-galactosyl-globoside) is derived by the addition of an α-galactosyl residue to globoside as a consequence of a single nucleotide change in *A4GALT*.

$$
\begin{array}{l}
\text{Gal} \\
\quad \big|\alpha 1\text{-}4 \quad \text{NOR} \\
\text{GalNAc} \\
\quad \big|\beta 1\text{-}3 \\
\text{Gal} \\
\quad \big|\alpha 1\text{-}4 \\
\text{Gal} \\
\quad \big|\beta 1\text{-}4 \\
\text{Glc} \\
\quad \big|\beta 1\text{-}1 \\
\text{Ceramide}
\end{array}
$$

Amino acid	Glu211
Nucleotide	G at bp 631 in exon 3 of the *A4GALT*
NOR– (wild type)	Gln211 and C at bp 631

Effect of enzymes and chemicals on NOR antigen on intact RBCs

Ficin/Papain	Resistant (enhanced)
Trypsin	Resistant (enhanced)

In vitro characteristics of alloanti-NOR

Immunoglobulin class	IgM
Optimal technique	RT
Neutralization	Hydatid cyst fluid; avian P1 glycoproteins

Clinical significance of alloanti-NOR

No data because transfusion of NOR+ blood is rare.

Comments

NOR+ RBCs are said to be polyagglutinable because they are agglutinated by most ABO-compatible human sera. It can be distinguished from Cad poly-agglutination by its non-reactivity with the lectins *G. soja*, *D. biflorus* or *S. horminum*[1].

The NOR phenotype is characterized by the presence of two unique neutral glycospingolipids (designated NOR1 and NOR2) that react strongly with *Griffonia simplicifolia* IB4 lectin (GSL-IB4)[5]. NOR2 is an extended NOR (NOR1) glycolipid that expresses NOR activity due to the sequential addition of β3GalNAc and α4Gal.

The Gln211Glu change in the 4-α-galactosyltransferase appears to alter its acceptor preferences so that it can add a Gal also to the P antigen, while retaining its capacity to synthesize the P^k and P1 antigens.

References

[1] Harris, P.A., et al., 1982. An inherited RBC characteristic, NOR, resulting in erythrocyte polyagglutination. Vox Sang 42, 134–140.

[2] Duk, M., et al., 2001. Structure of a neutral glycosphingolipid recognized by human antibodies in polyagglutinable erythrocytes from the rare NOR phenotype. J Biol Chem 276, 40574–40582.

[3] Suchanowska, A., et al., 2010. A possible genetic background of NOR polyagglutination [abstract]. Vox Sang 99 (Suppl. 1), 333.

[4] Suchanowska, A., et al., 2011. Genetic background of NOR polyagglutination [abstract]. Vox Sang 101 (Suppl. 1), 20.

[5] Duk, M., et al., 2002. Serologic identification of NOR polyagglutination with *Griffonia simplicifolia* IB4 lectin. Transfusion 42, 806–807.

Rh Blood Group System

Number of antigens   52

Polymorphic	D, C, E, c, e, f, Ce, G, hr^S, C^G, Rh26 (c-like), cE, hr^B, Rh41
Low prevalence	C^W, C^X, $V^\wedge$, E^W, $VS^\wedge$, CE, D^W, hr^H, Go^a, Rh32, Rh33, Rh35, Be^a, Evans, Tar, Rh42, Crawford, Riv, JAL, STEM, FPTT, BARC, JAHK, $DAK^\wedge$, LOCR, CENR
High prevalence	Hr_0, Hr, Rh29, Hr^B, Rh39, Nou, Sec, Dav, MAR, CEST, CELO, CEAG

$\wedge$ = polymorphic in populations with African ancestry.

Terminology

ISBT symbol (number)	RH (004)
CD number	CD240D (RhD); CD240CE (RhCcEe)
Obsolete name	*Rhesus*, which is obsolete because it is a genus of monkey
History	Antibodies, made in 1940 by Landsteiner and Wiener, in rabbits or guinea pigs in response to injected rhesus monkey (*Macacus rhesus*) RBCs, were thought to be the same specificity as the human antibody reported in 1939 and the antigen detected by them was named Rh.

Expression

Cord RBCs	Expressed
Tissues	Erythroid specific

Gene[1,2]

Chromosome	1p36.11
Name	*RHD, RHCE*

The Blood Group Antigen (3/e). DOI: http://dx.doi.org/ 10.1016/B978-0-12-415849-8.00006-5

Organization	*RHD* and *RHCE*, each with 10 exons, are distributed over 69 kbp of DNA in opposite orientation with their 3' ends facing each other. The genes are separated by a region of about 30 kbp of DNA that contains the TMEM50A gene (previously called *SMP1* for Small Membrane Protein 1). The 3' and 5' ends of *RHD* are flanked by two 9 kbp homologous regions of DNA named the *Rhesus boxes*[3].
Product	RhD polypeptide (obsolete names: Rh30; Rh30B; Rh30D; D_{30}) RhCE polypeptide (obsolete names: Rh30; Rh30A; Rh30C)

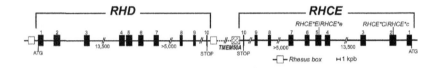

In diagrams representing *RH* exons, the information for *RHCE* is presented in the order of exon 1 to exon 10. The opposite orientation of *RHD* and *RHCE* and a putative "hairpin" formation allows homologous DNA segments to come into close proximity, and most gene recombination occurs through gene conversion rather than unequal crossover.

A third ancestral homologous gene (*RHAG*), located on chromosome 6, encodes the Rh-associated glycoprotein [RHAG **(030)**] and is essential for the expression of Rh antigens.

Database accession numbers

Gene	RHD	RHCE
GenBank	NG_007494 (gene) NM_016124 (mRNA)	NM_138618 (mRNA) NM_020485 (mRNA)
Accession number	L08429	DQ322275 (RHCE*01)
Entrez Gene ID	6007	6006
Allele names	*RHD*01*	*RHCE*01* or *RHCE*ce* *RHCE*02* or *RHCE*Ce* *RHCE*03* or *RHCE*cE* *RHCE*04* or *RHCE*CE*

Differences in nucleotides between *RHD* and *RHCE* and amino acids encoded

Differences between *RHD* and *RHCE*Ce* and *RHCE*cE* are not given.

Exon	Nucleotide #*RHD* > *RHCE*ce*	Amino acid RhD > Rhce
1	48G>C	Trp16Cys
2	150T>C	Silent
	178A>C	Ile60Leu
	201G>A	Silent
	203G>A	Ser68Asn
	307T>C	Ser103Pro
3	361T>A	Leu121Met
	380T>C	Val127Ala
	383A>G	Asp128Gly
	455A>C	Asn152Thr
4	505A>C	Met169Leu
	509T>G	Met170Arg
	514A>T	Ile172Phe
	544T>A	Ser182Thr
	577G>A	Glu193Lys
	594A>T	Lys198Asn
	602C>G	Thr201Arg
5	667T>G	Phe223Val
	697G>C	Glu233Gln
	712G>A	Val238Met
	733G>C	Val245Leu
	744C>T	Silent
	787G>A	Gly263Arg
	800A>T	Lys267Met
6	916G>A	Val306Ile
	932A>G	Tyr311Cys
7	941G>T	Gly314Val
	968C>A	Pro323His
	974G>T	Ser325Ile
	979A>G	Ile327Val
	985 G>C	Gly329His
	986 G>A	Gly329His
	989 A>C	Tyr330Ser
	992 A>T	Asn331Ile
	1025T>C	Ile342Thr
	1048G>C	Asp350His
	1053C>T	Silent
	1057G>T	Gly353Trp
	1059A>G	Gly353Trp
	1060G>A	Ala354Asn
	1061C>A	Ala354Asn
	1063G>T	Silent

(Continued)

Rh

(Continued)

Exon	Nucleotide #*RHD* > *RHCE**ce	Amino acid RhD > Rhce
8	No differences	No differences
9	1170T>C 1193A>T	Silent Glu398Val
10	No differences	No differences

Molecular bases of partial and weak partial D phenotypes

People with a partial D phenotype can make anti-D even though their RBCs are D+. Reference allele *RHD*01* (L08429) encodes D (RH1). Nucleotide differences from the reference allele, and amino acids affected, are given. In the following tables, when an exon(s) of *RHD* is substituted with the equivalent exon(s) of *RHCE*, the number of the substituted exon is given in parentheses. The nucleotide and amino acid changes between all *RHD* and *RHCE* exons are given in the previous table "Differences in nucleotides between *RHD* and *RHCE* and amino acids encoded."

Some of the variant D phenotypes in this table have not (yet) been associated with production of alloanti-D. They are included because of similarities to established partial D phenotypes, which may include the epitope profile and antigen site density when tested with monoclonal anti-D.

Allele encodes	Allele name	Exon	Nucleotide	Amino acid	Ethnicity (prevalence)
DII	*RHD*02 or RHD*DII*	7	1061C>A	Ala354Asp	(Few)
DIIIa	*RHD*03.01or RHD*DIIIa*	2	186G>T	Leu62Phe	Africans (Many)
DAK+ or RH:54		3	410C>T	Ala137Val	
(Previously DIIIa type 5)		4	455A>C	Asn152Thr	
		4	602C>G	Thr201Arg	
		5	667T>G	Phe223Val	
DIIIb− Caucasian	*RHD*03.02 or RHD*DIIIb*	2	D-CE(2)-D	See table for CE exon 2	Caucasians (Rare)
G− or RH:−12					
DIIIc	*RHD*03.03 or RHD*DIIIc*	3	D-CE(3)-D	See table for CE exon 3	Caucasians (Many)
DIII type 4	*RHD*03.04 or RHD*DIII.04*	2	186G>T	Leu62Phe	(Few)
		3	410C>T	Ala137Val	
		3	455A>C	Asn152Thr	
DIII type 6	*RHD*03.06 or RHD*DIII.06*	3	410C>T	Ala137Val	Africans (Many)
		3	455A>C	Asn152Thr	
		4	602C>G	Thr201Arg	
		5	667T>G	Phe223Val	
DIII type 7^ (likely the serologically defined (historic) DIIIb of Tippett and Sanger) G− or RH:−12	*RHD*03.07 or RHD*DIII.07*	2	Exon 2	See table	Africans (Few)
		3	410C>T	Ala137Val	
		3	455A>C	Asn152Thr	
		4	602C>G	Thr201Arg	
		5	667T>G	Phe223Val	

(Continued)

Rh

(Continued)

Allele encodes	Allele name	Exon	Nucleotide	Amino acid	Ethnicity (prevalence)
DIVa (previously DIVa.2) Go(a+) or RH:30	RHD*04.01 or RHD*DIVa.01	2 3 3 7	186G>T 410C>T 455A>C 1048G>C	Leu62Phe Ala137Val Asn152Thr Asp350His	Africans (Many)
DIVa type 2 (obsolete – same as original DIVa)					
DIVb Evans+ or RH:37	RHD*04.02 or RHD*DIVb	7 7 7 7 7 8–9	D-CE (part 7–9)-D 1048G>C 1057G>T 1059A>G 1060G>A 1061C>A See table	See table Asp350His Gly353Trp Gly353Trp Ala354Asn Ala354Asn See table	Europeans, Japanese (Many)
DIV type 3	RHD*04.03 or RHD* DIV.03	6, 7, 8, 9	D-CE(6–9)-D	See table	(Few)
DIV type 4	RHD*04.04 or RHD*DIV.04	7 7 7 7 7	1048G>C 1057G>T 1059A>G 1060G>A 1061C>A	Asp350His Gly353Trp Gly353Trp Ala354Asn Ala354Asn	(Few)
DIV type 5	RHD*04.05 or RHD*DIV.05	7, 8, 9	D-CE(7–9)-D	See table	(Few)
DV type 1 (KOU, FK) D^W+ or RH:23	RHD*05.01 or RHD*DV.01	5 5	667T>G 697G>C	Phe223Val Glu233Gln	Europeans, Japanese, (Many) Africans (Several)

			D-CE(5)-D	See table	
DV type 2 (Hus) D^W+ or RH:23	RHD*05.02 or RHD*DV.02	5	D-CE(5)-D	See table	Europeans, Japanese, (Many), Africans (Several)
DV type 3 (DBS0) E±	RHD*05.03 or RHD*DV.03	5 5 5 5	667T>G 676G>C 697G>C 712 G>A	Phe223Val Ala226Pro Glu233Gln Val238Met	(Few)
DV type 4 (SM) D^W+ or RH:23	RHD*05.04 or RHD*DV.04	5	697G>C	Glu233Gln	(Few)
DV type 5 (DHK, DYO) D^W+ or RH:23	RHD*05.05 or RHD*DV.05	5	697G>A	Glu233Lys	Japanese (Many), Austrians
DV type 6	RHD*05.06 or RHD*DV.06	5 5 5	667T>G 697G>C 712G>A	Phe223Val Glu233Gln Val238Met	Japanese
DV type 7 (DAL) D^W− or RH:−23	RHD*05.07 or RHD*DV.07	5 5 5 5 5	667T>G 697G>C 712G>A 733G>C 787G>A	Phe223Val Glu233Gln Val238Met Val245Leu Gly263Arg	European (Several)

(Continued)

Rh

(Continued)

Allele encodes	Allele name	Exon	Nucleotide	Amino acid	Ethnicity (prevalence)
DV type 8 (TT)	RHD*05.08 or RHD*DV.08	5 5 5 5	667T>G 697G>C 712G>A 733G>C	Phe223Val Glu233Gln Val238Met Val245Leu	Japanese (Several)
DV type 9 (TO)	RHD*05.09 or RHD*DV.09	5 5	697G>C 712G>A	Glu233Gln Val238Met	Japanese (Few)
DVI type 1 G±/− BARC− or RH:±/−12, −52	RHD*06.01 or RHD*DVI.01	4,5	D-CE(4-5)-D	See table	Europeans (Many)
DVI type 2 BARC+ or RH:52	RHD*06.02 or RHD*DVI.02	4,5,6	D-CE(4-6)-D	See table	Europeans (Many)
DVI type 3 BARC+ or RH:52	RHD*06.03 or RHD*DVI.03	3,4,5,6	D-CE(3-6)-D	See table	Germans, Chinese (Few)
DVI type 4 BARC+ or RH:52	RHD*06.04 or RHD*DVI.04	3, 4, 5	D-CE(3-5)-D	See table	Spanish (Many)
DVII Tar+ or RH:40	RHD*07.01 or RHD*DVII.01	2	329T>C	Leu110Pro	Europeans, Israelis (Many)
DVII type 2 Tar+ or RH:40	RHD*07.02 or RHD*DVII.02	2 2	307T>C 329 T>C	Ser103Pro Leu110Pro	(Few)
DFV	RHD*08.01 or RHD*DFV	5	667T>G	Phe223Val	Africans, Indians (Few)

Rh

Phenotype	Allele name	Exon	Nucleotide	Amino acid	Population
DAR1	*RHD*09.01 or RHD*DAR1*	4	602C>G	Thr201Arg	Africans (Many), Europeans (Few)
		5	667T>G	Phe223Val	
		7	1025T>C	Ile342Thr	
DAR1.1 Weak D 4.2.1 (silent change distinguish from DAR1)	*RHD*09.01.01 or RHD*DAR1.01*	4	602C>G	Thr201Arg	Africans (Many)
		5	667T>G	Phe223Val	
		7	957G>C	Silent	
		7	1025T>C	Ile342Thr	
DAR1.2 Weak D 4.2.2 (silent changes distinguish from DAR1)	*RHD*09.01.02 or RHD*DAR1.02*	4	602C>G	Thr201Arg	
		5	667T>G	Phe223Val	
		5	744C>T	Silent	
		7	957G>C	Silent	
		7	1025T>C	Ile342Thr	
DAR1.3 Weak D 4.2.3 (silent change distinguishes from DAR1)	*RHD*09.01.02 or RHD*DAR1.02*	4	602C>G	Thr201Arg	
		5	667T>G	Phe223Val	
		5	744C>T	Silent	
		7	1025T>C	Ile342Thr	
DAR2 (DARE)	*RHD*09.02 or RHD*DAR2*	4	602C>G	Thr201Arg	Ethiopians (Many)
		5	667T>G	Phe223Val	
		5	697G>C	Glu233Gln	
		7	1025T>C	Ile342Thr	
DAR3 Weak partial D 4.0.1	*RHD*09.03 or RHD*DAR3*	4	602C>G	Thr201Arg	Europeans (Many)
		5	667T>G	Phe223Val	
DAR3.1 Weak partial D 4.0 (silent change distinguishes from DAR3)	*RHD*09.03.01 or RHD*DAR3.01*	4	602C>G	Thr201Arg	Africans (Many)
		5	667T>G	Phe223Val	
		6	819G>A	Silent	

(Continued)

Rh

(Continued)

Allele encodes	Allele name	Exon	Nucleotide	Amino acid	Ethnicity (prevalence)
DAR4 Weak partial D 4.1	RHD*09.04 or RHD*DAR4	1 4 5 6	48G>C 602C>G 667T>G 819G>A	Trp16Cys Thr201Arg Phe223Val Silent	Africans (Many), Europeans (Few)
DAR5 Weak partial D 4.3 or DEL[†]	RHD*09.05 or RHD*DAR5	4 5 6 6	602C>G 667T>G 819G>A 872C>G	Thr201Arg Phe223Val silent Pro291Arg	Austrians
DAU0	RHD*10.00 or RHD*DAU0	8	1136C>T	Thr379Met	Africans (Many), Eurasians (Few)
DAU1	RHD*10.01 or RHD*DAU1	5 8	689G>T 1136C>T	Ser230Ile Thr379Met	Africans
DAU2	RHD*10.02 or RHD*DAU2	2 7 8	209G>A 998G>A 1136C>T	Arg70Gln Ser333Asn Thr379Met	Africans
DAU3	RHD*10.03 or RHD*DAU3	6 8	835G>A 1136C>T	Val279Met Thr379Met	Africans
DAU4	RHD*10.04 or RHD*DAU4	5 8	697G>A 1136C>T	Glu233Lys Thr379Met	Africans
DAU5	RHD*10.05 or RHD*DAU5	5 5 8	667T>G 697G>C 1136C>T	Phe223Val Glu233Gln Thr379Met	Africans, Canadians, Germans (Several)

	Allele name	Exon(s)	Nucleotide change	Amino acid change	Population
DAU6	RHD*10.06 or RHD*DAU6	7 8	998G>A 1136C>T	Ser333Asn Thr379Met	Africans
DAU7	RHD*10.07 or RHD*DAU7	6 7 8	835G>A 998G>A 1136C>T	Val279Met Ser333Asn Thr379Met	Africans (Rare)
Weak Partial D 11 (or Del)†	RHD*11 or RHD*Weak partial 11	6	885G>T	Met295Ile	European (Many)
DOL1 DAK+ or RH:54	RHD*12.01 or RHD*DOL1	4 5	509T>C 667T>G	Met170Thr Phe223Val	(Many)
DOL2 DAK+ or RH:54	RHD*12.02 or RHD*DOL2	4 5 8	509T>C 667T>G 1132C>G	Met170Thr Phe223Val Leu378Val	(Few)
DOL3	RHD*12.03 or RHD*DOL3	3 4 5	410C>T 509T>C 667T>G	Ala137Val Met170Thr Phe223Val	(Few)
DBS1	RHD*13.01 or RHD*DBS1	5 5	D-CE(5)-D 676G>C	See table Ala226Pro	(Few)
DBS2	RHD*13.02 or RHD*DBS2	5 5 5	667T>G 676G>C 697G>C	Phe223Val Ala226Pro Glu233Gln	Germans, Japanese (Few)
DBT1 Rh32+ or RH:32	RHD*14.01 or RHD*DBT1	5, 6, 7	D-CE(5-7)-D	See table	Caucasians, Japanese, Blacks, Moroccans (Several)

(Continued)

(Continued)

Allele encodes	Allele name	Exon	Nucleotide	Amino acid	Ethnicity (prevalence)
DBT2 Rh32+ or RH:32	RHD*14.02 or RHD*DBT2	5, 6, 7, 8, 9	D-CE(5–9)-D	See table	Japanese (Few)
Weak partial D 15	RHD*15 or RHD*Weak partial 15	6	845G>A	Gly282Asp	Asians (Many), Europeans (Many)
DCS1	RHD*16.01 or RHD*DCS1	5 5	667G>T 676G>C	Phe223Val Ala226Pro	Austrians (Several)
DCS2	RHD*16.02 or RHD*DCS2	5	676G>C	Ala226Pro	Germans, Chinese (Several)
DFR1 FPTT+ or RH:50	RHD*17.01 or RHD*DFR1	4 4 4	505A>C 509T>G 514A>T	Met169Leu Met170Arg Ile172Phe	Caucasians (Many)
DFR2	RHD*17.02 or RHD*DFR2	4	D-CE(4)-D	See table	(Few)
DFR3	RHD*17.03 or RHD*DFR3	4 4 4 4	505A>C 509T>G 514A>T 539G>C	Met169Leu Met170Arg Ile172Phe Gly180Ala	(Few)
DFR4	RHD*17.04 or RHD*DFR4	4 4	505A>C 509T>G	Met169Leu Met170Arg	(Few)
DFW	RHD*18 or RHD*DFW	4	497A>C	His166Pro	(Few)
DHMi	RHD*19 or RHD*DHMi	6	848C>T	Thr283Ile	Caucasians (Many)

DHO	RHD*20 or RHD*DHO	5	704A>C	Lys235Thr	Germans (Few)
Weak partial D 21	RHD*21 or RHD*weak partial 21	6	938C>T	Pro313Leu	Austrians, Germans (Few)
DHR	RHD*22 or RHD*DHR	5	686G>A	Arg229Lys	(Few)
DMH	RHD*23 or RHD*DMH	2	161T>C	Leu54Pro	Portuguese (Few)
DNAK	RHD*24 or RHD*DNAK	7	1070G>A	Gly357Asp	(Few)
DNB	RHD*25 or RHD*DNB	7	1063G>A	Gly355Ser	Swiss, Germans, Danish (Many)
DNU	RHD*26 or RHD*DNU	7	1057G>A	Gly353Arg	(Few)
DDE	RHD*27 or RHD*DDE	1	120T>A	Asp40Glu	(Few)
DFL	RHD*28 or RHD*DFL	4	494A>G	Tyr165Cys	Austrians, French (Few)
DYU (DQC)	RHD*29 or RHD*DYU	5	700A>T	Arg234Trp	(Few)
DTO	RHD*30 or RHD*DTO	5 5	667T>G 674C>T	Phe223Val Ser225Phe	(Few)
DVL1	RHD*31 or RHD*DVL1	5	deletion 684 to 686 GAG	Arg229del	(Few)
DVL2	RHD*32 or RHD*DVL2	5	deletion 705 to 707 GAA	Lys235del	Swiss (Many)
DWI (DWLLE)	RHD*33 or RHD*DWI	7	1073T>C	Met358Thr	(Few)

(Continued)

Rh

(Continued)

Allele encodes	Allele name	Exon	Nucleotide	Amino acid	Ethnicity (prevalence)
DIM(DIIeM)	*RHD*34* or *RHD*DIM*	6	854G>A	Cys285Tyr	(Few)
DMA	*RHD*35* or *RHD*DMA*	5	621G>C	Leu207Phe	(Few)
DLO	*RHD*36* or *RHD*DLO*	6	851C>T	Ser284Leu	(Few)
DUC2	*RHD*37* or *RHD*DUC2*	5	733G>C	Val245Leu	(Few)

a Weak D phenotype when *in cis* to *RHCE*ce* and Del phenotype when *in cis* to *RHCE*Ce*.

b The published molecular basis for DIIIb was determined using DNA from Caucasian probands who are G– and probably have a weak D phenotype, and thus are not DIIIb as defined by Tippett. It is likely that the DIII type 7 phenotype is the same as the originally serologically defined DIIIb phenotype of Tippett and Sanger.

Rh

Molecular bases of weak D phenotypes[4–6]

The weak D phenotype is a quantitative, and not a qualitative, polymorphism, and in most cases all immunogenic D epitopes are present. However, some rare individuals who were considered to have a weak D phenotype were later found to have made anti-D. There is debate about the clinical significance of the anti-D, and the possibility that they are autoanti-D may be difficult to rule out[7]. The D antigen with reduced expression is usually detected by the indirect antiglobulin test. If a person with a D+ phenotype that was previously reported as a weak D has made alloanti-D, the allele is now listed in the "Molecular bases of partial and weak partial RhD phenotypes" table.

Nucleotide differences from reference allele *RHD*01* (L08429), and amino acids affected, are given.

Allele encodes weak D	Allele name	Exon	Nucleotide	Amino acid	Ethnicity (prevalence)
Type 1	RHD*01W.01 or RHD*weak D type 1	6	809T>G	Val270Gly	European (Many)
Type 1.1	RHD*01W.01.01 or RHD*weak D type 1.1	1, 6	52C>G, 809G>A	Leu18Val, Val270Gly	Northern Germans (Many)
Type 2	RHD*01W.02 or RHD*weak D type 2	9	1154G>C	Gly385Ala	European (Many)
Type 2.1	RHD*01W.02.01 or RHD*Weak D type 2.1	2, 9	301T>A, 1154G>C	Phe101Ile, Gly385Ala	(Few)
Type 3	RHD*01W.03 or RHD*weak D type 3	1	8C>G	Ser3Cys	European (Many)
Types 4.0, 4.1, 4.2, 4.2.2, 4.3	See partial DAR or RHD*09.03				
Type 5	RHD*01W.05 or RHD*weak D type 5‡	3	446C>A	Ala149Asp	European (Several)
Type 6	RHD*01W.06	1	29G>A	Arg10Gln	Taiwanese, Germans (Few)
Type 7	RHD*01W.07	7	1016G>A	Gly339Glu	Germans (Few)
Type 8	RHD*01W.08	6	919G>A	Gly307Arg	Germans (Few)
Type 9	RHD*01W.09	6	880G>C	Ala294Pro	Germans (Few)
Type 10	RHD*01W.10	9	1177T>C	Trp393Arg	Germans (Few)
Type 11	See partial D RHD*11				
Type 12	RHD*01W.12	6	830G>A	Gly277Glu	(Few)

Type	Allele	Exon	Nucleotide	Amino acid	Population
Type 13	RHD*01W.13	6	826G>C	Ala276Pro	Austrians (Few)
Type 14	RHD*01W.14	4 4 4	544T>A 594A>T 602C>G	Ser182Thr Lys198Asn Thr201Arg	Austrians (Few)
Type 15	See partial RHD*15				
Type 16	RHD*01W.16	5	658T>C	Trp220Arg	(Few)
Type 17	RHD*01W.17	3	340C>T	Arg114Trp	(Few)
Type 18	RHD*01W.18	1	19C>T	Arg7Trp	(Few)
Type 19	RHD*01W.19	4	611T>C	Ile204Thr	(Few)
Type 20	RHD*01W.20	10	1250T>C	Phe417Ser	(Several)
Type 21	See partial RHD*21				
Type 22	RHD*01W.22	9	1224G>C	Trp408Cys	(Few)
Type 23	RHD*01W.23	4	634G>T	Gly212Cys	Japanese
Type 24	RHD*01W.24	7	1013T>C	Leu338Pro	Japanese
Type 25	RHD*01W.25	3	341G>A	Arg114Gln	(Several)
Type 26	RHD*01W.26	1	26T>A	Val9Asp	(Few)
Type 27	RHD*01W.27	5	661C>T	Pro221Ser	(Few)
Type 28	RHD*01W.28	8	(1152 A>C)	Thr384Thr splice site change	(Few)

(Continued)

Rh

(Continued)

Allele encodes weak D	Allele name	Exon	Nucleotide	Amino acid	Ethnicity (prevalence)
Type 29	RHD*01W.29	2	178A>C	Ile60Leu	(Few)
		2	201G>A	Ser68Asn	
		4	203G>A	Lys198Asn	
		5	594A>T	Phe223Val	
		7	1025T>C	Ile342Thr	
Type 30	RHD*01W.30	7	1018G>A	Glu340Met	(Few)
			1019A>T		
Type 31	RHD*01W.31	1	17C>T	Leu6Pro	(Few)
Type 32	RHD*01W.32	8	1121A>T	Ile374Asn	(Few)
Type 33 Probably a partial D; as alloanti-D has been made	RHD*01W.33	4	520G>A	Val174Met	Taiwanese, Europeans (Several)
Type 34	RHD*01W.34	6	809T>A	Val270Glu	Taiwanese (Few)
Type 35	RHD*01W.35	2	260G>A	Gly87Asp	(Few)
Type 36	RHD*01W.36	6	842T>G	Val281Gly	(Few)
Type 37	RHD*01W.37	3	399G>T	Lys133Asn	(Few)
Type 38	RHD*01W.38	6	833G>A	Gly278Asp	(Few)
Type 39	RHD*01W.39	7	1015G>A	Gly339Arg	(Few)

Rh

Type 40	RHD*01W.40	4	602C>G	Thr201Arg	(Few)
Type 41	RHD*01W.41	9	1193A>T	Glu398Val	(Few)
Type 42	RHD*01W.42	9	1226A>T	Lys409Met	(Few)
Type 43	RHD*01W.43	4	605T>C	Ala202Val	Koreans (Few)
Type 44	RHD*01W.44	5	728A>G	Tyr243Cys	(Few)
Type 45	RHD*01W.45	9	1195G>A	Ala399Thr	Austrians (Few)
Type 46	RHD*01W.46	9	1221C>A	Phe407Leu	(Few)
Type 47	RHD*01W.47	3	340C>G	Arg114Gly	(Few)
Type 48	RHD*01W.48	2	182G>T	Gly61Val	(Few)
Type 49	RHD*01W.49	5	770C>T	Ser257Phe	(Few)
Type 50	RHD*01W.50	5	727T>A	Tyr243Asn	Germans (Few)
Type 51	RHD*01W.51	4	594A>T 602C>G	Lys198Asn Thr201Arg	Chinese (Few)
Type 52	RHD*01W.52	1	92T>C	Phe31Ser	Chinese (Few)
Type 53	RHD*01W.53	5	740T>G	Val247Gly	Chinese (Few)
Type 54	RHD*01W.54	3	365C>T	Ser122Leu	(Few)
Type 55	RHD*01W.55	6	895C>G	Leu299Val	(Few)
Type 56	RHD*01W.56	1	65C>A	Ala22Glu	French (Few)

(Continued)

Rh

(Continued)

Allele encodes weak D	Allele name	Exon	Nucleotide	Amino acid	Ethnicity (prevalence)
Type 57	RHD*01W.57	5	640C>T	Leu214Phe	French (Few)
Type 58	RHD*01W.58	7	1006G>C	Gly336Arg	French (Few)
Type 59	RHD*01W.59	8	1148T>C	Leu383Pro	French (Few)
Type 60	RHD*01W.60	9	1219-1224 del TTCTGG	Phe407 Trp408del	French (Few)
Type 61	RHD*01W.61	1	28C>T	Arg10Trp	Chinese, Germans (Few)
Type 62	RHD*01W.62	5	661C>A	Pro221Thr	Germans (Few)
Type 63	RHD*01W.63	5	758T>A	Ile253Asn	Germans (Few)

Type 64	RHD*01W.64	6	881C>T	Ala294Val	Germans (Few)
Type 65	RHD*01W.65	1	68C>A	Ala23Asp	Germans (Few)
Type 66	RHD*01W.66	6	916G>A	Val306Ile	Austrians (Few)
Type 67	RHD*01W.67	5	722C>T	Thr241Ile	Germans (Few)
Type 68	RHD*01W.68	9	1213C>G	Gln405Glu	Germans (Few)
Type 69	RHD*01W.69	7	953G>A	Arg318Gln	Austrians (Few)
Type 70	RHD*01W.70	7	1012C>G	Leu338Val	Austrians (Few)
Type 71	RHD*01W.71	1	29G>C	Arg10Pro	Chinese
Type 72	RHD*01W.72	9	1212C>A	Asp404Glu	Chinese
Type 73	RHD*01W.73	10	1241C>T	Ala414Val	Chinese

† = This and subsequent weak *RHD* can also use the *RHD*weak D type #* designation (following the obvious pattern above); in the interest of space, this is not included.

Rh

Molecular bases of Del phenotype[8]

Very weakly expressed D antigen is detectable only by adsorption and elution of anti-D.

Nucleotide differences from *RHD*01* reference allele (L08429), and amino acids affected, are given for some of the reported Del.

Allele encodes	Allele name	Exon (intron)	Nucleotide	Amino acid	Ethnicity (prevalence)
Del	RHD*01EL.01 or RHD*DEL1	9	1227G>A	Lys409Lys Splice site change	Chinese, Koreans, Europeans (Several)
Del	RHD*01EL.02 or RHD*DEL2	1	3G>A Start codon lost	Met1Ile	Chinese
Del	RHD*01EL.03 or RHD*DEL3	1	53T>C	Leu18Pro	Chinese
Del	RHD*01EL.04 or RHD*DEL4	1	147delA, fs	fs, Stop	Germans (Few)
Del	RHD*01EL.05 or RHD*DEL5	(1)	+1g>a	Splice site change	Japanese (Few)
Del or weak D	RHD*01EL.06 or RHD*DEL6	2	251T>C	Leu84Pro	Chinese
Del or weak D	RHD*01EL.07 or RHD*DEL7	3	410C>A	Ala137Glu	Chinese
Del	RHD*01EL.08 or RHD*DEL8	(3)	+1g>a	Splice site change	Germans, Austrians, Slovenians (Few)
Del or D−	RHD*01EL.09 or RHD*DEL9	(3)	+2t>a	Splice site change	Germans (Few)
Del	RHD*01EL.10 or RHD*DEL10	9	1222T>C	Trp408Arg	Koreans (Few)
Del	RHD*01EL.11 or RHD*DEL11	10	1252 ins T	Stop418Leu (488 amino acids)	Austrians (Few)
Del	RHD*01EL.12 or RHD*DEL12	3	458T>C	Leu153Pro	Germans (Few)
Del or D−	RHD*01EL.13 or RHD*DEL13	5	785delA	fs, Stop	(Few)

Del phenotype also is associated with *RHD*11* and *RHD*09.05* (*RHD*DAR5*).

Molecular bases of D-negative phenotype[8–10]

Several molecular backgrounds result in the D-negative phenotype: deletion of *RHD* predominates in people of European descent; in Asian populations an intact but silenced *RHD* is common; in black African populations two-thirds of D-negative people have an inactive *RHD*, (the *RHD* pseudogene or *RHD*Ψ*) with a 37 bp internal duplication resulting in a premature stop codon. Alleles encoding D-negative phenotypes are designated with N (to represent the null of the *RHD*), and numbered according to the background allele on which the change has occurred.

RHD*01 is used when the changes are on the consensus sequence or if the derivation of the RHD allele is not known. The partial allele number is used when the changes are on a partial *RHD* background. Nucleotide differences from *RHD*01* reference allele (L08429), and amino acids affected, are given for some of the reported D−.

Rh

Allele encodes	Allele name	Exon (intron)	Nucleotide	Amino acid	Ethnicity (prevalence)
D–	RHD*01N.01	1–10	RHD deletion	No protein	Caucasians (Many)
D–	RHD*01N.01 or RHD*Pseudogene RHD*Ψ	(3) 4 5 5 5 6	37bp insert 609G>A 654G>C 667T>G 674C>T 807T>G	No protein	Africans (Many)
D–	RHD*01N.02	1–9	CE(1–9)-D	Hybrid	Germans (Few)
D–	RHD*01N.03	2–9	D-CE(2–9)-D	Hybrid	
D–	RHD*01N.04	3–9	D-CE(3–9)-D	Hybrid	Germans, Chinese, Koreans (Several)
D–	RHD*01N.05	2–7	D-CE(2–7)-D	Hybrid	(Few)
D–C+^{VW} or RH:–1, +^{VW}2	RHD*01N.06	3–7	D-CE(3–7)-D Type 2 hybrid (Part of r'^{S})	Hybrid	Africans (Several)
D–C+^{W} or RH:–1, +^{W}2	RHD*03N.01	4–7	DIIIa-CE(4–7)-D Type 1 hybrid (Part of r'^{S})	Hybrid	Africans (Many)
r'^{G} D– G+	RHD*01N.07	4–7	D-CE(4–7)-D	Hybrid	(Few)
D–	RHD*01N.08	1	48G>A	Trp16Stop	Germans (Few)

D–	*RHD*01N.09*	1 5 5 7	121C>T 643T>C 646T>C 988T>C	Gln41Stop	(Few)
D–	*RHD*01N.10*	2	270G>A	Trp90Stop	Chinese
D–	*RHD*01N.11*	2	325delA, fs	109fs>Stop	Chinese
D–	*RHD*01N.12*	3	449delT, fs	150fs>Stop	Austrians (Few)
D–	*RHD*01N.13*	4	487delACAG, fs	Met167Stop	Caucasians (Few)
D–	*RHD*01N.14*	4	554G>A	Trp185Stop	Koreans (Few)
D–	*RHD*01N.15*	5	635G>T, splice site change	Gly212Val	Germans (Few)
D–	*RHD*01N.16*	5	711delC, fs	238fs> 245Stop	Chinese (Few)
D–	*RHD*01N.17*	5	652delA, 653T>G, fs	228Stop	Chinese
D–	*RHD*01N.18*	6	807T>G	Tyr269Stop	Germans (Few)
D–	*RHD*01N.19*	6	933C>A	Tyr311Stop	Chinese (Few)
D–	*RHD*01N.20*	7	941G>T	Gly314Val	Japanese (Few)
D–	*RHD*01N.21*	7	990C>G	Tyr330Stop	Germans (Few)
D–	*RHD*01N.22*	9	1203T>A	Tyr401Stop	Russians (Few)

(Continued)

Rh

(Continued)

Allele encodes	Allele name	Exon (intron)	Nucleotide	Amino acid	Ethnicity (prevalence)
D–	*RHD*01N.23*	3	343delC, fs	115fs>Stop	Germans (Few)
D–	*RHD*01N.24*	(2)	+1g>a	splice site change	Chinese (Few)
D–	*RHD*01N.25*	(2)	–1g>a	splice site change	Koreans (Few)
D–	*RHD*01N.26*	(8)	+1g>a	splice site change	Germans (Few)
D–	*RHD*01N.27*	(6)	ins tggct+2del taag	fs and splice site change	Chinese (Few)
D–	*RHD*01N.28*	7 7	970del CAC, 976del TCCATCATGGGC TACA), fs	His324del, fs>352Stop	Chinese (Few)

Log on to the ISBT, dbRBC, and Rhesus Base, websites for more information, hyperlinks to original reports, and updates.

Molecular bases of phenotypes associated with RhCE

The RHCE common alleles are designated *RHCE*01* for ce, *RHCE*02* for Ce, *RHCE*03* for cE, and *RHCE*04* for CE. Serologically similar phenotypes can have different allelic backgrounds. Reported alterations in antigen expression are noted. As an alternative terminology, the nucleotides that differ from the reference allele (*RHCE*01* or *RHCE*ce*) may be listed, e.g., *RHCE*01.02* (*RHCE*ceTI*) can be written *RHCE*ce48C, 1025T*.

In the following tables, when an exon(s) of *RHCE* is substituted with the equivalent exon(s) of *RHD*, the number of the substituted exon is given in parentheses. The nucleotide and amino acid changes between all *RHD* and *RHCE* exons are given in the table "Differences in nucleotides between *RHD* and *RHCE* and amino acids encoded" (above).

Molecular bases of Rhce phenotypes

People who express a partial D on their RBCs can be immunized to make anti-D; however, the availability of D-negative blood precludes the need to name the specific anti-D made by each type of partial D phenotype. As a parallel scenario, people who are homozygous or hemizygous for alleles encoding partial c and/or partial e can produce alloantibody that is directed at conventional Rhce and appears as anti-Rh17 in that only RBCs with the D$--$ phenotype (or those expressing the same Rhce variant) do not react. These antibodies are not necessarily mutually compatible. In contrast to the availability of D-negative blood donors for transfusion, Rh17-negative blood donors are rare. Thus, some antibodies, especially those made by people with partial e antigens, have been given names to aid in communication and finding compatible blood.

Reference allele *RHCE*01* or *RHCE*ce* (Accession number DQ322275) encodes c (RH4), e (RH5), f (RH6), RH17, RH18, RH19, RH31, RH34, etc. Differences from *RHCE*01* reference allele are given in rows 2 and 3. These differences are also present in all other alleles in this table. Nucleotide differences from the reference allele, and amino acids affected, are given for some of the reported Rhce variants.

Allele encodes	Allele name	Exon	Nucleotide	Amino acid	Ethnicity (prevalence)
c+e+f+RH:4,5,6	RHCE*01 or RHCE*ce				
c+ or RH:4	RHCE*^	2	307C	Pro103	
e+ or RH:5	RHCE*^^	5	676G	Ala226	
e+ (weak with some monoclonal anti-e)	RHCE*01.01 or RHCE*ce.01	1	48G>C	Trp16Cys	Africans (Many) Europeans (Several)
Partial c, partial e	RHCE*01.02 or RHCE*ceTI	1 7	48G>C 1025C>T	Trp16Cys Thr342Ile	Africans (Several)
Partial e	RHCE*01.03 or RHCE*ceTI type 2	7	1025C>T	Thr342Ile	(Few)
Partial c, partial e V+^{w}VS– Hr– hr^S– or RH:+^{w}10,–18,–19,–20	RHCE*01.04 or RHCE*ceAR	1 5 5 5 5 6	48G>C 712A>G 733C>G 787A>G 800T>A 916A>G	Trp16Cys Met238Val Leu245Val Arg263Gly Met267Lys Ile306Val	Africans (Many)
Partial c, partial e Hr– hr^S– or RH:–18,–19	RHCE*01.05 or RHCE*ceEK	1 5 5 5	48G>C 712A>G 787A>G 800T>A	Trp16Cys Met238Val Arg263Gly Met267Lys	Africans (Several)
Partial e hr^B– CEAG– or RH:–31,–59	RHCE*01.06 or RHCE*ceAG	2	254C>G	Ala85Gly	Africans (Many)
Partial c, partial e hr^S– hr^B– or RH:–19,–31	RHCE*01.07 or RHCE*ceMO	1 5	48G>C 667G>T	Trp16Cys Val223Phe	Africans (Many)

		Exon	Nucleotide	Amino acid	Population
c+e± Hr− hrˢ− STEM+ or RH:−18,−19,49	RHCE*01.08 or RHCE*ceBI	1 5 6 8	48G>C 712A>G 818C>T 1132C>G	Trp16Cys Met238Val Ala273Val Leu378Val	Africans (Many)
c+e± Hr− hrˢ− STEM+ʷ or RH:−18,−19,49	RHCE*01.09 or RHCE*ceSM	1 5 6	48G>C 712A>G 818C>T	Trp16Cys Met238Val Ala273Val	Blacks (Many)
c+e+ʷ weakly reactive with some MAb anti-D	RHCE*01.10.01 or RHCE*ceSL	1 3	48G>C 365TC>T	Trp16Cys Ser122Leu	European (Few)
c+e+ʷ	RHCE*01.10.02	3	365TC>T	Ser122Leu	European (Rare)
c+e+ʷ weakly reactive with some MAb anti-D	RHCE*01.11 or RHCE*ceRT	3	461G>C	Arg154Thr	Japanese, Germans (Few)
c+e+ʷ (very weak)	RHCE*01.12 or RHCE*ceRA	1 4	48G>C 538G>C	Trp16Cys Gly180Arg	Indians (Few)
c+eᵛʷ CELO+ʷ or RH:+ʷ58	RHCE*01.13 or RHCE*ceBP	5	685-687 delAGA	Arg229del	Caucasians (Few)
c+ʷ, e+ʷ Be(a+) or RH:36	RHCE*01.14 or RHCE*ceBE	5	662C>G	Pro221Arg	Caucasians (Germans, Poles) (Several)
c+ʷ or e+ʷ Rh26− LOCR+ or RH:−26,55	RHCE*01.15 or RHCE*ceLOCR	2	286G>A	Gly96Ser	Dutch, Germans (Few)

(Continued)

Rh

(Continued)

Allele encodes	Allele name	Exon	Nucleotide	Amino acid	Ethnicity (prevalence)
Partial c, partial e V+VS+hr^B+^VW/– or RH:10,20,+^VW31	RHCE*01.20.01 or RHCE*ceVS.01	5	733C>G	Leu245Val	Africans (Many)
Partial c, partial e V+VS+hr^B– or RH:10,20,–31	RHCE*01.20.02 or RHCE* ceVS.02	1 5	48G>C 733C>G	Trp16Cys Leu245Val	Africans (Many)
Partial c, partial e V– VS+hr^B– or RH:-10,20,–31	RHCE*01.20.03 or RHCE* ceVS.03	1 5 7	48G>C 733C>G 1006G>T	Trp16Cys Leu245Val Gly336Cys	Africans (Many)
Partial e V+VS+or RH:10,20 Probably hr^B– or RH:–31	RHCE*01.20.04 or RHCE*ceVS.04	1 5 7	48G>C 733C>G 1025C>T	Trp16Cys Leu245Val Thr342Ile	Africans (Some)
Partial e V– VS+hr^B– or RH:-10,20,–31	RHCE*01.20.05 or RHCE*ceVS.05	5 7	733C>G 1006G>T	Leu245Val Gly336Cys	Africans (Several)

RHCE*01.20.06 or RHCE*ceCF RHCE*ceVS.06	Partial c, partial e hr^S+/−VS+hr^B−Crawford+ CELO− RH:+/−19,20,−31,43,−58 Strongly reactive with some MAb anti-D	1 5 5	48G>C 697C>G 733C>G	Trp16Cys Gln233Glu Leu245Val	Africans (Many)
RHCE*01.20.07 or RHCE*ceJAL RHCE*ceVS.07	Partial c, partial e RHVS+/^{W},hr^B+^{W}JAL+, CEST−:+^{W}/−19, +^{W}20,+^{W}/−31,48,−57	3 5 5	340C>T 733C>G	Arg114Trp Leu245Val	Africans (Many)
RHCE*01.20.08 or RHCE*ceVS.08	e+^{W} V+VS+or RH:10,20 Probably hr^B− (RH:−31)	1 5 5	48G>C 733C>G 748G>A	Trp16Cys Leu245Val Val250Met	Africans (Few)
RHCE*01.20.09 or RHCE*ceVS.09	e+^{W} V+VS+hr^B+^{W} or RH:10,20,+^{W}31	1 5 7 8	48G>C 733C>G 941T>C 1006G>T	Trp16Cys Leu245Val Val314Ala Gly336Cys	African (Several)
RHCE*01.21	e+^{W} JAL+ or RH:48	3	341G>A	Arg114Gln	Asian, Caucasians (Few)

(Continued)

Rh

(Continued)

Allele encodes	Allele name	Exon	Nucleotide	Amino acid	Ethnicity (prevalence)
e+^{W} partial D (D^{HAR}) Rh33+FPTT+ or RH:33,50 D+ with some anti-D	*RHCE*01.22* or *RHCE*ceHAR*	5	ce-D(5)-ce	See table	Caucasians (Germans) (Many)
e+^{W}	*RHCE*01.23*	5	649T>C	Trp217Arg	Caucasians (Germans) (Few)
e+^{W}	*RHCE*01.24*	4	512G>A	His171Arg	Caucasians (Germans) (Few)
e+^{W}	*RHCE*01.25*	5	730G>A	Ala244Thr	Caucasians (Germans) (Few)
e+^{W}	*RHCE*01.26*	6	872C>T	Pro291Leu	Caucasians (Germans) (Rare)
e+^{W}	*RHCE*01.27*	9	1154G>C	Gly385Ala	Caucasians (Germans) (Few)
c+^{W}	*RHCE*01.28*	10	1254A>C	Stop418Tyr	Caucasians (Germans) (Few)
C–E–c+e– (Dc–haplotype)	*RHCE*01.29* *RHCE*ceBOL*	4 to 9	ce-D(4–9)-ce	See table	(Few)
e+^{W}	*RHCE*01.30*	4	526G>A	Ala176Thr	African (Few)

^ = Can be used if testing is focused on only c.
^^ = Can be used if testing is focused on only e.

Molecular bases of RhCe phenotypes

People who are homozygous for alleles encoding partial C and/or partial e can produce alloanti-C and/or anti-e or an alloantibody that is directed at conventional RhCe and appears as anti-Rh17 in that only RBCs with the D− − phenotype (or those expressing the same RhCe variant) do not react. These antibodies are not necessarily mutually compatible. Donors who are Rh17-negative are rare. Thus, antibodies with broad RhCe specificity have been given names to aid in communication and finding compatible blood.

Differences between *RHCE*01* reference allele (Accession number DQ322275) and the *RHCE*02* (*RHCE*Ce*) are given in rows 2 and 3. These differences are also present in all other alleles in this table. Nucleotide differences and amino acids affected are given.

Allele encodes	Allele name	Exon	Nucleotide	Amino acid	Ethnicity (prevalence)
C+e+ Ce+ or RH:2,5,7	*RHCE*02* or *RHCE*Ce*	1 2			
C+ or RH:2	*RHCE*C^*	1 2	48G>C *RHC exon 2*	Trp16Cys See table	
e+ or RH:5	*RHCE*e^*	5	676G	Ala226	
C+W/−, e+W JAL+ or RH:48	*RHCE*02.01* or *RHCE*CeMA* or *RHCE*CeJAL* or or *RHCE*Ce.01*	3	340C>T	Arg114Trp	Caucasians (Many)
C+e+	*RHCE*02.02* or *RHCE*CeFV* or *RHCE*Ce.02*†	5 5 5	667G>T 697C>G 712A>G	Val223Phe Gln233Glu Met238Val	Caucasians (Few)
rᴳ C+Ｗe+Ｗ JAHK+ or RH:53	*RHCE*02.03* or *RHCE*CeJAHK*	3	365C>T	Ser122Leu	Europeans (Several)
Partial C	*RHCE*02.04* or *RHE*CeVA*	5 5	CE-D(5)-CE	See table	Caucasians (Few)
Partial C Cᵂ+ MAR− or RH:8,−51	*RHCE*02.08.01* or *RHCE*CeCW*	1	122A>G	Gln41Arg	Scandinavians (Many)
Cᵂ+, CENR− or RH:8,−56	*RHCE*02.08.02* or *RHCE*CeNR*	1 6 to 10	122A>G Ce-D(6−10)	Gln41Arg See table	Blacks (Few)

(Continued)

(Continued)

Allele encodes	Allele name	Exon	Nucleotide	Amino acid	Ethnicity (prevalence)
Partial C C^X+ MAR– or RH:9,−51	*RHCE*02.09* or *RHCE*CeCX*	1	106G>A	Ala36Thr	Scandinavians (Many)
R^N C+^{W}/–e+^{W} Rh32+ Sec– DAK+ or RH:32,−46,54	*RHCE*02.10.01* or *RHCE*CeRN.01*	4	Ce-D(4)-ce	See table	Africans (Many)
R^N C+^{W}/–e+^{W} Rh32+ Sec– DAK+ or RH:32,−46,54	*RHCE*02.10.02* or *RHCE*CeRN.02*	3 4	455C>A Ce-D(4)-Ce	Thr152Asn See table above	? exists
C+^{W}	*RHCE*02.11*	2	286G>A	Gly96Ser	Caucasians (Rare)
C+^{W}	*RHCE*02.12*	3	344T>G	Leu115Arg	Caucasians (Rare)
e+^{W}	*RHCE*02.13*	3	364T>C	Ser122Pro	Caucasians (Rare)
C+^{W}	*RHCE*02.14*	4	497A>T	His166Leu	Caucasians (Rare)
e+^{W}	*RHCE*02.15*	5	689G>C	Ser230Thr	Caucasians (Rare)
C+^{W} e+^{W}	*RHCE*02.16*	5	728A>G	Tyr243Cys	Caucasians (Rare)
e+^{W}	*RHCE*02.17*	5	800T>A	Met267Lys	Caucasians (Rare)
C+^{W} e+^{W}	*RHCE*02.18*	6	890T>C	Leu297Pro	Caucasians (Rare)
e+^{W}	*RHCE*02.19*	3 8	464T>G 1118C>T	Met155Arg Ala373Val	Caucasians (Rare)
C+^{W} e+^{W}	*RHCE*02.20*	1	79–81delCTC	Leu27del	Caucasians (Rare)
C+^{W}	*RHCE*02.21*	4	527C>T	Ala176Val	Caucasians (Rare)
C+^{W} e+^{W}/–	*RHCE*02.22*	5	667G>T	Val223Phe	Caucasians (Rare)

(Continued)

(Continued)

Allele encodes	Allele name	Exon	Nucleotide	Amino acid	Ethnicity (prevalence)
C+^{W}	RHCE*02.23	7	941T>C	Val314Ala	Caucasians (Rare)
C+^{W} e+^{W}	RHCE*02.24	7	1007G>A	Gly336Asp	Caucasians (Rare)
C+^{W}	RHCE*02.25	7	1007G>T	Gly336Val	Caucasians (Rare)

† = This and subsequent RHCE*Ce alleles can also be referred to by the RHCE*Ce.01, RHCE*Ce.02, etc., designation (following the pattern above); in the interest of space, this is not included.
^ = Can be used if testing is focussed on only C or e.

RBCs from people with the r'^{S} haplotypes (types 1 and 2) type C+ express a partial C, and when immunized by "normal" C frequently make alloanti-C. As the r'^{S} haplotypes (types 1 and 2) arise from altered *RHD* they are not included in this table.

Molecular bases of RhcE phenotypes[11,12]

Differences between *RHCE*01* reference allele (Accession number DQ322275) and the *RHCE*03* (*RHCE*cE*) are given in rows 2 and 3. These differences are also present in all other alleles in this table. Nucleotide differences and amino acids affected are given.

Allele encodes	Allele name	Exon	Nucleotide	Amino acid	Ethnicity (prevalence)
c+E+cE+ RH:3,4,27	RHCE*03 or RHCE*cE				
c+ or RH:4	RHCE*c^	2	307C	Pro103	
E+ or RH:3	RHCE*E^	5	676C	226Pro	
Partial E E type I c+E+^{W}/– E^W+ or RH:11	RHCE*03.01 or RHCE*cEEW or RHCE*cE.01†	4	500T>A	Met167Lys	Caucasians, Asians (Many)

(Continued)

(Continued)

Allele encodes	Allele name	Exon	Nucleotide	Amino acid	Ethnicity (prevalence)
E type II Partial E c+^{W}/−	RHCE*03.02 or RHCE*cEKK	1-3	D(1-3)-cE	See table	Caucasians, Japanese (Few)
E type III Partial E	RHCE*03.03 or RHCE*cEFM	5 5	697C>G 712A>G	Gln233Glu Met238Val	Japanese, Caucasians (Few)
E type IV E c+^{W}	RHCE*03.04	4	602G>C	Arg201Thr	Caucasians (Few)
Partial E, E type V Partial E c+^{W}	RHCE*03.05 or RHCE*cEKH	3	461G>C	Arg154Thr	Japanese (Few)
c+^{W}, E+/+^{W}	RHCE*03.06	1	28C>T	Arg10Trp	Caucasians (Rare)
E+^{W}	RHCE*03.07	3	344T>C	Leu115Pro	Caucasians (Rare)
E+^{W}	RHCE*03.08	3	356G>A	Ser119Asn	Caucasians (Rare)
c+^{W}, E+^{W}	RHCE*03.09	3	374T>A	Ile125Asn	Caucasians (Rare)
E+^{W}	RHCE*03.10	4	506T>A	Ile169Gln	Caucasians (Rare)
c+^{W}, E+^{W}	RHCE*03.11	6	908T>A	Leu303Gln	Caucasians (Rare)
E+^{W}	RHCE*03.12	33	464T>G 477T>G	Met155Arg Asn159Lys	Caucasians (Rare)
c+^{W}, E+^{W}	RHCE*03.13	5	728A>G	Try243Cys	Caucasians (Rare)
c+^{W}, E+^{WW}/−	RHCE*03.14	5	734T>C	Leu245Pro	Caucasians (Rare)
E+^{W}	RHCE*03.15.0 or RHCE*cEBA	3 3	380C>T 383G>A	Ala127Val Gly128Asp	Caucasians (Rare)
E+^{W}	RHCE*03.15.02 or RHCE*ceJU	3 3 3	361A>T 380C>T 383G>A	Met121Glu Ala127Val Gly128Asp	Caucasians (Rare)

† = This and subsequent RHCE*cE alleles can also be referred to by the RHCE*cE.01, RHCE*cE.02, etc., designation (following the pattern above); in the interest of space, this is not included.
^ = Can be used if testing is focussed on only c or E.

Molecular bases of RhCE phenotypes

Differences between *RHCE*01* reference allele (Accession number DQ322275) and the *RHCE*04* (*RHCE*CE*) are given in rows 2 and 3. These differences are also present in the other allele in this table. The nucleotide difference and amino acid affected are given.

Allele encodes	Allele name	Exon	Nucleotide	Amino acid	Ethnicity (prevalence)
C+E+CE+ or RH:2,3,22	RHCE*04 or RHCE*CE	1 2 5			South American Indians, Native Americans, (Many) Caucasians (Several)
C+ or RH:2	RHCE*C^	1 2	48G>C RHC exon 2	Trp16Cys See table	
E+ or RH:3	RHCE*E^	5	676C	226Pro	
C+^W E±	RHCE*04.01	5	722C>T	Thr241Ile	Caucasians (Few)

^ = Can be used if testing is focussed on only C or E.

Molecular bases of silencing of *RHCE*; on any background

Heterozygosity for a silenced allele is usually revealed by discrepancy between molecular and serologic testing. Homozygosity or compound heterozygosity leads to an RhCE$_{null}$ phenotype; if *in cis* to a deleted or silenced *RHD*, the outcome is an amorph type of Rh$_{null}$. If *in cis* to *RHD*, the outcome is a D− − haplotype without an exalted D antigen. The nucleotide differences from the *RHCE* parent (*RHCE*01, RHCE*02, RHCE*03,* or *RHCE*04*) reference allele, and amino acids affected, are given.

Phenotype	Allele name	Exon (intron)	Nucleotide	Amino acid	Ethnicity (prevalence)
c– e–	RHCE*01N.01 or RHCE*ceN.01	1	80–84 del	Lys31Stop	Caucasians (Few)
c– e–	RHCE*01N.02 or RHCE*ceN.02	7	960–963 del G	Gly321fs	Caucasians (Few)
c– e–	RHCE*01N.03 or RHCE*ceN.03	(4)	+1g>t	Disrupts slice site	Caucasians (Few)
C– e–	RHCE*02N.01 or RHCE*CeN.01	7	966–969del or nt change +del	Ile322, His323fs; Gly398Stop	Caucasians (Few)
c– E–	RHCE*03N.01 or RHCE*cEMI	3	350–358del	Arg120del, Met121del, Ser122del	Caucasians (Few)
c– E–	RHCE*03N.02 or RHCE*cE907del	6	907del C	fs, Leu303Stop	Hispanics (Several)

Rh

Molecular bases of hybrid haplotypes resulting in an altered or silenced RH alleles

RBCs with these phenotypes have an exalted expression of D.

Phenotype	Haplotype name	RHD allele	RHCE allele	Ethnicity (prevalence)
D+^{S} C– C^W+^{W} E– c– e–; DC^W–	DC^W– (AM)	RHD*D–CE(10)	RHCE*CeCW-D(2–10)	(Rare)
D+^{S} C– C^W+^{W} E– c– e–; DC^W–	DC^W– (GLO)	RHD*D–CE(10)	RHCE*CeCW-D(2–9)–CE	(Rare)
D+^{S} C– E– c– e– D– –	D– – (LM)	RHD*01W.033–CE(10)	RHCE*CE-DW33(2–8)–CE-D(10)	(Rare)
D+^{S} C– E– c– e– D– –	D– – (SH)	RHD*01	RHCE*CE-D(2–9)–CE	(Rare)
C+^{VW} Go(a+) Rh33+ Riv+ FPTT+ or RH:30,33,45,50	DIVa(C)–	RHD*DIVa	RHCE*CE-DIVa(2,3)-CE-D(5)-CE	Africans (Few)
D– C+^{W} V– VS+ hr^B– Hr^B–Rh42+ or RH:–1, +^{w}2,–10,20, –31,–34,42	r'^{S} type 1	RHD*DIIIa-ceVS.03(4–7)-D	RHCE*ceVS.03 RHCE*01.20.03	Africans (Many)
D–, C+^{VW} V– VS+ hr^B– Hr^B– Rh42– or RH:–1, +^{vw}2,–10, 20,–31,–34,–42	r'^{S} type 2	RHD*D-ceVS.03 (3–7)-D	RHCE*ceVS.03 RHCE*01.20.03	Africans (Several)

Rearranged *RHD* and *RHCE*[1,9,11,13]

White boxes show exons encoded by *RHCE*; black boxes show exons encoded by *RHD*; hatched boxes depict exons that are not expressed. Amino acid substitutions, rather than nucleotide substitutions are shown under the exons.

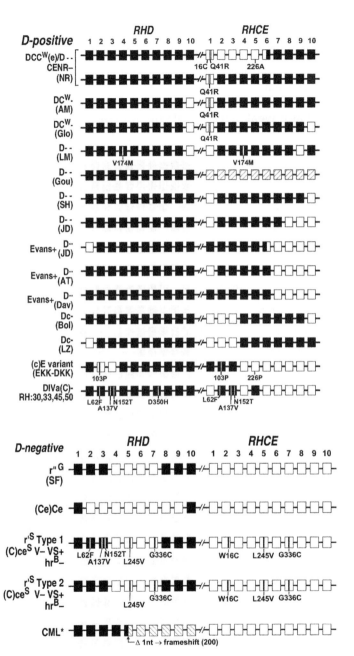

* *RHD* and *RHCE* identified in a D-positive patient, with chronic myeloid leukemia, who became D-negative

Rh

RHD alleles and associated information

Black boxes show exons encoded by *RHD*, white boxes show exons encoded by *RHCE*. Amino acid substitutions, rather than nucleotide substitutions, are shown under the exons.

Partial RhD and weak partial RhD phenotypes

Allele Name		Antigens
RHD*09.02	DAR2/DARE	
RHD*09.03	DAR3	
RHD*09.04	DAR4	
RHD*09.05	DAR5	
RHD*10.00	DAU0	
RHD*10.01	DAU1	
RHD*10.02	DAU2	
RHD*10.03	DAU3	
RHD*10.04	DAU4	
RHD*10.05	DAU5	
RHD*10.06	DAU6	
RHD*10.07	DAU7	
RHD*11	Weak Partial 11	
RHD*12.01	DOL1	DAK+
RHD*12.02	DOL2	DAK+
RHD*12.03	DOL3	
RHD*13.01	DBS1	
RHD*13.02	DBS2	
RHD*14.01	DBT1	Rh32+
RHD*14.02	DBT2	Rh32+
RHD*15	Weak Partial 15	
RHD*16.01	DCS1	
RHD*16.02	DCS2	
RHD*17.01	DFR1	FPTT+
RHD*17.02	DFR2	
RHD*17.03	DFR3	
RHD*17.04	DFR4	

For additional alleles and more information, see table
"Molecular bases of partial and weak partial D phenotypes."

RHCE alleles and associated information[14–16]

White boxes show exons encoded by *RHCE*; black boxes show exons encoded by *RHD*. Amino acid substitutions, rather than nucleotide substitutions, are shown under the exons. Associated low prevalence antigens and other relevant serological findings are given next to the exon diagram.

Partial and weak Rhce phenotypes

Differences from Rhce encoded by *RHCE*01 are given.

Allele Name		Substitutions (exons 1–10)	Antigens
*RHCE*01.01*	ce	W16C	e+/–
*RHCE*01.02*	ceTI Type 1	W16C; T342I	
*RHCE*01.03*	ceTI Type 2	T342I	
*RHCE*01.04*	ceAR	W16C; M238V/L245V/M267K/R263G; D36V	c+/– e+/– Rh18– hr^S–
*RHCE*01.05*	ceEK	W16C; M238V/R263G; M267K	c+/– e+/– Rh18– hr^S–
*RHCE*01.06*	ceAG	A85Q	e+/– hr^B– CEAG–
*RHCE*01.07*	ceMO	W16C; V223F	c+/– e+/– hr^S– hr^B–
*RHCE*01.08*	ceBI	W16C; M238V A273V; L379V	e+/– Rh18– hr^S– STEM+
*RHCE*01.09*	ceSM	W16C; M238V A273V	e+/– hr^S– STEM+
*RHCE*01.10.01*	ceSL	W16C; S122L	e+^{w} D+/–
*RHCE*01.11*	ceRT	R154T	D+/–
*RHCE*01.12*	ceRA	W16C; G180R	e+^{w}
*RHCE*01.13*	ceBP	R221del	e+^{w} CELO+^{w}
*RHCE*01.14*	ceBE	P221R	c+^{w} e+^{w} Be(a+)
*RHCE*01.15*	ceLOCR	Q66S	c+^{w} RH:–26 LOCR+
*RHCE*01.20.01*	ceVS.01	L245V	V+VS+ hr^B+^{w}/– c+/– e+/–
*RHCE*01.20.02*	ceVS.02	W16C; L245V	V+VS+ e+/– hr^B–
*RHCE*01.20.03*	ceVS.03	W16C; L245V G336C	c+/– e+/– V–VS+ hr^B–
*RHCE*01.20.04*	ceVS.04	W16C; L245V; T342I	e+/–, V+VS+
*RHCE*01.20.05*	ceVS.05	L245V; G336C	e+/– V–VS+ hr^B–
*RHCE*01.20.06*	ceVS.06	W16C; Q233E L245V	c+/– e+/– VS+ hr^B– Crawford+ CELO–
*RHCE*01.20.07*	ceVS.07	R114W; L245V	c+/– e+/– (hr^S) (V/VS) JAL+ CEST–
*RHCE*01.20.08*	ceVS.08	W16C; L245V/V250M	e+/– V+VS+
*RHCE*01.21*		R114Q	JAL+
*RHCE*01.22*	ceHAR	(exon 5 RHD)	D+/– e+^{w} Rh33+ FPTT+

( ) denotes reduced antigen expression
+/– positive with some antibodies (could be weak) negative with other antibodies
For additional alleles and more information, see table "Molecular bases of Rhce phenotypes"

Partial and weak RhCe phenotypes

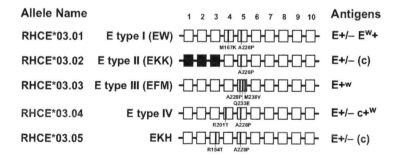

Allele Name		Antigens
RHCE*02.01	CeMA	C+ʷ/– e+ʷ JAL+
RHCE*02.02	CeFV	
RHCE*02.03	rᴳ	C+ʷ e+ʷ G+ʷ JAHK+
RHCE*02.04	CeVA	(C)(e) Rh33+ FPTT+
RHCE*02.08.01	CeCW	Cʷ+ MAR–
RHCE*02.09	CeCX	Cˣ+ MAR–
RHCE*02.10.01	CeRN.01	C+/– e+ʷ Rh32+ Rh46– DAK+
RHCE*02.10.02	CeRN.02	C+/– e+ʷ Rh32+ Rh46– DAK+
RHCE*02.11		C+ʷ
RHCE*02.12		C+ʷ
RHCE*02.15		eʷ
RHCE*02.16		C+ʷ e+ʷ
RHCE*02.18		C+ʷ e+ʷ
RHCE*02.19		e+ʷ
RHCE*02.20		C+ʷ e+ʷ

( ) denotes reduced antigen expression
+/– positive with some antibodies (could be weak)
    negative with other antibodies
For additional alleles and more information,
see table "Molecular bases of RhCe phenotypes."

Partial and weak RhcE Phenotypes

Allele Name			Antigens
RHCE*03.01	E type I (EW)	M167K A226P	E+/– Eʷ+
RHCE*03.02	E type II (EKK)	A226P	E+/– (c)
RHCE*03.03	E type III (EFM)	A226P M238V Q233E	E+w
RHCE*03.04	E type IV	R201T A226P	E+/– c+ʷ
RHCE*03.05	EKH	R154T A226P	E+/– (c)

( ) denotes reduced antigen expression
+/– positive with some antibodies (could be weak)
    negative with other antibodies.

For additional alleles and more information see table
"Molecular bases of RhcE phenotypes."

Amino acid sequence of RhCE and RhD[17,18]

The full sequence is the RhCE (C and E) protein. Differences in the sequence for c, e and D proteins are shown.

```
RhC:  MSSKYPRSVR  RCLPLCALTL  EAALILLFYF  FTHYDASLED  QKGLVASYQV   50
Rhc:                  W
RhD:                  W

RhC:  GQDLTVMAAI  GLGFLTSSFR  RHSWSSVAFN  LFMLALGVQW  AILLDGFLSQ  100
Rhc:           L           N
RhD:           I           S

RhC:  FPSGKVVITL  FSIRLATMSA  MSVLISAGAV  LGKVNLAQLV  VMVLVEVTAL  150
Rhc:           P
RhD:           S                       L     VD

RhC:  GTLRMVISNI  FNTDYHMNLR  HFYVFAAYFG  LTVAWCLPKP  LPKGTEDNDQ  200
RhD:  N                   MM          I           S           E   K

RhE:  RATIPSLSAM  LGALFLWMFW  PSVNSPLLRS  PIQRKNAMFN  TYYALAVSVV  250
Rhe:                                A
RhD:  T                               F  A        E   V         V

RhC:  TAISGSSLAH  PQRKISMTYV  HSAVLAGGVA  VGTSCHLIPS  PWLAMVLGLV  300
RhD:              G   K

RhC:  AGLISIGGAK  CLPVCCNRVL  GIHHISVMHS  IFSLLGLLGE  ITYIVLLVLH  350
RhD:  V           Y   G         P S I GY  N           I       D

RhC:  TVWNGNGMIG  FQVLLSIGEL  SLAIVIALTS  GLLTGLLLNL  KIWKAPHVAK  400
RhD:  GA                                              E

RhC:  YFDDQVFWKF  PHLAVGF                                         417
```

Carrier molecule

The assembly of the Rh proteins (RhD, RhCE) and the Rh-associated glycoprotein (RhAG) as a core complex in the RBC membrane appears to be essential for Rh-antigen expression. RhD and RhCE are multipass, acylated, palmitoylated, non-glycosylated proteins.

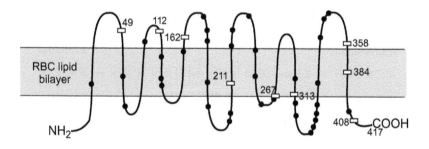

Circles indicate the amino acid positions that differ between RhD and RhCE. Depending on the RhCE haplotype, RhD differs from RhCE by 32 to 35 amino acids. The D antigen is unusual in that it is not derived from an amino acid polymorphism, but from the presence of the entire RhD protein. Expression of D antigen can vary qualitatively and quantitatively. Segments of

RhD and RhCE encoded by a particular exon are defined by numbered boxes, representing the start and finish of each exon.

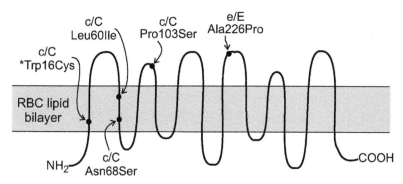

*Trp usually but not exclusively associated with c antigen
Cys usually but not exclusively associated with C antigen.

74% of C–c+black Americans with normal c have Cys16.

M_r (SDS-PAGE)	30,000–32,000	
Cysteine residues	4 in RhD; 6 in RhCE	
Palmitoylation sites	2 in RhD:	Cys12, Cys186
	3 in RhCE:	Cys12, Cys186, Cys311
Copies per RBC	100,000–200,000 for RhD and RhCE combined	

Function

The Rh membrane core complex interacts with band 3, GPA, GPB, LW, and CD47, and is associated with the RBC membrane skeleton via ankyrin and protein 4.2. This complex maintains erythrocyte membrane integrity, as demonstrated by the abnormal morphology and functioning of stomatocytic Rh$_{null}$ RBCs[9]. The Rh core proteins in the membrane may transport ammonia[19] and CO_2[20]. RhAG homologs are expressed in other tissues[1,21].

Disease association

Rh incompatibility is still the main cause of HDFN.
Compensated hemolytic anemia occurs in some individuals with Rh$_{null}$ or Rh$_{mod}$ RBCs. Reduced expression of Rh antigens and Rh mosaicism can occur in leukemia, myeloid metaplasia, myelofibrosis, and polycythemia. Rh and one form of hereditary spherocytosis are linked because both genes are on chromosome 1. Some Rh antigens are expressed weakly on South East Asian ovalocytes.

Rh

Phenotypes (% occurrence)

Haplotype	Caucasians	Blacks	Native Americans	Asians
DCe (R_1)	42	17	44	70
Ce (r')	2	2	2	2
DcE (R_2)	14	11	34	21
cE (r'')	1	0	6	0
Dce (R_0)	4	44	2	3
ce (r)	37	26	6	3
DCE (R_z)	0	0	6	1
CE (r^y)	0	0	0	0

Phenotype (alternative)	Caucasians	Blacks	Asians	D-antigen copy number
D-positive				
R_1R_1 (R_1r')	18.5	2.0	51.8	14,500–19,300
R_2R_2 (R_2r'')	2.3	0.2	4.4	15,800–33,300
R_1r (R_1R_0; R_0r')	34.9	21.0	8.5	9,900–14,600
R_2r (R_2R_0; R_0r'')	11.8	18.6	2.5	14,000–16,000
R_0r (R_0R_0)	2.1	45.8	0.3	12,000–20,000
R_zR_z (R_2r^y)	0.01	Rare	Rare	
R_1R_z (R_2r'; R_1r^y)	0.2	Rare	1.4	
R_2R_z (R_2r''; R_2r^y)	0.1	Rare	0.4	
R_1R_2 (R_1r''; R_2r'; R_zr; R_0R_z; R_0r^y)	13.3	4.0	30.0	23,000–36,000
D-negative				
r'r	0.8	Rare	0.1	
r'r'	Rare	Rare	0.1	
r''r	0.9	Rare	Rare	
r''r''	Rare	Rare	Rare	

(Continued)

(Continued)

Phenotype (alternative)	Caucasians	Blacks	Asians	D-antigen copy number
rr	15.1	6.8	0.1	
r'r" (r^{y}r)	0.05	Rare	Rare	
r'r^y; r"r^y; r^{y}r^y	Rare	Rare	Rare	
r'^{S}r	0	1–2	0	

Null: Rh$_{null}$
Unusual: Rh$_{mod}$; many variants.

Comparison of Rh$_{null}$ and Rh$_{mod}$ RBCs

Phenotype	Rh proteins/ antigens	RhAG	LW	CD47	GPB: S, s, and U antigens	Altered gene
Amorph Rh$_{null}$	Absent	Reduced (20%)	Absent	Reduced by 90%	Reduced by 50% S/s normal; U weak	RHCE (RHD deleted)
Regulator Rh$_{null}$	Absent	Absent	Absent	Reduced	Reduced by 70% S/s weak; U absent	RHAG
Rh$_{mod}$	Reduced (variable)	Absent or reduced (variable)	Absent or reduced	Reduced (variable)	Reduced (variable) S/s normal; U normal/ weak	RHAG

Comments

Useful websites are dbRBC (for *RHD, RHCE, RHAG*), Rhesus Base (for *RHD*), and NYBC (for *RHCE*) to obtain further details see hyperlinks to original papers, and updates.

References

[1] Huang, C.-H., et al., 2000. Molecular biologwy and genetics of the Rh blood group system. Semin Hematol 37, 150–165.

[2] Westhoff, C.M., 2007. The structure and function of the Rh antigen complex. Semin Hematol 44, 42–50.

[3] Wagner, F.F., Flegel, W.A., 2000. RHD gene deletion occurred in the Rhesus box. Blood 95, 3662–3668.

[4] Flegel, W.A., Wagner, F.F., 2002. Molecular biology of partial D and weak D: implications for blood bank practice. Clin Lab 48, 53–59.

[5] Müller, T.H., et al., 2001. PCR screening for common weak D types shows different distributions in three Central European populations. Transfusion 41, 45–52.

[6] Wagner, F.F., et al., 2000. Weak D alleles express distinct phenotypes. Blood 95, 2699–2708.

[7] Pham, B.N., et al., 2011. Anti-D investigations in individuals expressing weak D Type 1 or weak D Type 2: Allo- or autoantibodies? Transfusion 51, 2679–2685.

[8] Shao, C.P., et al., 2002. Molecular background of Rh D-positive, D-negative, D-el and weak D phenotypes in Chinese. Vox Sang 83, 156–161.

[9] Avent, N.D., Reid, M.E., 2000. The Rh blood group system: a review. Blood 95, 375–387.

[10] Wagner, F.F., et al., 2001. RHD positive haplotypes in D negative Europeans. BMC Genet 2, 10.

[11] Kashiwase, K., et al., 2001. E variants found in Japanese and c antigenicity alteration without substitution in the second extracellular loop. Transfusion 41, 1408–1412.

[12] Noizat-Pirenne, F., et al., 1998. Heterogeneity of blood group RhE variants revealed by serological analysis and molecular alteration of the RHCE gene and transcript. Br J Haematol 103, 429–436.

[13] Westhoff, C.M., et al., 2000. A new hybrid RHCE gene that is responsible for expression of a novel antigen [abstract]. Transfusion 40 (Suppl.), 7S.

[14] Daniels, G.L., et al., 1998. The VS and V blood group polymorphisms in Africans: a serological and molecular analysis. Transfusion 38, 951–958.

[15] Noizat-Pirenne, F., et al., 2002. Rare RHCE phenotypes in black individuals of Afro-Caribbean origin: identification and transfusion safety. Blood 100, 4223–4231.

[16] Noizat-Pirenne, F., et al., 2001. Two new alleles of the RHCE gene in Black individuals: the RHce allele ceMO and the RHcE allele cEMI. Br J Haematol 113, 672–679.

[17] Arce, M.A., et al., 1993. Molecular cloning of RhD cDNA derived from a gene present in RhD- positive, but not RhD-negative individuals. Blood 82, 651–655.

[18] Chérif-Zahar, B., et al., 1990. Molecular cloning and protein structure of a human blood group Rh polypeptide. Proc Natl Acad Sci USA 87, 6243–6247.

[19] Westhoff, C.M., et al., 2002. Identification of the erythrocyte Rh blood group glycoprotein as a mammalian ammonium transporter. J Biol Chem 277, 12499–12502.

[20] Soupene, E., et al., 2002. Rhesus expression in a green alga is regulated by CO_2. Proc Natl Acad Sci USA 99, 7769–7773.

[21] Huang, C.-H., Liu, P.Z., 2001. New insights into the Rh superfamily of genes and proteins in erythroid cells and nonerythroid tissues. Blood Cells Mol Dis 27, 90–101.

Rh

D Antigen

Terminology

ISBT symbol (number)	RH1 (004001 or 4.1)
Obsolete names	Rh_0
History	The original "Rh" antigen stimulated a transfusion reaction, which was investigated by Levine and Stetson in 1939. The reactions of this antibody paralleled those of the anti-"Rh" reported by Landsteiner and Wiener in 1940, but stimulated in animals. Some years later, upon recognition that the human and the animal anti-"Rh" did not react with the same antigen, the accumulation of publications about the clinically important human anti-"Rh" made a name change undesirable. Ultimately however, the antigen name switched to D and the system took the Rh name.

Occurrence

Caucasians	85%
Blacks	92%
Asians	99%
Native Americans	99%

Expression

Cord RBCs	Expressed
Altered	Partial and weak D phenotypes; exalted on D deletion phenotypes; appears exalted on GPA-deficient RBCs because of reduction of sialic acid

Number of D antigen sites per RBC

Common D phenotypes	10,000–33,000
Weak D phenotypes	<100–10,000
Exalted D phenotypes	75,000–200,000

Molecular basis of D antigen

See System pages.

Effect of enzymes and chemicals on D antigen on intact RBCs

Ficin/Papain	Resistant (markedly enhanced)
Trypsin	Resistant (enhanced)

Rh

α-Chymotrypsin	Resistant (enhanced)
DTT 200 mM	Resistant
Acid	Resistant

In vitro characteristics of alloanti-D

Immunoglobulin class	Most IgG, some IgM (IgA rare)
Optimal technique	IAT; enzymes
Complement binding	Extremely rarely

Clinical significance of alloanti-D

Transfusion reaction	Mild to severe/immediate or delayed
HDFN	Mild to severe

Autoanti-D

Yes	May appear as mimicking alloantibody

Partial D phenotypes: Qualitative variation of D

Tippett and Sanger studied the interactions of RBCs and serum from D+ people who had made anti-D. They observed a limited number of reaction patterns, and initially divided samples with partial D antigens into six categories (D^I to D^{VI}, which are now written without a superscript). The D antigen is a mosaic of different epitopes. People with RBCs lacking one or more of these epitopes (referred to as expressing a partial D antigen) can make alloanti-D directed at the missing D epitopes. Partial D phenotypes initially were classified into seven D categories (DI to DVII; DI later became obsolete) based on the interaction of the RBCs and sera of the D category members, and also by the reaction patterns with selected polyclonal anti-D. Low-prevalence marker antigens aided in their identification. Many other partial D were added later (e.g., DFR, DBT, DOL and DAU; see tables and figures in the Rh system pages). Monoclonal anti-D revealed different reaction patterns, and each reaction pattern recognizes a different epitope (epD) of the D mosaic. Seven reaction patterns were initially recognized and these were expanded to nine patterns with awareness that more epitopes would be identified.

Reactions with anti-D have shown that some partial D phenotypes have consistently strongly expressed D epitopes (e.g., DIII, DIVa), others have variable expression of the relevant epitopes, thereby demonstrating qualitative and quantitative alteration (e.g., DVa, DVII). Yet others have very weakly expressed epitopes (e.g., DVI Type 1, DAR) and these are referred to as "weak partial D phenotypes."

Epitope profiles of partial D antigens: The nine epitope model[1]

	Reactions with monoclonal anti-D							
	epD1	epD2	epD3	epD4	epD5	epD6/7	epD8	epD9
DII	+	+/0	+	0	+	+	+	0
DIIIa	+	+	+	+	+	+	+	+
DIIIb	+	+	+	+	+	+	+	+
DIIIc	+	+	+	+	+	+	+	+
DIVa	0	0	0	+	+	+	+	0
DIVb	0	0	0	0	+	+	+	0
DVa	0	+	+	+	0	+	+	+
DVI	0	0	+	+	0	0	0	+
DVII	+	+	+	+	+	+	0	+
DFR	+/0	+/0	+	+	+/0	+/0	0	+
DBT	0	0	0	0	0	+/0	+	0
DHAR	0	0	0	0	+/0	+/0	0	0

+ = Positive; +/0 = Positive with some anti-D, negative with other anti-D; 0 = Negative.

Recognition of new partial D phenotypes and use of hundreds of monoclonal anti-D has sub-split the nine epitopes. The nine epitope model, which was directly related to the original D categories, was expanded to accommodate the new reaction patterns. Sub-splits of the patterns by reactions observed with new unique partial D are being denoted by a dot followed by a second Arabic number, e.g., the sub-split of epD1 was defined by reactions with DFR cells: anti-epD1.1 are positive and anti-epD1.2 are negative with DFR cells. New reaction patterns defined with monoclonal anti-D have been assigned numbers above 9 (see table). These patterns were defined through multi-center ISBT workshops for a standardized and logical approach.

Rh

Epitope profiles of partial D antigens: the expanded 30 epitope model[2]

Anti-epD	Partial D phenotype																	
	DII	DIII	DIVa	DIVb	DV1	DV2	DV3/4	DV5	DVI	DVII	DFR	DBT	DHAR	DHMi	DNB	DAR	DNU	DOL
1.1	+	+	0	0	0	0	0	0	0	+	+	0	0	+	+	V	V	V
1.2	+	+	0	0	0	0	0	0	0	+	0	0	0	+	+			
2.1	+	+	0	0	+	+	+	0	0	+	+	0	0	+	+	+	+	+
2.2	+	+	0	0	+	+	+	0	0	+	0	0	0	0	+	0	+	+
3.1	+	+	0	0	+	+	+	+	+	+	+	0	0	+	V	+	+	+
4.1	0	+	+	0	0	0	+	+	+	+	+	0	0	+	+	+	+	+
5.1	+	+	+	+	+	0	0	0	0	+	+	0	+	+	+	+	+	+
5.2	+	+	+	+	+	+	0	0	0	+	+	0	0	+	+	+	0	+
5.3	+	+	+	+	0	0	0	0	0	+	0	0	0	+	+	0	+	0
5.4	+	+	+	+	+	0	0	0	0	+	0	0	0	+	+	+	+	V
5.5	+	+	+	+	+	+	+	+	0	+	0	0	0	0	+			
6.1	+	+	+	+	+	+	+	+	0	+	+	+	+	+	+	+	+	+
6.2	+	+	+	+	+	+	+	+	0	+	+	+	0	+	+	+	+	+
6.3	+	+	+	+	+	+	+	+	0	+	+	0	0	+	+	+	+	+

(Continued)

Rh

Cell	1	2	3	4	5	6	7	8	9	10	11	12	13
6.4	+	+	+	+	+	+	+	0	0	+	+	+	+
6.5	+	+	+	+	0	+	+	0	0	+	+	+	+
6.6	>	+	+	+	+	0	0	0	0	+	+	+	+
6.7	+	+	+	+	0	0	0	0	0	0	+	+	0
6.8	+	+	+	+	0	+	0	0	+	+	+	+	+
8.1	+	0	0	+	0	0	0	0	+	+	+	0	+
8.2	+	+	+	+	>	0	+	+	+	+	+	+	0
8.3	0	0	0	0	>	0	+	+	0	0	+	+	+
9.1	+	+	+	+	+	+	+	+	+	+	+	+	+
10.1	0	0	0	0	0	0	0	0	+	+	0	+	+
11.1	0	0	0	0	0	0	0	0	+	+	+	+	+
12.1	+	+	+	+	+	+	+	+	+	+	+	+	+
13.1	+	+	+	+	+	+	+	+	+	+	+	+	+
14.1	+	+	+	+	+	0	+	0	+	+	+	+	+
15.1	+	+	+	+	+	+	+	+	+	+	+	+	+
16.1	+	+	+	+	+	+	+	+	+	+	+	+	+

Nomenclature of partial D recommended by International Society of Blood Transfusion Working Party on Red Cell Immunogenetics and Blood Group Terminology[2,3]

D category phenotypes retain the original numbering system, but the historical superscript Roman numeral is now on the line, e.g., D^{VI} is written DVI. Subtypes of D categories are denoted by Arabic numerals, e.g., DVI type 1, DVI type 2, etc. Other (and new) partial D will be denoted by up to four upper case letters, e.g., DBT, DAR, DNU, DOL. Overall weak expression of D will be referred to as weak D (see later)[2].

Selected Partial D Phenotypes[1,4–7]

Partial D phenotype	Associated haplotype	Approximate number of D antigen sites	Number of probands	Ethnic origin	Made anti-D
DII	Ce	3,200	One	Caucasians	Yes
DIIIa	ce G+	12,300	Many	Blacks	Yes
DIIIb	ce G–		Few	Blacks	Yes
DIIIc	Ce G+	26,900	Many	Caucasians	Yes
DIII type 4		33,250	Few	Caucasians	Yes
DIVa	ce, [(C)–]	9,300	Many	Blacks	Yes
DIVb	Ce, cE	4,000	Many	Caucasians, Japanese	Yes
DIV type 3	Ce	600	One	Caucasians	
DIV type 4	Ce		Several		
DVa	ce, Ce, cE	9,400	Many	Caucasians, Japanese, Blacks	Yes
DVI type 1	cE	300–1,000	Many	Caucasians	Yes
DVI type 2	Ce	1,600–2,900	Many	Caucasians, Japanese	Yes
DVI type 3	Ce	14,500	Few	Caucasians	Yes
DVI type 4	Ce		One	Caucasians	
DVII	Ce	3,600–8,400	Many	Caucasians	Yes
DFR	Ce>cE	5,300	Many	Caucasians	Yes

(Continued)

(Continued)

Partial D phenotype	Associated haplotype	Approximate number of D antigen sites	Number of probands	Ethnic origin	Made anti-D
DBT type 1	Ce>(C)(e) and ce	4,300	Several	Caucasians, Japanese, Blacks	Yes
DBT type 2	Ce		Several	Japanese	
DHAR (ceHAR)	c(e) G–		Many	Caucasians	Yes
DHMi	cE	2,400	Several	Caucasians	Yes
DNB	Ce	6,000	Many	European (1 in 292 in Swiss)	Yes
DNU	Ce	10,000	Few	Caucasians	
DOL	ce	4,700	Several	Blacks	Yes
DAR	ce		Many	Blacks	Yes
Weak D type 4.2.2	ce	1,650	Few	Caucasians	Yes
DCS1	cE	3,000	One	Caucasians	
DCS-2	cE	800	Several	Caucasians	
DTI	cE		One	Japanese	
DBS	cE or ce		One	Asians	
DAL			Several	Caucasians	
DFW	Ce		One	Caucasians	
DHO	Ce	1,300		Caucasians	
DHR	cE	3,800		Caucasians	
DMH	ce			Caucasians	Yes
DIM	cE	200	One	Caucasians	
Weak D type 15	cE	300	Few	Caucasians	Yes
DAU0	ce	15,000	Many	Blacks (Caucasians)	
DAU1	ce	2,100	Several	Blacks	
DAU2	ce	370	Several	Blacks	Yes
DAU3	ce	10,880	Many	Blacks	Yes

Rh

Some partial D phenotypes in this table are not yet associated with production of alloanti-D; such phenotypes are included here because of their similarity to known partial D phenotypes as determined by molecular analysis or by the D epitope profile.

For additional partial D phenotypes see the system pages.

Weak D phenotypes: Quantitative variation of D

The weak D phenotype is a quantitative, not a qualitative, polymorphism and therefore all D epitopes are present. This reduced D antigen expression is usually detected by the indirect antiglobulin test, although some weak D phenotypes are directly agglutinated by MAb anti-D. For the molecular bases of weak D phenotypes see tables and figures in the system pages.

The different types of weak D defined at the molecular level, in accordance with ISBT nomenclature, are referred to as "type" with Arabic numerals, e.g., weak D type 1.

For molecular bases on weak D phenotypes, see system pages.

The weak D phenotypes shown in the table were those initially defined[7], but their number has greatly increased, as may be seen from the tables in the system pages.

Weak D phenotype	Associated haplotype	Approximate number of D antigen sites
Type 1	Ce	1,300
Type 2	cE	500
Type 3	Ce	1,900
Type 4.0	ce	2,300
Type 4.1		3,800
Type 5	cE	300
Type 6	Ce	1,000
Type 7	Ce	2,400
Type 8	Ce	1,000
Type 9	cE	250
Type 10	cE, Ce (majority)	1,200
Type 11	ce	200
Type 12	Ce	100
Type 13	Ce	1,000
Type 14	cE	
Type 16	cE	250
Type 17		60
Type 21	Ce	5,200

In European populations weak D types 1, 2, and 3 predominate[8].

Clinically relevant information about the D antigen

D phenotype	Amino acid changes in RhD	D Expression	Tests used to detect D	In patient			In donor
				Can make anti-D through transfusion or pregnancy	RBC suitable for transfusion	RhIgG prophylaxis recommended	Can immunize D– recipient
D+	None	Normal	Direct agglutination	No	D+†	No	Yes
Partial D	Usually extracellular	Altered (some D epitopes present, some absent)	Direct agglutination or IAT^	Yes	D– or matched partial D phenotype	Yes	Yes
Weak partial D	Usually extracellular	Altered (some D epitopes present, some absent) weak or variable	Direct agglutination or IAT^	Yes	D– or matched weak partial D phenotype	Yes	Possible
Weak D	Usually transmembrane or intracellular	Normal but weak	IAT (Direct agglutination for some)	No	D+ (or D–)	No	Possible
D–	RhD absent	Absent	IAT	Yes	D–^^	Yes	No

†Although cross-match compatible D– RBCs can be safely transfused, these RBCs/components should be reserved for use to D– patients.
^Depending on reagent used.
^^When suitable D– RBC components are not available, e.g., when the patient has made multiple additional alloantibodies or in times of blood shortage, D+ RBC components may be transfused until anti-D is made. The most suitable candidates for such a strategy are males or women unable to have children.

Rh

Comments

Expression of D may be weakened by a Ce, CE or (C)ce^S complex *in trans*.
A Rhesus Similarity Index[6] was devised to characterize the extent of qualitative changes in aberrant D antigens. Based on D epitope density profiles
ascertained by using FACS analysis with a panel of monoclonal anti-D, this
quantitative method may aid in the discrimination of normal D from partial D
and weak D.

References

[1] Tippett, P., et al., 1996. The Rh antigen D: partial D antigens and associated low incidence anti-
    gens. Vox Sang 70, 123–131.
[2] Scott, M., 2002. Section 1A: Rh serology. Coordinator's report. Transfus Clin Biol 9, 23–29.
[3] Storry, J.R., et al., 2011. International Society of Blood Transfusion Working Party on red cell
    immunogenetics and blood group terminology: Berlin report. Vox Sang 101, 77–82.
[4] Avent, N.D., Reid, M.E., 2000. The Rh blood group system: a review. Blood 95, 375–387.
[5] Müller, T.H., et al., 2001. PCR screening for common weak D types shows different distribu-
    tions in three Central European populations. Transfusion 41, 45–52.
[6] Wagner, F.F., et al., 2000. Weak D alleles express distinct phenotypes. Blood 95, 2699–2708.
[7] Wagner, F.F., et al., 2002. The *DAU* allele cluster of the *RHD* gene. Blood 100, 306–311.
[8] Flegel, W.A., Wagner, F.F., 2002. Molecular biology of partial D and weak D: implications for
    blood bank practice. Clin Lab 48, 53–59.

C Antigen

Terminology

ISBT symbol (number)	RH2 (004002 or 4.2)
Obsolete name	rh'
History	Reported in 1941 when it was recognized that, in addition to D, the Rh system had four other common antigens. Named because "C" was the next available letter in the alphabet.

Occurrence

Caucasians	68%
Blacks	27%
Asians	93%

Antithetical antigen

c (**RH4**)

Expression

Cord RBCs	Expressed
Altered	See System pages for unusual Rh complexes

Molecular basis associated with C antigen

Amino acid   Ser103; requirements for expression of C antigen are
                          not fully understood
Nucleotide   T at bp 307 in exon 2 of *RHCE*C*

See System pages for weak and partial C antigens.

Effect of enzymes and chemicals on C antigen on intact RBCs

Ficin/Papain   Resistant (markedly enhanced)
Trypsin   Resistant (enhanced)
α-Chymotrypsin   Resistant (enhanced)
DTT 200 mM   Resistant
Acid   Resistant

In vitro characteristics of alloanti-C

Immunoglobulin class   IgG; IgM
Optimal technique   IAT; enzymes

Clinical significance of alloanti-C

Transfusion reaction   Mild to severe/immediate or delayed/hemoglobinuria
HDFN   Mild

Autoanti-C

Yes, may be mimicking alloantibody.

Comments

Anti-C is often found in antibody mixtures, especially with anti-G (see **RH12**)
or anti-D (see **RH1**).
Apparent anti-C in Blacks may be anti-hr^B (see **RH31**).
Alloanti-C can be made by C+ individuals who express one of many partial
C phenotypes such as (C)ce^S (r$^{'S}$), C^W+, C^X+, and D(C)(e)/ce phenotypes (see
table "Molecular bases of RhCe phenotypes" in System pages).
C+RBCs express the G antigen (see **RH12**).
D(C)e RBCs carrying the low-prevalence antigen HOFM (**700050**) express
C weakly.

E Antigen

Terminology

ISBT symbol (number)   RH3 (004003 or 4.3)
Obsolete name   rh″

History	Reported in 1943 and named after the next letter in the alphabet when it was realized that the antigen was part the Rh system.

Occurrence

Caucasians	29%
Blacks	22%
Asians	39%

Antithetical antigen

e (**RH5**)

Expression

Cord RBCs	Expressed
Altered	See System pages for unusual Rh complexes

Molecular basis associated with E antigen

Amino acid	Pro226; requirements for expression of E antigen are not fully understood
Nucleotide	C at bp 676 in exon 5 of *RHCE*E*

See System pages for weak and partial E antigens.

Effect of enzymes and chemicals on E antigen on intact RBCs

Ficin/Papain	Resistant (markedly enhanced)
Trypsin	Resistant (enhanced)
α-Chymotrypsin	Resistant (enhanced)
DTT 200 mM	Resistant
Acid	Resistant

In vitro characteristics of alloanti-E

Immunoglobulin class	IgG and IgM
Optimal technique	RT; IAT; enzymes

Clinical significance of alloanti-E

Transfusion reaction	Mild to moderate/immediate or delayed/ hemoglobinuria
HDFN	Mild

Autoanti-E

Yes, may be mimicking alloantibody.

Comments

The E antigen is comprised of several epitopes as defined by monoclonal anti-E[1].
Anti-E is often present in sera containing anti-c.
Some examples of anti-E appear to be naturally-occurring.

Reference

[1] Noizat-Pirenne, F., et al., 1998. Heterogeneity of blood group RhE variants revealed by serological analysis and molecular alteration of the *RHCE* gene and transcript. Br J Haematol 103, 429–436.

c Antigen

Terminology

ISBT symbol (number)	RH4 (004004 or 4.4)
Obsolete names	hr'
History	Briefly reported in 1941 when it was recognized that, in addition to D, the Rh system had four other common antigens; named when the antithetical relationship to C was recognized.

Occurrence

Caucasians	80%
Blacks	98%
Asians	47%

Antithetical antigen

C (**RH2**)

Expression

Cord RBCs	Expressed
Altered	See System pages for unusual Rh complexes

Molecular basis associated with c antigen[1]

Amino acid	Pro103 (and Pro102[2]) requirements for expression of c antigen are not fully understood.

RhD with a substitution of Ser103Pro expresses a
weak c antigen.

Nucleotide   C at bp 307 in exon 2 of *RHCE*c*

See System pages for weak and partial c antigens.

Effect of enzymes and chemicals on c antigen on intact RBCs

Ficin/Papain	Resistant (markedly enhanced)
Trypsin	Resistant (enhanced)
α-Chymotrypsin	Resistant (enhanced)
DTT 200 mM	Resistant
Acid	Resistant

In vitro characteristics of alloanti-c

Immunoglobulin class	Most IgG; some IgM
Optimal technique	IAT; enzymes

Clinical significance of alloanti-c

Transfusion reaction	Mild to severe/immediate or delayed/ hemoglobulinuria
HDFN	Mild to severe

Autoanti-c

Yes, may be mimicking alloantibody.

References

[1] Faas, B.H.W., et al., 2001. Partial expression of RHc on the RHD polypeptide. Transfusion 41, 1136–1142.

[2] Westhoff, C.M., et al., 2000. Evidence supporting the requirement for two proline residues for expression of the "c" antigen. Transfusion 40, 321–324.

e Antigen

Terminology

ISBT symbol (number)	RH5 (004005 or 4.5)
Obsolete name	hr″
History	Named in 1945 when its antithetical relationship to E was recognized.

Occurrence

Caucasians	98%
Blacks	98%
Asians	96%

Antithetical antigen

E (**RH3**)

Expression

Cord RBCs   Expressed
Altered   See System pages for unusual Rh complexes
See table for reactions of monoclonal anti-e with unusual Rh complexes.

Molecular basis associated with e antigen

Amino acid   Ala226; requirements for expression of e antigen are
                   not fully understood
Nucleotide   G at bp 676 in exon 5 of *RHCE*e*
See System pages for weak and partial e antigens.

Effect of enzymes and chemicals on e antigen on intact RBCs

Ficin/Papain   Resistant (markedly enhanced)
Trypsin   Resistant (enhanced)
α-Chymotrypsin   Resistant (enhanced)
DTT 200 mM   Resistant
Acid   Resistant

In vitro characteristics of alloanti-e

Immunoglobulin class   Most IgG; some IgM
Optimal technique   IAT; enzymes

Clinical significance of alloanti-e

Transfusion reaction   Mild to moderate/delayed/hemoglobinuria
HDFN   Rare, usually mild

Autoanti-e

Common

Comments

Alloanti-e-like antibodies may be made by people with e+ RBCs lacking some
e epitopes. This occurs more frequently in Blacks than in Caucasians[1,2].
Many e variants, in people at risk of immunization against lacking Rhe epitopes,
have been defined with monoclonal anti-e and molecular studies[3,4].
The e antigen *in cis* with C and C^W (e.g., DCC^{W}e and CC^{W}e) is also a partial
antigen. This also applies in the presence of C and C^X.

Reaction of monoclonal anti-e with RBCs expressing e-variant phenotypes[4]

Phenotype	MS16	MS21	MS62/MS63	MS69	MS70
hr^S–, hr^B–, (ceMO)	W	W	0	0	0
hr^S– (ceAR)	+	+	0	W	+
hr^S– (ceEK)	+	+	+	+	+
hr^B– (ce^S) (ceVS, etc)	+	+	+	+	0
ce Cys16 (ce48C)	0	+	W	0	NT
CeRN$^\wedge$	W	0	0	W	0
RHCE*ceJAL	0	0	2+ to 3+	0	NT
ceCF	0	NT	3+	0	NT
ceSL	0	0	3+	NT	NT

$^\wedge$Previously referred to as $\overset{=N}{R}$.

References

[1] Chou, S.T., Westhoff, C.A., 2011. The role of molecular immunohematology in sickle cell disease. Transfus Apher Sci 44, 73–79.

[2] Issitt, P.D., 1991. An invited review: the Rh antigen e, its variants, and some closely related serological observations. Immunohematology 7, 29–36.

[3] Chou, S.T., Westhoff, C.M., 2010. The Rh and RhAG blood group systems. Immunohematology 26, 178–186.

[4] Noizat-Pirenne, F., et al., 2002. Rare RHCE phenotypes in black individuals of Afro-Caribbean origin: identification and transfusion safety. Blood 100, 4223–4231.

f Antigen

Terminology

ISBT symbol (number)	RH6 (004006 or 4.6)
Obsolete names	ce; hr
History	Reported in 1953 and named with the next letter of the alphabet when it was observed that c and e *in cis* were required for its expression.

Occurrence

Caucasians	65%
Blacks	92%
Asians	12%

Expression

Cord RBCs	Expressed
Altered	In some unusual Rh complexes, particularly in those with altered c and/or e expression

Molecular basis associated with f antigen

The f antigen is expressed on the Rhce protein, but the requirements for expression of the antigen are not understood.

Effect of enzymes and chemicals on f antigen on intact RBCs

Ficin/Papain	Resistant
Trypsin	Resistant
α-Chymotrypsin	Resistant
DTT 200 mM	Resistant
Acid	Resistant

In vitro characteristics of alloanti-f

Immunoglobulin class	Most IgG; some IgM
Optimal technique	RT; IAT; enzymes

Clinical significance of alloanti-f

Transfusion reaction	Mild/delayed/hemoglobinuria
HDFN	Mild

Autoanti-f

Yes

Comments

The f antigen is a compound antigen expressed on RBCs with c (**RH4**) and e (**RH5**) on the same protein (Rhce), e.g., on R_1r (DCe/ce), R_0R_0 (Dce/Dce) RBCs. The antigen is not expressed when c and e are on separate Rh proteins, e.g., on R_1R_2 (DCe/DcE) RBCs. The f antigen is expressed on RBCs of some people with the Dc− haplotype.

Anti-f is frequently a component of sera containing anti-c or anti-e, and can be made by people with partial c and e antigens. Anti-f is useful in distinguishing DCE/ce from DCe/cDE. Apparent anti-f in Blacks may be anti-hr^S (see **RH19**). Anti-f frequently fades *in vitro* and *in vivo*.

Ce Antigen

Terminology

ISBT symbol (number)	RH7 (004007 or 4.7)
Obsolete name	rh$_i$
History	Reported in 1958 when it was observed that C and e *in cis* were required for its expression.

Occurrence

Caucasians	68%
Blacks	27%
Asians	92%

Expression

Cord RBCs	Expressed

Molecular basis associated with Ce antigen

The Ce antigen is expressed on the RhCe protein, but the requirements for expression of the antigen are not understood.

Effect of enzymes and chemicals on Ce antigen on intact RBCs

Ficin/Papain	Resistant (markedly enhanced)
Trypsin	Resistant (enhanced)
α-Chymotrypsin	Resistant (enhanced)
DTT 200 mM	Resistant
Acid	Resistant

In vitro characteristics of alloanti-Ce

Immunoglobulin class	IgG more common than IgM
Optimal technique	IAT; enzymes

Clinical significance of alloanti-Ce

Transfusion reaction	Mild/delayed
HDFN	Mild

Comments

Ce is a compound antigen expressed on RBCs with C (**RH2**) and e (**RH5**) on the same protein (RhCe), e.g., on DCe/ce (R_1r) RBCs but not on DCE/ce (R_zr) RBCs.

Anti-Ce is usually found in sera containing anti-C. Apparent anti-Ce in a C+ Black may be anti-hr^B (see **RH31**).

C^W Antigen

Terminology

ISBT symbol (number)	RH8 (004008 or 4.8)
Obsolete names	Willis, rh^w
History	Reported in 1946 and named because of the association with C and "W" from "Willis," the first proband whose RBCs carried the antigen. For years C^W was thought to be antithetical to C. The weak C antigen on C^W+ RBCs is due to an altered expression of C rather than to "cross-reactivity" of anti-C^W.

Occurrence

Caucasians	2%
Blacks	1%
Finns	4%
Latvians	9%

Expression

Cord RBCs	Expressed
Altered	Weaker on DC^W–

See System pages for DC^W– phenotypes and unusual Rh complexes.

Molecular basis associated with C^W antigen[1]

Amino acid	Arg41
Nucleotide	G at bp 122 in exon 1 of *RHCE*
C^W– (wild type)	Gln41 and A at bp 122

Effect of enzymes and chemicals on C^W antigen on intact RBCs

Ficin/Papain	Resistant (markedly enhanced)
Trypsin	Resistant (enhanced)
α-Chymotrypsin	Resistant (enhanced)
DTT 200 mM	Resistant
Acid	Resistant

In vitro characteristics of alloanti-C^W

Immunoglobulin class	IgG and IgM
Optimal technique	RT; IAT; enzymes

Clinical significance of alloanti-C^W

Transfusion reaction	Mild to severe; immediate/delayed
HDFN	Mild to moderate

Comments

Anti-C^W are often naturally-occurring and found in multispecific sera.
Most C^W+ are C+; rare examples are C−. C^W has been associated with D(C)C^We, D(C)C^WE, (C)C^We, (C)C^WE, DC^W− and C^Wce haplotypes.
Alloanti-C can be made by individuals with the C+C^W+ phenotypes.
The e antigen *in cis* with C and C^W (e.g. DCC^{W}e and CC^{W}e) is also a partial antigen.
There is an association between C^W (RH9) and MAR (RH51) antigens.

Reference

[1] Mouro, I., et al., 1995. Molecular basis of the RhC^W (Rh8) and RhC^X (Rh9) blood group specificities. Blood 86, 1196–1201.

C^X Antigen

Terminology

ISBT symbol (number)	RH9 (004009 or 4.9)
Obsolete name	rh^X
History	Reported in 1954 and named because of the association with C and "X," because X was the next letter in the alphabet after W and the antigen had characteristics similar to C^W. C^X was thought to be antithetical to C. The weak C antigen on C^X+ RBCs is due to an altered expression of C rather than to "cross-reactivity" of anti-C^X.

Occurrence

Less than 0.01%; more common in Finns.

Expression

Cord RBCs   Expressed

Molecular basis associated with C^X antigen[1]

Amino acid   Thr36 on RhCe and rarely Rhce
Nucleotide   A at bp 106 in exon 1 of *RHCE*
C^X- (wild type)   Ala36 and G at bp 106

Effect of enzymes and chemicals on C^X antigen on intact RBCs

Ficin/Papain   Resistant (markedly enhanced)
Trypsin   Resistant (enhanced)
α-Chymotrypsin   Resistant (enhanced)
DTT 200 mM   Resistant
Acid   Resistant

In vitro characteristics of alloanti-C^X

Immunoglobulin class   IgG and IgM
Optimal technique   37°C; IAT; enzymes

Clinical significance of alloanti-C^X

Transfusion reaction   No to moderate; immediate/delayed
HDFN   Mild to moderate

Comments

Anti-C^X are often naturally-occurring and found in multispecific sera.
C^X+ are C+ except in the rare haplotype $C^X ce^S$ V–VS+ found in Somalia.
C^X has been associated with D(C)C^Xe, (C)C^Xe, and $C^X ce^S$ haplotypes.
Alloanti-C (and potentially alloanti-e) can be made by individuals with the $C+C^X+$ phenotypes.
There is an association between C^X (RH8) and MAR (RH51) antigens.

Reference

[1] Mouro, I., et al., 1995. Molecular basis of the RhC^W (Rh8) and RhC^X (Rh9) blood group specificities. Blood 86, 1196–1201.

V Antigen

Terminology

ISBT symbol (number)	RH10 (004010 or 4.10)
Obsolete names	ce^s; hr^V
History	Reported in 1955 and named after the first letter of the last name of the proband to make anti-V.

Occurrence

Caucasians	1%
Blacks	30%

Expression

Cord RBCs	Expressed

Molecular basis associated with V antigen[1]

The V antigen is associated with expression of VS antigen. For alleles encoding V see System pages.

Effect of enzymes and chemicals on V antigen on intact RBCs

Ficin/Papain	Resistant
Trypsin	Resistant
α-Chymotrypsin	Presumed resistant
DTT 200 mM	Resistant
Acid	Resistant

In vitro characteristics of alloanti-V

Immunoglobulin class	IgG
Optimal technique	IAT; enzyme

Clinical significance of alloanti-V

Transfusion reaction	Mild/delayed
HDFN	No

Comments

Anti-V frequently occurs in multispecific sera, particularly in sera containing anti-D.

Most V+ RBCs are also VS+ (**RH20**).

Reference

[1] Daniels, G.L., et al., 1998. The VS and V blood group polymorphisms in Africans: a serological and molecular analysis. Transfusion 38, 951–958.

E^W Antigen

Terminology

ISBT symbol (number)	RH11 (004011 or 4.11)
Obsolete name	rh^{W2}
History	Reported in 1955 as the cause of HDFN, and named after the affected family.

Occurrence

Less than 0.01%; more common in people of German ancestry.

Expression

Cord RBCs	Expressed

Molecular basis associated with E^W antigen

Amino acid	Lys167
Nucleotide	A at bp 500 in exon 4 of *RHCE*CE*
E^W– (wild type)	Met167 and T at bp 500

Effect of enzymes and chemicals on E^W antigen on intact RBCs

Ficin/Papain	Resistant (enhanced)
Trypsin	Presumed resistant
α-Chymotrypsin	Presumed resistant
DTT 200 mM	Presumed resistant

In vitro characteristics of alloanti-E^W

Immunoglobulin class	IgG
Optimal technique	IAT; enzymes

Clinical significance of alloanti-E^W

HDFN	Yes

Rh

Comments

E^W has only been found associated with the DcE^W haplotype. The E associated with expression of E^W is a partial antigen (category EI) that is detected by some, but not all anti-E.

Anti-E^W is a rare specificity.

G Antigen

Terminology

ISBT symbol (number)	RH12 (004012 or 4.12)
Obsolete name	rh^G
History	Reported in 1958 when a donor's D–C– RBCs were agglutinated by most anti-CD; given the next available letter in the alphabet.

Occurrence

Caucasians	84%
Blacks	92%
Asians	100%

Expression

Cord RBCs	Expressed
Altered	Weak on r^G and r''^G RBCs

See System pages for unusual Rh complexes

Molecular basis associated with G antigen[1]

Amino acid	Ser103 on Rh proteins expressing C or D
G–	Pro103 usually associated with D– phenotype and rarely with D+ phenotype[2].
Nucleotide	T at bp 307 in exon 2 of *RHD* or *RHCE*C*

See System pages.

Effect of enzymes and chemicals on G antigen on intact RBCs

Ficin/Papain	Resistant (markedly enhanced)
Trypsin	Resistant
α-Chymotrypsin	Resistant
DTT 200 mM	Resistant
Acid	Resistant

In vitro characteristics of alloanti-G

Immunoglobulin class	IgG
Optimal technique	IAT; enzymes

Clinical significance of alloanti-G

Transfusion reaction	No to severe/delayed
HDFN	No to severe

Comments

Anti-G is found as a component in sera from rr (ce/ce) people with anti-D (and/or anti-C), D+G– people with anti-C, and some DIIIb people with anti-D.

References

[1] Faas, B.H.W., et al., 1996. Involvement of Ser103 of the Rh polypeptides in G epitope formation. Transfusion 36, 506–511.

[2] Faas, B.H.W., et al., 2001. Partial expression of RHc on the RHD polypeptide. Transfusion 41, 1136–1142.

Hr_0 Antigen

Terminology

ISBT symbol (number)	RH17 (004017 or 4.17)
History	Anti-Hr_0 reported in 1958 and allocated Rh17 in 1962; defined by absorption/elution studies using sera from D−− probands. Hr_0 was considered to be a high-prevalence antigen expressed by all common Rh haplotypes.

Occurrence

All populations	100%

Expression

Cord RBCs	Expressed

Molecular basis of Hr_0 (Rh17)

See System pages.

Effect of enzymes and chemicals on Hr_0 (Rh17) antigen on intact RBCs

Ficin/Papain	Resistant (markedly enhanced)
Trypsin	Resistant (markedly enhanced)

α-Chymotrypsin	Resistant (markedly enhanced)
DTT 200 mM	Resistant
Acid	Resistant

In vitro characteristics of alloanti-Hr$_0$ (Rh17)

Immunoglobulin class	IgG
Optimal technique	IAT; enzymes

Clinical significance of alloanti-Hr$_0$ (-Rh17)

Transfusion reaction	No to severe
HDFN	No to severe

Autoanti-Hr$_0$ (Rh17)

Antibody with broad Rh specificity in patients with AIHA previously also known as anti-pdl.

Comments

Selected anti-Rh17 may be used to distinguish Rh_{mod} from Rh_{null} phenotypes. Anti-Rh17 is made by individuals with the following phenotypes: D$--$, D$\cdot\cdot$, Dc$-$, DC$^W-$.

Hr$_0$ (Rh17) appears to be composed of several epitopes, some of which may be lacking on RBCs with unusual Rh haplotypes, including those with partial C or c and/or e expression. People with phenotypes that have altered C or c and/or e can make an alloantibody that is directed at the conventional RhCE protein and initially appears to be anti-Rh17. Such antibodies, upon further testing, can be shown to have a precise specificity [see CEST (**RH57**), CELO (**RH58**), CEAG (**RH59**)].

Hr Antigen

Terminology

ISBT symbol (number)	RH18 (004018 or 4.18)
Obsolete names	Hr^S; Shabalala
History	Reported in 1960; two antibodies were distinguished in the serum of Mrs. Shabalala, the Bantu proband. One of the antibodies, anti-Hr, was removed by absorption with R_2R_2 (DcE/DcE) RBCs leaving anti-hr^S.

Occurrence

Most populations	100%

Hr– only found in Blacks.

Molecular basis associated with Hr antigen[1]

See Rh System pages.

Clinical significance of alloanti-Hr

Transfusion reaction	No to fatal
HDFN	Moderate[2]

Comments

Hr antigen is present on all RBCs except hr^S–, Rh$_{null}$, and RhCE-depleted phenotypes.

Anti-Hr is made by hr^S– people, and may be part of the immune response of people whose RBCs have Rh-depleted phenotypes. Several alleles encode the Hr– phenotype; see System pages.

References

[1] Noizat-Pirenne, F., et al., 2002. Rare RHCE phenotypes in black individuals of Afro-Caribbean origin: identification and transfusion safety. Blood 100, 4223–4231.

[2] Moores, P., 1994. Rh18 and hrS blood groups and antibodies. Vox Sang 66, 225–230.

hr^S Antigen

Terminology

ISBT symbol (number)	RH19 (004019 or 4.19)
Obsolete name	Shabalala
History	Reported in 1960. The name "hr" was from Wiener's terminology for e, and superscript "S" was from Shabalala, the e+ proband who made an apparent alloanti-e. See Rh18 (Hr).

Occurrence

All populations	98% (R$_2$R$_2$ RBCs lack hr^S)

RBCs of approximately 1% of Blacks are hr^S– as 1% of e+ Bantu people are hr^S–.

Expression

Cord RBCs	Expressed
Altered	Reduced on DC^{X}e and phenotypes with altered e antigens

Molecular basis associated with hr^S antigen[1]

See Rh System pages.

Rh

Effect of enzymes and chemicals on hr^S antigen on intact RBCs

Ficin/Papain	Resistant (markedly enhanced)
Trypsin	Resistant
α-Chymotrypsin	Resistant
DTT 200 mM	Resistant
Acid	Resistant

In vitro characteristics of alloanti-hr^S

Immunoglobulin class	IgG
Optimal technique	IAT; enzymes

Clinical significance of alloanti-hr^S

Transfusion reaction	No to fatal if with anti-Rh18
HDFN	Little evidence to indicate that anti-hr^S in the absence of anti-Hr has caused HDFN

Comments

Anti-hr^S reacts preferentially with haplotypes containing ce, and on initial test-ing may be mistaken for anti-f (see **RH6**). Antibodies made by hr^S– people are not necessarily anti-hr^S and, unless tested with appropriate rare e variant cells, are more correctly called anti-e-like.

cE haplotypes do not express hr^{S2,3}.

References

[1] Noizat-Pirenne, F., et al., 2002. Rare RHCE phenotypes in black individuals of Afro-Caribbean origin: identification and transfusion safety. Blood 100, 4223–4231.

[2] Issitt, P.D., 1991. An invited review: the Rh antigen e, its variants, and some closely related sero-logical observations. Immunohematology 7, 29–36.

[3] Moores, P., 1994. Rh18 and hrS blood groups and antibodies. Vox Sang 66, 225–230.

VS Antigen

Terminology

ISBT symbol (number)	RH20 (004020 or 4.20)
Obsolete name	e^s
History	Reported in 1960 and named after the initials of the first lady to make the antibody; the initial of her first name was used because of the association with the V antigen.

Occurrence

Blacks	26% to 40%
Other populations	<0.01%

Expression

Cord RBCs	Expressed
Altered	D(C)(e^S) FPTT+[1]; DC^{W}e/Dce^S (1 example, Inkelberger); Dce^S/DCe (1 example, Manday), and see System pages

Molecular basis associated with VS antigen[2]

Amino acid	Val245 in Rhce (several different alleles)
Nucleotide	G at bp 733 in exon 5 of *RHCE*
VS– (wild type)	Leu245 and C at bp 733

For RHCE*ceVS alleles expressing VS with or without V, see System pages.

Effect of enzymes and chemicals on VS antigen on intact RBCs

Ficin/Papain	Resistant
Trypsin	Resistant
α-Chymotrypsin	Presumed resistant
DTT 200 mM	Presumed resistant
Acid	Resistant

In vitro characteristics of alloanti-VS

Immunoglobulin class	IgG
Optimal technique	IAT; enzymes

Clinical significance of alloanti-VS

Transfusion reaction	Mild/delayed
HDFN	Positive DAT; no clinical HDFN

Comments

Anti-VS is often a component of sera with other specificities. Anti-VS are heterogeneous and may be naturally-occurring.

The majority of V+ RBCs are VS+ (**RH:20**). The majority of apparent hr^B– (**RH:–31**) RBCs are VS+[3].

References

[1] Bizot, M., et al., 1988. An antiserum identifying a red cell determinant expressed by Rh:33 and by some "new" depressed Rh phenotypes. Transfusion 28, 342–345.

[2] Daniels, G.L., et al., 1998. The VS and V blood group polymorphisms in Africans: a serological and molecular analysis. Transfusion 38, 951–958.

[3] Pham, B.N., et al., 2009. Heterogeneous molecular background of the weak C, VS+, hr B–, Hr B– phenotype in black persons. Transfusion 49, 495–504.

C^G Antigen

Terminology

ISBT symbol (number)	RH21 (004021 or 4.21)
History	Reported in 1961; considered to be the weak C antigen found on r^{G}r^G and r^{G}r RBCs. C^G is also made by all cells expressing C.

Occurrence

Caucasians	68%

Comments

There is no monospecific anti-C^G, but a minority of anti-C are anti-CC^G. Some consider that the C made by r$'^S$ is actually C^{G1}.

Reference

[1] Issitt, P.D., Anstee, D.J., 1998. Applied Blood Group Serology, fourth ed. Montgomery Scientific Publications, Durham, N.C.

CE Antigen

Terminology

ISBT symbol (number)	RH22 (004022 or 4.22)
Obsolete names	Jarvis
History	Reported in 1962 and named when it was observed that C and E *in cis* were required for its expression.

Occurrence

Less than 1% in most populations; 2% in Asians.

Expression

Cord RBCs	Expressed

Molecular basis associated with CE antigen

The CE antigen is expressed on the RhCE protein but the requirements for expression of CE are not understood.

Effect of enzymes and chemicals on CE antigen on intact RBCs

Ficin/Papain	Resistant
Trypsin	Presumed resistant
α-Chymotrypsin	Presumed resistant
DTT 200 mM	Presumed resistant

In vitro characteristics of alloanti-CE

Optimal technique   RT [Original anti-CE (Jarvis)]; 37°C

Clinical significance of alloanti-CE

No data are available because only two examples have been reported.

Comments

The two reported anti-CE appeared to be naturally-occurring, and were in sera that also contained anti-C.

This compound antigen is expressed on RBCs with C and E on the same protein (RhCE), e.g., on DCE (R_z) and CE (r^y) haplotypes.

D^W Antigen

Terminology

ISBT symbol (number)   RH23 (004023 or 4.23)
Obsolete name   Weil
History   Reported in 1962, and named after the first proband whose RBCs had this low prevalence antigen; shown to be an Rh antigen in 1965 and was associated with DVa.

Occurrence

All populations   <0.01%

Expression

Cord RBCs   Expressed

Molecular basis associated with D^W antigen

Associated with the partial D antigen encoded by several types of *RHD*DV*, in which all or part of exon 5 of *RHD* is replaced by the same exon from *RHCE*[1]. See System pages.

Effect of enzymes and chemicals on D^W antigen on intact RBCs

Ficin/Papain   Resistant (markedly enhanced)
Trypsin   Presumed resistant
α-Chymotrypsin   Presumed resistant
DTT 200 mM   Presumed resistant

In vitro characteristics of alloanti-D^W

Immunoglobulin class	IgG
Optimal technique	IAT; enzymes

Clinical significance of alloanti-D^W

HDFN	Moderate

Comments

Sera containing anti-D^W often contain anti-E.

Anti-D^W (anti-Rh23) is a rare specificity and has been found in multispecific sera. Some examples contain anti-Rh32, and these specificities are not separable. The molecular basis of the Rh haplotype in a person (NR) with D^W–, Rh32– RBCs that were agglutinated by one example of anti-Rh23/Rh32 is given in the system pages.

Reference

[1] Rouillac, C., et al., 1995. Transcript analysis of D category phenotypes predicts hybrid Rh D-CE-D proteins associated with alteration of D epitopes. Blood 85, 2937–2944.

Rh26 (c-like) Antigen

Terminology

ISBT symbol (number)	RH26 (004026 or 4.26)
Obsolete name	Deal
History	This variant of c was identified in 1964 when the serum of Mrs. Deal, considered to contain a potent anti-c, did not react with some c+ RBCs.

Occurrence

Expressed on the majority of c-positive RBCs.
The c+Rh26– phenotype has been found in Italians and Dutch.

Antithetical antigen

LOCR (**RH55**)

Molecular basis associated with Rh26 antigen[1]

Amino acid	Gly96 on Rhce
Nucleotide	G at bp 286 in exon 2 of *RHCE*ce*

Effect of enzymes and chemicals on Rh26 antigen on intact RBCs

Ficin/Papain	Resistant (enhanced)
Trypsin	Presumed resistant
α-Chymotrypsin	Presumed resistant
DTT 200 mM	Presumed resistant

In vitro characteristics of alloanti-Rh26

Immunoglobulin class	IgG
Optimal technique	37°C; IAT; enzymes

Clinical significance of alloanti-Rh26

No data are available.

Comments

One c– Rh26+ sample has been described. Rh26– RBCs have weak expression of f antigen.

Reference
[1] Faas, B.H.W., et al., 1997. Involvement of Gly96 in the formation of the Rh26 epitope. Transfusion 37, 1123–1130.

cE Antigen

Terminology

ISBT symbol (number)	RH27 (004027 or 4.27)
History	Reported in 1965 and named when it was observed that c and E *in cis* were required for its expression.

Occurrence

Caucasians	28%
Blacks	22%
Asians	38%

Molecular basis associated with cE antigen

The cE antigen is expressed on the RhcE protein, but the requirements for expression of cE are not understood.

Effect of enzymes and chemicals on cE antigen on intact RBCs

Ficin/Papain	Resistant
Trypsin	Presumed resistant
α-Chymotrypsin	Presumed resistant
DTT 200 mM	Presumed resistant

In vitro characteristics of alloanti-cE

Immunoglobulin class   IgG
Optimal technique   IAT; enzymes
Complement binding   Yes (one example)

Comments

Few examples of anti-cE have been reported. Expressed on RBCs having c [**RH4**] and E [**RH3**] antigens on the same protein (RhcE) e.g., R_2r (DcE/ce), r″r (cE/ce). The antigen is not expressed when c and E occur on separate haplotypes (in *trans*), e.g., R_zr (DCE/ce).

hr^H Antigen

Terminology

ISBT symbol (number)   RH28 (004028 or 4.28)
History   Reported in 1964. The antigen hr^H, primarily studied among South African Blacks, may be present on some RBCs that type V–VS+. hr^H has a complex relationship with VS (**RH20**).

Occurrence

All populations   <0.01%.

Rh29 Antigen

Terminology

ISBT symbol (number)   RH29 (004029 or 4.29)
Obsolete name   Total Rh
History   Reported in 1961 and given the next available number. The only Rh29– RBCs are Rh$_{null}$, which were originally called − − −/− − − when the first proband, an Australian Aboriginal woman, was identified.

Occurrence

All populations   100%

Expression

Cord RBCs   Expressed

Molecular basis of Rh29 antigen

For molecular basis of Rh29– (Rh$_{null}$), see RH and RHAG System pages.

Effect of enzymes and chemicals on Rh29 antigen on intact RBCs

Ficin/Papain	Resistant (markedly enhanced)
Trypsin	Resistant (markedly enhanced)
α-Chymotrypsin	Resistant (enhanced)
DTT 200 mM	Presumed resistant
Acid	Resistant

In vitro characteristics of alloanti-Rh29

Immunoglobulin class	IgG and IgM
Optimal technique	37°C; IAT; enzymes

Clinical significance of alloanti-Rh29

Transfusion reaction	No data available but potentially capable
HDFN	No to severe

Autoanti-Rh29

Antibody in AIHA with broad Rh specificity may be anti-Rh29 (previously also known as anti-dl).

Comments

Anti-Rh29 is the immune response of some Rh_{null} individuals (both amorph and regulator type). Some anti-Rh29 react with Rh_{mod} cells.

Go^a Antigen

Terminology

ISBT symbol (number)	RH30 (004030 or 4.30)
Obsolete names	Gonzales; D^{Cor}
History	Named after Mrs. Gonzales, the first maker of anti-Go^a. Reported briefly in 1962, and more extensively in 1967 when Go^a was shown to be an Rh antigen. In 1968, Go^a was confirmed to be a marker for D category IV (DIVa). Before partial D phenotypes were categorized, DIVa was called D^{Cor}.

Occurrence

Blacks	2%

Expression

Cord RBCs	Expressed

Rh

Molecular basis associated with Go^a antigen

Go^a is associated with the partial D antigen of category DIVa. See System pages.

Effect of enzymes and chemicals on Go^a antigen on intact RBCs

Ficin/Papain	Resistant (markedly enhanced)
Trypsin	Resistant
α-Chymotrypsin	Presumed resistant
DTT 200 mM	Presumed resistant
Acid	Resistant

In vitro characteristics of alloanti-Go^a

Immunoglobulin class	IgG
Optimal technique	37°C; IAT; enzymes

Clinical significance of alloanti-Go^a

Transfusion reaction	Moderate/delayed
HDFN	Mild to severe

Comments

Go^a is also expressed on RBCs with the rare DIVa(C)− Rh33+Riv+FPTT+ complex.

Anti-Go^a may be immune, but are often in multispecific sera, frequently with anti-Rh32 (see **RH32**) and/or anti-Evans (**RH37**); these Rh specificities are not separable by absorption/elution.

hr^B Antigen

Terminology

ISBT symbol (number)	RH31 (004031 or 4.31)
Obsolete name	Bastiaan
History	Reported in 1972. Named "hr" from Wiener's terminology for e and "B" from Bastiaan, the first antibody producer. See Hr^B (**RH34**).

Occurrence

All populations	98% [R_2R_2 (DcE/DcE) RBCs lack hr^B]
Blacks	97%, which includes numerous partial e

Expression

Cord RBCs	Expressed
Altered	Reduced on phenotypes with weak e antigens. See System pages

Molecular basis associated with hr^B antigen[1]

See System pages.

Effect of enzymes and chemicals on hr^B antigen on intact RBCs

Ficin/Papain	Resistant (markedly enhanced)
Trypsin	Resistant
α-Chymotrypsin	Presumed resistant
DTT 200 mM	Presumed resistant
Acid	Resistant

In vitro characteristics of alloanti-hr^B

Immunoglobulin class	IgG
Optimal technique	37°C; IAT; enzymes

Clinical significance of alloanti-hr^B

Transfusion reaction	Generally not clinically significant, but precise information is limited because anti-e-like antibodies are often incorrectly called anti-hr^B. However, the immune response of some hr^B– people may broaden to the clinically significant anti-Hr^B (**RH34**).
HDFN	Positive DAT; no clinical HDFN

Autoanti-hr^B

Yes, rare (often with transient suppression of antigen). Investigation of DNA from patients with apparent autoanti-hr^B has revealed the presence of partial e phenotypes, and suggests that some autoantibodies are alloantibodies.

Comments

cE haplotypes do not express hr^B. The majority of apparent e+ hr^B– RBCs are VS+[2].
Anti-hr^B can be mistaken for anti-Ce (see **RH7**).
The molecular basis of the hr^B– phenotype is heterogeneous, as are the anti-hr^B and e-like antibodies made by people with the hr^B– phenotype[1,3]. The fine specificity can often be determined by RH DNA typing of the patient, and testing the patient's plasma against RBCs characterized at the DNA level. Due to

limited availability of correctly characterized antibodies and RBC samples, prior to the use of RH DNA analysis many samples could only be partially characterized, and were (appropriately) labeled as anti-e-like.

References

[1] Pham, B.N., et al., 2009a. Heterogeneous molecular background of the weak C, VS+, hr B−, Hr B− phenotype in black persons. Transfusion 49, 495–504.

[2] Beal, C.L., et al., 1996. The *r'* gene is overrepresented in hr^B-negative individuals. Immunohematology 11, 74–77.

[3] Pham, B.N., et al., 2009b. Anti-Hr^B and anti-hr^b revisited. Transfusion 49, 2400–2405.

Rh32 Antigen

Terminology

ISBT symbol (number)	RH32 (004032 or 4.32)
Obsolete names	$\overset{=\text{N}}{R}$

History	Reported in 1971 after several years of investigation and was assigned the next Rh number in 1972.
	Incorrectly called $\overset{=\text{N}}{R}$, which is the name of the original (1960) haplotype with weak C and e antigens later shown to express Rh32. $\overset{=\text{N}}{R}$ is now referred to as R^N.

Occurrence

Blacks	1% ($\overset{=\text{N}}{R}$ phenotype)
Caucasians and Japanese	Rare (associated with the DBT partial D phenotype)

Antithetical antigen

Sec (**RH46**)

Expression

Cord RBCs	Expressed
Altered	May be slightly weaker on DBT phenotype RBCs and other rare variants

Molecular basis associated with Rh32 antigen[1,2]

R^N (formerly $\overset{=\text{N}}{R}$) phenotype: *RHCE*ceRN* hybrid in which exon 4 of *RHCE* is replaced by the corresponding exon of *RHD* [with or without nt 445C>A in exon 3 (Thr152Asn)].

Partial D phenotype *RHD*DBT* hybrid in which either exons 5 to 7 or exons 5 to 9 of *RHD* are replaced by the corresponding exons of *RHCE*. See System pages.

Effect of enzymes and chemicals on Rh32 antigen on intact RBCs

Ficin/Papain	Resistant (markedly enhanced)
Trypsin	Resistant
α-Chymotrypsin	Resistant
DTT 200 mM	Presumed resistant
Acid	Resistant

In vitro characteristics of alloanti-Rh32

Immunoglobulin class	IgG
Optimal technique	37°C; IAT; enzymes

Clinical significance of alloanti-Rh32

Transfusion reaction	None reported
HDFN	Mild to severe

Comments

RHCE*ceRN encodes Rh32 in combination with weakened expression of C (**RH2**) and e (**RH5**) antigens, and may be associated with normal or elevated expression of D antigen (**RH1**). It may be necessary to use sensitive techniques to detect the C antigen on some RBCs.

The RBCs of one proband with the DBT phenotype had weakened expression of C and e; another proband had weakened expression of C only.

Anti-Rh32 may be immune, but are often naturally-occurring in multispecific sera. Anti-Rh32 cannot be separated from anti-Go^a (see **RH30**) or anti-Evans (see **RH37**) by absorption/elution of sera containing these antibodies.

References

[1] Beckers, E.A.M., et al., 1996. The genetic basis of a new partial D antigen: D^{DBT}. Br J Haematol 93, 720–727.

[2] Rouillac, C., et al., 1996. Molecular basis of the altered antigenic expression of RhD in weak D (D^u) and RhC/e in R^N phenotypes. Blood 87, 4853–4861.

Rh33 Antigen

Terminology

ISBT symbol (number)	RH33 (004033 or 4.33)
Obsolete names	Har; R$_0$^{Har}; D^{Har}

History   Reported in 1971 and given the next Rh number.
                                 Although the complex expressing Rh33 was first
                                 detected on RBCs from a German donor, the
                                 complex was named R_0^{Har} after the name of an
                                 English donor with an informative family.

Occurrence

Less than 0.01%; Rh33 is more common in people of German ancestry.

Expression

Cord RBCs   Presumed expressed
Altered R_1^{Lisa1}

Molecular basis associated with Rh33 antigen

Encoded by *RHCE*ceHAR*, an RHCE*ce allele in which exon 5 is replaced
by exon 5 of *RHD*[2]. *RHCE*CeVA*, also a hybrid gene with exon 5 originating
from *RHD*, encodes Rh33 and weak C and e antigens. *RHCE*CeVA* may be
the allele encoding R_1^{Lisa3}.
See System pages.

Effect of enzymes and chemicals on Rh33 antigen on intact RBCs

Ficin/Papain   Resistant (markedly enhanced)
Trypsin   Presumed resistant
α-Chymotrypsin   Presumed resistant
DTT 200 mM   Presumed resistant

In vitro characteristics of alloanti-Rh33

Immunoglobulin class   IgM
Optimal technique   RT; enzymes

Clinical significance of alloanti-Rh33

No data are available.

Comments

*RHCE*ceHAR* encodes a partial D antigen, normal c (**RH4**), weak e (**RH5**),
weak f (**RH6**), and weak Hr_0 (**RH17**) antigens; it does not encode C (**RH2**),
E (**RH3**), G (**RH12**), hr^S (**RH19**) or Hr (**RH18**) antigens.
Rh33 is also expressed by the rare complexes DIVa(C)−, R_0^{JOH} and R_1^{Lisa}.
All Rh33+ RBCs also express the low prevalence antigen FPTT (**RH50**).

Anti-Rh33 is a rare specificity. Two examples were in serum also containing anti-D.

References

[1] Moores, P., et al., 1991. Rh33 in two of three German siblings with D+ C+ c+ E− e+red cells. Transfusion 31, 759–761.

[2] Beckers, E.A.M., et al., 1996. The R_0^{Har}Rh:33 phenotype results from substitution of exon 5 of the *RHCE* gene by the corresponding exon of the *RHD* gene. Br J Haematol 92, 751–757.

[3] Noizat-Pirenne, F., et al., 2002. Molecular background of *D(C)(e)* haplotypes within the white population. Transfusion 42, 627–633.

Hr^B Antigen

Terminology

ISBT symbol (number)	RH34 (004034 or 4.34)
Obsolete names	Bas; Baas; Bastiaan; Rh34
History	Reported in 1972. Anti-Hr^B initially described the total immune response of Mrs. Bastiaan (hence "B" in the name), a South African. Later, absorptions showed her serum contained two specificities: anti-hr^B (see **RH31**) and an antibody reacting with RBCs of all common phenotypes that was called anti-Hr^{B1}.

Occurrence

All populations	100%

Expression

Cord RBCs	Expressed

Molecular basis associated with Hr^B antigen

See Rh System pages for molecular basis of Hr^B− phenotypes.

Effect of enzymes and chemicals on Hr^B antigen on intact RBCs

Ficin/Papain	Resistant (markedly enhanced)
Trypsin	Presumed resistant
α-Chymotrypsin	Presumed resistant
DTT 200 mM	Presumed resistant

In vitro characteristics of alloanti-Hr^B

Immunoglobulin class	IgG
Optimal technique	IAT; enzymes

Clinical significance of alloanti-Hr^B

Transfusion reaction	No data available, presumed to be significant because of similarity to anti-RH18
HDFN	Positive DAT but no clinical HDFN[1]

Comments

Weak examples of anti-Hr^B resemble anti-C (see **RH3**) in that C+ RBCs give the strongest reactions; c+ RBCs give intermediate strength reactions; and DcE/DcE (R$_2$R$_2$) cells give the weakest reactions. Several alleles encode the Hr^B– phenotype; see System pages.

There was debate as to whether anti-Hr^B and anti-hr^B were separate specificities or two aspects of a single specificity. Anti-Hr^B and anti-hr^B are indeed separate specificities[2,3].

References

[1] Moores, P., Smart, E., 1991. Serology and genetics of the red blood cell factor Rh34. Vox Sang 61, 122–129.

[2] Pham, B.N., et al., 2009a. Anti-Hr^B and anti-hr^b revisited. Transfusion 49, 2400–2405.

[3] Pham, B.N., et al., 2009b. Heterogeneous molecular background of the weak C, VS+, hr B−, Hr B- phenotype in black persons. Transfusion 49, 495–504.

Rh35 Antigen

Terminology

ISBT symbol (number)	RH35 (004035 or 4.35)
Obsolete name	1114
History	Reported in 1971. Rh35 is produced by an Rh complex that produces weak C and e antigens and normal D antigen.

Occurrence

Less than 0.01%; Rh35 was originally found in people of Danish ancestry.

Expression

Cord RBCs	Presumed expressed

Molecular basis associated with Rh35 antigen

For the molecular basis of a phenotype with weak C and e expression (CeMA), which may be Rh35+, see System pages.

Effect of enzymes and chemicals on Rh35 antigen on intact RBCs

Ficin/Papain	Resistant (markedly enhanced)
Trypsin	Presumed resistant
α-Chymotrypsin	Presumed resistant
DTT 200 mM	Presumed resistant

In vitro characteristics of alloanti-Rh35

Immunoglobulin class	IgG
Optimal technique	Enzymes

Clinical significance of alloanti-Rh35

No data available because only one example of the antibody has been reported.

Be^a Antigen

Terminology

ISBT symbol (number)	RH36 (004036 or 4.36)
Obsolete name	Berrens
History	Reported in 1953, and named after the family in which HDFN occurred. Be^a is produced by a complex that produces weak c, e, and f antigens, and no D antigen. Family studies in 1974 confirmed it as an Rh antigen.

Occurrence

All populations   <0.1%.
Propositi were of German/Polish extraction from the Baltic region.

Expression

Cord RBCs	Expressed

Molecular basis associated with Be^a antigen[1]

Amino acid	Arg221 in Rhce
Nucleotide	G at bp 662 in exon 5 of *RHCE*ce*
Be(a–) (wild type)	Pro221 and C at bp 662

Effect of enzymes and chemicals on Be^a antigen on intact RBCs

Ficin/Papain	Resistant (markedly enhanced)
Trypsin	Presumed resistant
α-Chymotrypsin	Presumed resistant
DTT 200 mM	Presumed resistant
Acid	Resistant

In vitro characteristics of alloanti-Be^a

Immunoglobulin class	IgG
Optimal technique	37°C; IAT; enzymes

Clinical significance of alloanti-Be^a

Transfusion reaction	None reported
HDFN	Moderate to severe

Comments

Anti-Be^a is immune. Be^a appears to be highly immunogenic: the primary stimulus for production of anti-Be^a in 2 non-transfused women occurred during their first pregnancy and in each case the child of the second pregnancy had severe HDFN[1].

Reference

[1] Hue-Roye, K., et al., 2010. The low prevalence Rh antigen Be^a (Rh36) is associated with RHCE*ce 662C>G in exon 5, which is predicted to encode Rhce 221Arg. Vox Sang 98, e263–e268.

Evans Antigen

Terminology

ISBT symbol (number)	RH37 (004037 or 4.37)
History	Evans, identified in 1968, was named after the family in which HDFN occurred. Evans segregated with a D− − like complex (D··) in the family of the second Evans+ proband. Family studies, reported in 1978, confirmed Evans as an Rh antigen.

Occurrence

Less than 0.01%; may be more common in Welsh and Scots.

Expression

Cord RBCs	Expressed
Altered	Weak on DIVb RBCs

Molecular basis associated with Evans antigen[1,2]

Dav	*RHD(1–6)-RHCE(7–10)//RHD*
JD	*RHD(1–5 and part 6)-RHCE(part 6 and 6–10)// RHCE(1)-RHD(2–10)*
AT	*RHCE(1)-RHD(2–6)-RHCE(7–10)//RHD*
DIVb	*RHCE//RHD(1–6 and part of 7)-RHCE(part of 7–9)-RHD10*

RhD-CE-D hybrids with different proportions of RhCE into RhD. See System pages.

Effect of enzymes and chemicals on Evans antigen on intact RBCs

Ficin/Papain	Resistant (markedly enhanced)
Trypsin	Presumed resistant
α-Chymotrypsin	Presumed resistant
DTT 200 mM	Presumed resistant
Acid	Presumed resistant

In vitro characteristics of alloanti-Evans

Immunoglobulin class	IgM less common than IgG
Optimal technique	37°C; IAT; enzymes

Clinical significance of alloanti-Evans

Transfusion reaction	None reported
HDFN	Mild and moderate

Comments

The Rh complex D·· produces Evans antigen, elevated expression of D, normal expression of G, the high prevalence antigens Rh29 and Dav; C, c, E, and e antigens are not produced. However, a preliminary study suggested that RBCs from JD express a minute amount of e[3].

Anti-Evans may be naturally-occurring, and is often found in multispecific sera. Anti-Evans cannot be separated from anti-Go[a] (see **RH30**) or anti-Rh32 (see **RH32**) by absorption/elution of sera containing these antibodies.

References

[1] Avent, N.D., Reid, M.E., 2000. The Rh blood group system: a review. Blood 95, 375–387.

[2] Huang, C.-H., et al., 2000. Molecular biology and genetics of the Rh blood group system. Semin Hematol 37, 150–165.

[3] Lomas-Francis, C., et al., 2011. Surprising findings with RBCs expressing the low prevalence RH antigen Evans [abstract]. Transfusion 51 (Suppl.), 35A–36A.

Rh39 Antigen

Terminology

ISBT symbol (number)	RH39 (004039 or 4.39)
Obsolete name	C-like
History	Reported in 1979. Anti-Rh39 reacts more strongly with C+ than C– RBCs, and can be absorbed to exhaustion by all C+ and C– RBCs with common and uncommon Rh phenotypes except Rh_{null}.

Occurrence

All populations	100%

Autoanti-Rh39

Yes, always; made by some C– people.

Comments

One patient with this "mimicking" anti-C antibody proceeded to make alloanti-C.

Tar Antigen

Terminology

ISBT symbol (number)	RH40 (004040 or 4.40)
Obsolete name	Targett
History	Reported in 1975, and named after the proband whose RBCs expressed the antigen. When family studies in 1979 showed Tar to be an Rh antigen, it was awarded an Rh number. In 1986, Tar was established as a marker for the DVII partial D antigen.

Occurrence

All populations	<0.01%.

Expression

Cord RBCs   Expressed

Molecular basis associated with Tar antigen[1]

Amino acid   Pro110
Nucleotide   C at bp 329 in exon 2 of *RHD*
Tar− (wild type RhD)   Leu110 and T at bp 329

Effect of enzymes and chemicals on Tar antigen on intact RBCs

Ficin/Papain   Resistant (enhanced)
Trypsin   Presumed resistant
α-Chymotrypsin   Presumed resistant
DTT 200 mM   Resistant

In vitro characteristics of alloanti-Tar

Immunoglobulin class   IgG
Optimal technique   37°C; IAT; enzymes

Clinical significance of alloanti-Tar

HDFN   Moderate

Comments

In addition to the association with DVII *in cis* to Ce, Tar is expressed on a variant RhD protein that also expresses weak c[2]. Tar also was found on a D−− like complex, which produced weaker than usual D antigen.

Anti-Tar is a rare specificity; the antibody has been produced through pregnancy and has been found without known stimulus.

References

[1] Rouillac, C., et al., 1995. Leu110Pro substitution in the RhD polypeptide is responsible for the D^VII category blood group phenotype. Am J Hematol 49, 87–88.

[2] Faas, B.H.W., et al., 2001. Partial expression of RHc on the RHD polypeptide. Transfusion 41, 1136–1142.

Rh41 Antigen

Terminology

ISBT symbol (number)   RH41 (004041 or 4.41)
Obsolete name   Ce-like

| History | Reported in 1981 and given the next Rh number in 1990. The only example of anti-Rh41 reacted with RBCs that have C and e in the same haplotype. However, unlike anti-Ce, anti-Rh41 reacts with r'^S (C)ce^S RBCs, and does not react with C^W and e *in cis*[1]. |

Occurrence

| Caucasians | 70% |

Expression

| Cord RBCs | Presumed expressed |

Reference

[1] Svoboda, R.K., et al., 1981. Anti-Rh41, a new Rh antibody found in association with an abnormal expression of chromosome 1 genetic markers. Transfusion 21, 150–156.

Rh42 Antigen

Terminology

ISBT symbol (number)	RH42 (004042 or 4.42)
Obsolete names	Ce^S; Cce^S; rh^S; Thornton
History	Reported in 1980. It is a marker for the Cce^S V– VS+ haplotype.

Occurrence

| Caucasians | <0.1% |
| Blacks | 2% |

Expression

| Cord RBCs | Expressed |

Molecular basis associated with Rh42 antigen

Encoded by the RHD*DIIIa-CE(4–7)-D hybrid allele that encodes the type 1 but not the type 2 (C)ce^S (r$'^S$) haplotype[1]. See System pages.

Effect of enzymes and chemicals on Rh42 antigen on intact RBCs

| Ficin/Papain | Resistant (enhanced) |

In vitro characteristics of alloanti-Rh42

Immunoglobulin class	IgG
Optimal technique	37°C; IAT; enzymes

Clinical significance of alloanti-Rh42

Transfusion reaction	None reported
HDFN	Moderate

Comments

At least two examples of anti-Rh42 have been reported.

Reference

[1] Pham, B.N., et al., 2009. Heterogeneous molecular background of the weak C, VS+, hr B−, Hr B− phenotype in black persons. Transfusion 49, 495–504.

Crawford Antigen

Terminology

ISBT symbol (number)	RH43 (004043 or 4.43)
History	Reported in 1980, the only example of anti-Crawford was found in a reagent anti-D.

Occurrence

Blacks	0.1%

Expression

Cord RBCs	Expressed

Antithetical antigen

CELO (**RH58**)

Molecular basis associated with Crawford antigen[1]

Amino acids	16Cys, 233Glu, 245Val in Rhce
Nucleotides	C at bp 48, G at bp 697, and G at bp 733 in *RHCE*ce*

See System pages.

Effect of enzymes and chemicals on Crawford antigen on intact RBCs

Ficin/Papain   Resistant (enhanced)

In vitro characteristics of alloanti-Crawford

Immunoglobulin class   IgG
Optimal technique   37°C; IAT; enzymes

Comment

Crawford is encoded by an RHCE allele (*RHCE*ceCF*) that also encodes some D-specific amino acids. These D-specific amino acids are recognized by several potent MAb anti-D and (D−) Crawford+ RBCs have been errone-ously typed as D+.

Reference

[1] Flegel, W.A., et al., 2006. The RHCE allele ceCF: the molecular basis of Crawford (RH43). Transfusion 46, 1334–1342.

Nou Antigen

Terminology

ISBT symbol (number)   RH44 (004044 or 4.44)
History   The antigen was reported in 1969 and named after Mme Nou, from the Ivory Coast, who was homozygous for *DIVa(C)*−. Anti-Nou, reported in 1981, is a component of some anti-Hr_0 (see **RH17**) sera and can be separated by adsorption/elution with DIVa(C)−/DIVa(C)− cells; the antibody does not react with Rh_{null}, D− −, D··, DC^W− or Dc− cells.

Occurrence

All populations   100%

Expression

Cord RBCs   Expressed

Effect of enzymes and chemicals on Nou antigen on intact RBCs

Ficin/Papain   Resistant (enhanced)

Riv Antigen

Terminology

ISBT symbol (number) RH45 (004045 or 4.45)
History   Reported in 1983 and named for the Puerto Rican
                      family in which the antigen and antibody were
                      identified.

Occurrence

Six Riv+ probands are known.

Expression

Cord RBCs   Expressed

Molecular basis of Riv antigen[1]

Associated with a *RHCE*CE–DIVa(2–3)–CE–D(5)–CE* hybrid allele encoding a complex hybrid RhD-RhCE protein. See System pages "Rearranged *RHD* and *RHCE*."

Effect of enzymes and chemicals on Riv antigen on intact RBCs

Ficin/Papain   Resistant (enhanced)
Trypsin   Presumed resistant
α-Chymotrypsin   Presumed resistant
DTT 200 mM   Presumed resistant

In vitro characteristics of alloanti-Riv

Immunoglobulin class IgG
Optimal technique   37°C; IAT; enzymes

Clinical significance of alloanti-Riv

HDFN   Mild; caused by the only example of anti-Riv in a
                     serum which also contained anti-Go^a (see **RH30**)[2]

Comments

The Riv antigen is expressed by the rare Rh complex DIVa(C)–; this complex (which also expresses Go^a (**RH30**), Rh33 (**RH33**), FPTT (**RH50**), the D antigen (**RH1**) characteristic of category DIVa, G (**RH12**), Nou (**RH44**), and very weak C (**RH2**), but no c (**RH4**), E (**RH3**), e (**RH5**) or f (**RH6**) antigen), was shown to be encoded by *RHD*DIVa in cis* to an *RHCE*CE–DIVa(2–3)– CE–D(5)–CE* hybrid allele.

References

[1] Halter Hipsky, C., et al., 2011. Molecular basis of the rare gene complex, *DIV(C)–*, which encodes four low prevalence antigens in the Rh blood group system. Vox Sang (epub.).

[2] Delehanty, C.L., et al., 1983. Riv: a new low incidence Rh antigen [abstract]. Transfusion 23, 410.

Sec Antigen

Terminology

ISBT symbol (number)	RH46 (004046 or 4.46)
History	Described in 1989, given an Rh number in 1990, named after the first antibody producer.

Occurrence

All populations	100%

Antithetical antigen

Rh32 (**RH32**)

Expression

Cord RBCs	Expressed

Molecular basis associated with the Sec (RH46) antigen

See System pages.

Effect of enzymes and chemicals on Sec (RH46) antigen on intact RBCs

Ficin/Papain	Resistant (enhanced)
Trypsin	Presumed resistant
α-Chymotrypsin	Presumed resistant
DTT 200 mM	Presumed resistant

In vitro characteristics of alloanti-Sec (-RH46)

Immunoglobulin class	IgG
Optimal technique	37°C; IAT; enzymes

Clinical significance of alloanti-Sec (-RH46)

HDFN   No to severe

Comments

Immunized D(C)(e)/D(C)(e) people, homozygous for Rh32 and RH:−46, make anti-Sec.

Sec is expressed by RBCs of common Rh phenotype but is absent from Rh_{null} RBCs and not expressed by the following haplotypes: R^N, D−−, Dc−, DC^W−, and D··.

Dav Antigen

Terminology

ISBT symbol (number)   RH47 (004047 or 4.47)
History   Reported in 1982 and named after the first donor
                        with D·· RBCs. Anti-Dav is a component of some
                        anti-Hr_0 (see **RH17**) sera, and can be separated by
                        adsorption/elution with D··/D·· cells.

Occurrence

All populations   100%

Expression

Cord RBCs   Expressed

Effect of enzymes and chemicals on Dav antigen on intact RBCs

Ficin/Papain   Resistant (enhanced)

Comments

Anti-Dav reacts with cells of all common Rh phenotypes and with D·· cells, but not with Rh_{null}, DIVa(C)−, D−−, DC^W− and Dc− cells.

JAL Antigen

Terminology

ISBT symbol (number)   RH48 (004048 or 4.48)
Obsolete names   S.Allen; J.Allen
History   Reported and numbered in 1990 after more than
                        a decade of using the Allen serum; named after
                        J. Allen, whose RBCs possessed the antigen.

Rh

Occurrence

Less than 0.01%; found in English, French-speaking Swiss, Brazilians, and Blacks.

Antithetical antigen

CEST (**RH57**)

Expression

Cord RBCs	Expressed

Molecular basis of JAL antigen[1,2]

Amino acids	Trp114 and Val245 or Gln114 in Rhce in Blacks
	Trp114 in RhCe in Caucasians
Nucleotides	T at bp 340 and G at bp 733 or A at bp 341 in
	*RHCE*ce* in Blacks
	T at bp 340 in *RHCE*Ce* in Caucasians

Effect of enzymes and chemicals on JAL antigen on intact RBCs

Ficin/Papain	Resistant (enhanced)
Trypsin	Presumed resistant
α-Chymotrypsin	Presumed resistant
DTT 200 mM	Presumed resistant

In vitro characteristics of alloanti-JAL

Immunoglobulin class	IgG
Optimal technique	37°C; IAT; enzymes

Clinical significance of alloanti-JAL

HDFN	Positive DAT, no clinical HDFN

Comments

The JAL antigen is encoded by two different RH alleles: in Blacks JAL is associated with weak expression of c antigen (**RH4**) on Rhce while in Caucasians JAL is associated with a weak C antigen (**RH2**) on RhCe. JAL is variably associated with weak e (**RH5**) expression. Expression of JAL on RhCe appears to be stronger than JAL expressed on Rhce[1]. Three examples of anti-JAL are reported: only one is monospecific[3,4].

References

[1] Hustinx, H., et al., 2009. Molecular basis of the Rh antigen RH48 (JAL). Vox Sang 96, 234–239.

[2] Westhoff, C., et al., 2009. The JAL Antigen (RH48) is the result of a change in RHCE that encodes Arg114Trp. Transfusion 49, 725–732.

[3] Lomas, C., et al., 1990. A low-incidence red cell antigen JAL associated with two unusual Rh gene complexes. Vox Sang 59, 39–43.

[4] Poole, J., et al., 1990. The red cell antigen JAL in the Swiss population: family studies showing that JAL is an Rh antigen (RH48). Vox Sang 59, 44–47.

STEM Antigen

Terminology

ISBT symbol (number)	RH49 (004049 or 4.49)
Obsolete name	Stemper
History	Reported in 1993 (Rh number was assigned at the 1992 ISBT meeting), and named after the Black family in which the antibody/antigen was first identified.

Occurrence

Indians (in South Africa)	0.4%
Blacks	6%

Expression

Cord RBCs	Expressed
Altered	Variable expression among STEM+ [1]

Molecular basis of STEM antigen[2]

Amino acid	Cys16, Val238, Val273, Val378 in Rhce (RhceBI) Cys16, Val238, Val273 in Rhce (RhceSM)
Nucleotide	C at bp 48, G at 712, T at 818, G at 1132 in *RHCE*ce* (*RHCE*ceBI*) C at bp 48, G at 712, T at 818 in *RHCE*ce* (*RHCE*ceSM*)
STEM– (wild type)	Trp16, Met238, Val273, (with Leu 378 for ceBI) and G at bp 48, A at 712, C at 818 (with C at 1132 for ceBI)

Effect of enzymes and chemicals on STEM antigen on intact RBCs

Ficin/Papain	Resistant (enhanced)

Trypsin	Presumed resistant
α-Chymotrypsin	Presumed resistant
DTT 200 mM	Presumed resistant

In vitro characteristics of alloanti-STEM

Immunoglobulin class	IgG
Optimal technique	IAT; enzymes

Clinical significance of alloanti-STEM

HDFN	Mild

Comments

STEM may be associated with Dce haplotypes that do not produce hr^S (**RH19**)[1]. Approximately 65% of hr^S–Hr– RBCs and 30% of hr^B–Hr^B– RBCs are STEM+.

References

[1] Marais, I., et al., 1993. STEM, a new low-frequency Rh antigen associated with the e– variant phenotypes hr^S-(Rh: −18, −19) and hr^B-(Rh: −31, −34). Transf Med 3, 35–41.

[2] Halter-Hipsky, C., et al., 2009. Two alleles with RHCE*nt818C>T change encode the low prevalence Rh antigen STEM [abstract]. Blood 114 (Suppl.), 1226–1227.

FPTT Antigen

Terminology

ISBT symbol (number)	RH50 (004050 or 4.50)
Obsolete names	700048; Mol
History	Reported in 1988 and named after the "French Post Telegraph and Telecommunication" because several of the original probands worked and donated blood there. Achieved Rh antigen status in 1994.

Occurrence

All populations	<0.01%

Expression

Cord RBCs	Expressed
Altered	Strength varies with type of FPTT+ Rh complex

Molecular basis associated with FPTT antigen[1–3]

In the partial D phenotype DFR, FPTT is associated with a hybrid *RH(D–CE–D)* gene in which part of exon 4 of *RHD* is replaced by the same part of exon 4 of *RHCE*.

FPTT is also associated with a hybrid *RHCE*ce*, in which exon 5 of *RHCE* is replaced by exon 5 of *RHD*. See System pages.

Effect of enzymes and chemicals on FPTT antigen on intact RBCs

Ficin/Papain	Resistant (enhanced)
Trypsin	Presumed resistant
α-Chymotrypsin	Presumed resistant
DTT 200 mM	Presumed resistant

In vitro characteristics of alloanti-FPTT

Immunoglobulin class	IgG
Optimal technique	IAT; enzymes

Clinical significance of alloanti-FPTT

No data are available.

Comments

FPTT antigen is also associated with rare "depressed" Rh phenotypes that have depressed C (**RH2**) and/or e (**RH5**) antigens (one family had weakened expression of VS antigen [**RH20**])[4].

The rare haplotype DIVa(C)− is FPTT+. All Rh33+ RBCs are FPTT+, but not all FPTT+ are Rh33+.

The only reported example of anti-FPTT was in a multispecific serum (Mol.) from a woman who had not been transfused or pregnant.

References

[1] Beckers, E.A.M., et al., 1996. The R_0^{Har}Rh:33 phenotype results from substitution of exon 5 of the *RHCE* gene by the corresponding exon of the *RHD* gene. Br J Haematol 92, 751–757.

[2] Noizat-Pirenne, F., et al., 2002. Molecular background of *D(C)(e)* haplotypes within the white population. Transfusion 42, 627–633.

[3] Rouillac, C., et al., 1995. Transcript analysis of D category phenotypes predicts hybrid Rh D–CE–D proteins associated with alteration of D epitopes. Blood 85, 2937–2944.

[4] Bizot, M., et al., 1988. An antiserum identifying a red cell determinant expressed by Rh:33 and by some "new" depressed Rh phenotypes. Transfusion 28, 342–345.

Rh

MAR Antigen

Terminology

ISBT symbol (number)   RH51 (004051 or 4.51)
History   Reported in 1994 and named after the first antibody
                        producer, a Finnish woman with C^W+, C^X+ RBCs.

Occurrence

All populations   100%
Occurrence of MAR– phenotype in Finns is 0.2%.

Expression

Cord RBCs   Expressed
Altered   Weak on hr^B–; RH:32; DC^We/DC^We; DC^Xe/DC^Xe
                        RBCs

Molecular basis associated with MAR antigen

MAR is likely to be expressed in the vicinity of amino acid residues 36–41 of
the RhCe protein[1,2]. MAR– RBCs have the DC^We/DC^Xe phenotype and thus
express both C^W (**RH8**) and C^X (**RH9**) antigens.
See System pages.

Effect of enzymes and chemicals on MAR antigen on intact RBCs

Ficin/Papain   Resistant (enhanced)
Trypsin   Resistant
α-Chymotrypsin   Resistant
DTT 200 mM   Resistant

In vitro characteristics of alloanti-MAR

Immunoglobulin class   IgG
Optimal technique   37°C; IAT; enzymes

Clinical significance of alloanti-MAR

The only reported example of anti-MAR was found in the serum of a non-transfused DC^We/DC^Xe woman upon delivery of her second child[3,4].

Comments

Antibodies made by people with C^W+C^X+ RBCs detect a high-prevalence antigen (MAR) and are weakly reactive with C^W+/C^W+ or C^X+/C^X+ RBCs. Antibodies made by people with C^W+/C^W+, or C^X+/C^X+ RBCs are non-reactive with C^W+/C^X+ RBCs[3,4]. There is an association between MAR (**RH51**), C^W (**RH8**) and C^X (**RH9**) antigens.

References

[1] Mouro, I., et al., 1995. Molecular basis of the RhC^W (Rh8) and RhC^X (Rh9) blood group specificities. Blood 86, 1196–1201.

[2] Sistonen, P., et al., 1994. A novel high-incidence Rh antigen revealing the existence of an allelic sub-system including C^W (Rh8) and C^x (Rh9) with exceptional distribution in the Finnish population. Vox Sang 66, 287–292.

[3] O'Shea, K.P., et al., 2001. An anti-MAR-like antibody in a DC^We/DC^We person. Transfusion 41, 53–55.

[4] Poole, J., et al., 2001. Anti-Rh51-like in a rare C^WDe/C^WDe individual [abstract]. Transfus Med 11 (Suppl. 1), 32.

BARC Antigen

Terminology

ISBT symbol (number)	RH52 (004052 or 4.52)
History	Reported in 1989 as a low-prevalence antigen associated with some DVI RBCs. Named after the Badger American Red Cross, where the antibody was first found. Confirmed as an Rh antigen in 1996 and awarded the next number.

Occurrence

All populations	<0.01%

Expression

Cord RBCs	Presumed expressed
Altered	Correlation between strength of BARC antigen and partial D antigen. See Comments.

Rh

Molecular basis associated with BARC antigen[1–3]

BARC is associated with partial D category VI in a DVICe haplotype. There are three types of DVICe and each is encoded by a hybrid *RHD*D–CE–D*. See System pages.

Effect of enzymes and chemicals on BARC antigen on intact RBCs

Ficin/Papain	Resistant
Trypsin	Presumed resistant
α-Chymotrypsin	Presumed resistant
DTT 200 mM	Presumed resistant

In vitro characteristics of alloanti-BARC

Immunoglobulin class	IgG
Optimal technique	IAT; enzymes

Clinical significance of alloanti-BARC

No data are available.

Comments

BARC subdivides category DVI[4]. Almost all DVI *in cis* to Ce express BARC; DVI *in cis* to cE does not express BARC. DVI RBCs with a weak expression of D have a weak expression of BARC. RBCs with a stronger expression of D have a strong expression of BARC. Anti-BARC is separated from a multispecific serum (Horowitz) by absorption and elution.

References

[1] Mouro, I., et al., 1994. Rearrangements of the blood group RhD gene associated with the D[VI] category phenotype. Blood 83, 1129–1135.

[2] Wagner, F.F., et al., 2001. A D[V]-like phenotype is obliterated by A226P in the partial D DBS. Transfusion 41, 1052–1058.

[3] Wagner, F.F., et al., 1998. Three molecular structures cause rhesus D category VI phenotypes with distinct immunohematological features. Blood 91, 2157–2168.

[4] Tippett, P., et al., 1996. The Rh antigen D: partial D antigens and associated low incidence antigens. Vox Sang 70, 123–131.

Rh

JAHK Antigen

Terminology

ISBT symbol (number)	RH53 (004053 or 4.53)
History	First described in 1995 as a low-prevalence antigen associated with the r^G haplotype. Family studies reported in 2002 confirmed Rh antigen status, and an Rh number was allocated. Name extracted from the family name of the original antibody producer.

Occurrence

All populations	<0.01%

Expression

Cord RBCs	Presumed expressed

Molecular basis associated with JAHK antigen[1]

Amino acid	Leu122 in RhCe
Nucleotide	T at bp 365 in exon 3 of *RHCE*Ce*
JAHK−(wild type)	Ser122 and C at bp 365
See System pages.	

Effect of enzymes and chemicals on JAHK antigen on intact RBCs

Ficin/Papain	Resistant (enhanced)
Trypsin	Resistant
α-Chymotrypsin	Resistant
DTT 200 mM	Presumed resistant

In vitro characteristics of alloanti-JAHK

Immunoglobulin class	IgG
Optimal technique	37°C; IAT; enzymes

Clinical significance of alloanti-JAHK

Not known.

Rh

Comments

Present on RBCs with r^G phenotype but not with the r''^G phenotype[2]. Anti-JAHK is found in multispecific sera[2,3].

References
[1] Scharberg, E.A., et al., 2005. Molecular basis of the JAHK (RH53) antigen. Transfusion 45, 1314–1318.
[2] Green, C., et al., 2002. JAHK: a low frequency antigen associated with the r^G complex of the Rh blood group system. Transfus Med 12, 55–61.
[3] Kosanke, J., et al., 2002. Confirmation that the JAHK antigen is associated with the r^G haplotype. Immunohematology 18, 46–47.

DAK Antigen

Terminology

ISBT symbol (number)	RH54 (004054 or 4.54)
History	Described in 1999; named "D" for the D antigen, and "AK" from the initials of the original antibody producer. Confirmed as an Rh antigen in 2002.

Occurrence

Caucasians	<0.01%
Blacks	4%

Expression

Cord RBCs	Presumed expressed
Altered	Weak on R^N

Molecular basis associated with DAK antigen

Encoded by *RHD*DIIIa*, *RHD*DOL*, and *RHCE*CeRN*[1]. See System pages.

Effect of enzymes and chemicals on DAK antigen on intact RBCs

Ficin/Papain	Resistant (enhanced)
Trypsin	Resistant
α-Chymotrypsin	Resistant
DTT 200 mM	Presumed resistant

In vitro characteristics of alloanti-DAK

Immunoglobulin class	IgG
Optimal technique	37°C; IAT; enzymes

Rh

Clinical significance of alloanti-DAK

Transfusion reaction   Presumed significant
HDFN   Presumed significant

Comments

Many examples of anti-DAK exist in multispecific sera.

Reference

[1] Reid ME, et al., 2003. DAK, a new low-incidence antigen in the Rh blood group system. Transfusion;43: 1394–7.

LOCR Antigen

Terminology

ISBT symbol (number)   RH55 (004055 or 4.55)
Obsolete name   700053
History   Described in 1994, and became part of the Rh blood group system in 2002. Name was derived from two families in which HDFN occurred.

Occurrence

Only five LOCR+ probands, all European, have been reported.

Expression

Cord RBCs   Presumed expressed

Antithetical antigen

Rh26 (**RH26**)

Molecular basis of LOCR antigen[1]

Amino acid   Ser96 in Rhce
Nucleotide   A at bp 286 in exon 2 of *RHCE*ce*

Effect of enzymes and chemicals on LOCR antigen on intact RBCs

Ficin/Papain   Resistant (enhanced)
Trypsin   Resistant
α-Chymotrypsin   Resistant
DTT 200 mM   Presumed resistant
Acid   Resistant

Rh

In vitro characteristics of alloanti-LOCR

Immunoglobulin class   IgG
Optimal technique   37°C; IAT; enzymes

Clinical significance of alloanti-LOCR

Transfusion reaction   No data
HDFN   Moderate

Comments

Travels with ce, and the c or e may be weakened[2].

References
[1] Coghlan, G., et al., 2006. Molecular basis of the LOCR (Rh55) antigen. Transfusion 46, 1689–1692.
[2] Coghlan, G., et al., 1994. A "new" low-incidence red cell antigen, LOCR, associated with altered expression of Rh antigens. Transfusion 34, 492–495.

CENR Antigen

Terminology

ISBT symbol (number)   RH56 (004056 or 4.56)
History   Described in 2004, and named "CE" after *RHCE*/RhCE and "NR" from the initials of the proband.

Occurrence

Only one CENR+ proband, a Caucasian woman with a CENR+ daughter has been reported.

Expression

Cord RBCs   Presumed expressed

Molecular basis of CENR antigen[1]

Associated with *RHCE*Ce–D*(6–10) with 122A>G in exon 1 (Gln41Arg). See System pages.

Effect of enzymes and chemicals on CENR antigen on intact RBCs

Ficin/Papain	Resistant (enhanced)
Trypsin	Resistant
α-Chymotrypsin	Resistant
DTT 200 mM	Resistant
Acid	Resistant

In vitro characteristics of alloanti-CENR

Immunoglobulin class	IgG
Optimal technique	IAT

Clinical significance of alloanti-CENR

No data. Only one example of anti-CENR has been found.

Comments

Anti-CENR was identified in a serum containing anti-D^W (RH23) and anti-Rh32. An eluate, made by adsorbing the reactive anti-D^W/Rh32 serum onto CENR+ RBCs, was reactive with D^W+ and Rh32+ RBCs. However, the CENR+ RBCs were shown to be D^W− and Rh32− by tests with several examples of these specificities.

Reference
[1] Westhoff, C.M., et al., 2004. A new hybrid *RHCE* Gene (CeNR) is responsible for expression of a novel antigen. Transfusion 44, 1047–1051.

CEST Antigen

Terminology

ISBT symbol (number)	RH57 (004057 or 4.57)
History	Named in 2009 when it was shown to be antithetical to JAL: "CE" for RhCE, and "ST" from the name of the antigen-negative proband.

Occurrence

Most populations	100%

The CEST− phenotype was found in people with African ancestry.

Expression

Cord RBCs	Expressed

Antithetical antigen

JAL (**RH48**)

Molecular basis of CEST antigen[1,2]

Amino acid	Arg114 and Leu245
Nucleotide	C at bp 34 and C at bp 733

See System pages.

Effect of enzymes and chemicals on CEST antigen on intact RBCs

Ficin/Papain	Resistant
Trypsin	Resistant
α-Chymotrypsin	Resistant
DTT 200 mM	Resistant

In vitro characteristics of alloanti-CEST

Immunoglobulin class	IgG
Optimal technique	IAT

Clinical significance of alloanti-CEST

No data because antibody is rare.

References

[1] Lomas-Francis, C., et al., 2009. JAL (RH48) blood group antigen: serological observations. Transfusion 49, 719–724.

[2] Westhoff, C., et al., 2009. The JAL Antigen (RH48) is the result of a change in RHCE that encodes Arg114Trp. Transfusion 49, 725–732.

CELO Antigen

Terminology

ISBT symbol (number)	RH58 (004058 or 4.58)
History	Named in 2011 when it was shown to be antithetical to Crawford. "CE" from RhCE, and "LO" from the first two antigen-negative probands.

Occurrence

Three CELO– probands have been reported; they were of African or Hispanic ancestry.

Expression

Cord RBCs   Expressed

Antithetical antigen

Crawford (**RH43**)

Molecular basis of CELO antigen[1]

Amino acids   Trp16, Gln233, Leu245 in Rhce
Nucleotides   G at bp 48, C at bp 697, and C at bp 733 in
                       *RHCE*ce (RHCE*ceCF)*
See System pages.

Effect of enzymes and chemicals on CELO antigen on intact RBCs

Ficin/Papain   Resistant
Trypsin   Resistant
α-Chymotrypsin   Resistant
DTT 200 mM   Resistant

In vitro characteristics of alloanti-CELO

Immunoglobulin class   IgG
Optimal technique   IAT

Clinical significance of alloanti-CELO

No data available because the antibody specificity is rare.

Comments

CELO– RBCs are also hr^S+/– VS+ and hr^B–.

Reference

[1] Halter Hipsky, C., et al., 2011. *RHCE*ceCF* encodes partial c and partial e but not CELO an antigen antithetical to Crawford. Transfusion 51, 25–31.

CEAG Antigen

Terminology

ISBT symbol (number)   RH59 (004059 or 4.59)
History   Named in 2009: "CE" for RhCE, "A" for Ala, and
                       "G" for Gly.

Occurrence

The only CEAG– proband was of African ancestry.

Molecular basis of CEAG antigen[1,2]

Amino acid	Ala85 in Rhce
Nucleotide	C at bp 254 in exon 2 of *RHCE*ce*
CEAG–	Gly85 and G at bp 254

See System pages.

Expression

Cord RBCs	Expressed

Effect of enzymes and chemicals on CEAG antigen on intact RBCs

Ficin/Papain	Resistant
Trypsin	Resistant
α-Chymotrypsin	Resistant
DTT 200 mM	Resistant

In vitro characteristics of alloanti-CEAG

Immunoglobulin class	IgG
Optimal technique	IAT

Clinical significance of alloanti-CEAG

No data are available because only one anti-CEAG has been described.

Comments

CEAG– RBCs also lack hr^B (**RH31**) and have a partial e antigen[2].

References

[1] Vege, S., et al., 2009. A novel 254 G>C (Ala85 Gly) change associated with partial Rhe and alloanti-e [abstract]. Transfusion 49 (Suppl.) 15A–15A.

[2] Westhoff, C.M., et al., 2010. Frequency of *RHCE*ce(254G>C)* in African-American patients and donors [abstract]. Transfusion 50 (Suppl.), 145A–146A.

Rh

Lutheran Blood Group System

Number of antigens 20

Polymorphic | Lu^a, Au^a, Au^b
Low prevalence | Lu9, Lu14
High prevalence | Lub, Lu3, Lu4, Lu5, Lu6, Lu7, Lu8, Lu11, Lu12, Lu13, Lu16, Lu17, Lu20, Lu21, LURC

Terminology

ISBT symbol (number) | LU (005)
CD number | CD239
Other name | B-CAM (Basal-cell adhesion molecule)
History | The first Lutheran antigen was described in 1945, and should have been named Lutteran, after the first Lu(a+) donor, but the writing on the label of the blood sample was misread as Lutheran.

Expression

Soluble form | Not described in a natural form
Other blood cells | Not on lymphocytes, granulocytes, monocytes, platelets
Tissues | Predominantly in the basal layer of the epithelium and endothelium of blood vessels, and in a broad range of cells and tissues: brain, heart, kidney glomeruli, liver, lung, pancreas, placenta, skeletal muscle, arterial wall, tongue, trachea, skin, esophagus, cervix, ileum, colon, stomach, gall bladder[1]

Gene[2]

Chromosome | 19q13.32
Name | *LU*
Organization | 15 exons distributed over approximately 12 kbp of DNA

Lutheran

The Blood Group Antigen (3/e). DOI: http://dx.doi.org/10.1016/B978-0-12-415849-8.00007-7

| Product | Lutheran glycoprotein (597 amino acids) and B-CAM (557 amino acids) by alternative splicing of exon 13[1,3] |

Database accession numbers

| GenBank | NM_005581 (mRNA); X83425 (gene) |
| Entrez Gene ID | 4059 |

Molecular basis of Lutheran phenotypes

The reference allele, *LU*02* or *LU*B* (Accession number X83425) encodes Lu[b] (LU2), LU3, LU4, LU5, LU6, LU8, LU12, LU13, LU16, LU17, LU18, LU20, LU21, LURC. Nucleotide differences from this reference allele, and the amino acids affected, are given.

Allele encodes	Allele name	Exon	Nucleotide	Restriction enzyme	Amino acid	Ethnicity (prevalence)
Lu(a+) or LU:1	*LU*01* or *LU*A*	3	230G>A[†]	AciI–	Arg77His	Caucasian (Many), Blacks (Several)
Lu(a+) or LU:1	*LU*01.02* or *LU*A*	3 5	230G>A 586G>A		Arg77His Val196His	
LU:–4	*LU*02.–04.1*	5	524G>A	HpaII–	Arg175Gln	Caucasian (Rare)
LU:–4	*LU*02.–04.2*	5	524G>T	HpaII–	Arg175Leu	Caucasian (Rare)
LU:–5	*LU*02.–05*	3	326G>A	DraIII+	Arg109His	Caucasian (Several)
LU:–7	*LU*02.–07*	10	127A>C	MnlI–	Glu425Ala	Caucasian (Rare)
LU:–6,9	*LU*02.09*	7	824C>T	CfoI+	Ser275Phe	Iranian Jews (Rare)
LU:–8,14	*LU*02.14*	6	611T>A	NlaIII–	Met204Lys	Caucasian (Several)
LU:–12	*LU*02.–12.1*	2	99–104del GCGCTT	HspAI–	delArg34, Leu35	Caucasian (Rare)

(Continued)

(Continued)

Allele encodes	Allele name	Exon	Nucleotide	Restriction enzyme	Amino acid	Ethnicity (prevalence)
LU:−12	LU*02.−12.2	3	419G>A	BstUI−	Arg140Gln	Caucasian (Rare)
LU:−13	LU*02.−13	11 13	1340C>T 1742A>T	MspAI−	Ser447Leu Gln581Leu	(Rare)
LU:−16	LU*01.−16	6	679C>T	NlaIV−	Arg227Cys	Blacks (Rare)
LU:−17	LU*02.−17	3	340G>A	BbvCI−	Glu114Lys	Italians (Rare)
Au(a−b+) or LU:18,19	LU*02.19	12	1615A>G		Thr539Ala	Caucasian, Blacks (Many)
LU:−20	LU*02.−20	7	905C>T	AccI−	Thr302Met	Israeli (Rare)
LU:−21	LU*02.−21	3	282C>G	BssSI+	Asp94Glu	Israeli (Rare)
LURC− or LU:−22^	LU*01.−22	2	223C>T	SacI−	Arg75Cys	(Rare)

[†]Also associated with 586G>A in exon 5 (Val96Ile)[4].
[^]Expression of LURC is dependant on Arg77[5].

Molecular basis of silencing of *LU*

Homozygosity or compound heterozygosity leads to the recessive Lu(a−b−) phenotype. Nucleotide changes from the *LU*02* reference allele (Accession number X83425), and the amino acid affected, are given.

Allele name	Exon	Nucleotide	Restriction enzyme	Amino acid	Ethnicity (prevalence)
LU*02N.01	6	691C>T	DdeI+	Arg231Stop	Caucasian (Rare)
LU*02N.02	3 4	322intron2 to exon4del		Thr69-Glu168 del	Caucasian (Rare)
LU*02N.03	6	711C>A	DdeI+ and CfoI−	Cys237Stop	Japanese (Rare)
LU*02N.04	3	361C>T		Arg121Stop	Caucasian (Rare)
LU*02N.05	2	Ins123GG		42Gly-Arg-Stop[6]	Caucasian (Dutch)

Lutheran

Molecular basis of dominant Lu(a–b–) phenotype [In(Lu)]

KLF1 encodes erythroid Krüppel-like factor (EKLF). Heterozygosity for one of several nucleotide changes in this gene are responsible for the dominant Lu(a–b–) phenotype[7,8], which is also known as the *In(Lu)* phenotype, and is characterized by reduced expression of antigens in the Lutheran system and for P1, In[b], and AnWj antigens. The KLF1 gene is located at 19p13.13–p13.12 and has 3 exons; the initiation codon is in exon 1 and the stop codon is in exon 3. GenBank accession numbers are U37106 (gene) and NM_006563 (mRNA); Entrez Gene ID 10661. Nucleotide changes from the *KLF1*01* reference allele (Accession number NM_006563), and the amino acid affected, are given. Heterozygosity leads to the dominant Lu(a–b–) phenotype whilst homozygosity for is not thought to be compatible with life.

Allele name	Exon (intron)	Nucleotide	Amino acid	Ethnicity (prevalence)
*KLF1*BGM01*	Promoter	−124T>C		(Rare)
*KLF1*BGM02*	2	380T>A	Leu127Stop	(Rare)
*KLF1*BGM03*	2	569delC	Pro190LeufsStop	(Rare)
*KLF1*BGM04*	2	874A>T	Lys292Stop	(Rare)
*KLF1*BGM05*	2	895C>T	His299Tyr	(Rare)
*KLF1*BGM06*	3	954dupG	Arg319GlufsStop	(Rare)
*KLF1*BGM07*	3	983G>T	Arg328Leu	(Rare)
*KLF1*BGM08*	3	983G>A	Arg328His	(Rare)
*KLF1*BGM09*	3	991C>G	Arg331Gly	(Rare)
*KLF1*BGM11*	1	90G>A	Trp30Stop	(Rare)
*KLF1*BGM12*	2	304T>C	Ser102Pro	(Rare)

Molecular basis of X-linked Lu(a–b–) phenotype (Lu$_{mod}$ phenotype)

The X-borne transcription factor, GATA-1, is essential for erythroid and megakaryocyte development. A nucleotide change in this gene causes the X-linked Lu(a–b–) phenotype[9].

The gene (Accession number NG_008846; X17254; Entrez Gene ID 2623) is located at Xp11.23 and has 6 exons; the initiation codon is in exon 2 and the stop codon is in exon 6.

Nucleotide change from the *GATA1*01* reference allele (NM_002049), and the amino acid affected, are given.

Allele name	Exon	Nucleotide	Restriction enzyme	Amino acid	Ethnicity (prevalence)
*GATA1*BGM01*	6	1240T>C	*Bsp*HI	Stop414Arg^	Australian (Rare)

^ = The Stop codon becomes arginine, which leads to an elongated protein product.

Amino acid sequence[2]

```
MEPPDAPAQA  RGAPRLLLLA  VLLAAHPDAQ  AEVRLSVPPL  VEVMRGKSVI   50
LDCTPTGTHD  HYMLEWFLTD  RSGARPRLAS  AEMQGSELQV  TMHDTRGRSP  100
PYQLDSQGRL  VLAEAQVGDE  RDYVCVVRAG  AAGTAEATAR  LNVFAKPEAT  150
EVSPNKGTLS  VMEDSAQEIA  TCNSRNGNPA  PKITWYRNGQ  RLEVPVEMNP  200
EGYMTSRTVR  EASGLLSLTS  TLYLRLRKDD  RDASFHCAAH  YSLPEGRHGR  250
LDSPTFHLTL  HYPTEHVQFW  VGSPSTPAGW  VREGDTVQLL  CRGDGSPSPE  300
YTLFRLQDEQ  EEVLNVNLEG  NLTLEGVTRG  QSGTYGCRVE  DYDAADDVQL  350
SKTLELRVAY  LDPLELSEGK  VLSLPLNSSA  VVNCSVHGLP  TPALRWTKDS  400
TPLGDGPMLS  LSSITFDSNG  TYVCEASLPT  VPVLSRTQNF  TLLVQGSPEL  450
KTAEIEPKAD  GSWREGDEVT  LICSARGHPD  PKLSWSQLGG  SPAEPIPGRQ  500
GWVSSSLTLK  VTSALSRDGI  SCEASNPHGN  KRHVFHFGAV  SPQTSQAGVA  550
VMAVAVSVGL  LLLVVAVFYC  VRRKGGPCCR  QRREKGAPPP  GEPGLSHSGS  600
EQPEQTGLLM  GGASGGARGG  SGGFGDEC                             628
```

LU encodes a signal peptide of 31 amino acids which is cleaved off to form the mature protein found in the red cell membrane.

Carrier molecule

The predicted mature protein has five disulfide-bonded, extracellular, immunoglobulin superfamily (IgSF) domains (two variable-region (V) sets and three constant region (C) sets)[1].

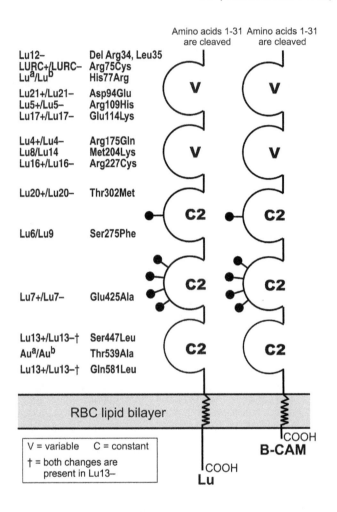

M_r (SDS-PAGE)	Lu: 85,000 (has a cytoplasmic tail)
	B-CAM: 78,000 (has no cytoplasmic tail)
CHO: N-glycan	5 potential sites (residues 290, 346, 352, 388, 408)
CHO: O-glycan	Present
Cysteine residues	10 extracellular
Copies per RBC	1,500 to 4,000

Function

Possibly has adhesion properties and may mediate intracellular signaling. The extracellular domains and the cytoplasmic domain contain consensus motifs for the binding of integrin and Src homology 3 domains, respectively, suggesting possible receptor and signal-transduction function[1]. Lutheran glycoprotein binds to laminin (particularly to isoforms that contain α5 chains), strongly suggesting that it is a membrane constituent that is involved in cell–cell and cell–matrix binding events, and may function as a laminin receptor during erythropoiesis[10]. IgSF domains 1 to 3, and possibly domain 5, are involved in laminin binding[11-13]. Lu-glycoproteins may be involved in facilitating movement of maturing erythroid cells from erythroblastic islands of the bone marrow to the peripheral circulation, and may play a role in the migration of erythroid progenitors from fetal liver to the bone marrow.

Disease association

Expression of B-CAM is increased in certain malignant tumors and cells. May mediate adhesion of sickle cells to vascular endothelium, and contribute to blockage of the vessels and painful episodes of vaso-occlusion. In polycythemia vera, the Lu-glycoproteins are phosphorylated which increases RBC adhesion, and may promote thrombosis in polycythemia vera.

Phenotypes (% occurrence)

Phenotype	Most populations
Lu(a+b–)	0.2
Lu(a–b+)	92.4
Lu(a+b+)	7.4
Lu(a–b–)	Rare
Null: Lu(a–b–) recessive type	
Unusual: Lu(a–b–) dominant type and X-linked type	

Comparison of three types of Lu(a–b–) phenotypes

Lu(a–b–) phenotype	Lutheran antigens	Make anti-Lu3	CD44	CD75	I/i antigen
Recessive	Absent	Yes	Normal	Normal	Normal/normal
Dominant	Weak	No, except in one case[1]	Weak (25–39% of normal)	Strong	Normal/weak
X-linked	Weak	No	Normal	Absent	Weak/strong

Comments

Lu(a–b–) dominant type RBCs have reduced expression of Lutheran, P1, AnWj, Indian, Knops, Cs^a, and MER2 blood group antigens. The X-linked type (*XS2*) of Lu(a–b–) has only been found in one family.

Lutheran, along with *Secretor*, provided the first example of autosomal linkage in humans. Some Lu(a–b–) RBCs are acanthocytic[14].

References

1   Parsons, S.F., et al., 1995. The Lutheran blood group glycoprotein, another member of the immunoglobulin superfamily, is widely expressed in human tissues and is developmentally regulated in human liver. Proc Natl Acad Sci USA 92, 5496–5500.

2   El Nemer, W., et al., 1997. Organization of the human LU gene and molecular basis of the Lu^a/Lu^b blood group polymorphism. Blood 89, 4608–4616.

3   Rahuel, C., et al., 1996. A unique gene encodes spliceoforms of the B-cell adhesion molecule cell surface glycoprotein of epithelial cancer and of the Lutheran blood group glycoprotein. Blood 88, 1865–1872.

4   Gowland, P., et al., 2005. A new polymorphism within the Lua/Lub blood group [abstract]. Transf Med Hemother 32 (Suppl. 1), 54–55.

5   Karamatic Crew, V., et al., 2009a. Two heterozygous mutations in an individual result in the loss of a novel high incidence Lutheran antigen Lurc [abstract]. Transfus Med 19 (Suppl. 1), 10.

6   Karamatic Crew, V., et al., 2009b. A novel LU mutation giving rise to a new example of the recessive type Lutheran-null phenotype [abstract]. Transfus Med 19 (Suppl. 1), 24.

7   Crowley, J., et al., 2010. Novel mutations in *EKLF/KLF1* encoding In(Lu) phenotype [abstract]. Transfusion 50 (Suppl. 47A).

8   Singleton, B.K., et al., 2008. Mutations in EKLF/KLF1 form the molecular basis of the rare blood group In(Lu) phenotype. Blood 112, 2081–2088.

9   Singleton, B.K., et al., 2009. A novel GATA-1 mutation (Ter414Arg) in a family with the rare X-linked blood group Lu(a–b–) phenotype [abstract]. Blood 114, 783.

10   El Nemer, W., et al., 1998. The Lutheran blood group glycoproteins, the erythroid receptors for laminin, are adhesion molecules. J Biol Chem 273, 16686–16693.

11   El Nemer, W., et al., 2001. Characterization of the laminin binding domains of the Lutheran blood group glycoprotein. J Biol Chem 276, 23757–23762.

12   Udani, M., et al., 1998. Basal cell adhesion molecule Lutheran protein: The receptor critical for sickle cell adhesion to laminin. J Clin Invest 101, 2550–2558.

13   Zen, Q., et al., 1999. Critical factors in basal cell adhesion molecule/Lutheran-mediated adhesion to laminin. J Biol Chem 274, 728–734.

14   Udden, M.M., et al., 1987. New abnormalities in the morphology, cell surface receptors, and electrolyte metabolism of In(Lu) erythrocytes. Blood 69, 52–57.

Lu^a Antigen

Terminology

ISBT symbol (number)	LU1 (005001 or 5.1)
History	Identified in 1945; named after the donor whose blood stimulated the production of anti-Lu^a in a patient with SLE who had received multiple transfusions.

Occurrence

Caucasians	8%
Blacks	5%

Antithetical antigen

Lu^b (**LU2**)

Expression

Cord RBCs	Weak

There is considerable variation in the strength of Lu^a expression on RBCs. This variation is inherited.

Molecular basis associated with Lu^a antigen[1,2]

Amino acid	His77 in IgSF domain 1
Nucleotide	A at bp 230 in exon 3

*LU*A* also encodes Val196Ile in IgSF domain 2 (586G>A in exon 5)[3].

Effect of enzymes and chemicals on Lu^a antigen on intact RBCs

Ficin/Papain	Resistant (may be weakened)
Trypsin	Sensitive
α-Chymotrypsin	Sensitive
Pronase	Sensitive
DTT 200 mM/50 mM	Sensitive/resistant (thus sensitive to WARM™ and ZZAP)
Acid	Resistant

In vitro characteristics of alloanti-Lu^a

Immunoglobulin class	IgM; IgG
Optimal technique	RT or IAT with characteristic "loose" agglutinates surrounded by unagglutinated RBCs; capillary
Complement binding	Rare

Clinical significance of alloanti-Lu^a

Transfusion reaction	No
HDFN	No to mild (rare). The presence of fetal Lutheran glycoprotein on placental tissue may result in absorption of maternal antibodies to Lutheran antigens.

Lutheran

Comments

Anti-Lu^a is not infrequently found in serum from patients following transfusion, and also may be naturally-occurring. Sera containing anti-Lu^a often also contain HLA antibodies.

References

[1] El Nemer, W., et al., 1997. Organization of the human LU gene and molecular basis of the Lu^a/Lu^b blood group polymorphism. Blood 89, 4608–4616.

[2] Parsons, S.F., et al., 1997. Use of domain-deletion mutants to locate Lutheran blood group antigens to each of the five immunoglobulin superfamily domains of the Lutheran glycoprotein: Elucidation of the molecular basis of the Lu^a/Lu^b and the Au^a/Au^b polymorphisms. Blood 89, 4219–4225.

[3] Gowland, P., et al., 2005. A new polymorphism within the Lu^a/Lu^b blood group [abstract]. Transf Med Hemother 32 (Suppl. 1), 54–55.

Lu^b Antigen

Terminology

ISBT symbol (number)	LU2 (005002 or 5.2)
History	Named because of its antithetical relationship to Lu^a; anti-Lu^b was identified in 1956.

Occurrence

All populations	99.8%

Antithetical antigen

Lu^a (**LU1**)

Expression

Cord RBCs	Weak
Altered	Weak on RBCs of the dominant type of Lu(a–b–)

There is considerable variation in the strength of Lu^b expression on RBCs. This variation is inherited.

Molecular basis associated with Lu^b antigen[1,2]

Amino acid	Arg77 in IgSF domain 1
Nucleotide	G at bp 230 in exon 3

Effect of enzymes and chemicals on Lu^b antigen on intact RBCs

Ficin/Papain	Resistant (may be weakened)
Trypsin	Sensitive
α-Chymotrypsin	Sensitive

Pronase	Sensitive
DTT 200 mM/50 mM	Sensitive/resistant (thus sensitive to WARM™ and ZZAP)
Acid	Resistant

In vitro characteristics of alloanti-Lu^b

Immunoglobulin class	IgG; IgM
Optimal technique	RT; IAT; capillary
Complement binding	Rare

Clinical significance of alloanti-Lu^b

| Transfusion reaction | Mild to moderate |
| HDFN | Mild; the presence of Lutheran glycoprotein on placental tissue may result in absorption of maternal antibodies to Lutheran antigens |

Comments

Weak expression of Lu^b on dominant type Lu(a–b–) RBCs is detectable by absorption/elution.

Siblings of patients with anti-Lu^b should be tested for compatibility, and the patient urged to donate blood for cryogenic storage when his/her clinical state permits.

References

[1] El Nemer, W., et al., 1997. Organization of the human LU gene and molecular basis of the Lu^a/Lu^b blood group polymorphism. Blood 89, 4608–4616.

[2] Parsons, S.F., et al., 1997. Use of domain-deletion mutants to locate Lutheran blood group antigens to each of the five immunoglobulin superfamily domains of the Lutheran glycoprotein: elucidation of the molecular basis of the Lu^a/Lu^b and the Au^a/Au^b polymorphisms. Blood 89, 4219–4225.

Lu3 Antigen

Terminology

ISBT symbol (number)	LU3 (005003 or 5.3)
Obsolete names	Lu^{ab}; Lu^{a}Lu^b
History	Reported in 1963; renamed Lu3 to be computer-friendly after Lu(a–b–) RBCs were shown to lack other high prevalence antigens in the Lutheran system. Dr. Crawford, a blood banker, found her own RBCs to be of the dominant type of Lu(a–b–).

Lutheran

Occurrence

All populations   100%

Expression

Cord RBCs	Weak
Altered	Weak or non-detectable by hemagglutination on RBCs of the dominant type of Lu(a–b–)

Molecular basis associated with the LU:–3 phenotype

See System pages for molecular basis of LU:–3 phenotype.

Effect of enzymes and chemicals on Lu3 antigen on intact RBCs

Ficin/Papain	Resistant
Trypsin	Sensitive
α-Chymotrypsin	Sensitive
Pronase	Sensitive
DTT 200 mM/50 mM	Sensitive/resistant (thus sensitive to WARM™ and ZZAP)
Acid	Resistant

In vitro characteristics of alloanti-Lu3

Immunoglobulin class	IgG
Optimal technique	IAT
Complement binding	Rare

Clinical significance of alloanti-Lu3

No data are available because anti-Lu3 is rare.

Comments

Anti-Lu3 is only made by immunized individuals of the rare recessive type Lu(a–b–). In these cases, Lu(a–b–) blood of the recessive or dominant type can be used for transfusion. One person with the dominant type Lu(a–b–) has made anti-Lu3[1].

Siblings of patients with anti-Lu3 should be tested for compatibility, and the patient urged to donate blood for cryogenic storage when his/her clinical state permits.

Reference

[1] Crowley, J., et al., 2010. Novel mutations in *EKLF/KLF1* encoding In(Lu) phenotype [abstract]. Transfusion 50 (Suppl. 47A).

Lu4 Antigen

Terminology

ISBT symbol (number)	LU4 (005004 or 5.4)
Obsolete name	Barnes
History	The first of a series of Lu(a–b+) people who made an antibody compatible only with Lu(a–b–) RBCs. Described in 1971.

Occurrence

Only one family with the LU:–4 phenotype has been reported.

Expression

Cord RBCs	Weak
Altered	Weak or non-detectable by hemagglutination on RBCs of the dominant type of Lu(a–b–)

Molecular basis associated with Lu4 antigen[1]

Amino acid	Arg175 in IgSF domain 2
Nucleotide	G at bp 524 in exon 5
Lu4–	Gln175 and A at bp 524

Effect of enzymes and chemicals on Lu4 antigen on intact RBCs

Ficin/Papain	Resistant
Trypsin	Sensitive
α-Chymotrypsin	Sensitive
Pronase	Sensitive
DTT 200 mM/50 mM	Sensitive/resistant (thus sensitive to WARM™ and ZZAP)

In vitro characteristics of alloanti-Lu4

Immunoglobulin class	IgG
Optimal technique	IAT

Clinical significance of alloanti-Lu4

Transfusion reaction	No data are available because only one anti-Lu4 has been described
HDFN	No in two infants born to one LU:–4 female. The presence of Lutheran glycoprotein on placental tissue may result in absorption of maternal antibodies to Lutheran antigens.

Lutheran

Comments

Siblings of patients with anti-Lu4 should be tested for compatibility, and the patient urged to donate blood for cryogenic storage when his/her clinical state permits.

Experts recommend transfusing Lu(a–b–) blood (dominant or recessive type) if specific antigen-negative blood is not available.

Reference

[1] Crew, V.K., et al., 2003. Molecular bases of the antigens of the Lutheran blood group system. Transfusion 43, 1729–1737.

Lu5 Antigen

Terminology

ISBT symbol (number)	LU5 (005005 or 5.5)
Obsolete names	Beal; Fox
History	Identified in 1972; given the next number in the series of Lu(a–b+) people who made an antibody compatible only with Lu(a–b–) RBCs.

Occurrence

All populations	100%

Expression

Cord RBCs	Weak
Altered	Non-detectable by hemagglutination on RBCs of the dominant type of Lu(a–b–)

Molecular basis associated with Lu5 antigen[1]

Amino acid	Arg109 in IgSF domain 1
Nucleotide	G at bp 326 in exon 3
Lu5–	His109 and A at bp 326

Effect of enzymes and chemicals on Lu5 antigen on intact RBCs

Ficin/Papain	Resistant
Trypsin	Sensitive
α-Chymotrypsin	Sensitive
Pronase	Sensitive
DTT 200 mM/50 mM	Sensitive/resistant (thus sensitive to WARM™ and ZZAP)

Lutheran

In vitro characteristics of alloanti-Lu5

Immunoglobulin class	IgG
Optimal technique	IAT

Clinical significance of alloanti-Lu5

Transfusion reaction	No (potentially significant by a chemiluminescent assay)
HDFN	No; the presence of Lutheran glycoprotein on placental tissue may result in absorption of maternal antibodies to Lutheran antigens

Comments

Several examples of immune anti-Lu5 have been reported.

Siblings of patients with anti-Lu5 should be tested for compatibility and the patient urged to donate blood for cryogenic storage when his/her clinical state permits.

Experts recommend transfusing Lu(a–b–) blood (dominant or recessive type) if specific antigen-negative blood is not available.

Reference

[1] Crew, V.K., et al., 2003. Molecular bases of the antigens of the Lutheran blood group system. Transfusion 43, 1729–1737.

Lu6 Antigen

Terminology

ISBT symbol (number)	LU6 (005006 or 5.6)
Obsolete names	Jan; Jankowski
History	Identified in 1972; given the next number in the series of Lu(a–b+) people who made an antibody compatible only with Lu(a–b–) RBCs.

Occurrence

All populations	100%

Three LU:–6 were Iranian Jews[1].

Antithetical antigen

Lu9 (**LU9**)

Expression

Cord RBCs	Weak
Altered	Non-detectable by hemagglutination on RBCs of the dominant type of Lu(a–b–)

Molecular basis associated with Lu6 antigen[2]

Amino acid	Ser275 in IgSF domain 3
Nucleotide	C at bp 824 in exon 7

Effect of enzymes and chemicals on Lu6 antigen on intact RBCs

Ficin/Papain	Resistant
Trypsin	Sensitive
α-Chymotrypsin	Sensitive
Pronase	Sensitive
DTT 200 mM/50 mM	Sensitive/Resistant (thus sensitive to WARM™ and ZZAP)

In vitro characteristics of alloanti-Lu6

Immunoglobulin class	IgG
Optimal technique	IAT

Clinical significance of alloanti-Lu6

Transfusion reaction	No to moderate
HDFN	No; the presence of Lutheran glycoprotein on placental tissue may result in absorption of maternal antibodies to Lutheran antigens

Comments

Siblings of patients with anti-Lu6 should be tested for compatibility, and the patient urged to donate blood for cryogenic storage when his/her clinical state permits.

Experts recommend transfusing Lu(a–b–) blood (dominant or recessive type) if specific antigen-negative blood is not available.

References

[1] Yahalom, V., et al., 2002. The rare Lu:–6 phenotype in Israel and the clinical significance of anti-Lu6. Transfusion 42, 247–250.

[2] Crew, V.K., et al., 2003. Molecular bases of the antigens of the Lutheran blood group system. Transfusion 43, 1729–1737.

Lu7 Antigen

Terminology

ISBT symbol (number)	LU7 (005007 or 5.7)
Obsolete name	Gary
History	Identified in 1972; given the next number in the series of Lu(a–b+) people who made an antibody compatible only with Lu(a–b–) RBCs.

Occurrence

Only two Lu7– probands have been reported.

Expression

Cord RBCs	Weak
Altered	Non-detectable by hemagglutination on RBCs of the dominant type of Lu(a–b–)

Molecular basis associated with Lu7 antigen

Amino acid	Glu425 in IgSF domain 4
Nucleotide	A at bp 1274 in exon 10
Lu7–	Ala425 and C at bp 1274

Effect of enzymes and chemicals on Lu7 antigen on intact RBCs

Ficin/Papain	Resistant
Trypsin	Sensitive
α-Chymotrypsin	Sensitive
Pronase	Sensitive
DTT 200mM/50mM	Presumed sensitive/resistant (thus sensitive to WARM™ and ZZAP)

In vitro characteristics of alloanti-Lu7

Immunoglobulin class	IgG
Optimal technique	IAT

Clinical significance of alloanti-Lu7

Transfusion reaction	No to mild
HDFN	No; the presence of Lutheran glycoprotein on placental tissue may result in absorption of maternal antibodies to Lutheran antigens

Comments

Only two examples of anti-Lu7 have been described.

Siblings of patients with anti-Lu7 should be tested for compatibility, and the patient urged to donate blood for cryogenic storage when his/her clinical state permits.

Experts recommend transfusing Lu(a–b–) blood (dominant or recessive type) if specific antigen-negative blood is not available.

Lutheran

Lu8 Antigen

Terminology

ISBT symbol (number)	LU8 (005008 or 5.8)
Obsolete names	Taylor; MT
History	Identified in 1972; given the next number in the series of Lu(a–b+) people who made an antibody compatible only with Lu(a–b–) RBCs.

Occurrence

All populations	100%

Several Lu8– probands have been described.

Antithetical antigen

Lu14 (**LU14**)

Expression

Cord RBCs	Weak
Altered	Non-detectable by hemagglutination on RBCs of the dominant type of Lu(a–b–)

Molecular basis associated with Lu8 antigen[1]

Amino acid	Met204 in IgSF domain 2
Nucleotide	T at bp 611 in exon 6

Effect of enzymes and chemicals on Lu8 antigen on intact RBCs

Ficin/Papain	Variable
Trypsin	Sensitive
α-Chymotrypsin	Sensitive
Pronase	Sensitive
DTT 200 mM/50 mM	Sensitive/resistant (thus sensitive to WARM™ and ZZAP)

In vitro characteristics of alloanti-Lu8

Immunoglobulin class	IgG
Optimal technique	IAT

Clinical significance of alloanti-Lu8

Transfusion reaction	No to mild
HDFN	No; the presence of Lutheran glycoprotein on placental tissue may result in absorption of maternal antibodies to Lutheran antigens

Comments

Siblings of patients with anti-Lu8 should be tested for compatibility, and the patient urged to donate blood for cryogenic storage when his/her clinical state permits.

Experts recommend transfusing Lu(a–b–) blood (dominant or recessive type) if specific antigen-negative blood is not available.

Reference

[1] Crew, V.K., et al., 2003. Molecular bases of the antigens of the Lutheran blood group system. Transfusion 43, 1729–1737.

Lu9 Antigen

Terminology

ISBT symbol (number)	LU9 (005009 or 5.9)
Obsolete name	Mull
History	Reported in 1973; given the next available number when its antithetical relationship to Lu6 was recognized.

Occurrence

Reported as 1 to 2%, but probably lower because the original anti-Lu9 contained anti-Bg^a.

Antithetical antigen

Lu6 (**LU6**)

Expression

Cord RBCs	Weak

Molecular basis associated with Lu9 antigen[1]

Amino acid	Phe275 in IgSF domain 3
Nucleotide	T at bp 824 in exon 7

Effect of enzymes and chemicals on Lu9 antigen on intact RBCs

Ficin/Papain	Resistant
Trypsin	Sensitive
α-Chymotrypsin	Sensitive
Pronase	Sensitive
DTT 200 mM/50 mM	Presumed sensitive/resistant (thus sensitive to WARM™ and ZZAP)
Acid	Resistant

Lutheran

In vitro characteristics of alloanti-Lu9

Immunoglobulin class	IgG
Optimal technique	IAT, capillary

Clinical significance of alloanti-Lu9

Transfusion reaction	No data; the second example was probably stimulated by transfusion[2]
HDFN	Positive DAT, but no clinical HDFN in the only case; the presence of Lutheran glycoprotein on placental tissue may result in absorption of maternal antibodies to Lutheran antigens

References

[1] Crew, V.K., et al., 2003. Molecular bases of the antigens of the Lutheran blood group system. Transfusion 43, 1729–1737.

[2] Champagne, K., et al., 1999. Anti-Lu9: the finding of the second example after 25 years. Immunohematology 15, 113–116.

Lu11 Antigen

Terminology

ISBT symbol (number)	LU11 (005011 or 5.11)
Obsolete name	Reynolds
History	Identified in 1974; given the next number in the series of Lu(a–b+) people who made an antibody compatible only with Lu(a–b–) RBCs.

Occurrence

All populations   100%

Several Lu11– probands have been described.

Expression

Cord RBCs	Weak
Altered	Non-detectable by hemagglutination on RBCs of the dominant type of Lu(a–b–)

Effect of enzymes and chemicals on Lu11 antigen on intact RBCs

Ficin/Papain	Resistant
Trypsin	Presumed sensitive
α-Chymotrypsin	Presumed sensitive
Pronase	Sensitive
DTT 200 mM/50 mM	Presumed sensitive/resistant (thus sensitive to WARM™ and ZZAP)

In vitro characteristics of alloanti-Lu11

Immunoglobulin class	IgM and IgG
Optimal technique	RT and IAT

Clinical significance of alloanti-Lu11

Transfusion reaction	No to mild (not much data)
HDFN	No (not much data)

Comments

Few anti-Lu11 have been reported and are typically very weakly reactive. There is no evidence that Lu11 is inherited or carried on the Lutheran glycoprotein.

Lu12 Antigen

Terminology

ISBT symbol (number)	LU12 (005012 or 5.12)
Obsolete names	Muchowski; Much
History	Identified in 1973; given the next number in the series of Lu(a–b+) people who made an antibody compatible only with Lu(a–b–) RBCs.

Occurrence

Only two Lu12– probands have been reported.

Expression

Cord RBCs	Weak
Altered	Non-detectable by hemagglutination on RBCs of the dominant type of Lu(a–b–)

Molecular basis associated with absence of Lu12 antigen[1]

Amino acid	Deletion of Arg34 and Leu35 in IgSF domain 1
Nucleotide	99_104delGCGCTT in exon 2

Effect of enzymes and chemicals on Lu12 antigen on intact RBCs

Ficin/Papain	Resistant
Trypsin	Sensitive
α-Chymotrypsin	Sensitive
Pronase	Sensitive
DTT 200 mM/50 mM	Sensitive/resistant (thus sensitive to WARM™ and ZZAP)

Lutheran

In vitro characteristics of alloanti-Lu12

Immunoglobulin class	IgG
Optimal technique	IAT

Clinical significance of alloanti-Lu12

No data are available because only two examples of anti-Lu12 have been reported.

Comments

Siblings of patients with anti-Lu12 should be tested for compatibility, and the patient urged to donate blood for cryogenic storage when his/her clinical state permits.

Experts recommend transfusing Lu(a–b–) blood (dominant or recessive type) if specific antigen-negative blood is not available.

Reference

[1] Crew, V.K., et al., 2003. Molecular bases of the antigens of the Lutheran blood group system. Transfusion 43, 1729–1737.

Lu13 Antigen

Terminology

ISBT symbol (number)	LU13 (005013 or 5.13)
Obsolete name	Hughes
History	Reported in 1983; given the next number in the series of Lu(a–b+) people who made an antibody compatible only with Lu(a–b–) RBCs.

Occurrence

All populations   100%

Only a few probands have been reported.

Expression

Cord RBCs	Weak
Altered	Non-detectable by hemagglutination on RBCs of the dominant type of Lu(a–b–)

Molecular basis associated with Lu13 antigen[1]

Amino acid	Ser447 in IgSF domain 5 and Gln581 in transmembrane domain
Nucleotide	C at bp 1340 in exon 11 and A at bp 1742 in exon 13
Lu13–	Leu447; Leu581, and T at bp 1340; T at bp 1742

Effect of enzymes and chemicals on Lu13 antigen on intact RBCs

Ficin/Papain	Resistant
Trypsin	Sensitive
α-Chymotrypsin	Sensitive
Pronase	Sensitive
DTT 200 mM/50 mM	Sensitive/resistant (thus sensitive to WARM™ and ZZAP)

In vitro characteristics of alloanti-Lu13

Immunoglobulin class	IgG
Optimal technique	IAT

Clinical significance of alloanti-Lu13

No data are available because only four examples of anti-Lu13 are known.

Comments

Siblings of patients with anti-Lu13 should be tested for compatibility, and the patient urged to donate blood for cryogenic storage when his/her clinical state permits.

Experts recommend transfusing Lu(a–b–) blood (dominant or recessive type) if specific antigen-negative blood is not available.

Reference
[1] Crew, V.K., et al., 2003. Molecular bases of the antigens of the Lutheran blood group system. Transfusion 43, 1729–1737.

Lu14 Antigen

Terminology

ISBT symbol (number)	LU14 (005014 or 5.14)
Obsolete name	Hofanesian
History	Reported in 1977; given the next available number when its antithetical relationship to Lu8 was recognized.

Occurrence

English	1.8%
Danes	1.5%

Antithetical antigen

Lu8 (**LU8**)

Expression

Cord RBCs	Presumed weak

Molecular basis associated with Lu14 antigen[1]

Amino acid	Lys204 in IgSF domain 2
Nucleotide	T at bp 611 in exon 6

Effect of enzymes and chemicals on Lu14 antigen on intact RBCs

Ficin/Papain	Variable
Trypsin	Presumed sensitive
α-Chymotrypsin	Presumed sensitive
Pronase	Sensitive
DTT 200 mM/50 mM	Sensitive/resistant (thus sensitive to WARM™ and ZZAP)

In vitro characteristics of alloanti-Lu14

Immunoglobulin class	IgG
Optimal technique	IAT

Clinical significance of alloanti-Lu14

Transfusion reaction	No data
HDFN	Positive DAT; HDFN in one case; the presence of Lutheran glycoprotein on placental tissue may result in absorption of maternal antibodies to Lutheran antigens

Reference

[1] Crew, V.K., et al., 2003. Molecular bases of the antigens of the Lutheran blood group system. Transfusion 43, 1729–1737.

Lu16 Antigen

Terminology

ISBT symbol (number)	LU16 (005016 or 5.16)
History	Reported in 1980 when three Lu(a+b–) black women were found to have an antibody to a high prevalence antigen in addition to anti-Lu[b].

Occurrence

Only four Lu16– probands have been reported, all were of African-American heritage.

Molecular basis associated with Lu16 antigen[1]

Amino acid	Arg227 in IgSF domain 2
Nucleotide	C at bp 679 in exon 6
Lu16–	Cys227 and T at bp 679

In vitro characteristics of alloanti-Lu16

Immunoglobulin class	IgG
Optimal technique	IAT

Clinical significance of alloanti-Lu16

Transfusion reaction	No data
HDFN	No

Comments

Siblings of patients with anti-Lu16 should be tested for compatibility, and the patient urged to donate blood for cryogenic storage when his/her clinical state permits.

Experts recommend transfusing Lu(a–b–) blood (dominant or recessive type) if specific antigen-negative blood is not available.

Reference

[1] Crew, V.K., et al., 2003. Molecular bases of the antigens of the Lutheran blood group system. Transfusion 43, 1729–1737.

Lu17 Antigen

Terminology

ISBT symbol (number)	LU17 (005017 or 5.17)
Obsolete names	Delcol, nee: Pataracchia
History	Reported in 1979; given the next number in the series of Lu(a–b+) people who made an antibody compatible only with Lu(a–b–) RBCs.

Occurrence

Only one Lu17– proband, Italian, has been reported.

Expression

Cord RBCs	Expressed
Altered	Non-detectable by hemagglutination on RBCs of the dominant type of Lu(a–b–)

Lutheran

Molecular basis associated with Lu17 antigen[1]

Amino acid	Glu114 in IgSF domain 1
Nucleotide	G at bp 340 in exon 3
Lu17–	Lys114 and A at bp 340

Effect of enzymes and chemicals on Lu17 antigen on intact RBCs

Ficin/Papain	Resistant
Trypsin	Sensitive
α-Chymotrypsin	Sensitive
Pronase	Sensitive
DTT 200 mM/50 mM	Sensitive/resistant (thus sensitive to WARM™ and ZZAP)

In vitro characteristics of alloanti-Lu17

Immunoglobulin class	IgG
Optimal technique	IAT

Clinical significance of alloanti-Lu17

Transfusion reaction	*In vivo* RBC survival study suggested that anti-Lu17 might cause modest destruction of transfused RBCs
HDFN	The only anti-Lu17 was made by a woman with four uneventful pregnancies; the presence of Lutheran glycoprotein on placental tissue may result in absorption of maternal antibodies to Lutheran antigens

Comments

Siblings of patients with anti-Lu17 should be tested for compatibility, and the patient urged to donate blood for cryogenic storage when his/her clinical state permits.

Experts recommend transfusing Lu(a–b–) blood (dominant or recessive type) if specific antigen-negative blood is not available.

Reference

[1] Crew, V.K., et al., 2003. Molecular bases of the antigens of the Lutheran blood group system. Transfusion 43, 1729–1737.

Au[a] Antigen

Terminology

ISBT symbol (number)	LU18 (005018 or 5.18)
Obsolete names	Auberger; 204001

| History | Named in 1961 after the first producer of anti-Au^a, a multi-transfused woman; placed into the Lutheran blood group system in 1990. |

Occurrence

| All populations | 90% (originally the prevalence was reported as 82%) |

Antithetical antigen

Au^b (**LU19**)

Expression

| Cord RBCs | Weak |
| Altered | Non-detectable by hemagglutination on RBCs of the dominant type of Lu(a–b–) |

There is considerable variation in the strength of Au^a expression on RBCs. This variation is inherited.

Molecular basis associated with Au^a antigen[1]

| Amino acid | Thr539 in IgSF domain 5 |
| Nucleotide | A at bp 1615 in exon 12 |

Effect of enzymes and chemicals on Au^a antigen on intact RBCs

Ficin/Papain	Resistant
Trypsin	Sensitive
α-Chymotrypsin	Sensitive
Pronase	Sensitive
DTT 200 mM/50 mM	Sensitive/resistant (thus sensitive to WARM™ and ZZAP)

In vitro characteristics of alloanti-Au^a

| Immunoglobulin class | IgG |
| Optimal technique | IAT |

Clinical significance of alloanti-Au^a

| Transfusion reaction | No to mild |
| HDFN | No; the presence of Lutheran glycoprotein on placental tissue may result in absorption of maternal antibodies to Lutheran antigens |

Comments

Only three examples of anti-Au^a have been reported, all in sera containing other antibodies[2].

Lutheran

Due to the scarcity of anti-Au^a, DNA analysis may be used to predict the antigen status.

References

[1] Crew, V.K., et al., 2003. Molecular bases of the antigens of the Lutheran blood group system. Transfusion 43, 1729–1737.

[2] Drachmann, O., et al., 1982. Serological characteristics of the third anti-Au^a. Vox Sang 43, 259–262.

Au^b Antigen

Terminology

ISBT symbol (number)	LU19 (005019 or 5.19)
Obsolete name	204002
History	Reported in 1989 and named because it is antithetical to Au^a.

Occurrence

Caucasians	51%
Blacks	68%

Antithetical antigen

Au^a (**LU18**)

Expression

Cord RBCs	Weak
Altered	Non-detectable by hemagglutination on RBCs of the dominant type of Lu(a–b–)

There is considerable variation in the strength of Au^b expression on RBCs. This variation is inherited.

Molecular basis associated with Au^b antigen[1]

Amino acid	Ala539 in IgSF domain 5
Nucleotide	G at bp 1615 in exon 12

Effect of enzymes and chemicals on Au^b antigen on intact RBCs

Ficin/Papain	Resistant
Trypsin	Sensitive
α-Chymotrypsin	Sensitive
Pronase	Sensitive
DTT 200 mM/50 mM	Presumed sensitive/resistant (thus sensitive to WARM™ and ZZAP)

In vitro characteristics of alloanti-Au^b

Immunoglobulin class	IgG
Optimal technique	IAT

Clinical significance of alloanti-Au^b

Transfusion reaction	No to mild
HDFN	No; the presence of Lutheran glycoprotein on placental tissue may result in absorption of maternal antibodies to Lutheran antigens

Comments

Four examples of anti-Au^b have been reported, all in sera also containing anti-Lu^a.

Due to the scarcity of anti-Au^b, DNA analysis may be used to predict the antigen status.

Reference

[1] Crew, V.K., et al., 2003. Molecular bases of the antigens of the Lutheran blood group system. Transfusion 43, 1729–1737.

Lu20 Antigen

Terminology

ISBT symbol (number)	LU20 (005020 or 5.20)
History	Reported in 1992, antibody made by an Israeli thalassemic; given the next number in the series of Lu(a–b+) people who made an antibody compatible only with Lu(a–b–) RBCs.

Occurrence

Only one Lu20– proband, an Israeli, has been reported.

Expression

Cord RBCs	Weak
Altered	Non-detectable by hemagglutination on RBCs of the dominant type of Lu(a–b–)

Molecular basis associated with Lu20 antigen[1]

Amino acid	Thr302 in IgSF domain 3
Nucleotide	C at bp 905 in exon 7
Lu20–	Met302 and T at bp 905

Effect of enzymes and chemicals on Lu20 antigen on intact RBCs

Ficin/Papain	Resistant
Trypsin	Sensitive
α-Chymotrypsin	Sensitive
DTT 200 mM/50 mM	Sensitive/resistant (thus sensitive to WARM™ and ZZAP)

In vitro characteristics of alloanti-Lu20

Immunoglobulin class	IgG
Optimal technique	37°C; IAT

Clinical significance of alloanti-Lu20

Not known since only one example of anti-Lu20 has been described.

Comments

Siblings of patients with anti-Lu20 should be tested for compatibility, and the patient urged to donate blood for cryogenic storage when his/her clinical state permits.

Experts recommend transfusing Lu(a–b–) blood (dominant or recessive type) if specific antigen-negative blood is not available.

Reference

[1] Crew, V.K., et al., 2003. Molecular bases of the antigens of the Lutheran blood group system. Transfusion 43, 1729–1737.

Lu21 Antigen

Terminology

ISBT symbol (number)	LU21 (005021 or 5.21)
History	Reported in 2002; given the next number in the series of Lu(a–b+) people who made an antibody compatible only with Lu(a–b–) RBCs.

Occurrence

Only one Lu21– proband, an Israeli, has been reported.

Expression

Cord RBCs	Weak
Altered	Non-detectable by hemagglutination on RBCs of the dominant type of Lu(a–b–)

Molecular basis associated with Lu21 antigen[1]

Amino acid	Asp94 in IgSF domain 1
Nucleotide	C at bp 282 in exon 3
Lu21–	Glu94 and G at bp 282

Effect of enzymes and chemicals on Lu21 antigen on intact RBCs

Ficin/Papain	Resistant
Trypsin	Sensitive
α-Chymotrypsin	Sensitive
DTT 200 mM/50 mM	Sensitive/resistant (thus sensitive to WARM™ and ZZAP)

In vitro characteristics of alloanti-Lu21

Immunoglobulin class	IgG; IgM
Optimal technique	IAT; RT and 37°C

Clinical significance of alloanti-Lu21

Transfusion reaction	No data
HDFN	No in the proband's 2nd, 3rd, and 4th pregnancies; the presence of Lutheran glycoprotein on placental tissue may result in absorption of maternal antibodies to Lutheran antigens

Comments

Siblings of patients with anti-Lu21 should be tested for compatibility, and the patient urged to donate blood for cryogenic storage when his/her clinical state permits.

Experts recommend transfusing Lu(a–b–) blood (dominant or recessive type) if specific antigen-negative blood is not available.

Reference

[1] Crew, V.K., et al., 2003. Molecular bases of the antigens of the Lutheran blood group system. Transfusion 43, 1729–1737.

LURC Antigen

Terminology

ISBT symbol (number)	LURC (005022 or 5.22)
History	Reported in 2009 and named "LU" for the system and "RC" for the amino acid change (Arg to Cys) in the antigen-negative phenotype.

Lutheran

Occurrence

Only one LURC– proband has been described.

Expression

Cord RBCs	Weak
Altered	Non-detectable by hemagglutination on RBCs of the dominant type of Lu(a–b–)

Molecular basis associated with LURC antigen[1]

Amino acid	Arg75 in IgSF domain 1
Nucleotide	C at bp 223 in exon 3
LURC–	Cys75 and T at bp 223

The LURC– proband was heterozygous for *LU*230A/G* (*LU*01/LU*02*) and *LU*586G/A* (a polymorphism described by Gowland, et al.[2], as well as for the *LU*223C/T* (*LU*02.22/LU*02.–22*)[1]. The presence of Lu^b (Arg77) is required for expression of LURC. LURC– RBCs have a weak expression of Lu^b and other high prevalence antigens[2].

Effect of enzymes and chemicals on LURC antigen on intact RBCs

Ficin/Papain	Resistant
Trypsin	Sensitive
α-Chymotrypsin	Sensitive
DTT 200 mM/50 mM	Sensitive/resistant (thus sensitive to WARM™ and ZZAP)

In vitro characteristics of alloanti-LURC

Immunoglobulin class	IgG
Optimal technique	IAT

Clinical significance of alloanti-LURC

No data as only one proband with anti-LURC has been reported.

Comments

Siblings of patients with anti-LURC should be tested for compatibility, and the patient urged to donate blood for cryogenic storage when his/her clinical state permits.

Experts recommend transfusing Lu(a–b–) blood (dominant or recessive type) if specific antigen-negative blood is not available.

References

[1] Karamatic Crew, V., et al., 2009. Two heterozygous mutations in an individual result in the loss of a novel high incidence Lutheran antigen Lurc [abstract]. Transfus Med 19 (Suppl. 1), 10.

[2] Gowland, P., et al., 2005. A new polymorphism within the Lu^a/Lu^b blood group [abstract]. Transf Med Hemother 32 (Suppl. 1), 54–55.

Kell Blood Group System

Number of antigens 34

Polymorphic	K
Low prevalence	Kpᵃ, Jsᵃ, Ulᵃ, K17, Kpᶜ, K23, K24, VLAN, VONG, KYO
High prevalence	k, Kpᵇ, Ku, Jsᵇ, K11, K12, K13, K14, K16, K18, K19, Km, K22, TOU, RAZ, KALT, KTIM, KUCI, KANT, KASH, KELP, KETI, KHUL

Terminology

ISBT symbol (number)	KEL (006)
CD number	CD238
History	Named in 1946 after the first antibody producer (Mrs. Kelleher) of anti-K that caused HDFN.

Expression

Other blood cells	Appears early in erythropoiesis, but may also be expressed on myeloid progenitors and possibly on megakaryocytes
Tissues	Primarily in bone marrow, fetal liver, testes; lesser amounts in other tissues including various parts of the brain, lymphoid organs, heart, and skeletal muscle.

Gene

Chromosome	7q34
Name	*KEL*
Organization	19 exons distributed over 21.5 kbp of gDNA
Product	Kell glycoprotein

The Blood Group Antigen (3/e). DOI: http://dx.doi.org/10.1016/B978-0-12-415849-8.00008-9

Database accession numbers

GenBank	M64934 (mRNA); AH008123 (gene) NM_000420 (mRNA)
Entrez Gene ID	3792

Molecular basis of Kell phenotypes

The reference allele, *KEL*02* (Accession number M64934) encodes k (KEL2) KEL4, KEL5, KEL7, KEL11, KEL12, KEL13, KEL14, KEL18, KEL19, KEL20, KEL22, KEL26, KEL27, KEL29, KEL30, KEL32, KEL33, KEL35, KEL36. Nucleotide differences from this reference allele, and the amino acids affected, are given.

Allele encodes	Allele name	Exon	Nucleotide[†]	Restriction enzyme	Amino acid	Ethnicity (prevalence)
K+k− or KEL:1,−2	KEL*01.01	6	578C>T	BsmI+	Thr193Met	Caucasians (9%), Blacks (2%), Iranian Jews (12%)
K+w or KEL:1weak	KEL*01.02	6	577A>T	BsmI as ref. allele	Thr193Ser	(Rare)
K+w, Kp(a+) or KEL:1weak,3	KEL*01.03	6,8	578C>T 841C>T		Thr193Met Arg281Trp	(Rare)[1]
Kp(a+b−c−) or KEL:3,−4,−21	KEL*02.03	8	841C>T	NlaIII+	Arg281Trp	Caucasians (2%)
Js(a+b−) or KEL:6,−7	KEL*02.06	17	1790T>C	MnlI−	Leu597Pro	Blacks (20%)
Ul(a+) or KEL:10	KEL*02.10	13	1481A>T	AccI+	Glu494Val	Finns > Japanese (Several)
K12− or KEL:−12	KEL*02.−12	15	1643A>G	NlaIII−	His548Arg	Caucasians (Rare)
K14− or KEL:−14,−24	KEL*02.−14.1	6	538C>T		Arg180Cys	Japanese (Rare)
K14− or KEL:−14	KEL*02.−14.2	6	539G>A		Arg180His	Japanese (Rare)
Wk(a+) K11− or KEL:−11,17	KEL*02.17	8	905T>C	HaeII+	Val302Ala	(Several)
K18− or KEL:−18	KEL*02.−18.1	4	388C>T	TaqII+	Arg130Trp	(Rare)
K18− or KEL:−18	KEL*02.−18.2	4	389G>A	Eco57+	Arg130Gln	(Rare)
K19− or KEL:−19	KEL*02.−19	13	1475G>A		Arg492Gln	(Rare)
Kp(c+) or KEL:−3,−4,21	KEL*02.21	8	842G>A	PvuII+	Arg281Gln	Japanese > Caucasians (Rare)
K22− or KEL:−22	KEL*02.−22	9	965C>T	Tsp45I+	Ala322Val	Iranian Jews (Rare)
K23+ or KEL:23	KEL*02.23	10	1145A>G	BcnI+	Gln382Arg	(Rare)
K24+ or KEL:−14,24	KEL*02.24	6	539G>C	HaeIII+	Arg180Pro	French Cajuns (Rare)

(Continued)

(Continued)

Allele encodes	Allele name	Exon	Nucleotide†	Restriction enzyme	Amino acid	Ethnicity (prevalence)
VLAN+VONG− or KEL:25,−28	KEL*02.25	8	743G>A	PspGI+^	Arg248Gln	(Rare)
TOU− or KEL:−26	KEL*02.−26	11	1217G>A		Arg406Gln	Native American, Hispanic (Rare)
RAZ− or KEL:−27	KEL*02.−27	8	745G>A	EcoRI+^	Glu249Lys	(Rare)
VLAN−VONG+ or KEL:−25,28	KEL*02.28	8	742C>T		Arg248Trp	Chinese (Rare)
KALT− or KEL:−29	KEL*02.−29	17	1868G>A	Tfil−^	Arg623Lys	(Rare)
KTIM− or KEL:−30	KEL*02.−30	8	913G>A	TaqI−	Asp305Asn	(Rare)
KYO+ or KEL:31	KEL*02.31	8	875G>A		Arg292Gln	Japanese (Rare)
KUCI− or KEL:−32	KEL*02.−32	11	1271C>T	FnuHI−	Ala424Val	American Indian (Rare)
KANT− or KEL:−33	KEL*02.−33	11	1283G>T		Arg428Leu	European (Rare)
KASH− or KEL:−34	KEL*02.−34	8	758A>G		Tyr253Cys	(Rare)
KELP− or KEL:−35	KEL*02.−35	8 / 18	780G>T / 2024G>A	GsaI− / TaqI−	Leu260Phe / Arg675Gln	(Rare)
KETI− or KEL:−36	KEL*02.−36^^	12	1391C>T	BsmAI−	Thr464Ile	European (Few)
KHUL− or KEL:−37	KEL*02.−37	8	877C>T		Arg293Trp	Asian (Rare)

†Nucleotide #1 is the first nucleotide of the translation/initiation codon, which is 120bp downstream from that given in early reports.
^Nucleotide change(s) introduced into primer(s).
^^Has been also found on an allele with the 905 T>C change associated with K11–K17+.

Kell

Molecular bases for silencing of *KEL*

Homozygosity or compound heterozygosity leads to the K_0 (Kell$_{null}$) phenotype. Nucleotide differences from *KEL*02* reference allele (Accession number M64934), and the amino acids affected, are given.

Allele name	Exon/ intron	Nucleotide[†]	Restriction enzyme	Amino acid	Ethnicity (prevalence)
*KEL*01N.01*	15	1678C>G		Pro560Ala	(Rare)
*KEL*02N.01*	Intron 3	IVS3+1g>c	*Dde*I+	Alternative splicing	Taiwanese (Rare)
*KEL*02N.02*	4 17	382C>T on a 1790T>C background		Arg128Stop Leu597Pro	Blacks (Rare)
*KEL*02N.03*	4	246T>A		Cys82Stop	Yugoslavians (Rare)
*KEL*02N.04*	9	1042C>T	*Tsp*45I –	Gln348Stop	Portuguese, Caucasians (Rare)
*KEL*02N.05*	18	2027G>A	*Alu*I–	Ser676Asn	Israeli (Rare)
*KEL*02N.06*	Intron 3	IVS3+1g>a	*Nla*III+	Alternative splicing	Reunion Islands (Few)
*KEL*020N.7*	6	574C>T		Arg192Stop	(Rare)
*KEL*02N.08*	Intron 5	IVS5–2a>g		Alternative splicing	Japanese (Rare)
*KEL*02N.09*	12	1377G>A		Trp459Stop	Japanese (Rare)
*KEL*02N.10*	13	1420C>T		Gln474Stop	Swedish (Rare)
*KEL*02N.11*	8	903delG		fs, Stop	Swedish (Rare)
*KEL*02N.12*	Intron 8	IVS8+1g>a		Alternative splicing	(Rare)
*KEL*02N.13*	Intron 8	IVS8+1g>t		Alternative splicing	(Rare)
*KEL*02N.14*	9	948G>A		Trp316Stop	(Rare)
*KEL*02N.15*	11	1216C>T		Arg406Stop	(Rare)
*KEL*02N.16*	13	1477C>T		Gln493Stop	(Rare)

(Continued)

(Continued)

Allele name	Exon/intron	Nucleotide[†]	Restriction enzyme	Amino acid	Ethnicity (prevalence)
KEL*02N.17	14	1546C>T		Arg516Stop	(Rare)
KEL*02N.18	Obsolete				
KEL*02N.19	18	2023C>T		Arg675Stop	(Rare)
KEL*02N.20	15	1596G>A		Trp532Stop	(Rare)
KEL*02N.21	18	1947C>G		Tyr649Stop	(Rare)
KEL*02N.22	Intron 7	IVS7–1g>c		Alternative splicing	(Rare)
KEL*02N.23	3	185insT		Glu239Stop	Chinese (Rare)

[†]Nucleotide #1 is the first nucleotide of the translation/initiation codon, which is 120 bp downstream from that given in early reports.

Molecular bases of weak Kell antigens

Homozygosity or compound heterozygosity leads to the Kell$_{mod}$ phenotype. Nucleotide differences from the *KEL*02* reference allele (Accession number M64934), and amino acids affected, are given. K$_{mod}$ is an umbrella term used to describe various phenotypes with very weak expression of Kell antigens and increased expression of Kx antigen. Classification of a mod phenotype may depend on the reagents and techniques used.

Allele name	Exon	Nucleotide[†]	Restriction enzyme	Amino acid	Ethnicity (prevalence)	Other
KEL*01M.01	6	578C>G		Thr193Arg	Taiwanese (Rare)	KEL:1 weak
KEL*02M.01	10	1088G>A	HaeIII+	Ser363Asn	Caucasian (Rare)	
KEL*02M.02	18	2030A>G		Tyr677Cys	Caucasian	
KEL*02M.03	9	986T>C		Leu329Pro	Caucasian (Rare)	KEL:–13
KEL*02M.04	19	2107G>A		Gly703Arg	Caucasian (Rare)	

(Continued)

(Continued)

Allele name	Exon	Nucleotide[†]	Restriction enzyme	Amino acid	Ethnicity (prevalence)	Other
KEL*02M.05	16	1719C>T		Gly573Gly	(Rare)	
KEL*02M.06	4	306C>A		Asp102Glu	(Rare)	
	11	1298C>T		Pro433Leu		
KEL*02M.07	16	1763A>G		Tyr588Cys	(Rare)	
KEL*02M.08	13	1490A>T		Asp497Val	(Rare)	
KEL*02M.09	16	1757T>G		Ile586Ser	(Rare)	
KEL*02M.10	8	787G>A		Gly263Arg	(Rare)	
KEL*02M.11	11	1268C>T		Ala423Val	Caucasian (Rare)	KEL:2 weak. KEL:7 weak.

[†]Nucleotide #1 is the first nucleotide of the translation/initiation codon, which is 120 bp downstream from that given in early reports.

Amino acid sequence[2]

```
MEGGDQSEEE  PRERSQAGGM  GTLWSQESTP  EERLPVEGSR  PWAVARRVLT   50
AILILGLLLC  FSVLLFYNFQ  NCGPRPCETS  VCLDLRDHYL  ASGNTSVAPC  100
TDFFSFACGR  AKETNNSFQE  LATKNKNRLR  RILEVQNSWH  PGSGEEKAFQ  150
FYNSCMDTLA  IEAAGTGPLR  QVIEELGGWR  ISGKWTSLNF  NRTLRLLMSQ  200
YGHFPFFRAY  LGPHPASPHT  PVIQIDQPEF  DVPLKQDQEQ  KIYAQIFREY  250
LTYLNQLGTL  LGGDPSKVQE  HSSLSISITS  RLFQFLRPLE  QRRAQGKLFQ  300
MVTIDQLKEM  APAIDWLSCL  QATFTPMSLS  PSQSLVVHDV  EYLKNMSQLV  350
EEMLLKQRDF  LQSHMILGLV  VTLSPALDSQ  FQEARRKLSQ  KLRELTEQPP  400
MPARPRWMKC  VEETGTFFEP  TLAALFVREA  FGPSTRSAAM  KLFTAIRDAL  450
ITRLRNLPWM  NEETQNMAQD  KVAQLQVEMG  ASEWALKPEL  ARQEYNDIQL  500
GSSFLQSVLS  CVRSLRARIV  QSFLQPHPQH  RWKVSPWDVN  AYYSVSDHVV  550
VFPAGLLQPP  FFHPGYPRAV  NFGAAGSIMA  HELLHIFYQL  LLPGGCLACD  600
NHALQEAHLC  LKRHYAAFPL  PSRTSFNDSL  TFLENAADVG  GLAIALQAYS  650
KRLLRHHGET  VLPSLDLSPQ  QIFFRSYAQV  MCRKPSPQDS  HDTHSPPHLR  700
VHGPLSSTPA  FARYFRCARG  ALLNPSSRCQ  LW                      732
```

Carrier molecule

Single-pass RBC membrane glycoprotein (type II) that is highly folded via disulfide bonds. In the RBC membrane Kell glycoprotein is covalently linked at Cys72 to the Cys347 of the XK protein.

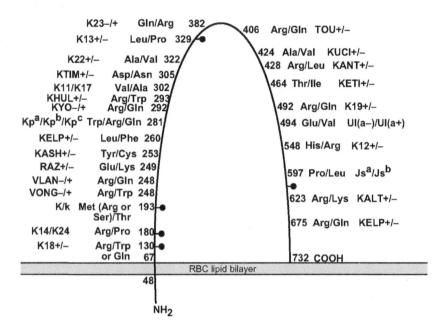

M_r (SDS-PAGE)	93,000; 79,000–80,000 after N-glycanase treatment
CHO: N-glycan	5 sites
Cysteine residues	16 (1 of which is in the membrane)
Copies per RBC	3,500–18,000[3]

Function

Kell glycoprotein is an endothelin-3-converting enzyme, preferentially cleaving big endothelin-3, creating bioactive endothelin-3, which is a potent vasoconstrictor. Kell glycoprotein is a member of the Neprilysin (M13) sub-family of zinc endopeptidases and, in common with all of them, shares a pentameric sequence, HEXXH, which is central to zinc binding and catalytic activity[2].

Disease association

In one study, 1 in 250 patients with AIHA had autoantibodies directed to Kell system antigens. Transiently depressed Kell system antigens have been associated with the presence of autoantibodies mimicking alloantibodies in AIHA, and with microbial infection. The Kell protein was reduced in one case of ITP[4]. Kell antigens are weak on RBCs from McLeod CGD (X-linked type) males. Antibodies to antigens in the Kell blood group system have caused HDFN, due both to immune destruction of RBCs and, more significantly, suppression of erythropoiesis[5]. This can result in severe anemia, which may be prolonged and without overt signs of hemolysis.

Phenotypes (% occurrence)

Phenotype	Caucasians	Blacks
K–k+	91	98
K+k–	0.2	Rare
K+k+	8.8	2
Kp(a+b–)	Rare	0
Kp(a–b+)	97.7	100
Kp(a+b+)	2.3	Rare
Js(a+b–)	0	1
Js(a–b+)	100	80
Js(a+b+)	Rare	19

Null: K_0; very rare, but a little less rare in Finland, Japan, and Reunion Islands

Unusual: K_{mod}, McLeod (see Kx blood group system [**XK, 019**]) and table below showing comparison of Kell phenotypes), Kp(a+b–), Leach and Gerbich types of Ge-negative

Comparison of Kell phenotypes[6]

Phenotype	Expression of antigen		Possible antibody in serum	RBC morphology
	Kell system	Kx		
Inherited Kell system phenotypes				
Common	Normal	Weak	Alloantibody	Discocytes
Kp(a+b–)	Slight/moderate reduction	Slight increase	Anti-Kp^b	Discocytes
K_{mod}	Marked reduction^	Moderate increase	Anti-Ku-like (not mutually compatible)	Discocytes
K_0 heterozygote	Normal	Moderate increase	None	Discocytes

(Continued)

(Continued)

Phenotype	Expression of antigen		Possible antibody in serum	RBC morphology
	Kell system	Kx		
K₀	None†	Marked increase††	Anti-Ku	Discocytes

Inherited Kx system phenotypes

Phenotype	Expression of antigen		Possible antibody in serum	RBC morphology
McLeod CGD	Marked reduction	None	Anti-KL (anti-Kx+ anti-Km)	Acanthocytes
McLeod non-CGD	Marked reduction	None	Anti-Km (anti-Kx in one case)	Acanthocytes
McLeod carriers‡	Normal to marked reduction	Not known	None	Discocytes and acanthocytes

Other

Phenotype	Expression of antigen		Possible antibody in serum	RBC morphology
Gerbich and Leach phenotypes	Slight decrease	Normal/ weak	Not Kell-related	Discocytes and elliptocytes in Leach phenotype
AIHA (Kell-related)	Normal to marked reduction	Slight increase (when Kell reduced)	"Kell-related" antibodies or non-specific	Discocytes or spherocytes (due to the hemolytic anemia)
Thiol-treated RBCs	Not detectable	Slight increase	Not applicable	Not known

^Will adsorb and elute antibody to inherited antigens in Kell system.
†Do not adsorb and elute.
††Xk protein is decreased; antigen may be more accessible.
‡The proportion of normal to McLeod phenotype RBCs varies in different carrier females. Only affected males present with 100% of RBCs having the McLeod phenotype.

Comparison of features of McLeod phenotype with normal and K_0 RBCs

Features	Normal Kell phenotype	K_0	McLeod non-CGD	McLeod CGD
Kell system antigens	++++	0	Weak	Weak
Kx antigen	+	++	0	0
Km antigen	++	0	0	0
Antibodies made	To lacking Kell antigens	Anti-Ku	Anti-Km (anti-Kx in one case)	Anti-Kx+ −Km
Creatine kinase level	Normal	Normal	Elevated	Normal or elevated
Blood for transfusion	Normal antigen-negative phenotype	K_0	McLeod or K_0	McLeod
Gene defect	Not applicable	Changes in *KEL*	Changes in *XK*	Deletion of *XK* and *CGD*^
Morphology	Discocytes	Discocytes	Acanthocytes	Acanthocytes
Pathology	None	None	Muscular and neurological defects	Muscular and neurological defects with CGD

^=The official name for the CDG gene is now *CBB*.

Comments

It is incorrect to refer to the K and k antigens as, respectively, K1 and K2; in the numerical terminology they should be referred to as KEL1 and KEL2. However, some Kell antigens were assigned a traditional numerical name, e.g., K11 and K14, and thus can be referred to as either K11 or KEL11, and K14 or KEL14.

Historically, no Kell system haplotype had more than one low-prevalence antigen. Recently, a novel *KEL*1,3* allele was reported that encoded K and Kp^a on the same molecule[1].

The Kp^a antigen *in cis* weakens the expression of Kell antigens (*cis*-modifying effect)[1,7]. K_{mod} is an umbrella term used to describe various phenotypes with very weak expression of Kell antigens and increased expression of Kx.

Kell antigens are sensitive to treatment by a mixture of α-chymotrypsin and trypsin or to sequential treatment of antigen-positive RBCs with these enzymes.

Antibodies produced by K_{mod} individuals are not necessarily mutually compatible.

References

[1] Körmöczi, G.F., et al., 2009. A novel KEL*1,3 allele with weak Kell antigen expression confirming the cis-modifier effect of KEL3. Transfusion 49, 733–739.

[2] Lee, S., 2007. The value of DNA analysis for antigens of the Kell and Kx blood group systems. Transfusion 47 (Suppl.), 32S–39S.

[3] Parsons, S.F., et al., 1993. Monoclonal antibodies against Kell glycoprotein: serology, immunochemistry and quantification of antigen sites. Transf Med 3, 137–142.

[4] Williamson, L.M., et al., 1994. Transient loss of proteins carrying Kell and Lutheran red cell antigens during constructive relapses of autoimmune thrombocytopenia. Br J Haematol 87, 805–812.

[5] Vaughan, J.I., et al., 1998. Inhibition of erythroid progenitor cells by anti-Kell antibodies in fetal alloimmune anemia. N Engl J Med 338, 798–803.

[6] Øyen, R., et al., 1997. Review: conditions causing weak expression of Kell system antigens. Immunohematology 13, 75–79.

[7] Yazdanbakhsh, K., et al., 1999. Identification of a defect in the intracellular trafficking of a Kell blood group variant. Blood 94, 310–318.

K Antigen

Terminology

ISBT symbol (number)	KEL1 (006001 or 6.1)
Obsolete names	Kell; K1
History	Named after first antibody producer (Mrs. Kelleher) of anti-K, which caused HDFN; reported in 1946.

Occurrence

Caucasians	9%
Blacks	2%
Asians	Rare
Iranian Jews	12%
Arabs	As high as 25%

Antithetical antigen

k (**KEL2**)

Expression

Cord RBCs	Expressed
Altered	See table showing Comparison of Kell Phenotypes on System pages
	Weak with Arg193 or Ser193

Molecular basis associated with K antigen[1,2,3]

Amino acid	Met193
Nucleotide	T at bp 578 (previously reported as 698) in exon 6
Weak expression of K	578C>G (Thr193Arg) and 577A>T (Thr193Ser)

All three changes disrupt the N-glycosylation motif, so Asn191 is not glycosylated.

Effect of enzymes and chemicals on K antigen on intact RBCs

Ficin/Papain	Resistant
Trypsin	Resistant
α-Chymotrypsin	Resistant*
DTT 200 mM/50 mM	Sensitive/sensitive (thus sensitive to WARM™ and ZZAP)
Acid	Sensitive (thus sensitive to EGA)

*May be weakened or sensitive if the enzyme preparation is contaminated with trypsin.

In vitro characteristics of alloanti-K

Immunoglobulin class	IgG more common than IgM
Optimal technique	IAT, sometimes RT; may not react well by LISS procedures
Complement binding	Rare

Clinical significance of alloanti-K

Transfusion reaction	Mild to severe/delayed/hemolytic
HDFN	Mild to severe (rare); often with anemia, which is sometimes delayed

Comments

Some bacteria elicit production of IgM anti-K. Expression of K can be acquired as a result of bacterial activity *in vivo* and *in vitro*.

References

[1] Lee, S., et al., 1995. Molecular basis of the Kell (K1) phenotype. Blood 85, 912–916.

[2] Poole, J., et al., 2006. A KEL gene encoding serine at position 193 of the Kell glycoprotein results in expression of KEL1 antigen. Transfusion 46, 1879–1885.

[3] Uchikawa, M., et al., 2000. Molecular basis of unusual Kmod phenotype with K+ᵂk−[abstract]. Vox Sang 78 (Suppl. 1), 0011.

k Antigen

Terminology

ISBT symbol (number)	KEL2 (006002 or 6.2)
Obsolete names	Cellano; K2
History	Identified in 1949 when an antibody was shown to recognize the antithetical antigen to K. Cellano, the original name, was derived by rearranging the proband's last name (Nocella).

Occurrence

Caucasians	99.8%
Blacks	100%

Antithetical antigen

K (**KEL1**)

Expression

Cord RBCs	Expressed
Altered	See table showing Comparison of Kell Phenotypes on System pages Weakened in rare genetic variants; weak expression with concomitant 423Val[1]

Molecular basis associated with k antigen[2]

Amino acid	Thr193
Nucleotide	C at bp 578 (previously reported as 698) in exon 6

Effect of enzymes and chemicals on k antigen on intact RBCs

Ficin/Papain	Resistant
Trypsin	Resistant
α-Chymotrypsin	Resistant (see K [**KEL1**])
DTT 200 mM/50 mM	Sensitive/sensitive (thus sensitive to WARM™ and ZZAP)
Acid	Sensitive (thus sensitive to EGA)

In vitro characteristics of alloanti-k

Immunoglobulin class	IgG more common than IgM
Optimal technique	IAT

Clinical significance of alloanti-k

Transfusion reaction	Mild to moderate/delayed
HDFN	Mild to severe (rare)

Comments

Siblings of patients with anti-k should be tested for compatibility, and the patient urged to donate blood for cryogenic storage when his/her clinical state permits.

References

[1] Lee, S., 1997. Molecular basis of Kell blood group phenotypes. Vox Sang 73, 1–11.
[2] Lee, S., et al., 1995. Molecular basis of the Kell (K1) phenotype. Blood 85, 912–916.

Kp^a Antigen

Terminology

ISBT symbol (number)	KEL3 (006003 or 6.3)
Obsolete names	Penny; K3
History	Identified in 1957; the antigen, which was shown to be related to the Kell System, took its name from "K" for "Kell" and "p" for the first letter of the antibody producer's name (Penny).

Occurrence

Caucasians	2%
Blacks	<0.01%

Antithetical antigens

Kp^b (**KEL4**), Kp^c (**KEL21**)

Expression

Cord RBCs	Expressed
Altered	See table showing Comparison of Kell Phenotypes on System pages

Molecular basis associated with Kp^a antigen[1]

Amino acid	Trp281
Nucleotide	T at bp 841 in exon 8 (previously reported as 961)

Effect of enzymes and chemicals on Kp^a antigen on intact RBCs

Ficin/Papain	Resistant
Trypsin	Resistant
α-Chymotrypsin	Resistant (see K [**KEL1**])
DTT 200 mM/50 mM	Sensitive/sensitive (thus sensitive to WARM™ and ZZAP)
Acid	Sensitive (thus sensitive to EGA)

In vitro characteristics of alloanti-Kp^a

Immunoglobulin class	IgG
Optimal technique	IAT

Clinical significance of alloanti-Kp^a

Transfusion reaction	Mild to moderate/delayed
HDFN	Mild to severe

Comments

In the presence of Kp^a, other inherited Kell system antigens are suppressed (*cis*-modifier effect) to varying degrees.

Until recently, in people with K+Kp(a+) RBCs, *K* was always *in trans* to *Kp^a*. However, Körmöczi et al. reported a novel *KEL*1,3* allele that encoded K and Kp^a on the same molecule[2].

Anti-Kp^a is often found with anti-K.

References

[1] Lee, S., et al., 1996. Point mutations characterize *KEL10*, the *KEL3*, *KEL4*, and *KEL21* alleles, and the *KEL17* and *KEL11* alleles. Transfusion 36, 490–494.

[2] Körmöczi, G.F., et al., 2009. A novel *KEL*1,3* allele with weak Kell antigen expression confirming the cis-modifier effect of KEL3. Transfusion 49, 733–739.

Kp^b Antigen

Terminology

ISBT symbol (number)	KEL4 (006004 or 6.4)
Obsolete names	Rautenberg; K4
History	Identified in 1958 and recognized to be antithetical to Kp^a.

Occurrence

All populations	100%

Antithetical antigens

Kp^a (**KEL3**); Kp^c (**KEL21**)

Expression

Cord RBCs   Expressed
Altered   Weak on RBCs from some patients with AIHA
See table showing Comparison of Kell phenotypes on System pages.

Molecular basis associated with Kp^b antigen[1]

Amino acid   Arg281
Nucleotide   C at bp 841 (previously reported as 961), G at bp
                         842 (previously reported as 962) in exon 8

Effect of enzymes and chemicals on Kp^b antigen on intact RBCs

Ficin/Papain   Resistant
Trypsin   Resistant
α-Chymotrypsin   Resistant (see K [**KEL1**])
DTT 200 mM/50 mM   Sensitive/sensitive (thus sensitive to WARM™ and
                         ZZAP)
Acid   Sensitive (thus sensitive to EGA)

In vitro characteristics of alloanti-Kp^b

Immunoglobulin class   IgG, rarely IgM
Optimal technique   IAT

Clinical significance of alloanti-Kp^b

Transfusion reaction   No to moderate/delayed
HDFN   Mild to moderate

Autoantibody

Yes. May appear as alloantibody on initial presentation due to suppression of Kp^b antigen.

Comments

Sera containing anti-Kp^b often contain anti-K (see **KEL1**).
Siblings of patients with anti-Kp^b should be tested for compatibility, and the patient urged to donate blood for cryogenic storage when his/her clinical state permits.

Reference

[1] Lee, S., et al., 1996. Point mutations characterize *KEL10*, the *KEL3*, *KEL4*, and *KEL21* alleles, and the *KEL17* and *KEL11* alleles. Transfusion 36, 490–494.

Ku Antigen

Terminology

ISBT symbol (number)	KEL5 (006005 or 6.5)
Obsolete names	Peltz; K5
History	Antibody in serum of K_0 [K–k–Kp(a–b–)] person identified in 1957; originally called anti-KkKp^a or anti-Peltz (after the proband); renamed anti-Ku (K for Kell, u for universal) in 1961.

Occurrence

All populations	100%

Expression

Cord RBCs	Expressed
Altered	See table showing Comparison of Kell phenotypes on System pages.

Molecular basis associated with Ku antigen

For the molecular basis associated with the K_0 phenotype, see Kell System pages.

Effect of enzymes and chemicals on Ku antigen on intact RBCs

Ficin/Papain	Resistant
Trypsin	Resistant
α-Chymotrypsin	Resistant (see K [**KEL1**])
DTT 200 mM/50 mM	Sensitive/sensitive (thus sensitive to WARM™ and ZZAP)
Acid	Sensitive (thus sensitive to EGA)

In vitro characteristics of alloanti-Ku

Immunoglobulin class	IgG
Optimal technique	IAT

Clinical significance of alloanti-Ku

Transfusion reaction	Mild to severe
HDFN	No to moderate

Autoantibody

Yes

Comments

Anti-Ku is made by K_0 people who may make additional antibodies directed at other Kell antigens and, rarely, make Kell system specificities without making anti-Ku. K_{mod} people make Ku-like antibodies that are not necessarily mutually compatible.

An antibody detected only in the presence of trimethoprim, found in co-trimoxazole (CTMX; a combination of two drugs, trimethoprim [the culprit] and sulfamethoxazole, present in the suspension medium of some reagent RBCs) was identified as anti-Ku[1].

Siblings of patients with anti-Ku should be tested for compatibility, and the patient urged to donate blood for cryogenic storage when his/her clinical state permits.

Reference

[1] Le Pennec, P., et al., 1999. Sulfamethoxazole and trimethoprim dependent antibodies with respective anti-H (H1) and anti-Ku (KEL5) specificity [abstract]. Transfusion 39 (Suppl.), 81S.

Js^a Antigen

Terminology

ISBT symbol (number)	KEL6 (006006 or 6.6)
Obsolete names	Sutter; K6
History	Described in 1958; "J" is from the first name (John) and "s" is from the last name (Sutter) of the first producer of the antibody. Js^a was shown to belong to the Kell System in 1965.

Occurrence

Caucasians	<0.01%
Blacks	20%

Antithetical antigen

Js^b (**KEL7**)

Expression

Cord RBCs	Expressed
Altered	See table showing Comparison of Kell phenotypes on System pages

Molecular basis associated with Js^a antigen[1]

Amino acid	Pro597
Nucleotide	C at bp 1790 in exon 17 (previously reported as 1910)

Effect of enzymes and chemicals on Js^a antigen on intact RBCs

Ficin/Papain	Resistant
Trypsin	Resistant
α-Chymotrypsin	Resistant (see K [**KEL1**])
DTT 200 mM/50 mM	Sensitive/sensitive (see Comments) (thus sensitive to WARM™ and ZZAP)
Acid	Sensitive (thus sensitive to EGA)

In vitro characteristics of alloanti-Js^a

Immunoglobulin class	IgG more common than IgM
Optimal technique	IAT

Clinical significance of alloanti-Js^a

Transfusion reaction	No to moderate/delayed
HDFN	Mild to severe

Comments

At least one example of "naturally-occurring" anti-Js^a has been reported in a Japanese woman.

Js^a is extremely sensitive to thiol reagents (it is sensitive to 2 mM DTT), most likely because it is located between two cysteine residues.

Reference

[1] Lee, S., et al., 1995. Molecular basis of the K:6, −7 [Js(a+b−)] phenotype in the Kell blood group system. Transfusion 35, 822–825.

Js^b Antigen

Terminology

ISBT symbol (number)	KEL7 (006007 or 6.7)
Obsolete names	Matthews; K7
History	Named in 1963 when it was found to be antithetical to Js^a; joined the Kell blood group system in 1965.

Occurrence

Caucasians	100%
Blacks	99%

Antithetical antigen

Js^a (**KEL6**)

Expression

Cord RBCs	Expressed
Altered	See table showing Comparison of Kell phenotypes on System pages

Molecular basis associated with Js^b antigen[1]

Amino acid	Leu597
Nucleotide	T at bp 1790 in exon 17 (previously reported as 1910)

Effect of enzymes and chemicals on Js^b antigen on intact RBCs

Ficin/Papain	Resistant (some enhanced)
Trypsin	Resistant
α-Chymotrypsin	Resistant (see K [**KEL1**])
DTT 200 mM/50 mM	Sensitive/sensitive (see Comments) (thus sensitive to WARM™ and ZZAP)
Acid	Sensitive (thus sensitive to EGA)

In vitro characteristics of alloanti-Js^b

Immunoglobulin class	IgG
Optimal technique	IAT

Clinical significance of alloanti-Js^b

Transfusion reaction	Mild to moderate/delayed
HDFN	Mild to severe (1 fatality)[2,3]

Comments

Js^b is extremely sensitive to thiol reagents (it is sensitive to 2 mM DTT), most likely because it is located between two cysteine residues.

Siblings of patients with anti-Js^b should be tested for compatibility, and the patient urged to donate blood for cryogenic storage when his/her clinical state permits.

References

[1] Lee, S., et al., 1995. Molecular basis of the K:6,−7 [Js(a+b−)] phenotype in the Kell blood group system. Transfusion 35, 822–825.

[2] Gordon, M.C., et al., 1995. Severe hemolytic disease of the newborn due to anti-Js^b. Vox Sang 69, 140–141.

[3] Stanworth, S., et al., 2001. Severe haemolytic disease of the newborn due to anti-Js^b. Vox Sang 81, 134–135.

Kell

Ul^a Antigen

Terminology

ISBT symbol (number)	KEL10 (006010 or 6.10)
Obsolete names	Karhula; K10
History	Described in 1968, and shown to be part of the Kell system in 1969. Named after the last letters of the antibody producer (Karhula).

Occurrence

Most populations	<0.01%
Finns	2.6% (higher in some regions)
Japanese	0.46%

Expression

Cord RBCs	Expressed

Molecular basis associated with Ul^a antigen[1]

Amino acid	Val494
Nucleotide	T at bp 1481 in exon 13 (previously reported as 1601)
Ul(a–) (wild type)	A at bp 1481 and Glu494

Effect of enzymes and chemicals on Ul^a antigen on intact RBCs

Ficin/Papain	Resistant
Trypsin	Resistant
α-Chymotrypsin	Resistant (see K [**KEL1**])
DTT 200 mM/50 mM	Presumed sensitive

In vitro characteristics of alloanti-Ul^a

Immunoglobulin class	IgG
Optimal technique	IAT

Clinical significance of alloanti-Ul^a

Transfusion reaction	No data but anti-Ul^a has been stimulated by transfusion
HDFN	One case[2]. 19 Ul(a–) mothers with Ul(a+) children did not make anti-Ul^a

Comments

Only a few examples of anti-Ul^a have been reported: two in Finland and two in Japan.

References

[1] Lee, S., et al., 1996. Point mutations characterize *KEL10*, the *KEL3*, *KEL4*, and *KEL21* alleles, and the *KEL17* and *KEL11* alleles. Transfusion 36, 490–494.

[2] Sakuma, K., et al., 1994. First case of hemolytic disease of the newborn due to anti-Ula antibodies. Vox Sang 66, 293–294.

K11 Antigen

Terminology

ISBT symbol (number)	KEL11 (006011 or 6.11)
Obsolete name	Côté
History	Found in 1971 in the serum of Mrs. Côté; the first of a series of K–k+ people who made an antibody compatible only with K_0 RBCs; a para-Kell antigen until proven to belong to Kell in 1974.

Occurrence

All populations	100%

Antithetical antigen

K17 (**KEL17**)

Expression

Cord RBCs	Expressed
Altered	See table showing Comparison of Kell phenotypes on System pages

Molecular basis associated with K11 antigen[1]

Amino acid	Val302
Nucleotide	T at bp 905 in exon 8 (previously reported as 1025)

Effect of enzymes and chemicals on K11 antigen on intact RBCs

Ficin/Papain	Resistant
Trypsin	Resistant
α-Chymotrypsin	Resistant (see K [**KEL1**])
DTT 200 mM/50 mM	Sensitive/sensitive (thus sensitive to WARM™ and ZZAP)
Acid	Sensitive (thus sensitive to EGA)

In vitro characteristics of alloanti-K11

Immunoglobulin class	IgG
Optimal technique	IAT

Clinical significance of alloanti-K11

Transfusion reaction	Mild to moderate (not much data)
HDFN	No to mild (not much data)

Comments

Siblings of patients with anti-K11 should be tested for compatibility, and the patient urged to donate blood for cryogenic storage when his/her clinical state permits.

Reference

[1] Lee, S., et al., 1996. Point mutations characterize *KEL10*, the *KEL3*, *KEL4*, and *KEL21* alleles, and the *KEL17* and *KEL11* alleles. Transfusion 36, 490–494.

K12 Antigen

Terminology

ISBT symbol (number)	KEL12 (006012 or 6.12)
Obsolete names	Bøc (Bøckman); Spears
History	Described in 1973; given the next number in the series of K–k+ people who made an antibody compatible only with K_0 RBCs.

Occurrence

The KEL:–12 phenotype has been reported only in four Caucasian families.

Expression

Cord RBCs	Presumed expressed
Altered	See table showing Comparison of Kell phenotypes on System pages RBCs from a KELP– (see **KEL35**) person were KEL:–12

Molecular basis associated with K12 antigen[1]

Amino acid	His548
Nucleotide	A at bp 1643 in exon 15 (previously reported as 1763)
K12–	Arg548 and G at bp 1643

Effect of enzymes and chemicals on K12 antigen on intact RBCs

Ficin/Papain	Resistant
Trypsin	Resistant
α-Chymotrypsin	Resistant (see K [**KEL1**])
DTT 200 mM/50 mM	Sensitive/sensitive (thus sensitive to WARM™ and ZZAP)

In vitro characteristics of alloanti-K12

Immunoglobulin class	IgG
Optimal technique	IAT

Clinical significance of alloanti-K12

Transfusion reaction	K12+ blood transfused to two patients (DL, MS) did not cause a transfusion reaction
HDFN	No data, although Mrs. Bøckman had at least two children

Comments

Siblings of patients with anti-K12 should be tested for compatibility, and the patient urged to donate blood for cryogenic storage when his/her clinical state permits.

Reference

[1] Lee, S., 1997. Molecular basis of Kell blood group phenotypes. Vox Sang 73, 1–11.

K13 Antigen

Terminology

ISBT symbol (number)	KEL13 (006013 or 6.13)
Obsolete name	SGRO
History	Described in 1974, given the next Kell System number. The K13– proband is a K_{mod}, thereby explaining the weak expression of Kell antigens in this phenotype[1].

Occurrence

K13– has been found in only one family.

Expression

Cord RBCs	Presumed expressed
Altered	See table showing Comparison of Kell phenotypes on System pages

Molecular basis associated with K13 antigen[1]

Amino acid	Leu329
Nucleotide	T at bp 986 in exon 9 (previously reported as 1106)
K13–	C at bp 1106 and Pro329

Effect of enzymes and chemicals on K13 antigen on intact RBCs

Ficin/Papain	Resistant
Trypsin	Resistant
α-Chymotrypsin	Resistant (see K [**KEL1**])
DTT 200 mM/50 mM	Sensitive/sensitive (thus sensitive to WARM™ and ZZAP)

In vitro characteristics of alloanti-K13

Immunoglobulin class	IgG
Optimal technique	IAT

Clinical significance of alloanti-K13

Transfusion reaction	No data are available because only one anti-K13 has been reported
HDFN	The proband's KEL:–13 sister had seven children, without making anti-K13

Comments

K13– RBCs express Kell antigens weakly; it is a K_{mod} phenotype.
Siblings of patients with anti-K13 should be tested for compatibility, and the patient urged to donate blood for cryogenic storage when his/her clinical state permits.

Reference

[1] Lee, S., et al., 2003. Mutations that diminish expression of Kell surface protein and lead to the $Kell_{mod}$ red cell phenotype. Transfusion 43, 1121–1125.

K14 Antigen

Terminology

ISBT symbol (number)	KEL14 (006014 or 6.14)
Obsolete names	San; Santini; Dp
History	Described in 1973, given the next number in the series of K–k+ people who made an antibody compatible only with K_0 RBCs.

Occurrence

K14– has been found in only three French-Cajun families.

Antithetical antigen

K24 (**KEL24**)

Expression

Cord RBCs   Expressed
Altered   See table showing Comparison of Kell phenotypes
                   on System pages

Molecular basis associated with K14 antigen[1,2]

Amino acid   Arg180
Nucleotide   G at bp 539 in exon 6 (previously reported as 659)
The presence of Cys, His or Pro at amino acid residue 180 has resulted in the
K14– phenotype (see table "Molecular basis of Kell phenotypes").

Effect of enzymes and chemicals on K14 antigen on intact RBCs

Ficin/Papain   Resistant
Trypsin   Resistant
α-Chymotrypsin   Resistant (see K [**KEL1**])
DTT 200 mM/50 mM Sensitive/sensitive (thus sensitive to WARM™ and
                   ZZAP)

In vitro characteristics of alloanti-K14

Immunoglobulin class   IgG
Optimal technique   IAT

Clinical significance of alloanti-K14

Transfusion reaction   No data are available because only three anti-K14
                   have been reported
HDFN   Mild in one case[3]

Comments

Siblings of patients with anti-K14 should be tested for compatibility, and the
patient urged to donate blood for cryogenic storage when his/her clinical state
permits.

References

[1] Lee, S., 1997. Molecular basis of Kell blood group phenotypes. Vox Sang 73, 1–11.

[2] Lee, S., et al., 1997. The *KEL24* and *KEL14* alleles of the Kell blood group system. Transfusion 37, 1035–1038.

[3] Wallace, M.E., et al., 1976. Anti-K14: an antibody specificity associated with Kell blood group system. Vox Sang 30, 300–304.

K16 Antigen

Terminology

ISBT symbol (number)	KEL16 (006016 or 6.16)
Obsolete names	Weak k; k-like
History	When anti-k was absorbed with McLeod RBCs, an antibody remained that was non-reactive with McLeod RBCs and reactive with all other k+ RBCs. In 1976, this antibody was named anti-K16 and the antigen K16. No further studies have been performed.

K17 (Wk^a) Antigen

Terminology

ISBT symbol (number)	KEL17 (006017 or 6.17)
Obsolete names	Weeks
History	Reported in 1974; given a Kell System number because the antigen had linkage disequilibrium to K, and in 1975 was shown to be antithetical to K11.

Occurrence

All populations	0.3%

Antithetical antigen

K11 (**KEL11**)

Expression

Cord RBCs	Presumed expressed
Altered	See table showing Comparison of Kell phenotypes on System pages

Molecular basis associated with K17 antigen[1]

Amino acid	Ala302
Nucleotide	C at bp 905 in exon 8 (previously reported as 1025)

Effect of enzymes and chemicals on K17 antigen on intact RBCs

Ficin/Papain	Resistant
Trypsin	Resistant
α-Chymotrypsin	Presumed resistant (see K [**KEL1**])
DTT 200 mM/50 mM	Sensitive/sensitive (thus sensitive to WARM™ and ZZAP)

In vitro characteristics of alloanti-K17

Immunoglobulin class	IgG
Optimal technique	IAT

Clinical significance of alloanti-K17

No data because the antigen and antibody are rare.

Reference

[1] Lee, S., et al., 1996. Point mutations characterize *KEL10*, the *KEL3*, *KEL4*, and *KEL21* alleles, and the *KEL17* and *KEL11* alleles. Transfusion 36, 490–494.

K18 Antigen

Terminology

ISBT symbol (number)	KEL18 (006018 or 6.18)
Obsolete names	V.M.; Marshall
History	Described in 1975, given the next number in the series of K–k+ people who made an antibody compatible only with K_0 RBCs.

Occurrence

K18– has been found only in three families.

Expression

Cord RBCs	Expressed
Altered	See table showing Comparison of Kell phenotypes on System pages

Molecular basis associated with K18 antigen[1]

Amino acid	Arg130
Nucleotide	C at bp 388 and G at bp 389 in exon 4 (previously reported as 508 and 509)
K18–: Type 1	130Trp and T at bp 388
K18–: Type 2	130Gln and A at bp 389

Effect of enzymes and chemicals on K18 antigen on intact RBCs

Ficin/Papain	Resistant
Trypsin	Resistant
α-Chymotrypsin	Resistant (see K [**KEL1**])
DTT 200 mM/50 mM	Sensitive/sensitive (thus sensitive to WARM™ and ZZAP)

In vitro characteristics of alloanti-K18

Immunoglobulin class   IgG
Optimal technique   IAT

Clinical significance of alloanti-K18

Transfusion reaction   Clinical significance is largely unknown. Chromium survival studies showed accelerated RBC destruction in one case[2]. Transfusion of an incompatible unit to a woman with anti-K18 resulted in shortened survival of transfused RBCs[3].

HDFN   Mild in the only reported case. Positive DAT; only phototherapy was required[3].

Comments

Siblings of patients with anti-K18 should be tested for compatibility, and the patient urged to donate blood for cryogenic storage when his/her clinical state permits.

References

[1] Lee, S., 1997. Molecular basis of Kell blood group phenotypes. Vox Sang 73, 1–11.

[2] Barrasso, C., et al., 1983. *In vivo* survival of K:18 red cells in a recipient with anti-K18. Transfusion 23, 258–259.

[3] O'Leary, M.F., et al., 2011. Anti-K18 causing hemolytic disease of the fetus and newborn [abstract]. Transfusion 51 (Suppl.), 156A.

K19 Antigen

Terminology

ISBT symbol (number)   KEL19 (006019 or 6.19)
Obsolete names   Sub; Sublett
History   Described in 1979; given the next number in the series of K–k+ people who made an antibody compatible only with K_0 RBCs.

Occurrence

All populations   100%

Expression

Cord RBCs   Presumed expressed
Altered   See table showing Comparison of Kell phenotypes on System pages

Molecular basis associated with K19 antigen[1]

Amino acid	Arg492
Nucleotide	G at bp 1475 in exon 13 (previously reported as 1595)
K19–	Gln492 and A at bp 1475

Effect of enzymes and chemicals on K19 antigen on intact RBCs

Ficin/Papain	Resistant
Trypsin	Resistant
α-Chymotrypsin	Resistant (see K [**KEL1**])
DTT 200 mM/50 mM	Sensitive/sensitive (thus sensitive to WARM™ and ZZAP)

In vitro characteristics of alloanti-K19

Immunoglobulin class	IgG
Optimal technique	IAT

Clinical significance of alloanti-K19

Transfusion reaction	Moderate/delayed/hemolytic in one case
HDFN	No data

Comments

Only two examples of anti-K19 have been described, one made by a woman (ethnic origin unknown) probably as a result of pregnancy, the other made by a black, multiply-transfused man.

Siblings of patients with anti-K19 should be tested for compatibility, and the patient urged to donate blood for cryogenic storage when his/her clinical state permits.

Reference

[1] Lee, S., 1997. Molecular basis of Kell blood group phenotypes. Vox Sang 73, 1–11.

Km Antigen

Terminology

ISBT symbol (number)	KEL20 (006020 or 6.20)
Obsolete name	K20
History	Reported in 1979. The suffix "m" denotes the association with the McLeod phenotype[1].

Occurrence

All populations   100%

Expression

Cord RBCs   Presumed expressed
Altered   See table showing Comparison of Kell phenotypes
                       on System pages

Effect of enzymes and chemicals on Km antigen on intact RBCs

Ficin/Papain   Resistant
Trypsin   Resistant
DTT 200 mM/50 mM   Not known but sensitive to AET

In vitro characteristics of alloanti-Km

Immunoglobulin class   IgG
Optimal technique   IAT

Clinical significance of alloanti-Km

Transfusion reaction   Delayed/hemolytic in one case[2]
HDFN   Not applicable; anti-Km has been made only by
                       McLeod males

Comments

Anti-Km is made by non-CGD McLeod males. Both McLeod and K_0 phenotype blood will be compatible. Anti-Km+anti-Kx (sometimes called anti-KL) is made by CGD McLeod males, and only McLeod blood will be compatible[3].

References
[1] Marsh, W.L., 1979. Anti-KL. Vox Sang 36, 375.
[2] Marsh, W.L., et al., 1979. Delayed hemolytic transfusion reaction caused by the second example of anti-K19. Transfusion 19, 604–608.
[3] Marsh, W.L., Redman, C.M., 1990. The Kell blood group system: a review. Transfusion 30, 158–167.

Kp^c Antigen

Terminology

ISBT symbol (number)   KEL21 (006021 or 6.21)
Obsolete names   Levay; K21
History   First reported in 1945; joined the Kell System in 1979
                       when it was shown to be antithetical to Kp^a and Kp^b.
                       Anti-Levay (anti-Kp^c) was the first antibody to a low-
                       prevalence ("private") antigen found.

Occurrence

Most populations	<0.01%
Japanese	Up to 0.32% (several *Kp*^{c} homozygotes reported)

Antithetical antigens

Kp^a (**KEL3**), Kp^b (**KEL4**)

Expression

Cord RBCs	Expressed
Altered	See table showing Comparison of Kell phenotypes on System pages

Molecular basis associated with Kp^c antigen[1]

Amino acid	Gln281
Nucleotide	A at bp 842 in exon 8 (previously reported as 962)

Effect of enzymes and chemicals on Kp^c antigen on intact RBCs

Ficin/Papain	Resistant
Trypsin	Resistant
DTT 200 mM/50 mM	Sensitive/sensitive (thus sensitive to WARM™ and ZZAP)

In vitro characteristics of alloanti-Kp^c

Immunoglobulin class	IgG; IgM
Optimal technique	IAT; saline RT

Clinical significance of alloanti-Kp^c

No data are available because the antigen and antibody are rare.

Comments

A Japanese Kp(a–b–) blood donor with anti-Kp^b was found to be Levay positive [Kp(c+)].

Reference

[1] Lee, S., et al., 1996. Point mutations characterize *KEL10*, the *KEL3*, *KEL4*, and *KEL21* alleles, and the *KEL17* and *KEL11* alleles. Transfusion 36, 490–494.

K22 Antigen

Terminology

ISBT symbol (number)	KEL22 (006022 or 6.22)
Obsolete names	N.I.; Ikar
History	Described in 1982; given the next number in the series of K–k+ people who made an antibody compatible only with K_0 RBCs.

Occurrence

K22– has been found in only two Iranian Jewish families.

Expression

Cord RBCs	Expressed
Altered	See table showing Comparison of Kell phenotypes on System pages

Molecular basis associated with K22 antigen[1]

Amino acid	Ala322
Nucleotide	C at bp 965 in exon 9 (previously reported as 1085)
K22–	Val322 and T at bp 965

Effect of enzymes and chemicals on K22 antigen on intact RBCs

Ficin/Papain	Resistant
Trypsin	Resistant
α-Chymotrypsin	Resistant (see K [**KEL1**])
DTT 200 mM/50 mM	Sensitive/sensitive (thus sensitive to WARM™ and ZZAP)

In vitro characteristics of alloanti-K22

Immunoglobulin class	IgG
Optimal technique	IAT

Clinical significance of alloanti-K22

Transfusion reaction	No data
HDFN	Mild to severe in one case

Comments

Siblings of patients with anti-K22 should be tested for compatibility, and the patient urged to donate blood for cryogenic storage when his/her clinical state permits.

Reference

[1] Lee, S., 1997. Molecular basis of Kell blood group phenotypes. Vox Sang 73, 1–11.

K23 Antigen

Terminology

ISBT symbol (number)	KEL23 (006023 or 6.23)
Obsolete name	Centauro
History	Reported in 1987; antibody identified in serum of a pregnant woman; assigned to Kell because the serum precipitated Kell glycoprotein from her husband's RBCs.

Occurrence

K23+ has been reported in only one Italian family.

Expression

Cord RBCs	Expressed

Molecular basis associated with K23 antigen[1]

Amino acid	Arg382
Nucleotide	G at bp 1145 in exon 10 (previously reported as 1265)
K23– (wild type)	Gln382 and A at bp 1145

Effect of enzymes and chemicals on K23 antigen on intact RBCs

Ficin/Papain	Resistant
Trypsin	Resistant
DTT 200 mM/50 mM	Sensitive/sensitive (thus sensitive to WARM™ and ZZAP)

In vitro characteristics of alloanti-K23

Immunoglobulin class	IgG
Optimal technique	IAT

Clinical significance of alloanti-K23

Transfusion reaction	No data because antigen and antibody are rare
HDFN	Positive DAT; no clinical HDFN

Reference

[1] Lee, S., 1997. Molecular basis of Kell blood group phenotypes. Vox Sang 73, 1–11.

Kell

K24 Antigen

Terminology

ISBT symbol (number)	KEL24 (006024 or 6.24)
Obsolete names	CL; Callais; Cls
History	Described in 1985 when it was shown to be antithetical to K14.

Occurrence

K24+ has been reported in only three French-Cajun families.

Antithetical antigen

K14 (**KEL14**)

Expression

Cord RBCs	Expressed
Altered	See table showing Comparison of Kell phenotypes on System pages

Molecular basis associated with K24 antigen[1]

Amino acid	Pro180
Nucleotide	C at bp 539 in exon 6 (previously reported as 659)

Effect of enzymes and chemicals on K24 antigen on intact RBCs

Ficin/Papain	Resistant
Trypsin	Resistant
α-Chymotrypsin	Presumed resistant (see K [**KEL1**])
DTT 200 mM/50 mM	Sensitive/sensitive (thus sensitive to WARM™ and ZZAP)

In vitro characteristics of alloanti-K24

Immunoglobulin class	IgG
Optimal technique	IAT

Clinical significance of alloanti-K24

Transfusion reaction	No data because only one example of anti-K24 has been reported
HDFN	Positive DAT; no clinical HDFN

Reference

[1] Lee, S., 1997. Molecular basis of Kell blood group phenotypes. Vox Sang 73, 1–11.

VLAN Antigen

Terminology

ISBT symbol (number)	KEL25 (006025 or 6.25)
History	Named in 1996 after the last name of the proband who's RBCs possessed the antigen.

Occurrence

VLAN+ has been reported in only one Dutch family[1].

Antithetical antigen

VONG (**KEL28**)

Expression

Cord RBCs	Presumed expressed

Molecular basis associated with VLAN antigen[2]

Amino acid	Gln248
Nucleotide	A at bp 743 in exon 8 (previously reported as 863)
VLAN– (wild type)	Arg248 and G at bp 743

Effect of enzymes and chemicals on VLAN antigen on intact RBCs

Ficin/Papain	Resistant
Trypsin	Presumed resistant
DTT 200 mM/50 mM	Sensitive/sensitive (thus sensitive to WARM™ and ZZAP)

In vitro characteristics of alloanti-VLAN

Immunoglobulin class	IgG
Optimal technique	IAT

Clinical significance of alloanti-VLAN

No data because only one example of anti-VLAN was found in a serum (named BUS) following an incompatible cross-match[1].

References

[1] Jongerius, J.M., et al., 1996. A new low-incidence antigen in the Kell blood group system: VLAN (KEL25). Vox Sang 71, 43–47.

[2] Lee, S., et al., 2001. Point mutations in KEL exon 8 determine a high incidence (RAZ) and a low incidence (KEL25, VLAN) antigen of the Kell blood group system. Vox Sang 81, 259–263.

Kell

TOU Antigen

Terminology

ISBT symbol (number)	KEL26 (006026 or 6.26)
History	Named in 1995 after the last name of the proband whose serum contained an antibody to a high-prevalence antigen; provisional assignment to Kell System ratified in 1998.

Occurrence

TOU– has been reported in only two families, one Native American and one Hispanic[1].

Expression

Cord RBCs	Presumed expressed
Altered	See table showing Comparison of Kell phenotypes on System pages

Molecular basis associated with TOU antigen[2]

Amino acid	Arg406
Nucleotide	G at bp 1217 in exon 11 (previously reported as 1337)
TOU–	Gln406 and A at bp 1217

Effect of enzymes and chemicals on TOU antigen on intact RBCs

Ficin/Papain	Resistant
Trypsin	Resistant
DTT 200 mM/50 mM	Sensitive/sensitive (thus sensitive to WARM™ and ZZAP)

In vitro characteristics of alloanti-TOU

Immunoglobulin class	IgG
Optimal technique	IAT

Clinical significance of alloanti-TOU

No data because only two examples of anti-TOU have been reported.

Comments

Siblings of patients with anti-TOU should be tested for compatibility, and the patient urged to donate blood for cryogenic storage when his/her clinical state permits.

References

[1] Jones, J., et al., 1995. A novel common Kell antigen, TOU, and its spatial relationship to other Kell antigens. Vox Sang 69, 53–60.

[2] Lee, S., 1997. Molecular basis of Kell blood group phenotypes. Vox Sang 73, 1–11.

RAZ Antigen

Terminology

ISBT symbol (number)	KEL27 (006027 or 6.27)
History	Named in 1994 after the proband whose serum contained an antibody to a high-prevalence antigen; provisional Kell System assignment ratified in 2002.

Occurrence

RAZ– has been found in only one Indian family (from Gujarat)[1].

Expression

Cord RBCs	Presumed expressed
Altered	See table showing Comparison of Kell phenotypes on System pages

Molecular basis associated with RAZ antigen[2]

Amino acid	Glu249
Nucleotide	G at bp 745 in exon 8 (previously reported as 865)
K27–	Lys249 and A at bp 745

Effect of enzymes and chemicals on RAZ antigen on intact RBCs

Ficin/Papain	Resistant
Trypsin	Resistant
α-Chymotrypsin	Weakened
DTT 200 mM/50 mM	Sensitive/sensitive (thus sensitive to WARM™ and ZZAP)

In vitro characteristics of alloanti-RAZ

Immunoglobulin class	IgG
Optimal technique	IAT

Clinical significance of alloanti-RAZ

No data because only one example of anti-RAZ has been reported.

Comments

Siblings of patients with anti-RAZ should be tested for compatibility, and the patient urged to donate blood for cryogenic storage when his/her clinical state permits.

References

[1] Daniels, G.L., et al., 1994. Demonstration by the monoclonal antibody-specific immobilization of erythrocyte antigens assay that a new red cell antigen belongs to the Kell blood group system. Transfusion 34, 818–820.

[2] Lee, S., et al., 2001. Point mutations in KEL exon 8 determine a high incidence (RAZ) and a low incidence (KEL25, VLAN) antigen of the Kell blood group system. Vox Sang 81, 259–263.

VONG Antigen

Terminology

ISBT symbol (number)	KEL28 (006028 or 6.28)
History	Described in 2003 and named after the VONG+ proband's name.

Occurrence

VONG+ has been reported in only one Chinese family from Timor.

Antithetical antigen

VLAN (**KEL25**)

Molecular basis associated with VONG antigen[1]

Amino acid	Trp248
Nucleotide	T at bp 742 in exon 8 (previously reported as 862)
VONG– (wild type)	Arg248 and C at bp 742

Effect of enzymes and chemicals on VONG antigen on intact RBCs[1]

Ficin/Papain	Resistant
Trypsin	Resistant
α-Chymotrypsin	Resistant (see K [**KEL1**])
DTT 200 mM/50 mM	Sensitive/sensitive (thus sensitive to WARM™ and ZZAP)

In vitro characteristics of alloanti-VONG

Immunoglobulin class	IgG
Optimal technique	IAT

Clinical significance of alloanti-VONG

Transfusion reaction	No data, only one example of anti-VONG has been described
HDFN	Mild[1]

Reference

[1] Grey, D., et al., 2003. Haemolytic disease of the newborn caused by a new Kell antigen [abstract]. Transfus Med 13 (Suppl. 1), 30.

KALT Antigen

Terminology

ISBT symbol (number)	KEL29 (006029 or 6.29)
History	Named in 2006, "K" for the System and "ALT" from the KALT– proband's name.

Occurrence

Only one KALT– proband, a Mexican woman with a history of pregnancy, and her KALT– sister have been reported.

Molecular basis associated with KALT antigen[1]

Amino acid	Arg623
Nucleotide	G at bp 1868 in exon 17 (previously reported as 1988)
KALT–	Lys623 and A at bp 1868

Effect of enzymes and chemicals on KALT antigen on intact RBCs

Ficin/Papain	Resistant
Trypsin	Sensitive (see Comments below)
α-Chymotrypsin	Resistant
DTT 200 mM/50 mM	Sensitive/sensitive (thus sensitive to WARM™ and ZZAP)

In vitro characteristics of alloanti-KALT

Immunoglobulin class	IgG
Optimal technique	IAT

Clinical significance of alloanti-KALT

Transfusion reaction	No data because only one anti-KALT has been described
HDFN	Positive DAT but no clinical HDFN

Comments

The KALT antigen is currently the only Kell antigen that is sensitive to trypsin treatment of intact RBCs. Anti-KALT recognizes the presence of Arg623, which is located on the C-terminal side of the catalytic domain of the Kell glycoprotein.

Siblings of patients with anti-KALT should be tested for compatibility, and the patient urged to donate blood for cryogenic storage when his/her clinical state permits.

Reference
[1] Lee, S., et al., 2006. Molecular basis of two novel high prevalence antigens in the Kell blood group system, KALT and KTIM. Transfusion 46, 1323–1327.

KTIM Antigen

Terminology

ISBT symbol (number)	KEL30 (006030 or 6.30)
History	Named in 2006, "K" from the System name and "TIM" from the KTIM– proband's name.

Occurrence

Only one KTIM– proband, a White American woman with a history of transfusion and pregnancy, has been described.

Molecular basis associated with KTIM antigen[1]

Amino acid	Asp305
Nucleotide	G at bp 913 in exon 8 (previously reported as 1033)
KTIM–	Asn305 and A at bp 913

Effect of enzymes and chemicals on KTIM antigen on intact RBCs

Ficin/Papain	Resistant
Trypsin	Resistant
α-Chymotrypsin	Resistant (see K [**KEL1**])
DTT 200 mM/50 mM	Sensitive/sensitive (thus sensitive to WARM™ and ZZAP)

In vitro characteristics of alloanti-KTIM

Immunoglobulin class	IgG
Optimal technique	IAT

Clinical significance of alloanti-KTIM

No data have been described.

Comments

Siblings of patients with anti-KTIM should be tested for compatibility, and the patient urged to donate blood for cryogenic storage when his/her clinical state permits.

Reference

[1] Lee, S., et al., 2006. Molecular basis of two novel high prevalence antigens in the Kell blood group system, KALT and KTIM. Transfusion 46, 1323–1327.

KYO Antigen

Terminology

ISBT symbol (number)	KEL31 (006031 or 6.31)
History	Named in 2006; "K" for the System and "YO" from the KYO+ proband's name

Occurrence

Most populations	<0.1%
Japanese	1.5%

Molecular basis associated with KYO antigen[1]

Amino acid	Gln292
Nucleotide	A at bp 875 in exon 8 (previously reported as 995)
KYO– (wild type)	Arg292 and G at bp 875

Effect of enzymes and chemicals on KYO antigen on intact RBCs

Ficin/Papain	Resistant
Trypsin	Resistant
DTT 200 mM/50 mM	Sensitive/sensitive (thus sensitive to WARM™ and ZZAP)

In vitro characteristics of alloanti-KYO

Immunoglobulin class	IgG
Optimal technique	IAT

Clinical significance of alloanti-KYO

No data because anti-KYO is rare.

Reference

[1] Uchikawa, M., et al., 2006. Molecular basis for a novel low-frequency antigen in the Kell blood group system, KYO [abstract]. Vox Sang 91 (Suppl. 3), 136.

KUCI Antigen

Terminology

ISBT symbol (number)	KEL32 (006032 or 6.32)
History	Named in 2007, "K" for the System and "UCI" from the name of the KUCI– proband.

Occurrence

KUCI– has been found in only one American Indian family.

Molecular basis associated with KUCI antigen[1]

Amino acid	Ala424
Nucleotide	C at bp 1271 in exon 11 (previously reported as 1391)
KUCI–	Val424 and T at bp 1271

Effect of enzymes and chemicals on KUCI antigen on intact RBCs

Ficin/Papain	Resistant
Trypsin	Resistant
α-Chymotrypsin	Resistant (see K [**KEL1**])
DTT 200 mM/50 mM	Sensitive/sensitive (thus sensitive to WARM™ and ZZAP)

In vitro characteristics of alloanti-KUCI

Immunoglobulin class	IgG
Optimal technique	IAT

Clinical significance of alloanti-KUCI

No data because only one example of anti-KUCI has been described.

Comments

Siblings of patients with anti-KUCI should be tested for compatibility, and the patient urged to donate blood for cryogenic storage when his/her clinical state permits.

KUCI– RBCs were non-reactive with plasma from the KANT– proband [see KANT (**KEL33**)].

Reference

[1] Velliquette, R.W., et al., 2007. Two novel and related high-prevalence antigens in the Kell blood group system [abstract]. Transfusion 47 (Suppl.), 164A–165A.

KANT Antigen

Terminology

ISBT symbol (number)	KEL33 (006033 or 6.33)
History	Named in 2007, "K" for the System and "ANT" from the KANT– proband's name.

Occurrence

KANT– has been found in only one French proband.

Molecular basis associated with KANT antigen[1]

Amino acid	Arg428
Nucleotide	G at bp 1283 in exon 11 (previously 1403)
KANT–	Leu428 and T at bp 1283

Effect of enzymes and chemicals on KANT antigen on intact RBCs

Ficin/Papain	Resistant
Trypsin	Resistant
α-Chymotrypsin	Resistant (see K [**KEL1**])
DTT 200 mM/50 mM	Sensitive/sensitive (thus sensitive to WARM™ and ZZAP)

In vitro characteristics of alloanti-KANT

Immunoglobulin class	IgG
Optimal technique	IAT

Clinical significance of alloanti-KANT

No data because only one example of anti-KANT has been described.

Comments

KUCI– RBCs were non-reactive with anti-KANT, whereas KANT– RBCs reacted very weakly with anti-KUCI. The change in the KANT– proband is predicted to be just four amino acids from that found in the KUCI– proband; this may provide an explanation for partial serological compatibility between the KUCI– and KANT– probands. KANT– RBCs express K11 weakly and KETI very weakly [see KETI (**KEL36**)].

Siblings of patients with anti-KANT should be tested for compatibility, and the patient urged to donate blood for cryogenic storage when his/her clinical state permits.

Reference

[1] Velliquette, R.W., et al., 2007. Two novel and related high-prevalence antigens in the Kell blood group system [abstract]. Transfusion 47 (Suppl.), 164A–165A.

KASH Antigen

Terminology

ISBT symbol (number)	KEL34 (006034 or 6.34)
History	Named in 2010, "K" for the System and "ASH" from the KASH– proband's name.

Occurrence

Only one KASH– proband and her KASH– brother have been reported.

Molecular basis associated with KASH antigen[1]

Amino acid	253Tyr
Nucleotide	A at bp 758 in exon 8
KASH–	Cys253 and G at bp 758

Effect of enzymes and chemicals on KASH antigen on intact RBCs

Ficin/Papain	Resistant
Trypsin	Resistant
DTT 200 mM/50 mM	Sensitive/sensitive (thus sensitive to WARM™ and ZZAP)

In vitro characteristics of alloanti-KASH

Immunoglobulin class	IgG
Optimal technique	IAT

Clinical significance of alloanti-KASH

No data, because only one example of anti-KASH has been described. The antibody maker had a history of pregnancy but no transfusions.

Comments

Lack of the KEL34 epitope prevents normal expression of the Kell antigens, and gives rise to a K_{mod} phenotype: the RBCs of the KASH– proband and those of her KASH– brother gave extremely weak reactions with some antibodies to high prevalence Kell antigens, and were negative with the majority of them, but their cells absorbed and eluted anti-k and anti-Kp^b.

Reference

[1] Karamatic Crew, V., et al., 2010. KASH (KEL34): a novel high incidence antigen in the Kell blood group system [abstract]. Vox Sang 99 (Suppl. 1), 357.

KELP Antigen

Terminology

ISBT symbol (number)	KEL35 (006035 or 6.35)
History	Named in 2010, "KE" from the System, "L" for leucine and "P" for the first letter of phenylalanine.

Occurrence

Only one KELP– proband has been reported.

Molecular basis associated with KELP antigen[1]

Amino acid	Leu260 and Arg675
Nucleotide	G at bp 780 in exon 8 and G at bp 2024 in exon 18
KELP–	Phe260 and Gln675, with T at bp 780 and A at bp 2024

Effect of enzymes and chemicals on KELP antigen on intact RBCs

Ficin/Papain	Resistant
Trypsin	Resistant
DTT 200 mM/50 mM	Sensitive/sensitive (thus sensitive to WARM™ and ZZAP)

In vitro characteristics of alloanti-KELP

Immunoglobulin class	IgG
Optimal technique	IAT

Clinical significance of alloanti-KELP

No data, because only one example of anti-KELP in a pregnant woman has been described.

Comments

Siblings of patients with anti-KELP should be tested for compatibility, and the patient urged to donate blood for cryogenic storage when his/her clinical state permits.

RBCs from the KELP– proband typed K12–, although DNA sequence analysis predicted the K12+ phenotype.

Reference

[1] Karamatic Crew, V., et al., 2010. KELP (KEL35): a new high incidence antigen in the Kell blood group defined by two homozygous missense mutations in KEL [abstract]. Transfus Med 20 (Suppl. 1), 30.

KETI Antigen

Terminology

ISBT symbol (number)	KEL36 (006036 or 6.36)
History	Named in 2011, "KE" from the System, "T" for Thr and "I" for Ile.

Occurrence

KETI– has been found in only a few European families.

Molecular basis associated with KETI antigen[1]

Amino acid	Thr464
Nucleotide	C at bp 1391 in exon 12
KETI–	Ile464 and T at bp 1391

Effect of enzymes and chemicals on KETI antigen on intact RBCs

Ficin/Papain	Resistant
Trypsin	Resistant
α-Chymotrypsin	Resistant
DTT 200 mM/50 mM	Sensitive/sensitive (thus sensitive to WARM™ and ZZAP)

In vitro characteristics of alloanti-KETI

Immunoglobulin class	IgG
Optimal technique	IAT

Clinical significance of alloanti-KETI

No data because anti-KETI is rare.

Comments

An example of KASH– RBCs was found to be compatible with the anti-KETI[1].

KETI– RBCs are K11+/– and KUCI+, KANT+.

Siblings of patients with anti-KETI should be tested for compatibility, and the patient urged to donate blood for cryogenic storage when his/her clinical state permits.

Reference

[1] Karamatic Crew, V., et al., 2011. KETI, a novel high incidence antigen in the Kell blood group system: a serological and molecular study [abstract]. Vox Sang 101 (Suppl. 1), 19.

KHUL Antigen

Terminology

ISBT symbol (number)	KEL37 (006037 or 6.37)
History	Named in 2011, "K" from the System and "HUL" from the KHUL– proband's name.

Occurrence

KHUL– has been found in only one Asian proband and her sister.

Molecular basis associated with KHUL antigen[1]

Amino acid	Arg293
Nucleotide	C at bp 877 in exon 8
KHUL–	Trp293 and T at bp 877

Effect of enzymes and chemicals on KHUL antigen on intact RBCs

Ficin/Papain	Resistant
Trypsin	Resistant
α-Chymotrypsin	Resistant
DTT 200 mM/50 mM	Sensitive/sensitive (thus sensitive to WARM™ and ZZAP)

In vitro characteristics of alloanti-KHUL

Immunoglobulin class	IgG
Optimal technique	IAT

Clinical significance of alloanti-KHUL

No data because anti-KHUL is rare.

Comments

KHUL is, surprisingly, independent of the low-prevalence antigen, KYO, which is associated with the adjacent amino acid, Arg292Gln.

Siblings of patients with anti-KHUL should be tested for compatibility, and the patient urged to donate blood for cryogenic storage when his/her clinical state permits.

Reference

[1] Vege, S., et al., 2011. A new high prevalence antigen (KHUL) in the Kell blood group system [abstract]. Transfusion 51 (Suppl.), 25A–26A.

Lewis Blood Group System

Number of antigens 6

Polymorphic	Le^a, Le^b, Le^{bH}, ALe^b, BLe^b
High prevalence	Le^{ab}

Terminology

ISBT symbol (number)	LE (007)
History	Discovered by Mourant in 1946; named after one of the two original donors in whom anti-Le^a was identified.

Expression

Soluble form	Glycoproteins in saliva and body fluids except CSF plus glycolipids in plasma; the concerted effect of Lewis, ABH, and secretor status determines, what if any, Lewis antigens are present
Other blood cells	Lymphocytes, platelets
Tissues	For instance, pancreas, mucosa of stomach, small and large intestine (large intestine has Le^a only), renal cortex, adrenal glands

Gene

Chromosome	19p13.3
Name	LE (FUT3)
Organization	3 exons distributed over approximately 8 kbp of gDNA
Product	$\alpha(1,3/4)$ fucosyltransferase

The Blood Group Antigen (3/e). DOI: http://dx.doi.org/10.1016/B978-0-12-415849-8.00009-0

Database accession numbers

GenBank	NM_000149; X53578 (mRNA)
Entrez Gene ID	2525

Molecular basis of Le phenotypes

Lewis determinants are carbohydrates on glycolipids and glycoproteins, and are built on type 1 precursor structures. RBC membranes acquire Lewis glycolipids from circulating lipoproteins. The synthesis of the Lewis antigens is dependent only on *FUT3* which encodes the α(1,3/4) fucosyltransferase (FUT3). However, the type of precursor FUT3 utilizes determines what type of Lewis antigen is made. In individuals **without** the *FUT2* (secretor *SE*) gene which encodes the α(1,2) fucosyltransferase (FUT2) (see H Blood Group System [**018**]) only Le^a can be made by the Lewis FUT3 enzyme. In individuals **with** both the Lewis FUT3 and secretor FUT2 the final products made are determined by their ABO type. If group O, then Le^b is made (with small amounts of Le^a remaining). If group A, then predominantly ALe^b will be made (from A antigen) followed by lesser amounts of Le^b, and trace amounts of Le^a. A similar process occurs for B and AB individuals, but with appropriate A/B antigens resulting. The Le^a antigen cannot be converted into Le^b, hence once made it will remain as Le^a. This is why Le^b positive individuals always have Le^a (which is undetectable with serologically formulated reagents), and why the Le(a+b+) phenotype exists in persons with inefficient FUT2 glycosyltransferases. In individuals not inheriting a functional FUT3, their RBCs will phenotype as Le(a−b−) regardless of the FUT2 status.

As Lewis antigens are adsorbed by RBCs from plasma (not intrinsic) the ISBT has not yet named the alleles in the Lewis system (see dbRBC).

Molecular basis of Le(a+b+) phenotype due to nucleotide changes in *FUT2*

The Le(a+b+) phenotype is caused by mutation(s) in secretor *FUT2*, which changes the generally highly efficient FUT2 enzyme to become less efficient[1]. As a consequence, the Lewis FUT3 enzyme becomes relatively more efficient, and is thus able to compete more effectively for precursor. More Le^a means less H type 1 is made, and less H type 1 results in less Le^b and ABH substances, hence the association of this phenotype with the partial/weak secretor phenotypes.

Molecular basis of Le(a−b−) phenotype due to nucleotide changes in *FUT3* and other genes

More than 30 null alleles at the *FUT3* locus are known (see dbRBC). Homozygosity or compound heterozygosity for these alleles will result in absence of both Le^a and Le^b antigens on RBCs. In addition, FUT6 deficiency is associated with the Le(a–b–) phenotype. The *FUT6* locus which encodes a plasma fucosyltransferase is closely linked to *FUT3* at 19p13.3, so there is a high degree of genetic linkage between these loci. Another very rare reason for this phenotype is mutations in the gene encoding the GDP-fucose transporter, see the H [018] Blood Group System.

Amino acid sequence of α(1,3/4) fucosyltransferase (FUT3)[2]

```
MDPLGAAKPQ  WPWRRCLAAL  LFQLLVAVCF  FSYLRVSRDD  ATGSPRAPSG   50
SSRQDTTPTR  PTLLILLWTW  PFHIPVALSR  CSEMVPGTAD  CHITADRKVY  100
PQADTVIVHH  WDIMSNPKSR  LPPSPRPQGQ  RWIWFNLEPP  PNCQHLEALD  150
RYFNLTMSYR  SDSDIFTPYG  WLEPWSGQPA  HPPLNLSAKT  ELVAWAVSNW  200
KPDSARVRYY  QSLQAHLKVD  VYGRSHKPLP  KGTMMETLSR  YKFYLAFENS  250
LHPDYITEKL  WRNALEAWAV  PVVLGPSRSN  YERFLPPDAF  IHVDDFQSPK  300
DLARYLQELD  KDHARYLSYF  RWRETLRPRS  FSWALDFCKA  CWKLQQESRY  350
QTVRSIAAWF  T                                               361
```

Lewis

Carrier molecule[3]

Lewis antigens are not a primary gene product, but are instead the result of action of the FUT3 enzyme on different precursors. Lewis antigens can exist on both glycoproteins and glycolipids, and have different carrier chains, although only those based on the type 1 precursors (Galβ1-3GlcNAc) are considered as Lewis antigens. Antigens (e.g., Le^x and Le^y) also formed by the FUT3 enzyme but on type 2 precursors (Galβ1-4GlcNAc) are not considered red cell antigens, although they may be present. Glycoprotein forms of the Lewis antigen are found primarily in bodily secretions such as saliva and milk, while only the glycolipid forms are adsorbed onto red cells from the plasma.

Function

There are no apparent pathological consequences in Le(a–b–) people.
Sialylated forms of Le^a and Le^b may serve as ligands for E-selectins, although their type 2 cousins, Le^x and Le^y, are more likely to be the antigens of biological significance.

Disease association

The Le^b antigen in the gastric mucosal epithelium is the receptor for *Helicobacter pylori*, a major causative agent of gastric ulcers[4,5].
Lewis antigens may be lost from RBCs as a result of infectious mononucleosis complicated with hemolysis, severe alcoholic cirrhosis, alcoholic pancreatitis, and pregnancy.

Patients with fucosidosis may have increased expression of Lewis antigens in their saliva and on their RBCs.

RBCs from patients with leukocyte adhesion deficiency (LADII) syndrome are Le(a–b–), and are Bombay phenotype, due to a mutation in the GDP-fucose transporter[6,7].

Phenotypes (% occurrence)

	Caucasians	Blacks	Japanese
Le(a+b–)	22	23	0.2
Le(a–b+)	72	55	73
Le(a+b+)	Rare	Rare	16.8
Le(a–b–)	6	22	10

Null: Le(a–b–)
Unusual: Le(a+b+), rare in European populations, is found in most Australasian populations (e.g., Australian Aborigines, Chinese in Taiwan, Japanese, and Polynesians) with an incidence of 10% to 40%. The phenotype is the result of a mutation in the *FUT2* (*SE*) gene: 385A>T; Ile129Phe. (See H [018] System)

Comments

Saliva is a good source of soluble Lewis antigens, and should be made isotonic (if not already) before use in hemagglutination tests. Monoclonal Lewis reagents should never be used in determining the presence of Lewis antigens in saliva. It is well-established that quality anti-Le[b] serological reagents are always neutralized by Le[a] saliva.

During pregnancy, expression of Lewis antigens on RBCs is often greatly reduced, although the level of antigen in plasma remains normal[8].

Le[x] (SSEA-1; CD15) and Le[y], products of *FUT3* on type 2 precursor chains, are only found in trace amounts on the RBC surface, and are not part of the Lewis blood group system[9]. Le[x] and Le[y] are isomers of, respectively, Le[a] and Le[b], and often occur in sialylated forms. Sialyl-Le[x] is a major neutrophil ligand for E-selectin[10]. Le[a], Le[x], their sialyl derivates, and also Le[b] and Le[y], accumulate in tumor tissues. Evidence indicates that adhesion of tumor cells to endothelial cells is mediated between the sialylated Le[a] and Le[x] antigens and

E-selectin, and represents an important factor in hematogenous metastasis of tumor cells.

Le^X

Gal
|β1–4
Fucα1-3—GlcNAc
|β1-3
Gal
|β1-4
Glc
|β1-1
Ceramide

Sialyl-Le^X

NeuAc $\xrightarrow{\alpha2–3}$ Gal
|β1–4
Fucα1-3—GlcNAc
|β1-3
Gal
|β1-4
Glc
|β1-1
Ceramide

References

[1] Henry, S., et al., 1996. Molecular basis for erythrocyte Le(a+b+) and salivary ABH partial-secretor phenotypes: expression of a FUT2 secretor allele with an A-->T mutation at nucleotide 385 correlates with reduced α(1,2)fucosyltransferase activity. Glycoconj J 13, 985–993.

[2] Kukowska-Latallo, J.F., et al., 1990. A cloned human cDNA determines expression of a mouse state-specific embryonic antigen and the Lewis blood group α(1,3/1,4)fucosyltransferase. Genes & Devel 4, 1288–1303.

[3] Hauser, R., 1995. Le^a and Le^b tissue glycosphingolipids. Transfusion 35, 577–581.

[4] Boren, T., et al., 1993. Attachment of *Helicobacter pylori* to human gastric epithelium mediated by blood group antigens. Science 262, 1892–1895.

[5] Boren, T., et al., 1994. *Helicobacter pylori*: molecular basis for host recognition and bacterial adherence [review]. Trends Microbiol 2, 221–228.

[6] Hirschberg, C.B., 2001. Golgi nucleotide sugar transport and leukocyte adhesion deficiency II. J Clin Invest 108, 3–6.

[7] Luhn, K., et al., 2001. The gene defective in leukocyte adhesion deficiency II encodes a putative GDP-fucose transporter. Nat Genet 28, 69–72.

[8] Henry, S., et al., 1996. A second nonsecretor allele of the blood group α(1, 2)fucosyltransferase gene (FUT2). Vox Sang 70, 21–25.

[9] Lowe, J.B., 1995. Biochemistry and biosynthesis of ABH and Lewis antigens: characterization of blood group-specific glycosyltransferases. In: Cartron, J.-P., Rouger, P. (Eds.), Molecular Basis of Human Blood Group Antigens. Plenum Press, New York, NY, pp. 75–115.

[10] Walz, G., et al., 1990. Recognition by ELAM-1 of the sialyl-Le^x determinant on myeloid and tumor cells. Science 250, 1132–1135.

Le^a Antigen

Terminology

ISBT symbol (number)	LE1 (007001 or 7.1)
History	Identified in 1946; named Lewis after one of the two original producers of anti-Le^a.

Occurrence

Caucasians	22%
Blacks	23%

Expression

Cord RBCs	Not expressed; although RBCs from some cord bloods will react with anti-Le^a by IAT
Reduced	Weak in Le(a+b+); often weakened during pregnancy and certain diseases

Molecular basis associated with Le^a antigen[1]

$$
\begin{array}{c}
\textbf{Gal} \\
|\beta\text{1-3} \\
\textbf{Fuc} \xrightarrow{\alpha\text{1-4}} \textbf{GlcNAc} \\
|\beta\text{1-3} \\
\textbf{Gal} \\
|\beta\text{1-4} \\
\textbf{Glc} \\
|\beta\text{1-1} \\
\textbf{Ceramide}
\end{array}
$$

Effect of enzymes and chemicals on Le^a antigen on intact RBCs

Ficin/Papain	Resistant (markedly enhanced)
Trypsin	Resistant (markedly enhanced)
α-Chymotrypsin	Resistant (markedly enhanced)
DTT 200 mM	Resistant
Acid	Resistant

In vitro characteristics of alloanti-Le^a

Immunoglobulin class	IgM more frequent than IgG
Optimal technique	RT; 37°C; IAT; enzymes
Neutralization	Plasma and isotonic saliva
Complement binding	Yes; some hemolytic

Clinical significance of alloanti-Le^a

Transfusion reaction	No (rare cases of hemolytic reactions)
HDFN	No (one mild case)

Comments

Anti-Le^a (in conjunction with anti-Le^b [see **LE2**]) is a frequent naturally-occurring antibody made by Le(a–b–) people, especially during pregnancy.

There are rare reports about Le(a–b+) individuals making anti-Le[a], but it is unclear if this only applies to a subgroup of individuals with this phenotype[2].

References

[1] Henry, S., et al., 1995. Lewis histo-blood group system and associated secretory phenotypes. Vox Sang 69, 166–182.
[2] Chan, Y.S., Lin, M., 2011. Anti-Lea in Le(a–b+) individuals [abstract]. Vox Sang 101 (Suppl. 2), 97–98.

Le[b] Antigen

Terminology

ISBT symbol (number)	LE2 (007002 or 7.2)
History	Anti-Le[b] was identified in 1948; initially appeared to detect an antigen antithetical to Le[a].

Occurrence

Caucasians	72%
Blacks	55%

Expression

Cord RBCs	Not expressed
Reduced	Weak in Le(a+b+); often weakened during pregnancy and certain diseases

Molecular basis associated with Le[b] antigen[1]

$$\text{Fuc} \xrightarrow{\alpha1-2} \text{Gal}$$
$$|\beta1-3$$
$$\text{Fuc} \xrightarrow{\alpha1-4} \text{GlcNAc}$$
$$|\beta1-3$$
$$\text{Gal}$$
$$|\beta1-4$$
$$\text{Glc}$$
$$|\beta1-1$$
$$\text{Ceramide}$$

Effect of enzymes and chemicals on Le[b] antigen on intact RBCs

Ficin/Papain	Resistant (markedly enhanced)
Trypsin	Resistant (markedly enhanced)
α-Chymotrypsin	Resistant (markedly enhanced)
DTT 200 mM	Resistant
Acid	Resistant

In vitro characteristics of alloanti-Le^b

Immunoglobulin class	IgM more frequent than IgG
Optimal technique	RT; 37°C; IAT; enzymes
Neutralization	Plasma and isotonic saliva from secretors
Complement binding	Yes; some hemolytic

Clinical significance of alloanti-Le^b

Transfusion reaction	No
HDFN	No (one mild case)

Comments

Le^b is a receptor for *Helicobacter pylori* in gastric mucosal epithelium[2].

Anti-Le^b (in conjunction with anti-Le^a [see **LE1**]) is a frequent naturally-occurring antibody made by Le(a–b–) people. There are also rare reports about Le(a+b–) individuals making anti-Le^b, but it is unclear if this only applies to a subgroup of individuals with this phenotype[3].

There are two kinds of anti-Le^b: anti-Le^{bH} **(LE4)**, reacting with group O and A$_2$ Le(b+) RBCs, and anti-Le^{bL}, reacting with all Le(b+) RBCs. Other antibodies react specifically with the compound antigens, e.g., ALe^b **(LE5)** and BLe^b **(LE6)**.

References

[1] Henry, S., et al., 1995. Lewis histo-blood group system and associated secretory phenotypes. Vox Sang 69, 166–182.

[2] Boren, T., et al., 1994. *Helicobacter pylori*: molecular basis for host recognition and bacterial adherence [review]. Trends Microbiol 2, 221–228.

[3] Chan, Y.S., Lin, M., 2011. Anti-Lea in Le(a–b+) individuals [abstract]. Vox Sang 101 (Suppl. 2), 97–98.

Le^{ab} Antigen

Terminology

ISBT symbol (number)	LE3 (007003 or 7.3)
Obsolete names	X; Le^x; Le^{abx}
History	Described in 1949 as the antigen reacting with anti-X; referred to as Le^x from the mid-1950s; formally assigned to Lewis and renamed Le^{ab} by ISBT in 1998.

Occurrence

All populations: on Le(a+b–) and Le(a–b+) RBCs from adults and on 90% of cord samples.

Expression

Cord RBCs   Expressed

Molecular basis associated with Le^{ab} antigen[1]

The binding site for anti-Le^{ab} comprises the disaccharide Fucα1→4GlcNAc→R which is shared by the Le^a and Le^b active structures, similar to the A,B antigen being shared by A and B in the ABO system. It is of note that all serological monoclonal anti-Le^b reagents are anti-Le^{ab} reagents, but cannot detect Le^a antigen when it is present on RBCs.

In vitro characteristics of alloanti-Le^{ab}

Immunoglobulin class   IgM
Optimal technique   RT; 37°C (rare)
Effect of enzymes   Enhanced

Clinical significance of alloanti-Le^{ab}

Transfusion reaction   No
HDFN   No

Comments

Anti-Le^{ab} is a fairly common specificity, and is frequently found with anti-Le^a and/or anti-Le^b; it occurs mainly in serum from Le(a–b–) secretors of blood group A, B, or AB. Anti-Le^{ab} is inhibited by saliva that contains Le^a, and is weakly inhibited by saliva that contains Le^b. Saliva from Le(a–) non-secretors may also have a very weak inhibitory effect.

Experts recommend transfusion of Le(a–b–) blood if antibody reacts at 37°C.

The practical value of categorizing the epitope recognized by anti-Le^{ab} as a blood group antigen has been questioned by some experts, because of uncertainty about its biochemical basis and cross-reactivity issues.

Reference

[1] Schenkel-Brunner, H., 2000. In: Human Blood Groups: Chemical and Biochemical Basis of Antigen Specificity, second ed. Springer-Verlag Wien, New York, NY.

Le^{bH} Antigen

Terminology

ISBT symbol (number)   LE4 (007004 or 7.4)
History   Antigen detected by the original anti-Le^b in 1948; named Le^{bH} in 1959 upon recognition of heterogeneity of anti-Le^b; allocated ISBT number in 1998.

Occurrence

Present on group O and A_2 Le(b+) RBCs, i.e., those with strong expression of H antigen. Group A_1 or B RBCs react weakly or not at all.

Expression

Cord RBCs	Not expressed
Altered	Weak in Le(a+b+), often weakened during pregnancy and certain diseases

Molecular basis associated with Le^{bH} antigen[1]

Anti-Le^{bH} appears to require access to the fucose residue of H type 1 (see H [018] System) on group O RBCs, where the structure is not blocked by the immunodominant blood group A or B sugars. The determinant must also involve the L-fucose added by the *FUT3* (*LE*)-specified transferase because Le(a–b–) RBCs from secretors (which carry H type 1) do not react with anti-Le^{bH}.

Effect of enzymes and chemicals on Le^{bH} antigen on intact RBCs

Refer to Le^b antigen (**LE2**).

In vitro characteristics of alloanti-Le^{bH}

Immunoglobulin class	IgM predominates
Optimal technique	RT; 37°C; enzymes
Neutralization	Isotonic saliva (inhibited by saliva that contains H, or H and Le^b)
Complement binding	Yes; some hemolytic

Clinical significance of alloanti-Le^{bH}

Transfusion reaction	No
HDFN	No

Comments

Anti-Le^{bH} is a more common specificity than anti-Le^{bL}. Anti-Le^{bH} is unlikely to cause incompatible cross-matches if ABO-identical blood is selected. The practical value of categorizing the epitope recognized by anti-Le^{bH} as a blood group antigen has been questioned by some experts, because of uncertainty about its biochemical basis and cross-reactivity issues.

Reference

[1] Henry, S., et al., 1995. Lewis histo-blood group system and associated secretory phenotypes. Vox Sang 69, 166–182.

ALe^b Antigen

Terminology

ISBT symbol (number)	LE5 (007005 or 7.5)
Obsolete name	A$_1$Le^b
History	Anti-A$_1$Le^b was identified in 1967 during cross-matching; name derived from the unusual antibody reactivity; received an ISBT number in 1998; name amended to ALe^b.

Occurrence

On all group A and AB RBCs which are also Le(b+), i.e., on RBCs of secretors of A who have a *FUT3* (*LE*) gene. It is of note that Le^b cannot be made into ALe^b.

All group O and group B Le(b+) RBCs and all Le(b–) people are antigen-negative.

Expression

Cord RBCs	Not expressed

Molecular basis associated with ALe^b antigen[1]

```
              GalNAc
                | α1-3
   Fuc  α1–2  Gal
                | β1-3
   Fuc  α1–4  GlcNAc
                | β1-3
              Gal
                | β1-4
              Glc
                | β1-1
           Ceramide
```

ALe^b is expressed when A type 1 is modified by the addition of *FUT3*-gene-specified L-fucose.

Effect of enzymes and chemicals on ALe^b antigen on intact RBCs

Refer to Le^b (**LE2**).

In vitro characteristics of alloanti-ALe^b

Refer to Le^b (**LE2**).

Lewis

Clinical significance of alloanti-ALe^b

Few examples of anti-ALe[b] have been reported; may be lymphocytotoxic.

Comments

ALe[b] is adsorbed onto RBCs and lymphocytes, and is the dominant Lewis antigen on group A Le(b+) RBCs. It is also of note that the anti-Le[b] does not react with ALe[b] antigen, hence group A individuals may have weaker Le[b] reactions than group O (and are often mistyped).

Anti-ALe[b] is a single specificity that cannot be separated into anti-A and anti-Le[b]. Monoclonal anti-ALe[b] has been produced.

Reference

[1] Henry, S., et al., 1995. Lewis histo-blood group system and associated secretory phenotypes. Vox Sang 69, 166–182.

BLe^b Antigen

Terminology

ISBT symbol (number)	LE6 (007006 or 7.6)
History	Allocated a Lewis system number by ISBT in 1998.

Occurrence

On all group B and AB RBCs that are also Le(b+), i.e., on RBCs of secretors of B who have a *FUT3* (*LE*). It is of note that Le[b] cannot be made into BLe[b]. All group O and group A Le(b+) and all Le(b−) people are antigen-negative.

Expression

Cord RBCs	Not expressed

Molecular basis associated with BLe^b antigen¹

$$
\begin{array}{c}
\textbf{Gal} \\
|\,\alpha 1\text{-}3 \\
\textbf{Fuc} \xrightarrow{\alpha 1\text{-}2} \textbf{Gal} \\
|\,\beta 1\text{-}3 \\
\textbf{Fuc} \xrightarrow{\alpha 1\text{-}4} \textbf{GlcNAc} \\
|\,\beta 1\text{-}3 \\
\textbf{Gal} \\
|\,\beta 1\text{-}4 \\
\textbf{Glc} \\
|\,\beta 1\text{-}1 \\
\textbf{Ceramide}
\end{array}
$$

BLe^b is expressed when B type 1 is modified by the addition of *FUT3*-gene-specified L-fucose.

Effect of enzymes and chemicals on BLe^b antigen on intact RBCs

Refer to Le^b antigen (**LE2**).

In vitro characteristics of alloanti-BLe^b

Refer to Le^b antigen (**LE2**).

Clinical significance of alloanti-BLe^b

Few examples of anti-BLe^b have been reported; unlikely to cause a transfusion reaction or HDFN; may be lymphocytotoxic.

Comments

BLe^b is adsorbed onto RBCs and lymphocytes, and is the dominant Lewis antigen on group B Le(b+) RBCs. It is also of note that the anti-Le^b does not react with BLe^b antigen, hence group B individuals may have weaker Le^b reactions than group O (and are often mistyped). Anti-BLe^b is a single specificity; it cannot be separated into anti-B and anti-Le^b.

Reference
[1] Henry, S., et al., 1995. Lewis histo-blood group system and associated secretory phenotypes. Vox Sang 69, 166–182.

Lewis

Duffy Blood Group System

Number of antigens 5

Polymorphic	Fy^a, Fy^b
High prevalence	Fy3, Fy5, Fy6

Terminology

ISBT symbol (number)	FY (008)
CD number	CD234
History	Named after the family of the first proband who made anti-Fy^a.

Expression

Other blood cells	Not on granulocytes, lymphocytes, monocytes, platelets
Tissues	Endothelial cells of capillary and postcapillary venules, the epithelial cells of kidney collecting ducts, lung alveoli, and Purkinje cells of the cerebellum

Gene

Chromosome	1q23.2
Name	FY (DARC)
Organization	2 exons distributed over 1.5 kbp of gDNA
Product	Major (β) Duffy glycoprotein (DARC)
	Minor (α) Duffy glycoprotein

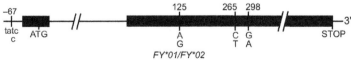

The Blood Group Antigen (3/e). DOI: http://dx.doi.org/10.1016/B978-0-12-415849-8.00010-7

361

Database accession numbers

GenBank   X85785 (gene); U01839 (mRNA); NM_002036
Entrez Gene ID   2532

Molecular basis of Duffy phenotypes

Two Duffy mRNA are translated from *FY*: a less abundant splice form (α) that encodes a protein of 338 residues that was discovered first and used for cloning (Accession number U01839, amino acid sequence MASSGYVLQAE. . .), and a more abundant splice form (β) that encodes a protein of 336 residues[1] (amino acid sequence MGNCLHRAE. . .). *FY*02* (Accession number U01839) is used as the reference allele, but the numbering by Iwamoto et al. is changed to conform with numbering from the codon for the initiator Met.

The reference allele *FY*02* or *FY*B* (Accession number U01839[1]) encodes Fy[b] (FY2), FY3, FY5, FY6. The nucleotide difference from this reference allele, and the amino acid affected, are given.

Allele encodes	Allele name	Exon	Nucleotide	Restriction enzyme	Amino acid	Ethnicity (prevalence)
Fy(a+b–) or FY:1	*FY*01* or *FY*A*	2	125A>G	*Ban*I+	Asp42Gly	Asians> Caucasians> Blacks (Common)

Molecular bases of silencing of *FY*

Homozygosity or compound heterozygosity leads to the FY$_{null}$ [FY:–3; Fy(a–b–)] phenotype. Nucleotide changes from the *FY*01* or *FY*02* backgrounds, and amino acids affected, are given.

Allele name	Exon	Nucleotide	Amino acid	Ethnicity (prevalence)
FY*01N.01	Promoter	–67t>c^	Protein absent from RBCs	Papua New Guinea (Rare)
FY*01N.02	2	281_295del	fs, Stop	Caucasians (Rare)
FY*01N.03	2	408G>A	Trp136Stop	(Rare)
FY*01N.04	2	287G>A	Trp96Stop	(Rare)

(Continued)

Duffy

(Continued)

Allele name	Exon	Nucleotide	Amino acid	Ethnicity (prevalence)
FY*01N.05	2	327delC	fs, Stop[2]	(Rare)
FY*02N.01	Promoter	−67t>c^	Protein absent from RBCs	Blacks (Common);Arabs, Jews, Romany (Several);Caucasians (Rare)
FY*02N.02	2	407G>A	Trp136Stop	(Rare)

^=This GATA-1 nucleotide change has been reported previously as −33 and −46. The gene is silenced only in erythroid cells.

Molecular bases of the FY$_{mod}$ (Fy^x) phenotype

Nucleotide changes from the *FY*02* background, and amino acids affected, are given.

Allele name	Exon	Nucleotide	Amino acid	Ethnicity (prevalence)
FY*02M.01	2	265C>T 298G>A	Arg89Cys Ala100Thr	Caucasians (Many), Blacks (Few)
FY*02M.02	2	145G>T 265C>T 298G>A	Ala49Ser Arg89Cys Ala100Thr	Caucasians (Few), Blacks (Rare)

Amino acid sequence of Fy^b [1,3]

Major product:

```
MGNCLHRAEL  SPSTENSSQL  DFEDVWNSSY  GVNDSFPDGD  YDANLEAAAP   50
CHSCNLLDDS  ALPFFILTSV  LGILASSTVL  FMLFRPLFRW  QLCPGWPVLA  100
QLAVGSALFS  IVVPVLAPGL  GSTRSSALCS  LGYCVWYGSA  FAQALLLGCH  150
ASLGHRLGAG  QVPGLTLGLT  VGIWGVAALL  TLPVTLASGA  SGGLCTLIYS  200
TELKALQATH  TVACLAIFVL  LPLGLFGAKG  LKKALGMGPG  PWMNILWAWF  250
IFWWPHGVVL  GLDFLVRSKL  LLLSTCLAQQ  ALDLLLNLAE  ALAILHCVAT  300
PLLLALFCHQ  ATRTLLPSLP  LPEGWSSHLD  TLGSKS                 336
```

At the N-terminus, the minor product of 338 amino acids has 9 amino acids, MASSGYVLQ, in place of the first 7 amino acids, MGNCLHR, of the major product.

Carrier molecule

A multipass membrane glycoprotein.

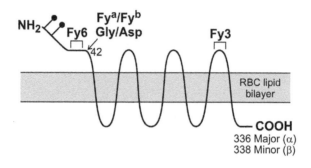

M_r (SDS-PAGE)	35,000–45,000
CHO: N-glycan	2 potential sites
CHO: O-glycan	No sites
Copies per RBC	6,000 to 13,000

Function

DARC is a promiscuous receptor for chemokines of both the C-X-C and C-C families, which have pleiotropic functions in innate and chronic immunobiology: IL-8 (interleukin-8), MGSA (melanoma growth stimulatory activity); MCP-1 (monocyte chemotactic protein 1); RANTES (regulated on activation, normal T-expressed and secreted). Clears proinflammatory peptides[4,5]. Fy(a–b–) RBCs do not bind chemokines, and RBCs with suppressed Fy^b antigen expression encoded by *FY*02M.01* bind reduced levels of chemokines.

Disease association

Receptor for *Plasmodium vivax*, *P. knowlesi* malarial parasites: Fy(a–b–) RBCs resist invasion.

The –67t>c change in *FY* (*DARC*) has been associated with a survival advantage in leukopenic HIV-infected persons, as well as low neutrophil counts, which in turn was associated with a three-fold higher risk of HIV susceptibility[6,7]. FY antigens may act as minor histocompatibility antigens in renal allograft rejection[8].

Duffy glycoprotein (DARC) has been reported to regulate prostate cancer growth, and its presence on vascular endothelium interacts with CD82 on cancer cells to inhibit the spread of cancer to remote sites and also may induce cancer cell senescence[9,10].

Phenotypes (% occurrence)

Phenotype	Caucasians	Blacks	Chinese	Japanese	Thai
Fy(a+b–)	17	9	90.8	81.5	69
Fy(a–b+)	34	22	0.3	0.9	3
Fy(a+b+)	49	1	8.9	17.6	28
Fy(a–b–)	Very rare	68	0	0	0
25% of Israeli Arabs and 4% of Israeli Jews have Fy(a–b–) RBCs					
Null: Fy(a–b–)					
Unusual: Fy^x expresses weak Fy^b antigen not detected by all anti-Fy^b					

Comments

Fy(a–b–) is also present in Arabs, Jews, Brazilians, and Romanies. The polymorphism 298G>A (Ala100Thr) occurs (without 265C>T) in ~15% of all *FY*02* alleles in Caucasians, and does not reduce expression of Fy^b.

References

[1] Iwamoto, S., et al., 1996. Identification of a novel exon and spliced form of Duffy mRNA that is the predominant transcript in both erythroid and postcapillary venule endothelium. Blood 87, 378–385.

[2] Tsuneyama, H., et al., 2000. Deletion in the Duffy gene of an apparently healthy individual with the Fy(a−b−) phenotype [abstract]. Transfusion 40 (Suppl.), 116S.

[3] Pogo, A.O., Chaudhuri, A., 2000. The Duffy protein: a malarial and chemokine receptor. Semin Hematol 37, 122–129.

[4] Darbonne, W.C., et al., 1991. Red blood cells are a sink for interleukin 8, a leukocyte chemotaxin. J Clin Invest 88, 1362–1369.

[5] Neote, K., et al., 1994. Functional and biochemical analysis of the cloned Duffy antigen: identity with the red blood cell chemokine receptor. Blood 84, 44–52.

[6] Kulkarni, H., et al., 2009. The Duffy-null state is associated with a survival advantage in leukopenic HIV-infected persons of African ancestry. Blood 114, 2783–2792.

[7] Ramsuran, V., et al., 2011. Duffy-null-associated low neutrophil counts influence HIV-1 susceptibility in high-risk South African black women. Clin Infect Dis 52, 1248–1256.

[8] Lerut, E., et al., 2007. Duffy and Kidd blood group antigens: minor histocompatibility antigens involved in renal allograft rejection? Transfusion 47, 28–40.

Duffy

[9] Bandyopadhyay, S., et al., 2006. Interaction of KAI1 on tumor cells with DARC on vascular endothelium leads to metastasis suppression. Nat Med 12, 933–938.

[10] Shen, H., et al., 2006. The Duffy antigen/receptor for chemokines (DARC) regulates prostate tumor growth. FASEB J 20, 59–64.

Fyᵃ Antigen

Terminology

ISBT symbol (number)	FY1 (008001 or 8.1)
History	Antibody identified in 1950 in the serum of Mr. Duffy, who was a multi-transfused hemophiliac. The last two letters of his name were used for the antigen name.

Occurrence

Caucasians	66%
Blacks	10%
Asians	99%
Thai	97%

Antithetical antigen

Fyᵇ (**FY2**)

Expression

Cord RBCs	Expressed

Molecular basis associated with Fyᵃ antigen[1–4]

Amino acid	Gly42 (44 in minor isoform)
Nucleotide	G at bp 125 in exon 2

Effect of enzymes and chemicals on Fyᵃ antigen on intact RBCs

Ficin/Papain	Sensitive
Trypsin	Resistant
α-Chymotrypsin	Sensitive
DTT 200 mM	Resistant
Acid	Resistant

In vitro characteristics of alloanti-Fyᵃ

Immunoglobulin class	IgG; IgM rarely
Optimal technique	IAT
Complement binding	Rare

Duffy

Clinical significance of alloanti-Fy^a

Transfusion reaction	Mild to severe (rare); immediate/delayed
HDFN	Mild to severe (rare)

Autoanti-Fy^a

Autoantibodies mimicking alloanti-Fy^a have been reported[5].

Comments

Fy^a has been demonstrated on fetal RBCs as early as 6 weeks gestation. Adult level of Fy^a expression is attained approximately 12 weeks after birth.

References

[1] Chaudhuri, A., et al., 1995. The coding sequence of Duffy blood group gene in humans and simians: restriction fragment length polymorphism, antibody and malarial parasite specificities, and expression in nonerythroid tissues in Duffy-negative individuals. Blood 85, 615–621.

[2] Iwamoto, S., et al., 1995. Genomic organization of the glycophorin D gene: Duffy blood group Fy^a/Fy^b alloantigen system is associated with a polymorphism at the 44-amino acid residue. Blood 85, 622–626.

[3] Mallinson, G., et al., 1995. Mutations in the erythrocyte chemokine receptor (Duffy) gene: the molecular basis of the Fy^a/Fy^b antigens and identification of a deletion in the Duffy gene of an apparently healthy individual with the Fy(a−b−) phenotype. Br J Haematol 90, 823–829.

[4] Tournamille, C., et al., 1995. Molecular basis and PCR-DNA typing of the Fy^a/Fy^b blood group polymorphism. Hum Genet 95, 407–410.

[5] Harris, T., 1990. Two cases of autoantibodies that demonstrate mimicking specificity in the Duffy blood group system. Immunohematology 6, 87–91.

Fy^b Antigen

Terminology

ISBT symbol (number)	FY2 (008002 or 8.2)
History	Named in 1951 when the antigen was shown to be antithetical to Fy^a.

Occurrence

Caucasians	83%
Blacks	23%
Chinese	9.2%
Asians	18.5%
Thai	31%

Duffy

Antithetical antigen

Fyª (**FY1**)

Expression

Cord RBCs	Expressed
Altered	Weak on RBCs with the Fyˣ phenotype

Molecular basis associated with Fyᵇ antigen[1–4]

Amino acid	Asp42 (44 in minor isoform)
Nucleotide	A at bp 125 in exon 2

Effect of enzymes and chemicals on Fyᵇ antigen on intact RBCs

Ficin/Papain	Sensitive
Trypsin	Resistant (weakened)
α-Chymotrypsin	Sensitive
DTT 200 mM	Resistant
Acid	Resistant

In vitro characteristics of alloanti-Fyᵇ

Immunoglobulin class	IgG; IgM rarely
Optimal technique	IAT
Complement binding	Rare

Clinical significance of alloanti-Fyᵇ

Transfusion reaction	Mild to severe (rare); immediate (rare); delayed
HDFN	Mild (rare)

Autoanti-Fyᵇ

Several examples of autoantibody mimicking alloanti-Fyᵇ have been reported; one caused AIHA.

Comments

Fyᵇ is a poor immunogen and has been estimated to be 20 times less immunogenic than Fyª.

Black individuals who have Fy(a–b–) RBCs invariably possess an allele that encodes Fyᵇ on cells other than RBCs. When the allele encoding RBC Fyᵇ antigen is silenced by a GATA box mutation (*FY*–67t>c*)[5], patients do not make anti-Fyᵇ.

References

[1] Chaudhuri, A., et al., 1995. The coding sequence of Duffy blood group gene in humans and simians: restriction fragment length polymorphism, antibody and malarial parasite specificities, and expression in nonerythroid tissues in Duffy-negative individuals. Blood 85, 615–621.

[2] Iwamoto, S., et al., 1995. Genomic organization of the glycophorin D gene: Duffy blood group Fy^a/Fy^b alloantigen system is associated with a polymorphism at the 44-amino acid residue. Blood 85, 622–626.

[3] Mallinson, G., et al., 1995. Mutations in the erythrocyte chemokine receptor (Duffy) gene: the molecular basis of the Fy^a/Fy^b antigens and identification of a deletion in the Duffy gene of an apparently healthy individual with the Fy(a−b−) phenotype. Br J Haematol 90, 823–829.

[4] Tournamille, C., et al., 1995a. Molecular basis and PCR-DNA typing of the Fy^a/Fy^b blood group polymorphism. Hum Genet 95, 407–410.

[5] Tournamille, C., et al., 1995b. Disruption of a GATA motif in the *Duffy* gene promoter abolishes erythroid gene expression in Duffy-negative individuals. Nature Genet 10, 224–228.

Fy3 Antigen

Terminology

ISBT symbol (number)	FY3 (008003 or 8.3)
Obsolete names	Fy^{ab}; Fy^{a}Fy^b
History	Anti-Fy3, found in 1971, was made by a pregnant Australian Fy(a–b–) woman who had been transfused. The specificity was named anti-Fy3 (and not anti-Fy^{ab}) because the antigenic determinant was resistant to enzyme treatment.

Occurrence

Caucasians	100% [Fy(a–b–) found in 4 Caucasians and one Cree Indian]
Blacks	32%
Asians	99.9%
Yemeni Jews	99%
Israeli Jews	96%
Israeli Arabs	75%

Expression

Cord RBCs	Expressed; increases after birth
Altered	Weak on RBCs with the Fy^x phenotype

Molecular basis associated with Fy3 antigen

See System pages for molecular basis associated with an absence of Fy3.
The third extracellular loop of the Duffy glycoprotein contains sequences necessary for binding of monoclonal anti-Fy3.

Duffy

Effect of enzymes and chemicals on Fy3 antigen on intact RBCs

Ficin/Papain	Resistant
Trypsin	Resistant
α-Chymotrypsin	Resistant
DTT 200 mM	Resistant
Acid	Resistant

In vitro characteristics of alloanti-Fy3

Immunoglobulin class	IgG
Optimal technique	IAT; enzymes
Complement binding	Rare

Clinical significance of alloanti-Fy3

Transfusion reaction	Mild to moderate; immediate (rare); delayed/ hemolytic
HDFN	Mild (rare)

Comments

The anti-Fy3 made by three non-black women reacted strongly with cord RBCs, whereas the anti-Fy3 made by black people does not react or reacts very weakly with cord RBCs.

Formation of anti-Fy3 is usually preceded by formation of anti-Fy^a. In spite of the high percentage of the Fy:–3 phenotype among blacks, anti-Fy3 is a rare specificity. To date, no black Fy(a–b–) individual has made anti-Fy^b, which is due to a GATA-box nucleotide change that silences *FY*B* only in the erythroid lineage.

Anti-Fy3 agglutinates Rh$_{null}$ RBCs, while anti-Fy5 does not.

Fy:–3 [Fy(a–b–)] RBCs resist invasion by *P. vivax* and *P. knowlesi* malarial parasites.

Fy5 Antigen

Terminology

ISBT symbol (number)	FY5 (008005 or 8.5)
History	Reported in 1973, and given the next Fy number when the antibody was shown to detect a novel antigen. Antibody was made by a black Fy(a–b–) boy with leukemia.

Occurrence

Blacks	32%
Most populations	99.9%

Expression

Cord RBCs	Expressed
Altered	Weak on Fy^x and D $-$ $-$ RBCs; absent from Rh$_{null}$ RBCs

Molecular basis associated with Fy5 antigen

Not known; possible interaction between Duffy and Rh proteins[1].

Effect of enzymes and chemicals on Fy5 antigen on intact RBCs

Ficin/Papain	Resistant
Trypsin	Not reported
α-Chymotrypsin	Not reported
DTT 200 mM	Resistant

In vitro characteristics of alloanti-Fy5

Immunoglobulin class	IgG
Optimal technique	IAT

Clinical significance of alloanti-Fy5

Transfusion reaction	Mild, delayed in one case
HDFN	No data

Comments

Several examples of anti-Fy5 have been found. All are in black, multiply-transfused (mostly because of sickle cell disease) Fy(a–b–) patients.
Fy(a–b–) RBCs from black individuals are FY:–3,–5; from a Caucasian (AZ; a.k.a. Findlay) are FY:–3,5, while Rh$_{null}$ RBCs are FY:3,–5.

Reference

[1] Colledge, K.I., et al., 1973. Anti-Fy5, an antibody disclosing a probable association between the Rhesus and Duffy blood group genes. Vox Sang 24, 193–199.

Fy6 Antigen

Terminology

ISBT symbol (number)	FY6 (008006 or 8.6)
History	Reported in 1987. This antigen, although numbered by the ISBT, has only been defined by murine monoclonal antibodies. No human anti-Fy6 has been described.

Occurrence

Blacks	32%
Most populations	100%

Expression

Cord RBCs	Expressed
Altered	Weak on RBCs with the Fy^x phenotype

Molecular basis associated with Fy6 antigen[1]

Anti-Fy6 binds to amino acid residues ^{19}Gln-Leu-Asp-Phe-Glu-Asp-Val-Trp^{26}.

Effect of enzymes and chemicals on Fy6 antigen on intact RBCs

Ficin/Papain	Sensitive
Trypsin	Resistant
α-Chymotrypsin	Sensitive
DTT 200 mM	Resistant
Acid	Resistant

Comments

Amino acid residues 8 (Ala) to 43 (Asp) are critical for *Plasmodium vivax* binding[2].

References

[1] Wasniowska, K., et al., 2002. Structural characterization of the epitope recognized by the new anti-Fy6 monoclonal antibody NaM 185-2C3. Transfus Med 12, 205–211.

[2] Pogo, A.O., Chaudhuri, A., 2000. The Duffy protein: a malarial and chemokine receptor. Semin Hematol 37, 122–129.

Duffy

Kidd Blood Group System

Number of antigens 3

Polymorphic	Jk^a, Jk^b
High prevalence	Jk3

Terminology

ISBT symbol (number)	JK (009)
History	Named in 1951 after the initials of the sixth child (John Kidd) of the first proband to make anti-Jk^a. John had hemolytic disease of the newborn.

Expression

Other blood cells	Not on lymphocytes, granulocytes, monocytes or platelets
Tissues	Kidney

Gene

Chromosome	18q12.3
Name	*JK (SLC14A1, HUT11A)*
Organization	11 exons distributed over 30 kbp of gDNA; exons 4 to 11 encode the mature protein
Product	Urea transporter UT-B, Kidd glycoprotein

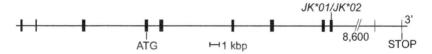

The Blood Group Antigen (3/e). DOI:http://dx.doi.org/10.1016/B978-0-12-415849-8.00011-9

Kidd

Database accession numbers

GenBank	NM_015865 (mRNA)
Entrez Gene ID	6563

Molecular basis of Kidd phenotype

The reference allele, *JK*02* or *JK*B* (Accession number NM_015865) encodes Jk^b (JK2), JK3. The nucleotide difference from this reference allele, and the amino acid affected, are given.

Allele encodes	Allele name	Exon	Nucleotide	Restriction enzyme	Amino acid	Ethnicity (prevalence)
Jk(a+b−) or JK:1,−2	*JK*01* or *JK*A*	9	838A>G	*Mnl*I+	Asn280Asp	Blacks> Whites> Asians (Common)

Molecular bases of weak or partial Kidd antigens

Homozygosity or compound herterozygosity leads to weak or partial phenotypes. Nucleotide changes from the *JK*01* or *JK*02* backgrounds, and the amino acids affected, are given. The term partial is used to classify an antigen-positive person who has the corresponding alloantibody in her/his plasma.

Allele encodes	Allele name	Exon	Nucleotide	Amino acid	Ethnicity (prevalence)	Antibody production^
Jk(a+^{w}b−)	*JK*01W.01*	4	130G>A	Glu44Lys	Many populations (Common)	−JK3 −Jk^a
Jk(a+^{w}b−)	*JK*01W.02*	7	511T>C	Trp171Arg	(Rare)	
Jk(a+^{w}b−)	*JK*01W.03*	4	28G>A	Val10Met	Black (Several)	−Jk^a
Jk(a+^{w}b−)	*JK*01W.04*	5	226G>A	Val76Ile	Black (Several)	−Jk^a
Jk(a+^{w}b−)	*JK*01W.05*	8	742G>A	Ala248Thr	Rare	
Jk(a−b+^{w})	*JK*02W.01*	7	548C>T	Ala183Val	Black (Rare)	

^Not all people with the allele have made the alloantibody.

Molecular bases of silencing of *JK*A* or *JK*B*

Homozygosity or compound heterozygosity leads to JK:–3 [Jk(a–b–)] phenotype.

Nucleotide changes from the *JK*01* or *JK*02* backgrounds, and the amino acids affected, are given.

Allele name	Exon/intron	Nucleotide	Amino acid	Ethnicity (prevalence)
JK*01N.01	4 & 5	Exons 4&5 deleted	Initiation Met absent	Tunisian, English, Bosnian (Rare)
JK*01N.02	5	202C>T	Gln68Stop	Caucasian, American (Rare)
JK*01N.03	7	582C>G	Tyr194Stop	Swiss, English (Few)
JK*01N.04	10	956C>T	Thr319Met	African American, (Rare)
JK*01N.05	7	561C>A	Tyr187Stop	African American (Rare) African Brazilian (Many)
JK*01N.06	Intron 5	IVS5–1g>a	Exon 6 skipped; in frame	Asian Indian (Rare)
JK*02N.01	Intron 5	IVS5–1g>a	Exon 6 skipped; in frame	Polynesian, Chinese (Several)
JK*02N.02	Intron 5	IVS5–1g>c	Exon 6 skipped; in frame	Chinese (Rare)
JK*02N.03	5	222C>A, 499A>G	Asn74Lys, Met167Val	Taiwanese (Rare)
JK*02N.04	Intron 7	IVS7+1g>t	Exon 7 skipped; frameshift→ Leu223Stop	French (Rare)
JK*02N.05	8	723delA	Frameshift→ Ile262Stop	Hispanic American (Rare)
JK*02N.06	9	871T>C	Ser291Pro	Finns (Several)
JK*02N.07	9	896G>A	Gly299Glu	Taiwanese (Rare)
JK*02N.08	10	956C>T	Thr319Met	Indian, Pakistani (Rare)
*JK*02N.09^*	5	191G>A	Arg64Gln	Black (Rare)

^Non-reactive with 1 monoclonal anti-Jk^b, positive with 1 polyclonal anti-Jk^b. Absorption and elution studies where not performed; may express some Jk protein.

Kidd

Amino acid sequence[1,2]

```
MEDSPTMVRV  DSPTMVRGEN  QVSPCQGRRC  FPKALGYVTG  DMKKLANQLK   50
DKPVVLQFID  WILRGISQVV  FVNNPVSGIL  ILVGLLVQNP  WWALTGWLGT  100
VVSTLMALLL  SQDRSLIASG  LYGYNATLVG  VLMAVFSDKG  DYFWWLLLPV  150
CAMSMTCPIF  SSALNSMLSK  WDLPVFTLPF  NMALSMYLSA  TGHYNPFFPA  200
KLVIPITTAP  NISWSDLSAL  ELLKSIPVGV  GQIYGCDNPW  TGGIFLGAIL  250
LSSPLMCLHA  AIGSLLGIAA  GLSLSAPFED  IYFGLWGFNS  SLACIAMGGM  300
FMALTWQTHL  LALGCALFTA  YLGVGMANFM  AEVGLPACTW  PFCLATLLFL  350
IMTTKNSNIY  KMPLSKVTYP  EENRIFYLQA  KKRMVESPL               389
```

Carrier molecule

Multi-pass glycoprotein.

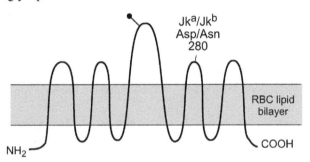

M_r (SDS-PAGE)	Predicted 43,000
CHO: N-glycan	2 potential sites (1 likely)
Cysteine residues	10
Copies per RBC	14,000

Function

A urea transporter in RBCs, UT-B plays a role in urea concentration by speeding up urea transport across the membrane as the RBCs pass through the descending and ascending vasa recta. Urea crosses the membrane of Jk(a–b–) RBCs about 1,000 times slower than in normal RBCs[3].

Disease association

Jk(a–b–) individuals have no clinical symptoms, but have their urine concentrating ability reduced by about one-third[3]. JK antigens may act as minor histocompatibility antigens in renal allograft rejection[4].

Phenotypes (% occurrence)

	Caucasians	Blacks	Asians
Jk(a+b–)	26.3	51.1	23.2
Jk(a–b+)	23.4	8.1	26.8
Jk(a+b+)	50.3	40.8	49.1
Jk(a–b–)	Rare	Rare	0.9 (Polynesians)

Null: Jk(a–b–)

Unusual: Jk(a–b–) [In(Jk)]; Several variants with altered (weakened or partial) expression of Jk^a or Jk^b (see table "Molecular bases of weak or partial Kidd phenotypes")

Comments

Jk(a–b–) RBCs resist lysis by 2 M urea[5].

Dominant type Jk(a–b–) [In(Jk)] RBCs have been found in Japanese.

Two transient Jk(a–b–) people have been described[6,7]. One was a Russian woman with myleofibrosis who made anti-Jk3 at the time her RBCs typed Jk(a–b–).

References

[1] Lucien, N., et al., 2002. Antigenic and functional properties of the human red blood cell urea transporter hUT-B1. J Biol Chem 277, 34101–34108.

[2] Olivès, B., et al., 1994. Cloning and functional expression of a urea transporter from human bone marrow cells. J Biol Chem 269, 31649–31652.

[3] Sands, J.M., et al., 1992. Urinary concentrating ability in patients with Jk(a−b−) blood type who lack carrier-mediated urea transport. J Am Soc Nephrol 2, 1689–1696.

[4] Lerut, E., et al., 2007. Duffy and Kidd blood group antigens: minor histocompatibility antigens involved in renal allograft rejection? *Transfusion*; 47, 28–40.

[5] Mougey, R., 1990. A review: the Kidd system. Immunohematology 6, 1–8.

[6] Issitt, P.D., et al., 1990. Temporary suppression of Kidd system antigen expression accompanied by transient production of anti-Jk3. Transfusion 30, 46–50.

[7] Obarski, G., et al., 1987. The Jk(a−b−) phenotype, probably occurring as a transient phenomenon [abstract]. Transfusion 27, 548.

Jk^a Antigen

Terminology

ISBT symbol (number)	JK1 (009001 or 9.1)
History	Reported in 1951. Name derived from the initials of the sixth child (John Kidd) of the antibody maker, Mrs. Kidd.

Occurrence

Caucasians	77%
Blacks	92%
Asians	72%

Antithetical antigen

Jk^b (**JK2**)

Expression

Cord RBCs	Expressed
Altered	See table: "Molecular basis of weak and partial Kidd phenotypes." When an altered allele is present, less Kidd glycoprotein is incorporated into the RBC membrane or epitope expression detected by monoclonal antibodies is reduced

Molecular basis associated with Jk^a antigen

Amino acid	Asp280
Nucleotide	G at bp 838 in exon 9

Effect of enzymes and chemicals on Jk^a antigen on intact RBCs

Ficin/Papain	Resistant (enhanced)
Trypsin	Resistant (enhanced)
α-Chymotrypsin	Resistant (enhanced)
Sialidase	Resistant
DTT 200 mM	Resistant
Acid	Resistant

In vitro characteristics of alloanti-Jk^a

Immunoglobulin class	IgG; many IgG plus IgM; IgM
Optimal technique	IAT; enzymes; PEG; CAT (gel)
Complement binding	Yes, provided that IgM is present; some hemolytic[1]

Clinical significance of alloanti-Jk^a

Transfusion reaction	No to severe; immediate or delayed/hemolytic
HDFN	Mild to moderate (rare)

Autoanti-Jk^a

Autoanti-Jk^a have been reported. With the discovery of weak or partial Jk^a phenotypes[2] it is possible some autoantibodies were actually alloantibodies.

Comments

Anti-Jk^a deteriorate *in vitro* and *in vivo*. Often found in multispecific sera. Anti-Jk^a may react more strongly with Jk(a+b–) than Jk(a+b+) RBCs (i.e., show dosage).

References

[1] Yates, J., et al., 1998. IgG anti-Jk^a/Jk^b antibodies are unlikely to fix complement. Transfus Med 8, 133–140.
[2] Wester, E.S., et al., 2011. Characterization of Jk(a+^{weak}): a new blood group phenotype associated with an altered JK*01 allele. Transfusion 51, 380–392.

Jk^b Antigen

Terminology

ISBT symbol (number)	JK2 (009002 or 9.2)
History	Found in 1953, and named for its antithetical relationship to Jk^a.

Occurrence

Caucasians	74%
Blacks	49%
Asians	76%

Antithetical antigen

Jk^a (**JK1**)

Expression

Cord RBCs	Expressed
Altered	See table: "Molecular basis of weak and partial Kidd phenotypes"

Molecular basis associated with Jk^b antigen

Amino acid	Asn280
Nucleotide	A at bp 838 in exon 9

Effect of enzymes and chemicals on Jk^b antigen on intact RBCs

Ficin/Papain	Resistant (enhanced)
Trypsin	Resistant (enhanced)
α-Chymotrypsin	Resistant (enhanced)
Sialidase	Resistant
DTT 200 mM	Resistant
Acid	Resistant

Kidd

In vitro characteristics of alloanti-Jk^b

Immunoglobulin class	IgG; many IgG plus IgM; IgM
Optimal technique	IAT; enzymes; PEG; CAT
Complement binding	Yes; provided that IgM is present; some hemolytic[1]

Clinical significance of alloanti-Jk^b

Transfusion reaction	No to severe; immediate or delayed/hemolytic
HDFN	No to mild (rare)

Autoanti-Jk^b

Autoanti-Jk^b have been reported. With the discovery of weak or partial Jk^b phenotypes it is possible that some autoantibodies were actually alloantibodies.

Comments

Anti-Jk^b deteriorate *in vitro* and *in vivo*. Often found in multispecific sera.

Reference

[1] Yates, J., et al., 1998. IgG anti-Jk^a/Jk^b antibodies are unlikely to fix complement. Transfus Med 8, 133–140.

Jk3 Antigen

Terminology

ISBT symbol (number)	JK3 (009003 or 9.3)
Obsolete names	Jk^{ab}; Jk^{a}Jk^b
History	Anti-Jk^{a}Jk^b was identified in 1959 and renamed anti-Jk3 by ISBT when numbers became popular.

Occurrence

Most populations	100%
Polynesians, Finns	>99%

Expression

Cord RBCs	Expressed
Altered	Very weak on Jk(a–b–) of the *In(Jk)* type (detected by absorption/elution); weak expression in the presence of certain alleles (see table: "Molecular basis of weak and partial Kidd phenotypes")

Molecular bases associated with Jk3 antigen

See System pages for molecular basis of Jk(a–b–) phenotype.

Effect of enzymes and chemicals on Jk3 antigen on intact RBCs

Ficin/Papain   Resistant (enhanced)
Trypsin   Resistant (enhanced)
α-Chymotrypsin   Resistant (enhanced)
Sialidase   Resistant
DTT 200 mM   Resistant
Acid   Resistant

In vitro characteristics of alloanti-Jk3

Immunoglobulin class   IgG more common than IgM
Optimal technique   IAT; PEG; Enzymes
Complement binding   Yes; some hemolytic

Clinical significance of alloanti-Jk3

Transfusion reaction   No to severe/immediate or delayed
HDFN   No to mild

Autoanti-Jk3

Rare

Comments

Anti-Jk3 has been found in a non-transfused male.

People with *In(Jk)* do not make anti-Jk3, and the presence of Jk antigens can be detected by absorption and elution.

Siblings of patients with anti-Jk3 should be tested for compatibility, and the patient urged to donate blood for cryogenic storage when his/her clinical state permits.

Kidd

Diego Blood Group System

Number of antigens 22

Low prevalence
Di^a, Wr^a, Wd^a, Rb^a, WARR, ELO, Wu, Bp^a, Mo^a, Hg^a, Vg^a, Sw^a, BOW, NFLD, Jn^a, KREP, Tr^a, Fr^a, SW1

High prevalence
Di^b, Wr^b, DISK

Terminology

ISBT symbol (number)
DI (010)

CD number
CD233

History
Named after the producer of the first anti-Di^a, discovered during the investigation of a case of HDFN in a Venezuelan family. Diego was described in 1955 by Layrisse et al; it had been mentioned briefly by Levine et al. in 1954.

Expression

Tissues
Intercalated cells of the distal and collecting tubules of the kidney
An isoform of band 3 is expressed in the distal nephron of the kidney

Gene

Chromosome
17q21.31

Name
DI (SLC4A1, AE1, EPB3)

Organization
20 exons distributed over 20 kbp of gDNA

Product
Band 3 (Anion Exchanger 1; Anion Transport Protein)

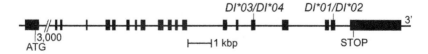

The Blood Group Antigen (3/e). DOI: http://dx.doi.org/10.1016/B978-0-12-415849-8.00012-0

Database accession numbers

GenBank   NM_000342 (mRNA); M27819 (mRNA)
Entrez Gene ID   6521

Molecular bases of Diego phenotypes[1]

Reference allele *DI*02* or *DI*B* (Accession number NM_000342) encodes Di^b
(DI2), DI4, DI22. Nucleotide differences, and amino acids affected, are given.

Allele encodes	Allele name	Exon	Nucleotide	Restriction enzyme	Amino acid	Ethnicity (prevalence)
Di(a+b−) or DI:1,−2	*DI*01* or *DI*A*	19	2561C>T	MspI−	Pro854Leu	South Americans, Native Americans, Japanese, Chinese, Poles (Many)
Wr(a+b−) or DI:3,−4	*DI*02.03*	16	1972G>A		Glu658Lys	(Many)
Wd(a+) or DI:5	*DI*02.05*	14	1669G>A	MslI+	Val557Met	Hutterites, Namibians (Rare)
Rb(a+) or DI:6	*DI*02.06*	14	1643C>T	EcoNI+	Pro548Leu	Caucasians (Rare)
WARR+ or DI:7	*DI*02.07*	14	1654C>T	BbsI −	Thr552Ile	Native Americans (Rare)
ELO+ or DI:8	*DI*02.08*	12	1294C>T	MspI −	Arg432Trp	Caucasians (Rare)
Wu+, DISK− or DI:9,−22	*DI*02.09*	14	1694G>C	ApaI −	Gly565Ala	Scandinavians, Dutch, Blacks, Irish (Several)
Bp(a+) or DI:10	*DI*02.10*	14	1707C>A	Tth2 −	Asn569Lys	English, Italians (Rare)
Mo(a+) or DI:11	*DI*02.11*	16	1967G>A	MsmI+	Arg656His	Belgians, Norwegians (Rare)
Hg(a+) or DI:12	*DI*02.12*	16	1966C>T	Cac8I+	Arg656Cys	Welsh, Australians (Rare)

(Continued)

(Continued)

Allele encodes	Allele name	Exon	Nucleotide	Restriction enzyme	Amino acid	Ethnicity (prevalence)
Vg(a+) or DI:13	DI*02.13	14	1663T>C	DraIII +	Tyr555His	Australians (Rare)
Sw(a+) or DI:14	DI*02.14	16	1937G>A		Arg646Gln	Caucasians (Rare)
BOW+ or DI:15	DI*02.15	14	1681C>T	BanI − BstEIII +	Pro561Ser	Caucasians (Rare)
NFLD+ or DI:16	DI*02.16	12; 14	1287A>T 1681C>G		Glu429Asp Pro561Ala	French Canadians, Japanese (Few)
Jn(a+) or DI:17	DI*02.17	14	1696C>T		Pro566Ser	Poles, Slovaks (Rare)
KREP+ or DI:18	DI*02.18	14	1696C>G		Pro566Ala	Poles (Rare)
Tr(a+) or DI:19	DI*02.19	14	1653G>C	BbsI −	Lys551Asn	English (Rare)
Fr(a+) or DI:20	DI*02.20	13	1438G>A	BsaI − BsmAI −	Glu480Lys	Mennonites (Several)
SW1 or DI:21	DI*02.21	16	1936C>T		Arg646Trp	Caucasians (Rare)

Molecular basis for silencing of *DI*

The nucleotide difference from *DI*02* reference allele (Accession number NM_000342), and the amino acid affected, are given.

Allele name	Exon	Nucleotide	Restriction enzyme	Amino acid	Ethnicity (prevalence)
DI*02N.01	13	1462G>A	NlaIII +	Val488Met[1]	Portuguese (Rare)

Diego

Amino acid sequence

```
MEELQDDYED  MMEENLEQEE  YEDPDIPESQ  MEEPAAHDTE  ATATDYHTTS   50
HPGTHKVYVE  LQELVMDEKN  QELRWMEAAR  WVQLEENLGE  NGAWGRPHLS  100
HLTFWSLLEL  RRVFTKGTVL  LDLQETSLAG  VANQLLDRFI  FEDQIRPQDR  150
EELLRALLLK  HSHAGELEAL  GGVKPAVLTR  SGDPSQPLLP  QHSSLETQLF  200
CEQGDGGTEG  HSPSGILEKI  PPDSEATLVL  VGRADFLEQP  VLGFVRLQEA  250
AELEAVELPV  PIRFLFVLLG  PEAPHIDYTQ  LGRAAATLMS  ERVFRIDAYM  300
AQSRGELLHS  LEGFLDCSLV  LPPTDAPSEQ  ALLSLVPVQR  ELLRRRYQSS  350
PAKPDSSFYK  GLDLNGGPDD  PLQQTGQLFG  GLVRDIRRRY  PYYLSDITDA  400
FSPQVLAAVI  FIYFAALSPA  ITFGGLLGEK  TRNQMGVSEL  LISTAVQGIL  450
FALLGAQPLL  VVGFSGPLLV  FEEAFFSFCE  TNGLEYIVGR  VWIGFWLILL  500
VVLVVAFEGS  FLVRFISRYT  QEIFSFLISL  IFIYETFSKL  IKIFQDHPLQ  550
KTYNYNVLMV  PKPQGPLPNT  ALLSLVLMAG  TFFFAMMLRK  FKNSSYFPGK  600
LRRVIGDFGV  PISILIMVLV  DFFIQDTYTQ  KLSVPDGFKV  SNSSARGWVI  650
HPLGLRSEFP  IWMMFASALP  ALLVFILIFL  ESQITTLIVS  KPERKMVKGS  700
GFHLDLLLVV  GMGGVAALFG  MPWLSATTVR  SVTHANALTV  MGKASTPGAA  750
AQIQEVKEQR  ISGLLVAVLV  GLSILMEPIL  SRIPLAVLFG  IFLYMGVTSL  800
SGIQLFDRIL  LLFKPPKYHP  DVPYVKRVKT  WRMHLFTGIQ  IICLAVLWVV  850
KSTPASLALP  FVLILTVPLR  RVLLPLIFRN  VELQCLDADD  AKATFDEEEG  900
RDEYDEVAMP  V                                              911
```

Carrier molecule[2,3]

A multipass glycoprotein.

Band 3 on intact RBCs is cleaved by α-chymotrypsin at residues 553 and 555 in the third extracellular loop.

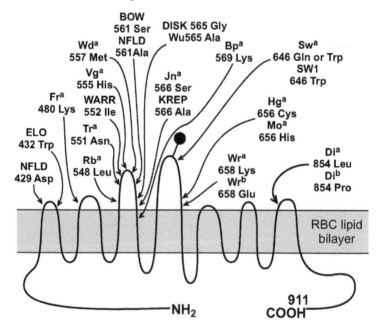

M_r (SDS-PAGE)	95,000–105,000
CHO: N-glycan	One (Asn642) in the 4th extracellular loop (carries more than half the ABH antigens on the RBC)
Copies per RBC	1,000,000

Function

Band 3 makes up 20% of the RBC membrane proteins; it has two functionally independent domains and numerous roles.

N-terminus cytoplasmic domain (residues 1–359):

Anchored to the membrane skeleton via ankyrin and protein 4.2, and contributes to maintaining the structural integrity of the RBC; interacts with several glycolytic enzymes, hemoglobin, catalase, and hemichromes (the oxidation products of denatured hemoglobin).

C-terminus membrane domain (residues 360–911):

Anion exchange (HCO_3^- and Cl^-) across the RBC membrane; contributes to the stability of the lipid bilayer through interaction with adjacent phospholipid molecules.

Band 3 may be involved in the removal of senescent or defective RBCs from the circulation and sequestration of RBCs infected with *Plasmodium falciparum*.

Disease association[2]

A severely hydropic baby, who lacked band 3 and protein 4.2, had to be resuscitated and kept alive by transfusions.

Products of variant alleles of band 3 have been implicated in the pathogenesis of South East Asian ovalocytosis, congenital acanthocytosis, hereditary spherocytosis, and distal renal tubular acidosis.

Band 3 has a role in the attachment of malarial parasites to the surface of RBCs, and in the adhesion of parasitized cells to the vascular epithelium.

Phenotypes (% occurrence)

Phenotype	Caucasians	Blacks	Asians	South American Indians
Di(a+b−)	<0.01	<0.01	<0.01	<0.1
Di(a−b+)	>99.9	>99.9	90	64
Di(a+b+)	<0.1	<0.1	10	36
Null: Di(a−b−) in one, transfusion-dependent case				
Unusual: Weak Di(b+)				

Diego

Comments

Band 3 and glycophorin A (GPA) interact during biosynthesis and within the RBC membrane: GPA appears to be a chaperone to aid the correct folding and efficient transport of band 3 to the RBC membrane. Lack of GPA in the RBC membrane results in failure to express Wr^b (**DI4**) antigen. An altered form of band 3 present in South East Asian ovalocytes is due to a deletion of amino acid residues 400 to 408[4].

References

[1] Zelinski, T., 1998. Erythrocyte band 3 antigens and the Diego blood group system. Transfus Med Rev 12, 36–45.

[2] Bruce, L.J., Tanner, M.J., 1999. Erythroid band 3 variants and disease. Baillieres Best Pract Res Clin Haematol 12, 637–654.

[3] Schofield, A.E., et al., 1994. The structure of the human red blood cell anion exchanger (EPB3, AE1, band 3) gene. Blood 84, 2000–2012.

[4] Tanner, M.J., 1993. Molecular and cellular biology of the erythrocyte anion exchanger (AE1). Semin Hematol 30, 34–57.

Di^a Antigen

Terminology

ISBT symbol (number)	DI1 (010001 or 10.1)
Obsolete name	Diego
History	Named after Mrs. Diego, producer of the first example of anti-Di^a; reported in detail in 1955; identified as a result of HDFN.

Occurrence

Most populations	0.01%
South American Indians	From 2% in Caracas Indians to 54% in Kainganges Indians
Japanese	12%
Chippewa Indians (Canada)	11%
Chinese	5%
Hispanics	1%
Poles	0.47%

Antithetical antigen

Di^b (**DI2**)

Expression

Cord RBCs	Expressed

Molecular basis associated with Di[a] antigen[1]

Amino acid	Leu854
Nucleotide	T at bp 2561 in exon19

Di[a] is predominantly associated with band 3 Memphis variant II, which has a mutation of Lys56 to Glu in the cytoplasmic domain. RBCs with the band 3 Memphis variant II bind stilbene disulfonate (H_2DIDS) more readily than do RBCs with Memphis variant I or common type band 3. Di[a] with Lys56 has been found in Amazonian Indians[2], but is not associated with antigen expression.

Effect of enzymes and chemicals on Di[a] antigen on intact RBCs

Ficin/Papain	Resistant
Trypsin	Resistant
α-Chymotrypsin	Resistant
DTT 200 mM	Resistant
Acid	Resistant

In vitro characteristics of alloanti-Di[a]

Immunoglobulin class	IgG (often IgG1 and IgG3)
Optimal technique	IAT
Complement binding	Some

Clinical significance of alloanti-Di[a]

Transfusion Reaction	None to severe/delayed
HDFN	Mild to severe

Comments

In contrast to many of the Diego system antibodies, anti-Di[a] is usually found as a single specificity; only occasionally does it occur in sera containing multiple antibodies to low-prevalence antigens. Examples of agglutinating anti-Di[a] and naturally-occurring anti-Di[a] exist, but are rare.

References

[1] Bruce, L.J., et al., 1994. Band 3 Memphis variant II. Altered stilbene disulfonate binding and the Diego (Di[a]) blood group antigen are associated with the human erythrocyte band 3 mutation Pro[854]-- > Leu. J Biol Chem 269, 16155–16158.

[2] Baleotti Jr., W., et al., 2003. A novel *DI* *A* allele without the Band 3-Memphis mutation in Amazonian Indians. Vox Sang 84, 326–330.

Diego

Di^b Antigen

Terminology

ISBT symbol (number)	DI2 (010002; 10.2)
Obsolete name	Luebano
History	Anti-Di^b identified in 1967; detected an antigen of high prevalence that is antithetical to Di^a.

Occurrence

Most populations	100%
Native Americans	99%

Antithetical antigen

Di^a (**DI1**)

Expression

Cord RBCs	Expressed
Altered	Weak on South East Asian ovalocytes, and on the Di(a−b+) and Di(a+b+) RBCs of some Hispanic-Americans

Molecular basis associated with Di^b antigen[1]

Amino acid	Pro854
Nucleotide	C at bp 2561 in exon 19

Effect of enzymes and chemicals on Di^b antigen on intact RBCs

Ficin/Papain	Resistant
Trypsin	Resistant
α-Chymotrypsin	Resistant
DTT 200 mM	Resistant
Acid	Resistant

In vitro characteristics of alloanti-Di^b

Immunoglobulin class	IgG
Optimal technique	IAT
Complement binding	No

Clinical significance of alloanti-Di^b

Transfusion reaction	None to moderate/delayed
HDFN	Mild

Autoanti-Di^b

Yes

Comments

Anti-Di^b demonstrates dosage, reacting more strongly with Di(a–b+) than with Di(a+b+) RBCs.

Siblings of patients with anti-Di^b should be tested for compatibility, and the patient urged to donate blood for cryogenic storage when his/her clinical state permits.

Reference

[1] Bruce, L.J., et al., 1994. Band 3 Memphis variant II. Altered stilbene disulfonate binding and the Diego (Di^a) blood group antigen are associated with the human erythrocyte band 3 mutation Pro^{854}--> Leu. J Biol Chem 269, 16155–16158.

Wr^a Antigen

Terminology

ISBT symbol (number)	DI3 (010003 or 10.3)
Obsolete names	Wright; 700001; 211001
History	Identified in 1953 as the cause of HDFN in the Wright family; assigned to the Diego blood group system in 1995.

Occurrence

All populations	<0.01%

Antithetical antigen

Wr^b (DI4)

Expression

Cord RBCs	Expressed

Molecular basis associated with Wr^a antigen[1]

Amino acid	Lys658
Nucleotide	A at bp 1972 in exon 16

Effect of enzymes and chemicals on Wr^a antigen on intact RBCs

Ficin/Papain	Resistant
Trypsin	Resistant
α-Chymotrypsin	Resistant
DTT 200mM	Resistant
Acid	Resistant

Diego

In vitro characteristics of alloanti-Wr^a

Immunoglobulin class	IgM; IgG
Optimal technique	RT; IAT

Clinical significance of alloanti-Wr^a

Transfusion reaction	None to severe/immediate or delayed/hemolytic
HDFN	Mild to severe

Comments

Alloanti-Wr^a is often an apparent naturally-occurring antibody and is found in the serum of 1% to 2% of blood donors. It is frequently found in multispecific sera and is a common specificity in patients with AIHA.

Reference

[1] Bruce, L.J., et al., 1995. Changes in the blood group Wright antigens are associated with a mutation at amino acid 658 in human erythrocyte band 3: a site of interaction between band 3 and glycophorin A under certain conditions. Blood 85, 541–547.

Wr^b Antigen

Terminology

ISBT symbol (number)	DI4 (010004 or 10.4)
Obsolete names	Fritz; MF; 901010; 900024; 211002
History	Anti-Wr^b identified in 1971; detected an antigen of high prevalence antithetical to Wr^a; assigned to Diego blood group system in 1995.

Occurrence

All populations	100%

Antithetical antigen

Wr^a (**DI3**)

Expression

Cord RBCs	Expressed
Altered	On ENEP– (HAG+), ENAV– (MARS+)[1], and ENEV– RBCs[2]

Molecular basis associated with Wr^b antigen[3]

Amino acid	Glu658
Nucleotide	G at bp 1972 in exon 16

Diego

For expression, Wr^b antigen requires the presence of amino acid residues 78 (previously 59) to 95 (previously 76) of GPA^4.

Effect of enzymes and chemicals on Wr^b antigen on intact RBCs

Ficin/Papain	Resistant (one example); sensitive (one example)[5]
Trypsin	Resistant
α-Chymotrypsin	Resistant
DTT 200 mM	Resistant
Acid	Resistant

In vitro characteristics of alloanti-Wr^b

Immunoglobulin class	IgM plus IgG
Optimal technique	IAT

Clinical significance of alloanti-Wr^b

Transfusion reaction	Not known because only three individuals with Wr(a+b–) RBCs and alloanti-Wr^b have been described, but chemiluminescence assay suggests that anti-Wr^b may cause accelerated destruction of transfused incompatible RBCs[1]
HDFN	Positive DAT but not clinical HDFN

Autoanti-Wr^b

Yes; fairly common specificity in patients with AIHA. Some autoanti-Wr^b appear to be benign while others are not: in two cases autoanti-Wr^b, reactive at 37°C, resulted in fatal intravascular hemolysis.

Comments

Alloanti-Wr^b can be a component in the plasma of immunized En(a–) people. RBCs that lack GPA, i.e., with the En(a–) or M^kM^k phenotype, type as Wr(a–b–). Some hybrid glycophorin molecules, e.g., En^a.UK, GP.Mur, GP.Hop, GP.Hil, GP.Bun, GP.HF, GP.JL, GP.SAT, GP.TK or GP.Dantu do not express Wr^b. All these hybrids have glutamic acid at residue 658 of band 3, but lack the required amino acids from GPA^6. GP.HAG [Gln82 (previously 63) Lys of GPA], GP.MARS [Ala84 (previously 65) Pro of GPA], and GP.ENEV [Val81 (previously 62) Gly of GPA] have an altered expression of Wr^b.

References

[1] Poole, J., 2000. Red cell antigens on band 3 and glycophorin A. Blood Rev 14, 31–43.

[2] Velliquette, R.W., et al., 2010. Novel single nucleotide change in *GYP*A* in a person who made an alloantibody to a new high prevalence MNS antigen called ENEV. Transfusion 50, 856–860.

[3] Bruce, L.J., et al., 1995. Changes in the blood group Wright antigens are associated with a mutation at amino acid 658 in human erythrocyte band 3: a site of interaction between band 3 and glycophorin A under certain conditions. Blood 85, 541–547.

Diego

[4] Reid, M.E., 1999. Contribution of MNS to the study of glycophorin A and glycophorin B. Immunohematology 15, 5–9.

[5] Storry, J.R., et al., 2001. A new Wr(a−b−) proband with anti-Wr[b] recognizing a ficin sensitive antigen [abstract]. Transfusion 41 (Suppl.), 23S.

[6] Huang, C.-H., et al., 1996. Human red blood cell Wright antigens: a genetic and evolutionary perspective on glycophorin A-band 3 interaction. Blood 87, 3942–3947.

Wd[a] Antigen

Terminology

ISBT symbol (number)	DI5 (010005 or 10.5)
Obsolete names	Waldner; 700030
History	Reported in 1983; first described in the Waldner family; identified when RBCs were being typed with a serum known to contain anti-Fr[a]; assigned to Diego blood group system in 1996.

Occurrence

Found only in two Schmiedeleut Hutterite families, one family in Holland with probable Hutterite connections, and two Namibian sisters.

Expression

Cord RBCs	Expressed

Molecular basis associated with Wd[a] antigen[1,2]

Amino acid	Met557
Nucleotide	A at bp 1669 in exon 14
Wd(a–) (wild type)	Val557 and G at bp 1669

Effect of enzymes and chemicals on Wd[a] antigen on intact RBCs

Ficin/Papain	Resistant
Trypsin	Resistant
α-Chymotrypsin	Sensitive
DTT 200 mM	Resistant
Acid	Resistant

In vitro characteristics of alloanti-Wd[a]

Immunoglobulin class	IgM; IgG (at least one IgG example reported)
Optimal technique	RT; IAT

Diego

Clinical significance of alloanti-Wd^a

Anti-Wd^a was not made by any of six Wd(a–) women, who between them gave birth to 30 Wd(a+) children[3]. In the same study, anti-Wd^a was found in the serum of 1 of 358 pregnant women. No other data are available.

Comments

Anti-Wd^a is a common specificity in multispecific sera.

References
[1] Bruce, L.J., et al., 1996. The low-incidence blood group antigen, Wd^a, is associated with the substitution Val$_{557}$--> Met in human erythrocyte band 3 (AE1). Vox Sang 71, 118–120.

[2] Jarolim, P., et al., 1997. Blood group antigens Rb^a, Tr^a, and Wd^a are located in the third ectoplasmic loop of erythroid band 3. Transfusion 37, 607–615.

[3] Lewis, M., Kaita, H., 1981. A "new" low incidence "Hutterite" blood group antigen Waldner (Wd^a). Am J Hum Genet 33, 418–420.

Rb^a Antigen

Terminology

ISBT symbol (number)	DI6 (010006 or 10.6)
Obsolete names	Redelberger; 700027
History	Reported in 1978; found on the RBCs of Mr. Redelberger, a donor and donor recruiter; assigned to Diego blood group system in 1996.

Occurrence

Found in three families.

Expression

Cord RBCs	Expressed

Molecular basis associated with Rb^a antigen[1]

Amino acid	Leu548
Nucleotide	T at bp 1643 in exon 14
Rb(a–) (wild type)	Pro548 and C at bp 1643

Effect of enzymes and chemicals on Rb^a antigen on intact RBCs

Ficin/Papain	Variable
Trypsin	Resistant
α-Chymotrypsin	Variable
Pronase	Variable

In vitro characteristics of alloanti-Rb^a

Immunoglobulin class   IgM (predominantly); IgG from the original report
Optimal technique   RT, IAT

Clinical significance of alloanti-Rb^a

Five Rb(a−) women, who gave birth to Rb(a+) children, did not make anti-Rb^a. No other data are available.

Comments

Common specificity in multispecific sera.

Reference

[1] Jarolim, P., et al., 1997. Blood group antigens Rb^a, Tr^a, and Wd^a are located in the third ectoplasmic loop of erythroid band 3. Transfusion 37, 607–615.

WARR Antigen

Terminology

ISBT symbol (number)   DI7 (010007 or 10.7)
Obsolete names   Warrior; 700055
History   Identified in 1991 as a result of HDFN in the Warrior family; assigned to Diego blood group system in 1996.

Occurrence

Found in two kindred, both with Native American heritage. The eldest WARR+ member in the Warrior kindred was of Absentee Shawnee ancestry[1].

Expression

Cord RBCs   Expressed

Molecular basis associated with WARR antigen[2]

Amino acid   Ile552
Nucleotide   T at bp 1654 in exon 14
WARR− (wild type)   Thr552 and C at bp 1654

Effect of enzymes and chemicals on WARR antigen on intact RBCs

Ficin/Papain   Resistant
Trypsin   Resistant

α-Chymotrypsin	Sensitive
Pronase	Sensitive
DTT 200 mM	Resistant

In vitro characteristics of alloanti-WARR

| Immunoglobulin class | IgG |
| Optimal technique | IAT |

Clinical significance of alloanti-WARR

| Transfusion reaction | No data are available |
| HDFN | Mild |

Comments

It is a common specificity in multispecific sera and immune anti-WARR exists.

References

[1] Coghlan, G., et al., 1995. A "new" low-incidence red cell antigen, WARR: unique to native Americans? Vox Sang 68, 187–190.

[2] Jarolim, P., et al., 1997. A Thr$_{552}$--> Ile substitution in erythroid band 3 gives rise to the warrior blood group antigen. Transfusion 37, 398–405.

ELO Antigen

Terminology

ISBT symbol (number)	DI8 (010008 or 10.8)
Obsolete name	700051
History	Antigen recognized in 1979; reported in detail in 1993; named after the first name of the original proband; assigned to Diego blood group system in 1998.

Occurrence

| All populations | <0.01% |

Expression

| Cord RBCs | Expressed |

Molecular basis associated with ELO antigen[1]

Amino acid	Trp432
Nucleotide	T at bp 1294 in exon 12
ELO– (wild type)	Arg432 and C at bp 1294

Diego

Effect of enzymes and chemicals on ELO antigen on intact RBCs

Ficin/Papain	Resistant
Trypsin	Resistant
α-Chymotrypsin	Variable
Pronase	Variable
DTT 200 mM	Resistant

In vitro characteristics of alloanti-ELO

Immunoglobulin class	IgG
Optimal technique	IAT

Clinical significance of alloanti-ELO

Transfusion reaction	No data are available
HDFN	Mild to severe

Comments

Several examples of immune monospecific anti-ELO exist and it is often found in multispecific sera.

Reference

[1] Zelinski, T., 1998. Erythrocyte band 3 antigens and the Diego blood group system. Transfus Med Rev 12, 36–45.

Wu Antigen

Terminology

ISBT symbol (number)	DI9 (010009 or 10.9)
Obsolete names	Wulfsberg (700013); Hov (700038); Haakestad
History	Identified in 1967; named after the original Wu+ donor; assigned to Diego blood group system in 1998.

Occurrence

Less than 0.01% (Dutch, Danish, and Norwegian ancestry; also in one Irish and one Black proband).

Antithetical antigen

DISK (**DI22**)

Expression

Cord RBCs	Expressed

Molecular basis associated with Wu antigen[1]

Amino acid	Ala565
Nucleotide	C at bp 1694 in exon 14

Effect of enzymes and chemicals on Wu antigen on intact RBCs

Ficin/Papain	Resistant
Trypsin	Resistant
α-Chymotrypsin	Sensitive
Pronase	Sensitive
DTT 200 mM	Resistant

In vitro characteristics of alloanti-Wu

Immunoglobulin class	IgM; IgG (few)
Optimal technique	RT; IAT

Clinical significance of alloanti-Wu

No data are available.

Comments

Anti-Wu is often found in multispecific sera and may be naturally-occurring. Several members in one family of Dutch descent are likely to be homozygous for *Wu*, and an Irish woman homozygous for *Wu* (*DI*02.09* allele) was reported. The serological relationship with NFLD and BOW[2] cannot (yet) be explained by the molecular knowledge.

References

[1] Zelinski, T., 1998. Erythrocyte band 3 antigens and the Diego blood group system. Transfus Med Rev 12, 36–45.
[2] Kaita, H., et al., 1992. A serologic relationship among the NFLD, BOW, and Wu red cell antigens. Transfusion 32, 845–847.

Bp^a Antigen

Terminology

ISBT symbol (number)	DI10 (010010 or 10.10)
Obsolete names	Bishop; 700010
History	Antigen discovered in 1964 on the RBCs of Mr. Bishop; assigned to Diego blood group system in 1998.

Occurrence

There are two probands (English and Italian).

Diego

Expression

Cord RBCs   Presumed expressed

Molecular basis associated with Bp^a antigen[1]

Amino acid	Lys569
Nucleotide	A at bp 1707 in exon 14
Bp(a–) (wild type)	Asn569 and C at bp 1707

Effect of enzymes and chemicals on Bp^a antigen on intact RBCs[2]

Ficin/Papain	Sensitive
Trypsin	Sensitive
α-Chymotrypsin	Sensitive
Pronase	Sensitive
DTT 200 mM	Presumed resistant

In vitro characteristics of alloanti-Bp^a

Immunoglobulin class	IgM
Optimal technique	RT

Clinical significance of alloanti-Bp^a

Transfusion reaction	No data are available
HDFN	No

Comments

Alloanti-Bp^a is often found in sera containing multiple naturally-occurring antibodies (including anti-Wr^a), and in sera from patients with AIHA. The band 3 amino acid substitution associated with expression of Bp^a antigen is likely to be located within the RBC lipid bilayer; thus, the enzyme sensitivity of Bp^a is somewhat surprising and may indicate the interaction with another enzyme sensitive component in the formation of the Bp^a epitope.

References

[1] Zelinski, T., 1998. Erythrocyte band 3 antigens and the Diego blood group system. Transfus Med Rev 12, 36–45.

[2] Jarolim, P., et al., 1998. Characterization of seven low incidence blood group antigens carried by erythrocyte band 3 protein. Blood 92, 4836–4843.

Mo^a Antigen

Terminology

ISBT symbol (number)	DI11 (010011 or 10.11)
Obsolete names	Moen; 700022

History   Antigen found in 1972 when random donors were
                          screened for Jn^a; assigned to Diego blood group
                          system in 1998.

Occurrence

Three probands have been reported: one from Norway and two from Belgium.

Antithetical antigen

Hg^a (**DI12**)

Expression

Cord RBCs   Presumed expressed

Molecular basis associated with Mo^a antigen[1]

Amino acid   His656
Nucleotide   A at bp 1967 in exon 16
Mo(a–) (wild type)   Arg656 and G at bp 1967

Effect of enzymes and chemicals on Mo^a antigen on intact RBCs

Ficin/Papain   Resistant
Trypsin   Resistant
α-Chymotrypsin   Resistant
Pronase   Resistant
DTT 200 mM   Resistant

In vitro characteristics of alloanti-Mo^a

Immunoglobulin class   IgM; IgG
Optimal technique   RT; IAT

Clinical significance of alloanti-Mo^a

No data are available because the antigen is rare.

Comments

Anti-Mo^a may be naturally-occurring, and is found in multispecific sera.

Reference

[1] Zelinski, T., 1998. Erythrocyte band 3 antigens and the Diego blood group system. Transfus
Med Rev 12, 36–45.

Diego

Hgª Antigen

Terminology

ISBT symbol (number)	DI12 (010012 or 10.12)
Obsolete names	Hughes; 700034; Tarplee; Tarp
History	The Hgª antigen was described in 1983. Its name was derived from the maiden name (Hughes) of the original Hg(a+) panel donor identified during routine antibody screening tests. Hgª joined the Diego blood group system in 1998.

Occurrence

Reported in three Welsh families and an Australian donor from New South Wales.

Antithetical antigen

Moª (DI11)

Expression

Cord RBCs	Expressed

Molecular basis associated with Hgª antigen[1]

Amino acid	Cys656
Nucleotide	T at bp 1966 in exon 16
Hg(a–) (wild type)	Arg656 and C at bp 1966

Effect of enzymes and chemicals on Hgª antigen on intact RBCs[2]

Ficin/Papain	Resistant
Trypsin	Resistant
α-Chymotrypsin	Resistant
Pronase	Resistant
DTT 200 mM	Presumed resistant

In vitro characteristics of alloanti-Hgª

Immunoglobulin class	IgM; IgG
Optimal technique	RT; IAT

Clinical significance of alloanti-Hgª

No data are available because the antigen is rare.

Comments

Anti-Hg^a is found in multispecific sera; anti-Hg^a as a single specificity has not been reported.

References

[1] Zelinski, T., 1998. Erythrocyte band 3 antigens and the Diego blood group system. Transfus Med Rev 12, 36–45.

[2] Jarolim, P., et al., 1998. Characterization of seven low incidence blood group antigens carried by erythrocyte band 3 protein. Blood 92, 4836–4843.

Vg^a Antigen

Terminology

ISBT symbol (number)	DI13 (010013 or 10.13)
Obsolete names	VanVugt; 700029
History	Antigen reported in 1981; found while screening Australian donors with anti-Wu; named after the first antigen positive donor Miss Van Vugt; assigned to Diego blood group system in 1998.

Occurrence

Only one family reported.

Expression

Cord RBCs	Presumed expressed

Molecular basis associated with Vg^a antigen[1]

Amino acid	His555
Nucleotide	C at bp 1663 in exon 14
Vg(a–) (wild type)	Tyr555 and T at bp 1663

Effect of enzymes and chemicals on Vg^a antigen on intact RBCs[2]

Ficin/Papain	Resistant
Trypsin	Resistant
α-Chymotrypsin	Sensitive
Pronase	Sensitive
DTT 200 mM	Presumed resistant

In vitro characteristics of alloanti-Vg^a

Immunoglobulin class	IgM and rarely IgG
Optimal technique	RT (IAT)

Diego

Clinical significance of alloanti-Vg^a

No data are available.

Comments

Anti-Vg^a is a relatively common antibody (11 examples of anti-Vg^a were found among 1669 donor sera), and is found in multispecific sera that frequently also contain anti-Wr^a.

References

[1] Zelinski, T., 1998. Erythrocyte band 3 antigens and the Diego blood group system. Transfus Med Rev 12, 36–45.

[2] Jarolim, P., et al., 1998. Characterization of seven low incidence blood group antigens carried by erythrocyte band 3 protein. Blood 92, 4836–4843.

Sw^a Antigen

Terminology

ISBT symbol (number)	DI14 (010014 or 10.14)
Obsolete names	Swann; 700004
History	Antigen identified in 1959, when serum from an AIHA patient was cross-matched against RBCs from donor Donald Swann; assigned to Diego blood group system in 1998.

Occurrence

All populations	<0.01%

Expression

Cord RBCs	Presumed expressed

Molecular basis associated with Sw^a antigen[1]

Amino acid	Gln or Trp646
Nucleotide	T at bp 1936 or A at bp 1937 in exon 16
Sw(a–) (wild type)	Arg646 and C at bp 1936 or G at bp 1937

Effect of enzymes and chemicals on Sw^a antigen on intact RBCs

Ficin/Papain	Resistant
Trypsin	Resistant
α-Chymotrypsin	Resistant
DTT 200 mM	Presumed resistant

In vitro characteristics of alloanti-Sw[a]

Immunoglobulin class   IgM; IgG
Optimal technique   RT (IAT)

Clinical significance of alloanti-Sw[a]

No data are available because the antigen is rare.

Comments[2,3]

Anti-Sw[a] is often found in AIHA and in multispecific sera. Anti-Sw[a] and anti-Fr[a], when present in the same serum, show cross-reactivity and cannot be separated by absorption.

Anti-Sw[a] also react with SW1+ (see **DI21**) RBCs: RBCs may be Sw(a+), SW1– (Gln646, the more common type) or Sw(a+) SW1+ (Trp646). Sw(a–), SW1+ RBCs have not been found.

References

[1] Zelinski, T., et al., 2000. Distinctive Swann blood group genotypes: molecular investigations. Vox Sang 79, 215–218.

[2] Contreras, M., et al., 1987. Swa: a subdivision. Vox Sang 52, 115–119.

[3] Lewis, M., et al., 1988. The Swann phenotype 700:4,-41: genetic studies. Vox Sang 54, 184–187.

BOW Antigen

Terminology

ISBT symbol (number)   DI15 (010015 or 10.15)
Obsolete names   Bowyer; 700046
History   Antigen reported in 1988; identified on the RBCs
                        of a donor (Bowyer) during an incompatible cross-
                        match; assigned to Diego blood group system in
                        1998.

Occurrence

Only a few probands have been reported.

Antithetical antigen

NFLD (**DI16**) (at amino acid 561, see Comments).

Expression

Cord RBCs   Presumed expressed

Molecular basis associated with BOW antigen[1]

Amino acid   Ser561
Nucleotide   T at bp 1681 in exon 14
BOW–, NFLD–   Pro561 and C at bp 1681
(wild type)

Effect of enzymes and chemicals on BOW antigen on intact RBCs

Ficin/Papain   Resistant
Trypsin   Resistant
α-Chymotrypsin   Sensitive
Pronase   Sensitive
DTT 200 mM   Resistant

In vitro characteristics of alloanti-BOW

Immunoglobulin class   IgG; some IgM
Optimal technique   IAT; RT

Clinical significance of alloanti-BOW

No data are available because the antigen is rare.

Comments

Several examples of immune monospecific anti-BOW exist and it is often found in multispecific sera. Molecular analysis showed that both BOW and NFLD are associated with a substitution at amino acid residue 561: serine is present when BOW is expressed and alanine is present when NFLD is expressed. Thus, BOW and NFLD can be considered antithetical, even though NFLD has a second mutation of Glu429Asp.

The serological relationship to Wu[2] cannot be explained by the molecular knowledge, although the critical residue (565Ala) for Wu is relatively close to residue 561 of band 3.

References

[1] McManus, K., et al., 2000. Amino acid substitutions in human erythroid protein band 3 account for the low-incidence antigens NFLD and BOW. Transfusion 40, 325–329.

[2] Kaita, H., et al., 1992. A serologic relationship among the NFLD, BOW, and Wu red cell antigens. Transfusion 32, 845–847.

Diego

NFLD Antigen

Terminology

ISBT symbol (number)	DI16 (010016 or 10.16)
Obsolete names	Newfoundland; 700037
History	Antigen reported in 1984; was found on the RBCs of a French Canadian in Newfoundland; assigned to Diego system in 1998.

Occurrence

Only a few probands (two French Canadian and two Japanese families) have been reported.

Antithetical antigen

BOW (**DI15**) (at amino acid 561, see Comments).

Expression

Cord RBCs	Presumed expressed

Molecular basis associated with NFLD antigen[1]

Amino acid	Asp429 and Ala561
Nucleotide	T at bp 1287 in exon 12 and G at bp 1681 in exon 14

NFLD–, BOW– (wild type) Glu429 with A at bp 1287, and Pro561 with C at bp 1681

Effect of enzymes and chemicals on NFLD antigen on intact RBCs

Ficin/Papain	Resistant
Trypsin	Resistant
α-Chymotrypsin	Sensitive
Pronase	Sensitive
DTT 200 mM	Resistant
Acid	Presumed resistant

In vitro characteristics of alloanti-NFLD

Immunoglobulin class	IgM and IgG
Optimal technique	RT (with albumin); IAT

Clinical significance of alloanti-NFLD

A Japanese NFLD– woman gave birth to three NFLD+ children without making anti-NFLD. No other data are available.

Diego

Comments

Anti-NFLD is found in multispecific sera. Molecular analysis showed that both NFLD and BOW are associated with a substitution at amino acid residue 561: alanine is present when NFLD is expressed and serine is present when BOW is expressed. Thus, NFLD and BOW can be considered antithetical at this residue. However, NFLD has a second mutation of Glu429Asp. The epitope defining NFLD may be created through an association and/or interaction between the first (residue 429) and third (residue 561) extracellular loops of band 3.

The serological relationship to Wu[2] cannot be explained by the molecular knowledge although the critical residue (565Ala) for expression of Wu is relatively close to residue 561 of band 3.

References

[1] McManus, K., et al., 2000. Amino acid substitutions in human erythroid protein band 3 account for the low-incidence antigens NFLD and BOW. Transfusion 40, 325–329.

[2] Kaita, H., et al., 1992. A serologic relationship among the NFLD, BOW, and Wu red cell antigens. Transfusion 32, 845–847.

Jnᵃ Antigen

Terminology

ISBT symbol (number)	DI17 (010017 or 10.17)
Obsolete names	Nunhart; JN; 700014
History	Antigen described in 1967; identified on the RBCs of Mr. J.N. during a study of the incidence of Wrᵃ in the Prague population; assigned to Diego blood group system in 1998.

Occurrence

Two probands (one of Polish, the other of Slovakian descent)[1].

Antithetical antigen

KREP (**DI18**)

Expression

Cord RBCs	Presumed expressed

Molecular basis associated with Jnᵃ antigen[1]

Amino acid	Ser566
Nucleotide	T at bp 1696 in exon 14
Jn(a−) (wild type)	Pro566 and C at bp 1696

Effect of enzymes and chemicals on Jn^a antigen on intact RBCs

Ficin/Papain	Resistant
Trypsin	Resistant
α-Chymotrypsin	Sensitive
DTT 200 mM	Resistant

In vitro characteristics of alloanti-Jn^a

Immunoglobulin class	IgM (no data regarding presence of an IgG component)
Optimal technique	RT

Clinical significance of alloanti-Jn^a

No data are available because the antigen is rare.

Comments

Anti-Jn^a is found in multispecific sera that also contain anti-KREP. The majority are naturally-occurring.

Reference
[1] Poole, J., 1999. The Diego blood group system – an update. Immunohematology 15, 135–143.

KREP Antigen

Terminology

ISBT symbol (number)	DI18 (010018 or 10.18)
Obsolete name	IK
History	Found in 1997 during investigation of the second Jn(a+) proband; named after the antigen-positive donor; assigned to Diego blood group system in 1998.

Occurrence

One Polish proband (IK)[1].

Antithetical antigen

Jn^a (**DI17**)

Expression

Cord RBCs	Presumed expressed

Molecular basis associated with KREP antigen[1]

Amino acid	Ala566
Nucleotide	G at bp 1696 in exon 14
KREP– (wild type)	Pro566 and C at bp 1696

Effect of enzymes and chemicals on KREP antigen on intact RBCs

Ficin/Papain	Resistant
Trypsin	Resistant
α-Chymotrypsin	Sensitive
DTT 200 mM	Resistant

In vitro characteristics of alloanti-KREP

Immunoglobulin class	IgM (no data regarding presence of an IgG component)
Optimal technique	RT

Clinical significance of alloanti-KREP

No data are available, but unlikely because anti-KREP have been naturally-occurring.

Comments

Anti-KREP is naturally-occurring and is present in multispecific sera. Among 13 sera tested, 12 contained anti-Jn[a] and anti-KREP, and only one serum contained anti-KREP without anti-Jn[a].

Reference
[1] Poole, J., 1999. The Diego blood group system – an update. Immunohematology 15, 135–143.

Tr[a] Antigen

Terminology

ISBT symbol (number)	DI19 (010019 or 10.19)
Obsolete names	Traversu; Lanthois; 700008
History	Antigen found in the 1960s during random testing of English blood donors with a multispecific serum, that also contained anti-Wr[a]; named after the first positive donor, Traversu.

Occurrence

Only found in two English probands.

Expression

Cord RBCs	Presumed expressed

Molecular basis associated with Tr^a antigen[1]

Amino acid	Asn551
Nucleotide	G at bp 1653 in exon 14
Tr(a–) (wild type)	Lys551 and C at bp 1653

Effect of enzymes and chemicals on Tr^a antigen on intact RBCs

Ficin/Papain	Resistant
Trypsin	Resistant
α-Chymotrypsin	Sensitive
DTT 200 mM	Presumed resistant

In vitro characteristics of alloanti-Tr^a

Immunoglobulin class	IgM and IgG
Optimal technique	RT; IAT

Clinical significance of alloanti-Tr^a

No data are available because the antigen is rare.

Comments

Anti-Tr^a was found as a separable specificity in 12 of 18 plasma samples that contained anti-Wr^a.

Anti-Tr^a is found in multispecific sera and in plasma from patients with AIHA.

Reference

[1] Jarolim, P., et al., 1997. Blood group antigens Rb^a, Tr^a, and Wd^a are located in the third ectoplasmic loop of erythroid band 3. Transfusion 37, 607–615.

Fr^a Antigen

Terminology

ISBT symbol (number)	DI20 (010020 or 10.20)
Obsolete names	Froese; 700026
History	Reported in 1978; named after the family (Froese) in which it was first recognized; assigned to Diego blood group system in 2000.

Occurrence

The reported Fr(a+) probands originate from three Mennonite kindred in Manitoba, Canada.

Diego

Expression

Cord RBCs   Expressed

Molecular basis associated with Fr^a antigen[1]

Amino acid   Lys480
Nucleotide   A at bp 1438 in exon 13
Fr(a–) (wild type)   Glu480 and G at bp 1438

Effect of enzymes and chemicals on Fr^a antigen on intact RBCs

Ficin/Papain   Resistant
Trypsin   Presumed resistant
α-Chymotrypsin   Presumed resistant
DTT 200 mM   Presumed resistant

In vitro characteristics of alloanti-Fr^a

Immunoglobulin class   IgG; IgM (few)
Optimal technique   IAT; RT

Clinical significance of alloanti-Fr^a

Transfusion reaction   No data are available
HDFN   Positive DAT, but no clinical HDFN

Comments

Several examples of immune monospecific anti-Fr^a exist, and it is often found in multispecific sera. Anti-Fr^a and anti-Sw^a, when present in the same serum show cross-reactivity and cannot be separated by absorption[2].

References

[1] McManus, K., et al., 2000. An amino acid substitution in the putative second extracellular loop of RBC band 3 accounts for the Froese blood group polymorphism. Transfusion 40, 1246–1249.
[2] Contreras, M., et al., 1987. Swa: a subdivision. Vox Sang 52, 115–119.

SW1 Antigen

Terminology

ISBT symbol (number)   DI21 (010021 or 10.21)
Obsolete name   700041
History   SW1 was documented in 1987; revealed by
                        heterogeneity among sera containing anti-Sw^a;
                        assigned to Diego blood group system in 2000.

Occurrence

Most populations   <0.01%

Expression

Cord RBCs   Presumed expressed

Molecular basis associated with SW1 antigen[1]

Amino acid   Trp646
Nucleotide   T at bp 1936 in exon 16
SW1–, Sw(a+)   Gln646 and A at bp 1936
SW1–, Sw(a–) (wild type) Arg646 and C at bp 1936

Effect of enzymes and chemicals on SW1 antigen on intact RBCs

Ficin/Papain   Resistant
Trypsin   Resistant
α-Chymotrypsin   Resistant
DTT 200 mM   Presumed resistant

In vitro characteristics of alloanti-SW1

Immunoglobulin class   IgM; IgG
Optimal technique   RT; IAT

Clinical significance of alloanti-SW1

No data are available because the antigen is rare.

Comments

Examples of anti-SW1 exist that do not react with Sw(a+) RBCs, but all anti-Sw[a] react with SW1+ RBCs[1]. See Sw[a] (**DI14**) for more details.

Reference

[1] Zelinski, T., et al., 2000. Distinctive Swann blood group genotypes: molecular investigations. Vox Sang 79, 215–218.

DISK Antigen

Terminology

ISBT symbol (number)   DI22 (010022 or 10.22)
History   Named in 2010 when an antibody to a high prevalence ficin-resistant, α-chymotrypsin-sensitive antigen was found in an untransfused Irish female and shown to be antithetical to Wu.

Occurrence

One Irish DISK– proband has been reported, and members of the Dutch family suspected to express a double dose of Wu antigen are also anticipated to be DISK–

Antithetical antigen

Wu (**DI9**)

Expression

Cord RBCs   Expressed

Molecular basis associated with DISK antigen[1]

Amino acid   Gly565
Nucleotide   G at bp 1694 in exon 14

Effect of enzymes and chemicals on DISK antigen on intact RBCs

Ficin/Papain   Resistant
Trypsin   Resistant
α-Chymotrypsin   Sensitive
DTT 200 mM   Presumed resistant

In vitro characteristics of alloanti-DISK

Immunoglobulin class   IgM; IgG
Optimal technique   RT; 37C; IAT

Clinical significance of alloanti-DISK

No data. However, serological characteristics of the only anti-DISK suggest a highly clinically significant antibody.

Comment

RBCs from the proband's brother reacted more weakly with anti-DISK, suggesting that the antibody exhibits dosage.
Siblings of patients with anti-DISK should be tested for compatibility, and the patient urged to donate blood for cryogenic storage when his/her clinical state permits.

Reference

[1] Poole, J., et al., 2010. Novel high incidence antigen in the Diego blood group system (DISK) and clinical significance of anti-DISK. Vox Sang 99 (Suppl. 1), 54–55.

Yt Blood Group System

Number of antigens   2

High prevalence   Yt^a
Polymorphic   Yt^b

Terminology

ISBT symbol (number)   YT (011)
Obsolete name   Cartwright
History   Named after the high prevalence antigen, Yt^a; became a system in 1964 after discovery of the antithetical antigen.

Expression

Other blood cells   Not on lymphocytes, granulocytes or monocytes
Tissues   Brain, muscle, nerves

Gene

Chromosome   7q22.1
Name   *YT (ACHE)*
Organization   6 exons distributed over 2.2 kbp of gDNA (exons 5 and 6 are alternatively spliced)
Product   Acetylcholinesterase (AChE)

Database accession numbers

GenBank   M55040 (mRNA); L42812 (DNA)
Entrez Gene ID   43

The Blood Group Antigen (3/e). DOI: http://dx.doi.org/10.1016/B978-0-12-415849-8.00013-2
415

Molecular bases of Yt phenotypes

The reference allele, *YT*01* or *YT*A* (Accession number M55040) encodes Yt^a (YT1). Nucleotide differences from this reference allele, and the amino acids affected, are given.

Allele encodes	Allele name	Exon	Nucleotide	Amino acid	Ethnicity (prevalence)
Yt(a–b+) or YT:–1,2	*YT*02.01* or *YT*B*	2	1057C>A	His353Asn	Israeli Jews, Israeli Arabs, Druse > Blacks > Europeans (Common)
Yt(a–b+) or YT:–1,2	*YT*02.02* or *YT*B*	2 5	1057C>A 1775C>G	His353Asn Pro592Arg	Israeli Arabs, Druse, and Jews

Amino acid sequence[1]

```
MRPPQCLLHT  PSLASPLLLL  LLWLLGGGVG  AEGREDAELL  VTVRGGRLRG   50
IRLKTPGGPV  SAFLGIPFAE  PPMGPRRFLP  PEPKQPWSGV  VDATTFQSVC  100
YQYVDTLYPG  FEGTEMWNPN  RELSEDCLYL  NVWTPYPRPT  SPTPVLVWIY  150
GGGFYSGASS  LDVYDGRFLV  QAERTVLVSM  NYRVGAFGFL  ALPGSREAPG  200
NVGLLDQRLA  LQWVQENVAA  FGGDPTSVTL  FGESAGAASV  GMHLLSPPSR  250
GLFHRAVLQS  GAPNGPWATV  GMGEARRRAT  QLAHLVGCPP  GGTGGNDTEL  300
VACLKTRPAQ  VLVNHEWHVL  PQESVFRFSF  VPVVDGDFLS  DTPEALINAG  350
DFHGLQVLVG  VVKDEGSYFL  VYGAPGFSKD  NESLISRAEF  LAGVRVGVPQ  400
VSDLAAEAVV  LHYTDWLHPE  DPARLREALS  DVVGDHNVVC  PVAQLAGRLA  450
AQGARVYAYV  FEHRASTLSW  PLWMGVPHGY  EIEFIFGIPL  DPSRNYTAEE  500
LIFAQRLMRY  WANFARTGDP  NEPRDPKAPQ  WPPYTAGAQQ  YVSLDLRPLE  550
VRRGLRAQAC  AFWNRFLPKL  LSATASEAPS  TCPGFTHGEA  APRPGLPLPL  600
LLLHQLLLLF  LSHLRRL                                         617
```

YT encodes a signal peptide of 31 amino acids, which is cleaved from the membrane bound protein. The carboxyl-terminal 29 amino acids are cleaved from the RBC GPI-linked form.

Carrier molecule

GPI-linked glycoprotein that probably exists as a dimer in the RBC membrane.

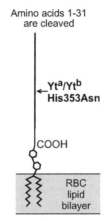

M_r (SDS-PAGE)	160,000 (72,000 as monomer under reducing conditions)
CHO: N-glycan	3 sites
CHO: O-glycan	Present
Cysteine residues	8
Copies per RBC	7,000–10,000 (or 3,500–5,000 dimers)

Function

AChE terminates nerve impulse transmission. AChE is in many tissues in various forms as a result of alternative splicing and post-translational modification. Function in RBC unknown.

Disease association

PNH III RBCs are deficient in AChE. Levels are reduced in myelodysplasias associated with chromosome 7 abnormalities and in some cases of SLE.

Phenotypes (% occurrence)

Phenotype	Most populations	Israelis
Yt(a+b−)	91.9	74.4
Yt(a+b+)	7.8	23.7
Yt(a−b+)	0.3	1.9
Null: Inherited Yt(a−b−) phenotype not found		
Unusual: One example of transient Yt(a−b−) RBCs reported[2]		

References

[1] Bartels, C.F., et al., 1993. Mutation at codon 322 in the human acetylcholinesterase (ACHE) gene accounts for YT blood group polymorphism. Am J Hum Genet 52, 928–936.

[2] Rao, N., et al., 1993. Human erythrocyte acetylcholinesterase bears the Yt^a blood group antigen and is reduced or absent in the Yt(a−b−) phenotype. Blood 81, 815–819.

Yt^a Antigen

Terminology

ISBT symbol (number)	YT1 (011001 or 11.1)
Obsolete name	Cartwright
History	In 1956 when the antibody to this high-prevalence antigen was found, most letters in the patient's name (Cartwright) had been taken by other antigens. The authors thought "why not T?" but to avoid confusion with T polyagglutination, they then said "why T" or "Yt". The Yt^a antigen achieved system status in 1964 after discovery of the antithetical antigen.

Occurrence

Most populations	>99.8%
Israeli Jews	98.6%
Israeli Arabs	97.6%
Israeli Druse	97.4%

Antithetical antigen

Yt^b (**YT2**)

Expression

Cord RBCs	Weak
Altered	Weak or absent from PNH III RBCs

Molecular basis associated with Yt^a antigen[1]

Amino acid	His353
Nucleotide	C at bp 1057 in exon 2

Effect of enzymes and chemicals on Yt^a antigen on intact RBCs

Ficin/Papain	Sensitive (variable)
Trypsin	Resistant
α-Chymotrypsin	Sensitive
DTT 200 mM/50 mM	Sensitive/weakened (thus sensitive to WARM™ and ZZAP)
Acid	Resistant

In vitro characteristics of alloanti-Yt[a]

Immunoglobulin class	IgG (some are IgG4)
Optimal technique	IAT
Complement binding	Some

Clinical significance of alloanti-Yt[a]

Transfusion reaction	No to moderate (rare)/delayed
HDFN	No

Comments

A report of an apparent alloanti-Yt[a] in a Yt(a+) person suggests the possibility of heterogeneity of the Yt[a] antigen[2].

Experts agree that anti-Yt[a] are often benign and antigen-negative blood may not need to be transfused.

References

[1] Bartels, C.F., et al., 1993. Mutation at codon 322 in the human acetylcholinesterase (ACHE) gene accounts for YT blood group polymorphism. Am J Hum Genet 52, 928–936.

[2] Mazzi, G., et al., 1994. Presence of anti-Yt[a] antibody in a Yt(a+) patient. Vox Sang 66, 130–132.

Yt[b] Antigen

Terminology

ISBT symbol (number)	YT2 (011002 or 11.2)
History	Identified in 1964 and named when its antithetical relationship to Yt[a] was recognized.

Occurrence

Europeans	8%
Israeli Jews	21.3%
Israeli Arabs	23.5%
Israeli Druse	26%

Not found in Japanese.

Antithetical antigen

Yt[a] (**YT1**)

Expression

Cord RBCs	Weak
Altered	Weak or absent from PNH III RBCs

Yt

Molecular basis associated with Yt^b antigen[1]

Amino acid	Asn353
Nucleotide	A at bp 1057 in exon 2

Effect of enzymes and chemicals on Yt^b antigen on intact RBCs

Ficin/Papain	Sensitive (variable)
Trypsin	Resistant
α-Chymotrypsin	Sensitive
DTT 200 mM/50 mM	Sensitive/weakened (thus sensitive to WARM™ and ZZAP)

In vitro characteristics of alloanti-Yt^b

Immunoglobulin class	IgG
Optimal technique	IAT

Clinical significance of alloanti-Yt^b

Transfusion reaction	No
HDFN	No

Comments

Anti-Yt^b is rare and usually occurs in sera with other antibodies. The second example of anti-Yt^b was made by a patient with PNH.

Reference

[1] Bartels, C.F., et al., 1993. Mutation at codon 322 in the human acetylcholinesterase (ACHE) gene accounts for YT blood group polymorphism. Am J Hum Genet 52, 928–936.

Xg Blood Group System

Number of antigens 2

Polymorphic   Xg^a
High prevalence   CD99

Terminology

ISBT symbol (number)   XG (012)
History   The Xg system was established in 1962 when it was found that Xg^a antigen expression was controlled by the X chromosome.

Expression

Other blood cells   Xg^a: expression may be restricted to RBCs
CD99: lymphocytes (27,000 sites), platelets (4,000 sites)[1]
Tissues   CD99: fibroblasts, fetal liver, lymph nodes, spleen, thymus, pancreatic islet cells, ovarian granulosa cells, sertoli cells, fetal adrenal, adult bone marrow[2]. Most abundant expression is in the most immature stages of the B cell, T cell, and granulocyte lineages

Gene

XG
Chromosome   Xp22.33
Name   *XG (PBDX)*[2]
Organization 10 small exons distributed over approximately 60 kbp of gDNA. Exon 1 to exon 3 are present in the pseudoautosomal region of the X and Y chromosomes. Exon 4 to exon 10 are only on the X chromosome[2]
Product   Xg^a glycoprotein

The Blood Group Antigen (3/e). DOI: http://dx.doi.org/10.1016/B978-0-12-415849-8.00014-4

CD99

Chromosome	Xp22.2 and Yp11.2
Name	*MIC2 (CD99)*
Organization	10 exons distributed over 52 kbp of gDNA
Product	CD99

X chromosome

Y chromosome

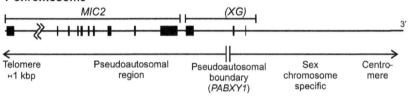

Database accession numbers

	XG	CD99
GenBank	NM_175569, AF380356 (mRNA)	M16279
Entrez Gene ID	7499	4267

Molecular basis of Xg phenotype

Molecular basis of Xg(a–) and CD99– phenotypes have not been determined.

Xgᵃ amino acid sequence

```
MESWWGLPCL  AFLCFLMHAR  GQRDFDLADA  LDDPEPTKKP  NSDIYPKPKP   50
PYYPQPENPD  SGGNIYPRPK  PRPQPQPGNS  GNSGGYFNDV  DRDDGRYPPR  100
PRPRPPAGGG  GGGYSSYGNS  DNTHGGDHHS  TYGNPEGNMV  AKIVSPIVSV  150
VVVTLLGAAA  SYFKLNNRRN  CFRTHEPENV                          180
```

XG encodes a putative leader sequence of 21 amino acids.

CD99 amino acid sequence

```
MARGAALALL  LFGLLGVLVA  APDGGFDLSD  ALPDNENKKP  TAIPKKPSAG   50
DDFDLGDAVV  DGENDDPRPP  NPPKPMPNPN  PNHPSSSGSF  SDADLADGVS  100
GGEGKGGSDG  GGSHRKEGEE  ADAPGVIPGI  VGAVVVAVAG  AISSFIAYQK  150
KKLCFKENAE  QGEVDMESHR  NANAEPAVQR  TLLEK                   185
```

MIC2 encodes a signal peptide of 22 amino acids.

Carrier molecule

A single pass type 1 membrane glycoprotein.

	Xg^a glycoprotein	CD99
M_r (SDS-PAGE)	22,000–29,000	32,500
CHO: N-glycan	No sites	No sites
CHO: O-glycan	11 potential sites[2]	11 potential sites[3]
Cysteine residues	3	1
Copies per RBC	9,000 (polyclonal anti-Xg^a) 18–450 (monoclonal anti-Xg^a)	200 to 2,000[3]

Function

CD99 is a cell surface glycoprotein involved in leukocyte migration, T-cell adhesion, and transmembrane protein transport, and also in T-cell death by a caspase-independent pathway. It may have the ability to rearrange the actin cytoskeleton, and may also act as an oncosuppressor in osteosarcoma. Cyclophilin A binds to CD99, and may act as a signaling regulator of CD99[2]. In RBCs the function of CD99 is not known.

Disease association

XG is linked to genes responsible for ichthyosis (*STS*), ocular albinism (*OAI*), and retinoschisis (*RS*).
High levels of CD99 are found in Ewing's sarcoma, some neuroectodermal tumors, lymphoblastic lymphoma, and acute lymphoblastic leukemia[2].

Phenotypes (% occurrence)

Phenotype	Male	Female
Xg(a+)	65.6	88.7
Xg(a−)	34.4	11.3

Phenotypic relationship of Xgᵃ and CD99 antigens

	Xgᵃ type	CD99 level
Male	Xg(a+)	High
	Xg(a–)	High or low
Female	Xg(a+)	High
	Xg(a+ᵂ)	High
	Xg(a–)	Low

Comments

First blood group system to be assigned to the X chromosome. Family studies with anti-Xgᵃ helped to define the mechanism responsible for various sex-chromosome aneuploides.

Xgᵃ and CD99 escape X chromosome inactivation.

XG transcripts were detected in thymus, bone marrow, and fetal liver, and in several non-erythroid tissues: heart, placenta, skeletal muscle, prostate, thyroid, spinal cord, trachea[4].

References

[1] Latron, F., et al., 1987. Immunochemical characterization of the human blood cell membrane glycoprotein recognized by the monoclonal antibody 12E7. Biochem J 247, 757–764.

[2] Tippett, P., Ellis, N.A., 1998. The Xg blood group system: a review. Transfus Med Rev 12, 233–257.

[3] Fouchet, C., et al., 2000a. Quantitative analysis of XG blood group and CD99 antigens on human red cells. Immunogenetics 51, 688–694.

[4] Fouchet, C., et al., 2000b. A study of the coregulation and tissue specificity of *XG* and *MIC2* gene expression in eukaryotic cells. Blood 95, 1819–1826.

Xgᵃ Antigen

Terminology

ISBT symbol (number)	XG1 (012001 or 12.1)
History	Discovered in 1962 when serum from multiply-transfused Mr. And detected an antigen with a higher prevalence in females than in males; encoded by a locus on the X chromosome. Named after the X chromosome and "g" from "Grand Rapids," where the patient was treated.

Occurrence

Females	89%
Males	66%

Expression

Cord RBCs	Weak
Altered	Weak expression on RBCs from adult females heterozygous for Xg^a. Weak expression on RBCs from adult males is rare.

Effect of enzymes and chemicals on Xgᵃ antigen on intact RBCs

Ficin/Papain	Sensitive
Trypsin	Sensitive
α-Chymotrypsin	Sensitive
DTT 200 mM	Resistant

In vitro characteristics of alloanti-Xgᵃ

Immunoglobulin class	IgG more common than IgM
Optimal technique	RT; IAT; capillary
Complement binding	Some

Clinical significance of alloanti-Xgᵃ

Transfusion reaction	No
HDFN	No

Autoanti-Xgᵃ

One example has been reported.

Comments

An uncommon antibody; occurs mostly in monospecific rather than multispe-cific sera. Some anti-Xgᵃ are naturally-occurring. Xgᵃ is a poor immunogen.

For the phenotypic relationship between Xg^a and CD99, see System pages. Xg^a escapes X chromosome inactivation.

CD99 Antigen

Terminology

ISBT symbol (number)	XG2 (012002 or 12.2)
Obsolete names	12E7; MIC2; E2; HuLy-m6; FMC29; HEC
History	Became part of the Xg blood group system in 2000 because *MIC2* and *XG* are adjacent, homologous genes, and two CD99-negative people were found with the alloantibody.

Occurrence

The only two CD99-negative probands that have been described are Japanese[1].

Expression

Cord RBCs	Weak

Effect of enzymes and chemicals on CD99 antigen on intact RBCs

Ficin/Papain	Sensitive
Trypsin	Sensitive
α-Chymotrypsin	Sensitive
DTT 200 mM	Resistant

In vitro characteristics of alloanti-CD99

Immunoglobulin class	IgG
Optimal technique	IAT

Clinical significance of alloanti-CD99

There are no data because the antibody is rare.

Comments

CD99 escapes X chromosome inactivation.
CD99 has a phenotypic relationship to Xg^a; see System pages.

Reference

[1] Uchikawa, M., et al., 1995. An alloantibody to 12E7 antigen detected in 2 healthy donors [abstract]. Transfusion 35 (Suppl.), 23S.

Scianna Blood Group System

Number of antigens   7

Low prevalence   Sc2, Rd
High prevalence   Sc1, Sc3, STAR, SCER, SCAN

Terminology

ISBT symbol (number)   SC (013)
History   Established in 1974; named after the family of the
                        first maker of anti-Sc1.

Expression

Other blood cells   Weakly expressed on leukocytes[1]
Tissues   Fetal liver, thymus, lymph nodes, spleen, and bone
                        marrow in adults[1]

Gene

Chromosome   1p34.2
Name   *SC (ERMAP)*
Organization   12 exons spanning 27.89 kbp of gDNA
Product   Sc glycoprotein [Erythroid membrane associated
                        protein (ERMAP)]

Database accession numbers

GenBank   NG_008749 (gene); NM_001017922 (mRNA)
Entrez Gene ID   114625

The Blood Group Antigen (3/e). DOI: http://dx.doi.org/10.1016/B978-0-12-415849-8.00015-6
427

Molecular bases of Sc phenotypes[2]

The reference allele SC*01 (Accession number NM_001017922) encodes Sc1 (SC1), SC3, SC5, SC6, SC7. Nucleotide differences, and amino acids affected, are given.

Allele encodes	Allele name	Exon^	Nucleotide	Restriction enzyme	Amino acid	Ethnicity (prevalence)
Sc1– Sc2+ or SC:–1,2	SC*02	4^	169G>A	SmaI –	Gly57Arg	Mennonites, Others (Several)
Rd+ or SC:4	SC*01.04	4^	178C>G		Pro60Ala	Danes, Jews, Canadians, Blacks (Several) Others (Rare)
STAR– or SC:–5	SC*01.–05	4^	139G>A		Glu47Lys	English, Irish (Rare)
SCER– or SC:–6	SC*01.–06	4^	242G>A		Arg81Gln	German (Rare)
SCAN– or SC:–7	SC*01.–07	4^	103G>A		Gly35Ser	German English- Native American heritage (Rare)

Note: A change of SC*54C>T in exon 3 (previously exon 2; silent; Leu18) and SC*76C>T in exon 3 (previously exon 2) are polymorphic, and more common in Caucasians than Blacks[3]. SC*76C>T encodes His26Tyr in the leader sequence of Sc glycoprotein, and thus is not in the RBC membrane.
^In 2011, the gene encoding ERMAP was shown to have 12 exons and not 11 as previously published. As the additional exon is upstream of the initiation codon, the exon that harbors nucleotide changes that affect the expression of a blood group is +1 from that given in the original publications, while the nucleotide and amino acid numbers remain the same.

Molecular bases of silencing of SC[2]

Homozygosity or compound heterozygosity leads to the Sc_{null} (SC:–1–2–3) phenotype. The nucleotide difference from SC*01 reference allele (Accession number NM_001017922), and the amino acid affected, are given.

Allele name	Exon	Nucleotide	Restriction enzyme	Amino acid	Ethnicity (prevalence)
SC*01N.01	4^	307_308delGA		113fs Stop	Saudi Arabian (Rare)
SC*01N.02	12^	994C>T@	Tsp45I+	Arg332Stop	Marshallese (Rare)

^See comment under previous table.
@This allele also may have SC*1514C>T in the 3'-UTR portion of exon 12 (previously published as exon 11).

Amino acid sequence[1,4]

```
MEMASSAGSW  LSGCLIPLVF  LRLSVHVSGH  AGDAGKFHVA  LLGGTAELLC   50
PLSLWPGTVP  KEVRWLRSPF  PQRSQAVHIF  RDGKDQDEDL  MPEYKGRTVL  100
VRDAQEGSVT  LQILDVRLED  QGSYRCLIQV  GNLSKEDTVI  LQVAAPSVGS  150
LSPSAVALAV  ILPVLVLLIM  VCLCLIWKQR  RAKEKLLYEH  VTEVDNLLSD  200
HAKEKGKLHK  AVKKLRSELK  LKRAAANSGW  RRARLHFVAV  TLDPDTAHPK  250
LILSEDQRCV  RLGDRRQPVP  DNPQRFDFVV  SILGSEYFTT  GCHYWEVYVG  300
DKTKWILGVC  SESVSRKGKV  TASPANGHWL  LRQSRGNEYE  ALTSPQTSFR  350
LKEPPRCVGI  FLDYEAGVIS  FYNVTNKSHI  FTFTHNFSGP  LRPFFEPCLH  400
DGGKNTAPLV  ICSELHKSEE  SIVPRPEGKG  HANGDVSLKV  NSSLLPPKAP  450
ELKDIILSLP  PDLGPALQEL  KAPSF                               475
```

SC encodes a signal peptide of 29 amino acids.

Carrier molecule[1,2,5]

Single pass type 1 membrane glycoprotein, a member of the immunoglobulin superfamily (IgSF) with one IgV domain.

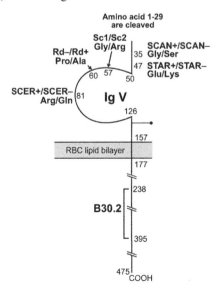

M_r (SDS-PAGE)   60,000–68,000
CHO: N-glycan   4 sites
Cysteine residues   11
Copies per RBC   Not determined

Function

Human ERMAP is an erythroid transmembrane adhesion/receptor protein.

Disease association

Not known

Phenotypes (% occurrence)

Phenotype	Caucasians	Blacks
SC:1,−2	99	100
SC:1,2	1	0
SC:−1,2	Very rare	0
SC:1,−2, Rd+	Very rare	Very rare
SC:1,2, Rd+	Very rare	0
Null: SC:−1,−2,−3		

Comments

The extracellular IgV domain of ERMAP is homologous with the butyrophilin family of milk proteins, autoantigens, and avian blood group antigens[1].
The intracellular B30.2 domain is highly homologous with a similar domain in a diverse group of proteins, including butyrophilin, pyrin, and MID1[1].

References

[1] Su, Y.Y., et al., 2001. Human ERMAP: an erythroid adhesion/receptor transmembrane protein. Blood Cells Mol Dis 27, 938–949.

[2] Brunker, P.A.R., Flegel, W.A., 2011. Scianna: the lucky 13th blood group system. Immunohematology 27, 41–57.

[3] Fuchisawa, A., et al., 2009. The polymorphism nt 76 in exon 2 of SC is more frequent in Whites than in Blacks. Immunohematology 25, 18–19.

[4] Xu, H., et al., 2001. Cloning and characterization of human erythroid membrane-associated protein, human ERMAP. Genomics 76, 2–4.

[5] Wagner, F.F., et al., 2003. The Scianna antigens including Rd are expressed by ERMAP. Blood 101, 752–757.

Scianna

Sc1 Antigen

Terminology

ISBT symbol (number)	SC1 (013001 or 13.1)
Obsolete name	Sm
History	Identified in 1962; name changed from Sm to Sc1 in 1974 when the Scianna system was established. Named for the first maker of anti-Sc1.

Occurrence

All populations	>99%

Antithetical antigen

Sc2 (**SC2**)

Expression

Cord RBCs	Expressed

Molecular basis associated with Sc1 antigen[1]

Amino acid	Gly57
Nucleotide	G at bp 169 in exon 4 (previously published as exon 3)

Effect of enzymes and chemicals on Sc1 antigen on intact RBCs[2]

Ficin/Papain	Resistant
Trypsin	Resistant
α-Chymotrypsin	Resistant
DTT 200 mM/50 mM	Resistant/resistant (thus resistant to WARM™ and ZZAP)

In vitro characteristics of alloanti-Sc1

Immunoglobulin class	IgG
Optimal technique	IAT

Clinical significance of alloanti-Sc1

Transfusion reaction	Not reported
HDFN	Positive DAT but no clinical HDFN

Autoanti-Sc1

Yes[3]. Some examples are reactive in tests using serum but not plasma[4].

Scianna

Comments

Siblings of patients with anti-Sc1 should be tested for compatibility, and the patient urged to donate blood for cryogenic storage when his/her clinical state permits.

References

[1] Wagner, F.F., et al., 2003. The Scianna antigens including Rd are expressed by ERMAP. Blood 101, 752–757.

[2] Velliquette, R.W., et al., 2011. The effect of proteases or DTT on Scianna antigens, revisited [abstract]. Transfusion 51 (Suppl.), 146A.

[3] Owen, I., et al., 1992. Autoimmune hemolytic anemia associated with anti-Sc1. Transfusion 32, 173–176.

[4] Tregellas, W.M., et al., 1979. An example of autoanti-Sci demonstrable in serum but not in plasma [abstract]. Transfusion 19, 650.

Sc2 Antigen

Terminology

ISBT symbol (number)	SC2 (013002 or 13.2)
Obsolete names	Bu^a; Bullee
History	Identified in 1962 and named Bu^a; renamed Sc2 in 1974 when it was shown to be antithetical to Sm (now Sc1).

Occurrence

1% in people of European ancestry; more common in Mennonites.

Antithetical antigen

Sc1 (**SC1**)

Expression

Cord RBCs	Expressed

Molecular basis associated with Sc2 antigen[1]

Amino acid	Arg57
Nucleotide	A at bp 169 in exon 4 (previously published as exon 3)

Effect of enzymes and chemicals on Sc2 antigen on intact RBCs[2]

Ficin/Papain	Resistant
Trypsin	Variable
α-Chymotrypsin	Variable
DTT 200 mM/50 mM	Variable/resistant (thus variable with WARM™ and ZZAP)

In vitro characteristics of alloanti-Sc2

Immunoglobulin class	IgG
Optimal technique	IAT

Clinical significance of alloanti-Sc2

Transfusion reaction	No
HDFN	Positive DAT but no clinical HDFN to mild[3]

Comments

Sc2 antigen has variable expression among different people.

References

[1] Wagner, F.F., et al., 2003. The Scianna antigens including Rd are expressed by ERMAP. Blood 101, 752–757.

[2] Velliquette, R.W., et al., 2011. The effect of proteases or DTT on Scianna antigens, revisited [abstract]. Transfusion 51 (Suppl), 146A.

[3] DeMarco, M., et al., 1995. Hemolytic disease of the newborn due to the Scianna antibody, anti-Sc2. Transfusion 35, 58–60.

Sc3 Antigen

Terminology

ISBT symbol (number)	SC3 (013003 or 13.3)
History	Named in 1980 when a person with SC:–1,–2 RBCs made an antibody to a high-prevalence antigen.

Occurrence

Most SC:–1,–2,–3 people have originated from the Marshall Islands or other Pacific Islands, including Papua New Guinea.

Expression

Cord RBCs	Expressed

Molecular basis associated with Sc3 antigen

For molecular bases of the SC:–1,–2,–3 (the null phenotype) see System pages.

Scianna

Effect of enzymes and chemicals on Sc3 antigen on intact RBCs[1]

Ficin/Papain	Resistant (enhanced)
Trypsin	Variable
α-Chymotrypsin	Variable
DTT 200 mM/50 mM	Resistant/resistant (thus resistant to WARM™ and ZZAP)

In vitro characteristics of alloanti-Sc3

Immunoglobulin class	IgG
Optimal technique	IAT

Clinical significance of alloanti-Sc3

Transfusion reaction	No to mild/delayed
HDFN	Mild

Autoanti-Sc3

Autoanti-Sc3-like antibody in two patients with suppressed Sc antigens (one patient with lymphoma; one with Hodgkins disease)[2].

References

[1] Velliquette, R.W., et al., 2011. The effect of proteases or DTT on Scianna antigens, revisited [abstract]. Transfusion 51 (Suppl.), 146A.

[2] Peloquin, P., et al., 1989. Anti-Sc3 as an apparent autoantibody in two patients [abstract]. Transfusion 29 (Suppl.), 49S.

Sc4 Antigen

Terminology

ISBT symbol (number)	Rd (013004 or 13.4)
Obsolete names	Radin; Rd^a; 700015
History	Named after the first family in which the antibody caused HDFN. Became part of the SC system when the associated polymorphism in human ERMAP was identified.

Occurrence

All populations	Less than 0.01%
Danes	0.5%
Jews, Canadians	0.1%
African Blacks	0.1%

Expression

Cord RBCs   Expressed

Molecular basis associated with Rd antigen[1]

Amino acid	Ala60
Nucleotide	G at bp 178 in exon 4 (previously published as exon 3)
Rd-negative	Pro60 and C at bp 178

Effect of enzymes and chemicals on Rd antigen on intact RBCs[2]

Ficin/Papain	Resistant
Trypsin	Variable, but often sensitive
α-Chymotrypsin	Variable, but often sensitive
DTT 200 mM/50 mM	Resistant/resistant (thus resistant to WARM™ and ZZAP)

In vitro characteristics of alloanti-Rd

Immunoglobulin class	IgG
Optimal technique	IAT

Clinical significance of alloanti-Rd

Transfusion reaction	No
HDFN	Mild to severe

References

[1] Wagner, F.F., et al., 2003. The Scianna antigens including Rd are expressed by ERMAP. Blood 101, 752–757.

[2] Velliquette, R.W., et al., 2011. The effect of proteases or DTT on Scianna antigens, revisited [abstract]. Transfusion 51 (Suppl.), 146A.

STAR Antigen

Terminology

ISBT symbol (number)	SC5 (013005 or 13.5)
History	Named in 2005 after the STAR– proband.

Occurrence

The only STAR– proband was of English-Irish heritage.

Molecular basis associated with STAR antigen[1]

Amino acid	Glu47
Nucleotide	G at bp 139 in exon 4 (previously published as exon 3)
SCAN–	Lys47 and A at bp 139

Effect of enzymes and chemicals on STAR antigen on intact RBCs

Ficin/papain	Resistant (Enhanced)

In vitro characteristics of alloanti-STAR

Immunoglobulin class	IgG
Optimal technique	IAT

Clinical significance of alloanti-STAR

No information because antibody is rare.

Comments

Siblings of patients with anti-STAR should be tested for compatibility, and the patient urged to donate blood for cryogenic storage when his/her clinical state permits.

Reference

[1] Hue-Roye, K., et al., 2005. STAR: a novel high prevalence antigen in the Scianna blood group system. Transfusion 45, 245–247.

SCER Antigen

Terminology

ISBT symbol (number)	SC6 (013006 or 13.6)
History	Named in 2005 after "SC" for Scianna and "ER" from the second and third letters of the SCER– proband's name.

Occurrence

The only SCER– proband was of German heritage.

Molecular basis associated with SCER antigen[1]

Amino acid	Arg81
Nucleotide	G at 242 in exon 4 (previously published as exon 3)
SCER–	Gln81 and A at bp 242

Effect of enzymes and chemicals on SCER antigen on intact RBCs

Ficin/papain	Resistant
Trypsin	Resistant
DTT 200/50 mM	Resistant/resistant (thus resistant to WARM™ and ZZAP)

In vitro characteristics of alloanti-SCER

Immunoglobulin class	IgG
Optimal technique	IAT

Clinical significance of alloanti-SCER

Transfusion reaction	No information because the antibody is rare, but an *in vivo* ^{51}Cr-labelled RBC survival study indicated reduced survival of antigen-positive RBCs
HDFN	No information because antibody is rare

Comments

Siblings of patients with anti-SCER should be tested for compatibility, and the patient urged to donate blood for cryogenic storage when his/her clinical state permits.

Reference

[1] Flegel, W.A., et al., 2005. SCER and SCAN: two novel high-prevalence antigens in the Scianna blood group system. Transfusion 45, 1940–1944.

SCAN Antigen

Terminology

ISBT symbol (number)	SC7 (013007 or 13.7)
History	Named in 2005 after "SC" for Scianna and "AN" from the second and third letters of the SCAN– proband's name.

Occurrence

The only SCAN– proband was of German, English, and Native American heritage.

Molecular basis associated with SCAN antigen[1]

Amino acid	Gly35
Nucleotide	G at bp 103 in exon 4 (previously published as exon 3)
SCAN–	Ser35 and A at bp 103

The SCAN– proband also had the *SC*54C>T* silent nucleotide change and *SC*76C>T* (His26Tyr).

Effect of enzymes and chemicals on SCAN antigen on intact RBCs

Ficin/papain   Resistant
DTT 200/50 Mm   Resistant/resistant (thus resistant to WARM™ and
                        ZZAP)

In vitro characteristics of alloanti-SCAN

Immunoglobulin class   IgG
Optimal technique   IAT

Clinical significance of alloanti-SCAN

Transfusion reaction   Delayed (only one case reported)
HDFN   No information because antibody is rare

Comments

Siblings of patients with anti-SCAN should be tested for compatibility, and the patient urged to donate blood for cryogenic storage when his/her clinical state permits.

Reference

[1] Flegel, W.A., et al., 2005. SCER and SCAN: two novel high-prevalence antigens in the Scianna blood group system. Transfusion 45, 1940–1944.

Dombrock Blood Group System

Number of antigens 8

Polymorphic	Doª, Doᵇ
High prevalence	Gyª, Hy, Joª, DOYA, DOMR, DOLG

Terminology

ISBT symbol (number)	DO (014)
CD number	CD297
History	Named after the producer of the first anti-Doª; identified in 1965.

Expression

Other blood cells	Lymphocytes
Tissues	Primarily in adult bone marrow and fetal liver; also in spleen, lymph nodes, intestine, ovary, testes, and fetal heart

Gene[1–3]

Chromosome	12p12.3
Name	*DO (ART4)*
Organization	3 exons distributed over 14 kbp of gDNA
Product	Do glycoprotein

Database accession numbers

GenBank	NM_021071; AF290204 (mRNA)
Entrez Gene ID	420

Dombrock

The Blood Group Antigen (3/e). DOI: http://dx.doi.org/10.1016/B978-0-12-415849-8.00016-8

Molecular bases of Dombrock phenotypes

Reference allele *DO*01* or *DO*A* (Accession number AF290204) encodes Do^a (DO1), DO3, DO4, DO5, DO6, DO7, DO8. Nucleotide differences from reference allele, and amino acids affected, are given.

Allele encodes	Allele name	Exon	Nucleotide	Restriction enzyme	Amino acid	Ethnicity (prevalence)
Do(b+) or DO:2	*DO*02* or *DO*B*	2	793A>G	*BseRI*+	Asn265Asp	Caucasians, Blacks, Asians, Thais (Common)
Hy– or DO:–4	*DO*02.–04.01* or *DO*HY1*	2 2 3	323G>T 793A>G 898C>G	*BsaJI* – *BseRI*+ *BsmAI*+	Gly108Val Asn265Asp Leu300Val	Blacks (Several)
Hy– or DO:–4	*DO*02.–04.02* or *DO*HY2*	2 2	323G>T 793A>G	*BsaJI* – *BseRI*+	Gly108Val Asn265Asp	Blacks (Several)
Jo(a–) or DO:–5	*DO*01.–05* or *DO*JO1*	2 2	350C>T	*XcmI* –	Thr117Ile	Blacks (Several)
Jo(a–) or DO:– 5	*DO*02.–05* or *DO*JO2*	2 2	350C>T 793A>G	*XcmI* – *BseRI*+	Thr117Ile Asn265Asp	Malis (Few)
DOYA– or DO:–6	*DO*01.–06*	2	547T>G	*BtgZI*+	Tyr183Asp[4]	Turkish Kurds (Rare)
DOMR– or DO:–7	*DO*02.–07*^	2 2 2 3	431C>A 432C>A 793A>G 898C>G	*AgsI*+ *BstNI* – *BseRI*+ *BsmAI*+	Ala144Glu Asn265Asp Leu300Val[5]	Brazilian Blacks (Rare)
DOLG– or DO:–8	*DO*01.–08*	2	674T>A		Leu225Gln[6]	Sri Lankan (Rare)

^The background for this allele is actually *DO*B-WL* (378T, 624C, 793G), but as 378 and 624 are silent changes, they are not listed.

Molecular bases of silencing *DO*

Homozygosity or compound heterozygosity leads to Do$_{null}$ [Gy(a–)] phenotype.

Nucleotide differences from *DO*01* reference allele (Accession number AF290204), and amino acids affected, are given.

Allele name	Exon	Nucleotide	Amino acid	Ethnicity (prevalence)
*DO*01N.01*^	2	442C>T^	Gln148Stop	(Rare)
*DO*01N.02*	2	343–350del	114 fs; premature Stop	(Rare)
*DO*02N.01*	2	IVS1 –2 a>g; 793A>G	Exon 2 skip Asn265Asp	(Rare)
*DO*02N.02*	2	IVS1+2 t>c; 793A>G	Exon 2 skip Asn265Asp	(Rare)
*DO*02N.03*	2	185T>C 793A>G	Phe62Ser[7] Asn265Asp	(Rare)

^ = The background for this allele is *DO*A-HA* (378T, 624T, 793A), but as nucleotides 378 and 624 are silent changes, they are not listed.

DO*A alleles, including those that do not express novel Do antigens

Allele name	nt (aa)	nt (aa)	nt^	nt^	nt (aa)	nt (aa)	nt (aa)
	323 (108)	350 (117)	378	624	793 (265)	898 (300)	Other
*DO*A*	G (Gly)	C (Thr)	C	T	A (Asn)	C (Leu)	
*DO*JO1*	G (Gly)	T (Ile)	T	T	A (Asn)	C (Leu)	
*DO*DOYA*	G (Gly)	C (Thr)	C	T	A (Asn)	C (Leu)	547T>G Tyr183Asp
*DO*DOLG*	G (Gly)	C (Thr)	C	T	A (Asn)	C (Leu)	674T>A Leu225Gln
*DO*A-HA*	G (Gly)	C (Thr)	T	T	A (Asn)	C (Leu)	
*DO*A-SH*	G (Gly)	C (Thr)	C	C	A (Asn)	C (Leu)	
*DO*A-WL*	G (Gly)	C (Thr)	C	T	A (Asn)	G (Val)	

nt = nucleotide; aa = amino acid.
^ = As nts 378 and 624 are silent changes, the amino acids are not listed.

Dombrock

DO*B alleles, including those that do not express novel Do antigens

Allele name	nt (aa)	nt (aa)	nt^	nt^	nt (aa)	nt (aa)	nt (aa)
	323 (108)	350 (117)	378	624	793 (265)	898 (300)	Other
DO*B	G (Gly)	C (Thr)	T	C	G (Asp)	C (Leu)	
DO*HY1	T (Val)	C (Thr)	C	C	G (Asp)	G (Val)	
DO*HY2	T (Val)	C (Thr)	C	C	G (Asp)	C (Leu)	
DO*JO2	G (Gly)	T (Ile)	T	C	G (Asp)		
DO*DOMR	G (Gly)	C (Thr)	T	C	G (Asp)	G (Val)	431C>A & 432C>A Ala144Glu
DO*B-SH	G (Gly)	C (Thr)	C	C	G (Asp)	C (Leu)	
DO*B-SH-Q149K	G (Gly)	C (Thr)	C	C	G (Asp)	C (Leu)	445C>A Gln149Lys
DO*B-WL	G (Gly)	C (Thr)	T	C	G (Asp)	G (Val)	
DO*B-I175N	G (Gly)	C (Thr)	T	C	G (Asp)	C (Leu)	524T>A; Ile175Asn

nt = nucleotide; aa = amino acid.
^ = As nts 378 and 624 are silent changes, the amino acids are not listed.

Amino acid sequence[1]

Signal peptide: Amino acids 1 to 44 or, more likely, 22 to 44 if initiation occurs at the second AUG[8].

```
MGPLINRCKK  ILLPTTVPPA  TMRIWLLGGL  LPFLLLLSGL  QRPTEGSEVA   50
IKIDFDFAPG  SFDDQYQGCS  KQVVEKLTQG  DYFTKDIEAQ  KNYFRMWQKA  100
HLAWLNQGKV  LPQNMTTTHA  VAILFYTLNS  NVHSDFTRAM  ASVARTPQQY  150
ERSFHFKYLH  YYLTSAIQLL  RKDSIMENGT  LCYEVHYRTK  DVHFNAYTGA  200
TIRFGQFLST  SLLKEEAQEF  GNQTLFTIFT  CLGAPVQYFS  LKKEVLIPPY  250
ELFKVINMSY  HPRGNWLQLR  STGNLSTYNC  QLLKASSKKC  IPDPIAIASL  300
SFLTSVIIFS  KSRV                                           314
```

GPI-anchor motif: Amino acids 298 to 314 or, more likely, 285 to 314[8].

Carrier molecule

GPI-linked glycoprotein.

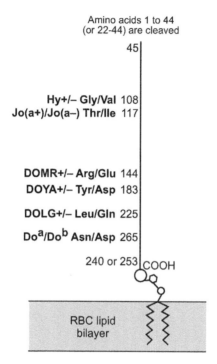

Amino acids 1 to 44
(or 22-44) are cleaved

45

Hy+/− Gly/Val 108
Jo(a+)/Jo(a−) Thr/Ile 117

DOMR+/− Arg/Glu 144
DOYA+/− Tyr/Asp 183
DOLG+/− Leu/Gln 225
Do^a/Do^b Asn/Asp 265

240 or 253 COOH

RBC lipid
bilayer

M_r (SDS-PAGE)	47,000–58,000
CHO: N-glycan	5 potential sites
Cysteine residues	4 or 5 in membrane-bound protein

Function

Its function in RBCs is not known. Dombrock is ADP-ribosyltransferase (ART) 4[1–3,9].

Disease association

Dombrock glycoprotein is absent from PNH III RBCs.

Dombrock

Phenotypes (% occurrence)

RBC phenotype	Doᵃ	Doᵇ	Gyᵃ	Hy	Joᵃ	DOYA	DOMR	DOLG	Whites	Blacks	Japanese	Thai
Do(a+b−)	+	0	+	+	+	+	+	+	18	11	1.5	0.5
Do(a+b+)	+	+	+	+	+	+	+	+	49	44	22	13
Do(a−b+)	0	+	+	+	+	+	+	+	33	45	76.5	86.5
Hy−	0	+ʷ	+ʷ	0	0/+ʷ	+ʷ	0/+ʷ	+ʷ	None	Rare	None	None
Do(a+b+) Jo(a−)^^	+ʷ	+ʷ	+	+ʷ	0	NT	NT	NT	None	Rare	None	None
Do(a+b−) Jo(a−)	+ʷ	0	+	+ʷ	0	+ʷ	+ʷ	+	None	Rare	None	None
Do(a−b+ʷ) Jo(a−)	0	+ʷ	+	+ʷ	0	NT	NT	NT	None	Rare	None	None
Gy(a−)	0	0	0	0	0	0	0	0	Rare	Rare	None	None
DOYA−	0	0	+ʷ	+ʷ	+ʷ	0	0/+ʷ	+ʷ	Rare	Rare	None	None
DOMR−	0	+ʷ	+ʷ	+ʷ	+ʷ	+ʷ	0	NT	None	None	None	None
DOLG−	+	0	+	+	+	NT	NT	0	None	None	None	None[#]
Null:	Gregory negative [Gy(a−)]											

NT = Not tested.
^ = Expression of Gyᵃ marginally reduced.
^^Associated with the compound heterozygote DO*HY/DO*JO.
[#]The only reported proband was from Sri Lanka.

Dombrock

References

[1] Gubin, A.N., et al., 2000. Identification of the Dombrock blood group glycoprotein as a polymorphic member of the ADP-ribosyltransferase gene family. Blood 96, 2621–2627.

[2] Koch-Nolte, F., 1999. Erratum (to Koch-Nolte et al., Genomics, 39;370-376, 1997). Genomics 55, 130.

[3] Koch-Nolte, F., et al., 1997. Two novel human members of an emerging mammalian gene family related to mono-ADP-ribosylating bacterial toxins. Genomics 39, 370–376.

[4] Mayer, B., et al., 2010. New antigen in the Dombrock blood group system, DOYA, ablates expression of Do^a and weakens expression of Hy, Jo^a and Gy^a antigens. Transfusion 50, 1295–1302.

[5] Costa, F., et al., 2010. Absence of DOMR, a new antigen in the Dombrock blood group system that weakens expression of Do^b, Gy^a, Hy, Jo^a, and DOYA antigens. Transfusion 50, 2026–2031.

[6] Karamatic Crew, V., et al., 2011. DOLG, a novel high incidence antigen in the Dombrock blood group system [abstract]. Vox Sang 101 (Suppl 1), 263.

[7] Westhoff, C., et al., 2007. A *DOB* allele encoding an amino acid substitution (Phe62Ser) resulting in a Dombrock null phenotype. Transfusion 47, 1356–1362.

[8] Reid, M.E., 2003. The Dombrock blood group system: a review. Transfusion 43, 107–114.

[9] Grahnert, A., et al., 2002. Mono-ADP-ribosyltransferases in human monocytes: regulation by lipopolysaccharide. Biochem J 362, 717–723.

Do^a Antigen

Terminology

ISBT symbol (number)	DO1 (014001 or 14.1)
Obsolete name	Dombrock
History	Named after the proband who made anti-Do^a; reported in 1965.

Occurrence

Caucasians	67%
Blacks	55%
Japanese	24%
Thais	14%

Antithetical antigen

Do^b (**DO2**)

Expression

Cord RBCs	Expressed
Altered	Absent from PNH III RBCs; absent from Hy– and DOYA– RBCs, and weak on Jo(a–) RBCs

Dombrock

Molecular basis associated with Do^a antigen[1]

Amino acid	Asn265
Nucleotide	A at bp 793 and C at bp 378 (silent 126Tyr); T at bp 624 (silent 208Leu); all in exon 2

Effect of enzymes and chemicals on Do^a antigen on intact RBCs

Ficin/Papain	Resistant (enhanced)
Trypsin	Sensitive
α-Chymotrypsin	Weakened
Pronase	Sensitive (weakened)
DTT 200 mM/50 mM	Sensitive/resistant (thus sensitive to WARM™ and ZZAP)
Acid	Variable

In vitro characteristics of alloanti-Do^a

Immunoglobulin class	IgG
Optimal technique	IAT; PEG; enzyme IAT

Clinical significance of alloanti-Do^a

Transfusion reaction	Delayed and acute/hemolytic
HDFN	Positive DAT but no clinical HDFN

Comments

Anti-Do^a is notorious for disappearing *in vivo*.
Do^a is a poor immunogen and anti-Do^a is rarely found as a single specificity. Due to the scarcity of monospecific anti-Do^a, DNA analysis may be used to predict the antigen status.

Reference

[1] Gubin, A.N., et al., 2000. Identification of the Dombrock blood group glycoprotein as a polymorphic member of the ADP-ribosyltransferase gene family. Blood 96, 2621–2627.

Do^b Antigen

Terminology

ISBT symbol (number)	DO2 (014002 or 14.2)
History	Named when it was recognized to be antithetical to Do^a; reported in 1973.

Occurrence

Caucasians	82%
Blacks	89%

Antithetical antigen

Do^a (**DO1**)

Expression

Cord RBCs	Expressed
Altered	Absent from PNH III RBCs. Weak on Hy– and DOMR–, and absent or weak on Jo(a–) RBCs.

Molecular basis associated with Do^b antigen[1]

Amino acid	Asp265
Nucleotide	G at bp 793 and T at bp 378 (silent 126Tyr); C at bp 624 (silent 208Leu); all in exon 2

Effect of enzymes and chemicals on Do^b antigen on intact RBCs

Ficin/Papain	Resistant (enhanced)
Trypsin	Sensitive
α-Chymotrypsin	Weakened
Pronase	Sensitive (weakened)
DTT 200 mM/50 mM	Sensitive/resistant (thus sensitive to WARM™ and ZZAP)
Acid	Variable

In vitro characteristics of alloanti-Do^b

Immunoglobulin class	IgG
Optimal technique	IAT; PEG; enzyme IAT

Clinical significance of alloanti-Do^b

Transfusion reaction	Acute and delayed
HDFN	Positive DAT but no clinical HDFN

Comments

Do^b is a poor immunogen and anti-Do^b is rarely found as a single specificity. Due to the scarcity of monospecific anti-Do^b, DNA analysis may be used to predict the antigen status.

Reference

[1] Gubin, A.N., et al., 2000. Identification of the Dombrock blood group glycoprotein as a polymorphic member of the ADP-ribosyltransferase gene family. Blood 96, 2621–2627.

Gyª Antigen

Terminology

ISBT symbol (number)   DO3 (014003 or 14.3)
Obsolete names   Gregory; GY1; 206001; 900005
History   Named in 1967 after the last name of the first
                        producer of the antibody. Placed in the Dombrock
                        system in 1992 when it was recognized that Gy(a–)
                        was the null phenotype of Do[1].

Occurrence

Most populations   100%
Eastern European   Greater than 99%
(Romany)
Japanese   Greater than 99%
Blacks   One proband[2]

Expression

Cord RBCs   Weak
Altered   Absent from PNH III RBCs; weak on Hy–, DOYA–,
                and DOMR– RBCs; marginally reduced on
                DOLG– RBCs

Molecular basis associated with Gyª antigen

Molecular bases underlying the Gy(a–) phenotype (see table on System pages)
result in an absence of Do glycoprotein in the membrane.

Effect of enzymes and chemicals on Gyª antigen on intact RBCs

Ficin/Papain   Resistant (enhanced)
Trypsin   Sensitive
α-Chymotrypsin   Weakened
Pronase   Sensitive (weakened)
DTT 200 mM/50 mM   Sensitive/resistant (thus sensitive to WARM™ and
                        ZZAP)
Acid   Variable

In vitro characteristics of alloanti-Gyª

Immunoglobulin class   IgG
Optimal technique   IAT

Clinical significance of alloanti-Gy^a

Transfusion reaction	No to moderate/delayed
HDFN	Positive DAT, but no clinical HDFN

Autoanti-Gy^a

Yes, may appear to be an alloantibody due to transient suppression of Gy^a antigen.

Comments

Gy(a–) RBCs are also Do(a–b–) Hy– Jo(a–) DOYA– DOMR– DOLG–.

Siblings of patients with anti-Gy^a should be tested for compatibility, and the patient urged to donate blood for cryogenic storage when his/her clinical state permits.

Experts agree that Hy– blood may be used for transfusion when Gy(a–) blood is not available.

References

[1] Banks, J.A., et al., 1995. Evidence that the Gy^a, Hy and Jo^a antigens belong to the Dombrock blood group system. Vox Sang 68, 177–182.

[2] Smart, E.A., et al., 2000. The first case of the Dombrock-null phenotype reported in South Africa [abstract]. Vox Sang 78 (suppl 1), P015.

Hy Antigen

Terminology

ISBT symbol (number)	DO4 (014004 or 14.4)
Obsolete names	Holley; GY2; 206002; 900011
History	Reported in 1967, and named after the proband who made anti-Hy. Joined the Dombrock system in 1995.

Occurrence

Most populations	100%
Blacks	Greater than 99%

Expression

Cord RBCs	Weak
Altered	Absent from PNH III RBCs; weak on Jo(a–), DOYA–, DOMR–, and DOLG– RBCs

Molecular basis associated with Hy antigen[1]

Amino acid	Gly108
Nucleotide	G at bp 323 in exon 2

Dombrock

Hy–	Val108 and Asp265; T at bp 323 and G at bp 793 in exon 2 (HY2 allele). HY1 allele also has 898C>G in exon 3 (Leu300Val)

Effect of enzymes and chemicals on Hy antigen on intact RBCs

Ficin/Papain	Resistant (enhanced)
Trypsin	Sensitive
α-Chymotrypsin	Weakened
Pronase	Sensitive (weakened)
DTT 200 mM/50 mM	Sensitive/resistant (thus sensitive to WARM™ and ZZAP)
Acid	Variable

In vitro characteristics of alloanti-Hy

Immunoglobulin class	IgG
Optimal technique	IAT

Clinical significance of alloanti-Hy

Transfusion reaction	No to moderate/delayed
HDFN	Positive DAT, but no clinical HDFN

Comments

Hy– RBCs are Do(a–b+^{W}) Gy(a+^{W}) Jo(a–) DOYA+^{W} DOMR+^{W}/– DOLG+^{W}. Siblings of patients with anti-Hy should be tested for compatibility, and the patient urged to donate blood for cryogenic storage when his/her clinical state permits.

Reference

[1] Reid, M.E., 2003. The Dombrock blood group system: a review. Transfusion 43, 107–114.

Jo^a Antigen

Terminology

ISBT symbol (number)	DO5 (014005 or 14.5)
Obsolete names	Joseph; 901004; 900010
History	Reported in 1972, and named after the proband who was reported to have made anti-Jo^a. Joined the Dombrock system in 1992. The original and 2nd probands were later shown to be Hy–! The 3rd proband had a JO allele and made anti-Jo^a. This explains some of the confusion in differentiating Hy and Jo^a antigens and antibodies.

Occurrence

| Most populations | 100% |
| Blacks | Greater than 99% |

Expression

| Cord RBCs | Weak |
| Altered | Absent or weak on Hy– and PNH III RBCs; weak on DOYA– and DOMR– RBCs. |

Molecular basis associated with Jo[a] antigen[1]

Amino acid	Thr117
Nucleotide	C at bp 350 in exon 2
Jo(a–)	Ile117 and Asn265; T at bp 350 and A at bp 793 in exon 2. Can also occur on Hy–, and on Do(b+) backgrounds (see tables above)

Effect of enzymes and chemicals on Jo[a] antigen on intact RBCs

Ficin/Papain	Resistant (enhanced)
Trypsin	Sensitive
α-Chymotrypsin	Weakened
Pronase	Sensitive (weakened)
DTT 200 mM/50 mM	Variable (thus variable to WARM™ and ZZAP)
Acid	Variable

In vitro characteristics of alloanti-Jo[a]

| Immunoglobulin class | IgG |
| Optimal technique | IAT |

Clinical significance of alloanti-Jo[a]

| Transfusion reaction | No to moderate/delayed |
| HDFN | No |

Comments

Jo(a–) RBCs are Do(a+^{W}b+^{W}/–) Gy(a+^{W}) Hy+^{W} DOYA+^{W} DOMR+ DOLG+.

In Malis, the Jo(a–) phenotype has been found on a *DO*B* background (Moulds JM, personal communication).

Siblings of patients with anti-Jo[a] should be tested for compatibility, and the patient urged to donate blood for cryogenic storage when his/her clinical state permits.

Dombrock

Reference

[1] Reid, M.E., 2003. The Dombrock blood group system: a review. Transfusion 43, 107–114.

DOYA Antigen

Terminology

ISBT symbol (number)	DO6 (014006 or 14.6)
History	Named in 2010, "DO" for the system and "YA" from the name of the DOYA– proband.

Occurrence

The only DOYA– proband reported was a Turkish Kurd; she had a DOYA– sister.

Expression

Cord RBCs	Weak
Altered	Absent on PNH III RBCs and weak on Hy–, Jo(a–), and DOMR– RBCs.

Molecular basis associated with DOYA antigen[1]

Amino acid	Tyr183
Nucleotide	T at bp 547 in exon 2
DOYA–	Asp183 and G at bp 547 on a *DO*A* background

Effect of enzymes and chemicals on DOYA antigen on intact RBCs

Ficin/Papain	Resistant (enhanced)
Trypsin	Sensitive
α-Chymotrypsin	Weakened
DTT 200 mM/50 mM	Variable (thus variable to WARM™ and ZZAP)

In vitro characteristics of DOYA

Immunoglobulin class	IgG
Optimal technique	IAT

Clinical significance of alloanti-DOYA

Transfusion reaction	The only producer of anti-DOYA was transfused on two occasions with incompatible blood that was tolerated with pretransfusion medication of antihistamine and steroids.
HDFN	No

Dombrock

Comments

DOYA– RBCs type Do(a–b–) Gy(a+^{W}) Hy+^{W} Jo(a+^{W}) DOMR+^{W}.

Reference

[1] Mayer, B., et al., 2010. New antigen in the Dombrock blood group system, DOYA, ablates expression of Do^a and weakens expression of Hy, Jo^a and Gy^a antigens. Transfusion 50, 1295–1302.

DOMR Antigen

Terminology

ISBT symbol (number)	DO7 (014007 or 14.7)
History	Named in 2010, "DO" from the system and "MR" from the name of the DOMR– proband.

Occurrence

The only DOMR– proband reported was African Brazilian.

Expression

Cord RBCs	Weak
Altered	Absent on PNH III RBCs and weak on Hy–, Jo(a–) and DOYA– RBCs

Molecular basis associated with DOMR antigen[1]

Amino acid	Ala144
Nucleotide	C at bp 431 and C at bp 432 in exon 2
DOMR–	Glu144 and A at bp 431 and A at bp 432 on a *DO*B-WL* background

Effect of enzymes and chemicals on DOMR antigen on intact RBCs

Ficin/Papain	Resistant (enhanced)
Trypsin	Sensitive
α-Chymotrypsin	Weakened
DTT 200 mM/50 mM	Variable (thus variable to WARM™ and ZZAP)

In vitro characteristics of alloanti-DOMR

Immunoglobulin class	IgG
Optimal technique	IAT

Clinical significance of alloanti-DOMR

Transfusion reaction	No data; only one example of anti-DOMR reported
HDFN	No

Dombrock

Comments

DOMR– RBCs type Do(a–b+^{W}) Gy(a+^{W}) Hy+^{W} Jo(a+^{W}) DOYA+^{W}.

Reference

[1] Costa, F., et al., 2010. Absence of DOMR, a new antigen in the Dombrock blood group system that weakens expression weakens expression of Do^b, Gy^a, Hy, Jo^a, and DOYA antigens. Transfusion 50, 2026–2031.

DOLG Antigen

Terminology

ISBT symbol (number)	DO8 (014008 or 14.8)
History	Named in 2011, "DO" from the system and "LG" for the Leu/Gln amino acid change found in the DOLG– proband[1].

Occurrence

The only DOLG– proband reported was from Sri Lanka.

Expression

Cord RBCs	Presumed weak
Altered	Presumed absent on PNH III RBCs and weak on Hy– RBCs

Molecular basis associated with DOLG antigen[1]

Amino acid	Leu225
Nucleotide	T at bp 674 in exon 2
DOLG–	Gln225 and A at bp 674 on a *DO*A* background

Effect of enzymes and chemicals on DOLG antigen on intact RBCs

No information reported.

In vitro characteristics of alloanti-DOLG

Immunoglobulin class	IgG
Optimal technique	IAT

Clinical significance of alloanti-DOLG

Transfusion reaction	No data because only one example of anti-DOLG reported
HDFN	No

Comments

DOLG– RBCs type Do(a+b–) Gy(a+^{W}) Hy+ Jo(a+).

Reference

[1] Karamatic Crew, V., et al., 2011. DOLG, a novel high incidence antigen in the Dombrock blood group system [abstract]. Vox Sang 101 (Suppl 1), 263.

Colton Blood Group System

Number of antigens 4

Polymorphic Co^b
High prevalence Co^a, Co3, Co4

Terminology

ISBT symbol (number)   CO (015)
History   Named in 1967 for the first of the three original
                         producers of anti-Co^a. Should have been named
                         Calton, but the handwriting on the tube was misread.

Expression

Tissues   Kidney (apical surface of proximal tubules,
                         basolateral membranes, subpopulation of collecting
                         ducts in cortex, descending tubules in medulla), liver
                         bile ducts, gall bladder, eye (epithelium, cornea,
                         lens, choroid plexus), hepatobilliary epithelia,
                         capillary endothelium[1]

Gene

Chromosome   7p14.3
Name   *CO [AQP1 (Aquaporin-1)]*
Organization   4 exons distributed over 11.6 kbp of gDNA
Product   Channel-forming integral protein (CHIP); Aquaporin
                         1 (AQP1); CHIP-1; CHIP28[1]

The Blood Group Antigen (3/e). DOI: http://dx.doi.org/10.1016/B978-0-12-415849-8.00017-X

Colton

Database accession numbers

GenBank	AY953319 (gene), M77829 (mRNA)
Entrez Gene ID	358

Molecular bases of Colton phenotypes

Reference allele, *CO*01* (M77829) encodes Co[a] (CO1), CO3, CO4. Differences from this allele are given.

Allele encodes	Allele name	Exon	Nucleotide	Restriction Enzyme	Amino acid	Ethnicity (prevalence)
Co(a–b+) or CO:–1,2	*CO*02.01* or *CO*B*	1	134C>T	*PflMI*+	Ala45Val	(Many)
Co(a–b+) or CO:–1,2	*CO*02.02*	1	133G>A		Ala45Thr[2]	(Rare)
^Co(a–b–) CO:3,–4 or CO:–1,–2, 3,–4	*CO*01. –04*	1	140A>G		Gln47Arg	Caucasians, Turkish (Rare)

^ = When *in trans* to *CO*02*, expression of Co[b] on RBCs is weakened.

Molecular bases of silencing of *CO*

Homozygosity or compound heterozygosity leads to Co_{null} [Co(a–b–) Co3–, CO:–3,–4] phenotype. Differences from *CO*01* reference allele (M77829) are given.

Allele name	Exon	Nucleotide	Amino acid	Ethnicity (prevalence)
*CO*01N.01*	1	del all or part exon 1	No protein	Northern European (Rare)
*CO*01N.02*	1	307insT	Gly104 fs →Stop	French (Rare)
*CO*01N.03*	3	576C>A	Asn192Lys	Portuguese (Rare)
*CO*01N.04*	1	232delG	Ala78 fs→119Stop	Indian (Rare)
*CO*01N.05*	1	113C>T	Pro38Ser[3]	Polish (Rare)
*CO*01N.06*	3	601delG	fs Val201Stop	Caucasian (Rare)

Molecular bases of weak expression of Co antigens

KLF1 encodes erythroid Krüppel-like factor (EKLF). Several nucleotide changes in this gene are responsible for *In(Lu)* (see Lutheran). *KLF1* has 3 exons; the initiation codon is in exon 1, and the stop codon is in exon 3. GenBank accession numbers are U37106 (gene) and NM_006563 (mRNA). Differences from *CO*01* reference allele (M77829) or *KLF1*01* reference allele (Accession number NM_006563) are given.

Allele name	Exon	Nucleotide change	Amino acid change	Ethnicity (prevalence)
*CO*01M.01*^	1	112C>T	Pro38Leu	Northern European (Rare)
*KLF1*BGM10*	3	973G>A^^	Glu325Lys	(Rare)

^ = When *in trans* to *CO*02*, expression of Co^b on RBCs is weakened.
^^ = Heterozygosity for this nucleotide change in a patient with dyserythropoietic anemia caused suppression of antigens in CO, IN, and LW blood group systems[4,5].

Amino acid sequence[1]

```
MASEFKKKLF  WRAVVAEFLA  TTLFVFISIG  SALGFKYPVG  NNQTAVQDNV   50
KVSLAFGLSI  ATLAQSVGHI  SGAHLNPAVT  LGLLLSCQIS  IFRALMYIIA  100
QCVGAIVATA  ILSGITSSLT  GNSLGRNDLA  DGVNSGQGLG  IEIIGTLQLV  150
LCVLATTDRR  RRDLGGSAPL  AIGLSVALGH  LLAIDYTGCG  INPARSFGSA  200
VITHNFSNHW  IFWVGPFIGG  ALAVLIYDFI  LAPRSSDLTD  RVKVWTSGQV  250
EEYDLDADDI  NSRVEMKPK                                       269
```

Carrier molecule[1]

A multipass membrane glycoprotein.

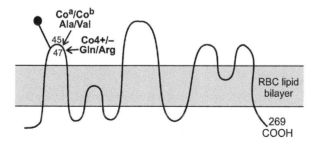

M_r (SDS-PAGE)	28,000 unglycosylated form
	40,000–60,000 glycosylated form
CHO: N-glycan	Polylactosaminoglycan that carries ABH determinants at residue 42

Colton

Cysteine residues   4
Copies per RBC   120,000–160,000 molecules arranged in tetramers

Function

Water transport. AQP1 accounts for 80% of water reabsorption in kidneys, and is a determinant of vascular permeability in the lung[6]. The ^{76}Asn-Pro-Ala^{78} (NPA) motif is essential for this function[6,7].

Disease association

Co^a is expressed weakly in Monosomy 7 due to certain chromosome 7 rearrangements that also cause acute leukemia.

One patient with dyserythropoietic anemia had suppression of CO, IN, and LW antigens[4,8]. Other examples have been found.

Phenotypes (% occurrence)

Phenotypes	Most populations
Co(a+b–)	90
Co(a–b+)	0.5
Co(a+b+)	9.5
Co(a–b–)	<0.01
Null: Co(a–b–)	
Unusual: Co(a–b+^{W}); Co(a–b–) CO:3,–4	

Comments

In RBCs, AQP1 exists in the membrane as a dimer, and accounts for 2.4% of the total membrane protein[7,9].

References

[1] Preston, G.M., Agre, P., 1991. Isolation of the cDNA for erythrocyte integral membrane protein of 28 kilodaltons: Member of an ancient channel family. Proc Natl Acad Sci USA 88, 11110–11114.

[2] Arnaud, L., et al., 2010. A functional *AQP1* allele producing a Co(a–b–) phenotype revises and extends the Colton blood group system. Transfusion 50, 2106–2116.

[3] Karpasitou, K., et al., 2010. A silenced allele in the Colton blood group system. Vox Sang 99, 158–162.

[4] Parsons, S.F., et al., 1994. A novel form of congenital dyserythropoietic anemia associated with deficiency of erythroid CD44 and a unique blood group phenotype [In(a−b−), Co(a−b−)]. Blood 83, 860–868.

[5] Singleton, B.K., et al., 2009b. A novel GATA-1 mutation (Ter414Arg) in a family with the rare X-linked blood group Lu(a–b–) phenotype [abstract]. Blood 114, 783.

[6] King, L.S., et al., 2002. Decreased pulmonary vascular permeability in aquaporin-1-null humans. Proc Natl Acad Sci USA 99, 1059–1063.

[7] Kozono, D., et al., 2002. Aquaporin water channels: Atomic structure and molecular dynamics meet clinical medicine. J Clin Invest 109, 1395–1399.

[8] Singleton, B.K., et al., 2009a. A novel EKLF mutation in a patient with dyserythropoietic anemia: The first association of EKLF with disease in man [abstract]. Blood 114, 72.

[9] Agre, P., et al., 2002. Aquaporin water channels – from atomic structure to clinical medicine. J Physiol London 542, 3–16.

Co^a Antigen

Terminology

ISBT symbol (number)	CO1 (015001 or 15.1)
Obsolete name	Colton
History	Named in 1967, after the first antibody producer. Should have been named Ca^a from Calton, but the handwriting on the tube was misread.

Occurrence

All populations	99.5%

Antithetical antigen

Co^b (CO2)

Expression

Cord RBCs	Expressed

Molecular basis associated with Co^a antigen[1]

Amino acid	Ala45
Nucleotide	C at bp 134 (and G at bp 133) in exon 1 The nucleotide at bp 133 was shown by transfectant studies to be important for expression of Co^a antigen

A change of Gln47 to Arg (in the Co4− phenotype) inhibits expression of Co^a but not Co3.

Colton

Effect of enzymes and chemicals on Co^a antigen on intact RBCs

Ficin/Papain	Resistant
Trypsin	Resistant
α-Chymotrypsin	Resistant
DTT 200 mM	Resistant
Acid	Resistant

In vitro characteristics of alloanti-Co^a

Immunoglobulin class	IgG (Rare IgM reported)[2]
Optimal technique	IAT
Complement binding	Some

Clinical significance of alloanti-Co^a

Transfusion reaction	No to moderate/delayed; immediate/hemolytic[3]
HDFN	Mild to severe[4] (rare)

Autoanti-Co^a

One example.

Comments

Siblings of patients with anti-Co^a should be tested for compatibility, and the patient urged to donate blood for cryogenic storage when his/her clinical state permits.

References

[1] Smith, B.L., et al., 1994. Human red cell aquaporin CHIP. I. Molecular characterization of ABH and Colton blood group antigens. J Clin Invest 94, 1043–1049.

[2] Kurtz, S.R., et al., 1982. Survival of homozygous Co^a (Colton) red cells in a patient with anti-Co^a. Vox Sang 43, 28–30.

[3] Covin, R.B., et al., 2001. Acute hemolytic transfusion reaction caused by anti-Co^a. Immunohematology 17, 45–49.

[4] Simpson, W.K.H., et al., 1973. Anti-Co^a and severe haemolytic disease of the newborn. S Afr Med J 47, 1302–1304.

Co^b Antigen

Terminology

ISBT symbol (number)	CO2 (015002 or 15.2)
History	Named in 1970 when it was shown to be antithetical to Co^a.

Occurrence

All populations   10%

Antithetical antigen

Co^a (**CO1**)

Expression

Cord RBCs	Expressed
Altered	Co(a–) RBCs with weak expression of Co^b exist. See System pages

Molecular basis associated with Co^b antigen[1]

Amino acid	Val45 or Thr45
Nucleotide	T at bp 134 or A at bp 133 in exon 1
	The nucleotide at bp 133, with 134T, was shown by transfectant studies to be important for expression of Co^b antigen

Effect of enzymes and chemicals on Co^b antigen on intact RBCs

Ficin/Papain	Resistant
Trypsin	Resistant
α-Chymotrypsin	Resistant
DTT 200 mM	Resistant
Acid	Resistant

In vitro characteristics of alloanti-Co^b

Immunoglobulin class	IgG
Optimal technique	IAT
Complement binding	Rare

Clinical significance of alloanti-Co^b

Transfusion reaction	No to moderate/delayed/hemolytic
HDFN	Mild

Comments

Co^b is a poor immunogen, and anti-Co^b is rarely found as a single specificity.

Reference

[1] Smith, B.L., et al., 1994. Human red cell aquaporin CHIP. I. Molecular characterization of ABH and Colton blood group antigens. J Clin Invest 94, 1043–1049.

Colton

Co3 Antigen

Terminology

ISBT symbol (number)   CO3 (015003 or 15.3)
Obsolete names Co^{ab}
History   Reported in 1974, when an antibody to a common antigen (then called anti-Co^aCo^b) was found in a patient whose RBCs typed Co(a–b–).

Occurrence

All populations   Greater than 99.9%

Expression

Cord RBCs   Expressed

Molecular basis associated with lack of Co3 antigen

Refer to System pages.

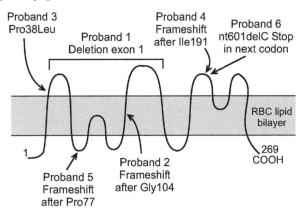

Effect of enzymes and chemicals on Co3 antigen on intact RBCs

Ficin/Papain   Resistant
Trypsin   Resistant
α-Chymotrypsin   Resistant
DTT 200 mM   Resistant
Acid   Resistant

In vitro characteristics of alloanti-Co3

Immunoglobulin class   IgG
Optimal technique   IAT
Complement binding   Yes

Colton

Clinical significance of alloanti-Co3

Transfusion reaction	Mild hemolytic
HDFN	Severe

Autoanti-Co3

One example described as mimicking anti-Co3 made by a patient with non-Hodgkins lymphoma.

Comments

RBCs from a baby with congenital dyserythropoietic anemia (CDA) were Co(a–b–), In(a–b–), AnWj–, and had a weak expression of LW[1,2]. More cases have been reported.

Siblings of patients with anti-Co3 should be tested for compatibility, and the patient urged to donate blood for cryogenic storage when his/her clinical state permits.

References

[1] Agre, P., et al., 1994. Human red cell Aquaporin CHIP. II. Expression during normal fetal development and in a novel form of congenital dyserythropoietic anemia. J Clin Invest 94, 1050–1058.

[2] Parsons, S.F., et al., 1994. A novel form of congenital dyserythropoietic anemia associated with deficiency of erythroid CD44 and a unique blood group phenotype [In(a−b−), Co(a−b−)]. Blood 83, 860–868.

Co4 Antigen

Terminology

ISBT symbol (number)	CO4 (015004 or 15.4)
History	First identified in 2002, and named with next available number in the Colton system in 2010 after a second Co4– proband (Turkish) was found with functional AQP1. Both cases had anti-Co4 in their plasma.

Occurrence

All populations	Only three Co4– probands have been reported

Expression

Cord RBCs	Presumed expressed

Colton

Molecular basis associated with Co4 antigen[1,2]

Amino acid	Gln47
Nucleotide	A at bp 140 in exon 1
CO:−4	Arg47 and G at bp 140

Effect of enzymes and chemicals on Co4 antigen on intact RBCs

Ficin/Papain	Resistant
Trypsin	Resistant
α-Chymotrypsin	Presumed resistant
DTT 200 mM	Presumed resistant

In vitro characteristics of alloanti-Co4

| Immunoglobulin class | IgG |
| Optimal technique | IAT |

Clinical significance of alloanti-Co4

No data because the antibody is rare.

Comments

Anti-Co4 is reactive with Co(a+b−) and Co(a−b+), but not Co(a−b−) RBCs; CO:−4 RBCs type Co(a−b−) but Co3+ and have functional AQP1[1].

Siblings of patients with anti-Co4 should be tested for compatibility and the patient urged to donate blood for cryogenic storage when his/her clinical state permits.

References

[1] Arnaud, L., et al., 2010. A functional *AQP1* allele producing a Co(a−b−) phenotype revises and extends the Colton blood group system. Transfusion 50, 2106–2116.

[2] Wagner, F.F., Flegel, W.A., 2002. A clinically relevant Co(a)-like allele encoded by *AQP1* (Q47R) [abstract]. Transfusion 42 (Suppl.), 24S–25S.

Landsteiner-Wiener Blood Group System

Number of antigens 3

High prevalence LW^a, LW^{ab}
Low prevalence LW^b

Terminology

ISBT symbol (number)	LW (016)
CD number	CD242
History	Anti-LW, or "anti-Rh" as it was called, was produced in 1940. However, the phenotypic relationship between LW and the RhD antigen delayed recognition that LW was an independent blood group system until 1963, when it was named to honor Landsteiner and Wiener who made anti-LW in rabbits and guinea pigs after immunizing them with blood from *Macacus rhesus*. In 1982, it became a three-antigen system. The LW_1, LW_2, LW_3, and LW_4 terminology was changed when it was realized that anti-Ne^a (now called anti-LW^b) detects an antigen antithetical to that recognized by anti-LW made by LW_3 people (now called anti-LW^a)[1].

Expression

Tissues   May be found in placenta

Gene[2]

Chromosome	19p13.2
Name	*LW (ICAM4, CD242)*
Organization	3 exons distributed over 2.6 kbp of gDNA
Product	LW glycoprotein, ICAM-4

The Blood Group Antigen (3/e). DOI: http://dx.doi.org/10.1016/B978-0-12-415849-8.00018-1

LW*05/LW*07

ATG   ⊢——⊣ 100 bp   STOP

Database accession numbers

GenBank   X93093 (gene), NM_001544 (mRNA), S78852

Entrez Gene ID   3386

Molecular basis of Landsteiner-Wiener phenotypes

The reference allele, *LW*05* or *LW*A* (Accession number S78852), encodes LW^a (LW5), LW6. Nucleotide differences from this reference allele, and amino acids affected, are given.

Allele encodes	Allele name	Exon	Nucleotide	Restriction enzyme	Amino acid†	Ethnicity (prevalence)
LW(a–b+) or LW:–5,7	*LW*07* or *LW*B*	1	299A>G	*Pvu*II –	Gln100 Arg	Estonians> Finns> Latvians> Lithuanians> Poles> Russians (Several), Others (Rare)

†Change from historical counting of #1 as Ala of the mature membrane-bound protein to #1, as Met results in all amino acid numbers being increased by 30. Thus, the LW5/LW7 polymorphism used to be amino acid number 70 and is now 100.

Molecular basis of silencing of *LW*

Homozygosity leads to LW$_{null}$ [LW:–5,–6,–7; LW(a–b–)] phenotype. Nucleotide changes from *LW*05* reference allele (Accession number S78852), and amino acids affected, are given.

Allele name	Exon	Nucleotide	Amino acid†	Ethnicity (prevalence)
*LW*05N.01*	1	346–355del	116del;fs,118Stop	Canadian (Rare)

†Change from historical counting of #1 as Ala of the mature (membrane-bound protein) results in all amino acid numbers increasing by 30.

Molecular basis of weak LW antigens

KLF1 encodes erythroid Krüppel-like factor (EKLF). Several nucleotide changes in this gene are responsible for the dominant Lu(a–b–) phenotype encoded by *In(Lu)* (see Lutheran blood group system)[3,4]. *KLF1* has 3 exons; the initiation codon is in exon 1 and the stop codon is in exon 3. GenBank accession numbers are U37106 (gene) and NM_006563 (mRNA).

Nucleotide difference from the *KLF1*01* reference allele (Accession number NM_006563), and amino acid affected, are given.

Allele name	Exon	Nucleotide	Amino acid	Ethnicity (prevalence)
KLF1*BGM10	3	973G>A^	Glu325Lys	(Rare)

^ = Heterozygosity for this change caused dyserythropoietic anemia and suppression of CO, IN, and LW antigens[5,6].

Amino acid sequence

```
MGSLFPLSLL  FFLAAAYPGV  GSALGRRTKR  AQSPKGSPLA  PSGTSVPFWV   50
RMSPEFVAVQ  PGKSVQLNCS  NSCPQPQNSS  LRTPLRQGKT  LRGPGWVSYQ  100
LLDVRAWSSL  AHCLVTCAGK  TRWATSRITA  YKPPHSVILE  PPVLKGRKYT  150
LRCHVTQVFP  VGYLVVTLRH  GSRVIYSESL  ERFTGLDLAN  VTLTYEFAAG  200
PRDFWQPVIC  HARLNLDGLV  VRNSSAPITL  MLAWSPAPTA  LASGSIAALV  250
GILLTVGAAY  LCKCLAMKSQ  A                                   271
```

LW encodes a signal peptide of 30 amino acids.

Carrier molecule[7]

A single pass type I membrane glycoprotein with two IgSF domains. A secreted form also has been described[8].

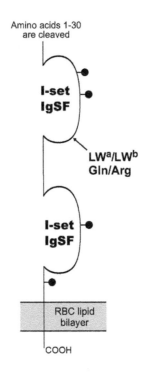

<table>
<tr><td>M_r (SDS-PAGE)</td><td>37,000–43,000</td></tr>
<tr><td>CHO: N-glycan</td><td>4 potential sites at residue 68, 78, 190, and 223</td></tr>
<tr><td>CHO: O-glycan</td><td>Present</td></tr>
<tr><td>Cysteine residues</td><td>3 pairs at residues 69/113, 73/117, and 153/210</td></tr>
<tr><td>Copies per RBC</td><td>D+ 4,400 (Adult); 5,100 (cord)
D− 2,800 (Adult); 3,600 (cord)</td></tr>
</table>

Function

The LW glycoprotein is an intercellular adhesion molecule (ICAM-4), and a ligand for integrins. LW has 30% sequence identity with other ICAMs. ICAM-4 binds to β_2 integrins including the LFA-1 and Mac-1 leukocyte integrins and VLA-4 ($\alpha_4\beta_1$) in haemopoietic tissue and also $\alpha_v\beta_1$, $\alpha_v\beta_5$, and maybe $\alpha_v\beta_3$[9,10]. Possible marker for lymphocyte maturation or differentiation. May assist in stabilizing erythroblastic islands during erythropoiesis[11]. May be involved in removal of senescent RBCs[7].

Disease association

LW antigens may be depressed during pregnancy and in some diseases, e.g., Hodgkin's disease, lymphoma, leukemia, and sarcoma[7]. Autoanti-LW is common in patients with warm AIHA.

Expression of ICAM-4 is elevated on sickle RBCs, and interaction between ICAM-4 and vascular endothelial cells may be involved in microvascular occlusions during painful crises of SCD[12].

Phenotypes (% occurrence)

Phenotype	Europeans	Finns
LW(a+b−)	97	93.9
LW(a+b+)	3	6
LW(a−b+)	Rare	0.1

Null: LW(a−b−); Rh$_{null}$ RBCs type LW(a−b−) although *LW* is normal

There is a phenotypic relationship between LW and D antigens: in adults, D− RBCs have lower expression of LW antigens than D+ RBCs (ratio 1:1.5). In cord RBCs, LW is strongly expressed in D− and D+ RBCs.

Obsolete compared to current phenotype names

Obsolete	Obsolete	Current
LW+, D+	LW$_1$	LW(a+b−) or LW(a+b+)
LW+, D−	LW$_2$	LW(a+b−) or LW(a+b+)
LW−, D+ or D−	LW$_3$	LW(a−b+)
LW−, D+ or D−	LW$_4$	LW(a−b−)
LW−, Rh$_{null}$	LW$_0$	LW(a−b−)

Comments

LW antigens require intramolecular disulfide bonds and the presence of divalent cations, notably Mg^{2+}, for expression[13].

References

[1] Sistonen, P., Tippett, P., 1982. A "new" allele giving further insight into the LW blood group system. Vox Sang 42, 252–255.

[2] Hermand, P., et al., 1996. Characterization of the gene encoding the human LW blood group protein in LW$^+$ and LW$^−$ phenotypes. Blood 87, 2962–2967.

[3] Singleton, B.K., et al., 2009. A novel GATA-1 mutation (Ter414Arg) in a family with the rare X-linked blood group Lu(a–b–) phenotype [abstract]. Blood 114, 783.

[4] Singleton, B.K., et al., 2008. Mutations in EKLF/KLF1 form the molecular basis of the rare blood group In(Lu) phenotype. Blood 112, 2081–2088.

[5] Arnaud, L., et al., 2010. A dominant mutation in the gene encoding the erythroid transcription factor KLF1 causes a congenital dyserythropoietic anemia. Am J Hum Genet 87, 721–727.

[6] Parsons, S.F., et al., 1994. A novel form of congenital dyserythropoietic anemia associated with deficiency of erythroid CD44 and a unique blood group phenotype [In(a–b–), Co(a–b–)]. Blood 83, 860–868.

[7] Parsons, S.F., et al., 1999. Erythroid cell adhesion molecules Lutheran and LW in health and disease. Baillieres Clin Haematol 12, 729–745.

[8] Lee, G., et al., 2003. Novel secreted isoform of adhesion molecule ICAM-4: potential regulator of membrane-associated ICAM-4 interactions. Blood 101, 1790–1797.

[9] Bailly, P., et al., 1995. The red cell LW blood group protein is an intercellular adhesion molecule which binds to CD11/CD18 leukocyte integrins. Eur J Immunol 25, 3316–3320.

[10] Spring, F.A., et al., 2001. Intercellular adhesion molecule-4 binds $\alpha_4\beta_1$ and α_V-family integrins through novel integrin-binding mechanisms. Blood 98, 458–466.

[11] Chasis, J.A., Mohandas, N., 2008. Erythroblastic islands: Niches for erythropoiesis. Blood 112, 470–478.

[12] Zennadi, R., et al., 2004. Epinephrine acts through erythroid signaling pathways to activate sickle cell adhesion to endothelium via LW-alphavbeta3 interactions. Blood 104, 3774–3781.

[13] Bloy, C., et al., 1990. Surface orientation and antigen properties of Rh and LW polypeptides of the human erythrocyte membrane. J Biol Chem 265, 21482–21487.

LW^a Antigen

Terminology

ISBT symbol (number)	LW5 (016005 or 16.5)
Obsolete names	LW; LW$_1$; LW$_2$
History	Named LW^a in 1982 when the antithetical relationship to Ne^a (LW^b) was recognized. At that time, the antigen names LW1 to LW4 were not used, because they had been used to designate phenotypes.

Occurrence

All populations	100%

Antithetical antigen

LW^b (**LW7**)

Expression

Cord RBCs	Well expressed on D+ and D–
Altered	Weak on D– RBCs from adults
	Weak or absent on RBCs stored in EDTA

Molecular basis associated with LW^a antigen[1]

Amino acid	Gln100 (previously reported as 70)
Nucleotide	A at bp 299 in exon 1

Effect of enzymes and chemicals on LW^a antigen on intact RBCs

Ficin/Papain	Resistant
Trypsin	Resistant
α-Chymotrypsin	May be weakened
Pronase	Sensitive
DTT 200 mM/50 mM	Sensitive/sensitive (thus sensitive to WARM™ and ZZAP)
Acid	Resistant

In vitro characteristics of alloanti-LW^a

Immunoglobulin class	IgG (usually); IgM
Optimal technique	IAT or RT

Clinical significance of alloanti-LW^a

Transfusion reaction	No to mild/delayed [Rare; D–, LW(a+) RBCs survive well]
HDFN	No to mild (very rare)

Autoanti-LW^a

Autoanti-LW^a with suppression of LW antigens has been reported.
Sometimes observed in plasma of patients with warm AIHA.

Comments

Testing pronase or DTT treated D+ RBCs is a useful way to differentiate anti-D from anti-LW; anti-D will be reactive while anti-LW will not.
Siblings of patients with alloanti-LW^a should be tested for compatibility, and the patient urged to donate blood for cryogenic storage when his/her clinical state permits, in cases where D–, LW(a+) RBCs do not survive well.
Antigen expression requires Mg^{2+} (may be weak in EDTA samples).

Reference

[1] Hermand, P., et al., 1995. Molecular basis and expression of the LW^a/LW^b blood group polymorphism. Blood 86, 1590–1594.

LW^{ab} Antigen

Terminology

ISBT symbol (number)	LW6 (016006 or 16.6)
Obsolete names	Bigelow; Big; LW; LW$_4$
History	LW$_4$ was renamed LW^{ab} when the LW blood group system was established in 1982.

Occurrence

All populations	100%

Expression

Cord RBCs	Well expressed on D+ and D–
Altered	Weak on D– RBCs from adults
	Weak or absent on RBCs stored in EDTA

Molecular basis associated with LW^{ab} antigen

See System pages for molecular basis of LW(a–b–) phenotype.

Effect of enzymes and chemicals on LW^{ab} antigen on intact RBCs

Ficin/Papain	Resistant
Trypsin	Resistant
α-Chymotrypsin	May be weakened
Pronase	Sensitive
DTT 200 mM/50 mM	Sensitive/sensitive (thus sensitive to WARM™ and ZZAP)
Acid	Resistant

In vitro characteristics of alloanti-LW^{ab}

Immunoglobulin class	IgG; IgM
Optimal technique	37°C; IAT

Clinical significance of alloanti-LW^{ab}

Transfusion reaction	No data
HDFN	Mild; an autoanti-LW^{ab} has been reported to cause HDFN[1]

Autoanti-LW^{ab}

Autoanti-LW^{ab} with suppression of LW antigens occurs[2].
Sometimes observed in plasma of patients with warm AIHA.

Comments

Siblings of patients with anti-LW[ab] should be tested for compatibility, and the patient urged to donate blood for cryogenic storage when his/her clinical state permits.

If random units have shortened survival, experts agree that D– RBCs are the component of choice.

Only one alloanti-LW[ab] has been described in an LW(a–b–) person (Bigelow). Her brother also had the LW(a–b–) phenotype.

When LW antigens are suppressed, the anti-LW[ab] may mimic an alloantibody and is a more common specificity than autoanti-LW[a].

Antigen expression requires Mg^{2+} (may be weak in EDTA samples).

References

[1] Davies, J., et al., 2009. Haemolytic disease of the foetus and newborn caused by auto anti-LW. Transfus Med 19, 218–219.

[2] Storry, J.R., 1992. Review: the LW blood group system. Immunohematology 8, 87–93.

LW[b] Antigen

Terminology

ISBT symbol (number)	LW7 (016007 or 16.7)
Obsolete names	Ne[a]; LW_3
History	Name changed from Ne[a] when the antithetical relationship to LW[a] was recognized in 1982.

Occurrence

Most populations	Rare
Estonians	8%
Finns	6%
Latvians and Lithuanians	5%
Poles and Russians	2%
Other Europeans	<1%[1]

Antithetical antigen

LW[a] (**LW5**)

Expression

Cord RBCs	Well expressed on D+ and D–
Altered	Weak on D– RBCs from adults
	Weak or absent on RBCs stored in EDTA

Molecular basis associated with LW^b antigen[2]

Amino acid	Arg100 (previously reported as 70)
Nucleotide	G at bp 299 in exon 1

Effect of enzymes and chemicals on LW^b antigen on intact RBCs

Ficin/Papain	Resistant (enhanced)
Trypsin	Resistant (enhanced)
α-Chymotrypsin	May be weakened
Pronase	Sensitive
DTT 200 mM/50 mM	Sensitive/sensitive (thus sensitive to WARM™ and ZZAP)
Acid	Resistant

In vitro characteristics of alloanti-LW^b

Immunoglobulin class	IgG; IgM
Optimal technique	37°C; IAT

Clinical significance of alloanti-LW^b

Transfusion reaction	No to mild
HDFN	No to mild

Comments

Antigen expression requires Mg^{2+} (may be weak in EDTA samples).

References

[1] Sistonen, P., et al., 1999. The LW^b blood group as a marker of prehistoric Baltic migrations and admixture. Hum Hered 49, 154–158.

[2] Hermand, P., et al., 1995. Molecular basis and expression of the LW^a/LW^b blood group polymorphism. Blood 86, 1590–1594.

Chido/Rodgers Blood Group System

Number of antigens 9

Polymorphic | WH
High prevalence | Ch1, Ch2, Ch3, Ch4, Ch5, Ch6, Rg1, Rg2

Terminology

ISBT symbol (number) | CH/RG (017)
History | Named after the first antibody producers, Chido and Rodgers. Anti-Ch was reported in 1967, and when anti-Rg was described in 1976 there were obvious similarities between them. Ch and Rg appeared to be RBC antigens and were given blood group system status. However, the antigens were later located on the fourth component of complement (C4), which becomes bound to RBCs from the plasma.

Expression

Soluble form | In plasma or serum
Altered | GPA-deficient RBCs have a weak expression of Ch and Rg antigens

Gene

Chromosome | 6p21.32
Name | *CH (C4B)*; *RG(C4A)*
Organization | *C4A* 41 exons distributed over 22 kpb of gDNA
C4B 41 exons distributed over 22 kbp or 16 kbp of gDNA after loss of a 6.8 kbp intron
Product | C4A complement component (Rg)
C4B complement component (Ch)

Database accession numbers

GenBank   NM_001002029 *(CH)*; K02403.1 *(RG)*
Entrez Gene ID   720 *(RG)*; 721 *(CH)*

Molecular basis of Ch/Rg phenotypes

Ch and Rg antigens are not intrinsic to the RBC membrane, but are adsorbed from the plasma. The molecular bases of selected phenotypes are given below.

Chido

The reference allele [Accession number NM_001002029 (mRNA)] encodes Ch1 (CH1), CH2, CH3, CH4, CH5, and CH6. Differences from this allele are given.

Allele encodes	Allotype	Exon	Nucleotide^	Amino acid
Ch+Rg– or CH:1,2,–3,4,5,–6. RG:–1,–2	C4B*1	27	3527G>A	Ser1176Asn
Ch+Rg– or CH:1,–2,3,4,5,6 RG:–1,–2	C4B*2	25	3218G>A	Gly1073Asp
Ch+Rg+WH+ or CH:–1,–2,–3,4,–5,6 RG:1,–2	C4B*5	25, 28	3620C>T; 3629G>T; 3630G>C	Gly1073Asp; Ala1207Val; Arg1210Val

^Nucleotide numbers start with A of ATG, which is 52 bp into the reference sequence. The allele names are from the complement community and not the ISBT.

Rodgers

The reference allele [Accession number K02403.1 (mRNA) for *C4A*3*] encodes Rg1 (RG1) and RG2. Differences from this allele are given.

Allele encodes	Allotype	Nucleotide change^	Amino acid change
Ch+Rg– or CH:–1,–2,3,4,5,6 RG:–1,–2	*C4A*1*	3567A>G; 3660T>C; 3669T>G; 3670C>G	Asp1054Gly; Asn1157Ser; Val1188Ala; Leu1191Arg
Ch–Rg+WH+ or CH:–1,–2,–3,–4,–5,6 RG:1,–2	*C4A*3.WH*	3567A>G	Asn1157Ser

^The site of ATG is not clear. The allele names are from the complement community and not the ISBT.

Carrier molecule

C4A and C4B are glycoproteins which are adsorbed onto the RBC membrane from the plasma. C4A binds preferentially to protein, and C4B to carbohydrate. C4A migrates more quickly in electrophoresis than C4B. C4A and C4B are 99% identical in their amino acid sequences.

Ch and Rg antigens are located in the C4d region of C4B or C4A, respectively. C4d is a tryptic fragment of C4.

C4 molecule (adapted from Daniels)[1].

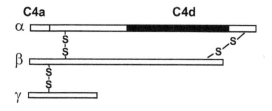

Chido/Rodgers

Amino acid residues associated with CH/RG phenotypes[2]

Allotype	Ch/Rg type	Phenotype	Amino acid residue										
			1054	1101	1102	1105	1106	1157	1188	1191			
C4A*3	Ch–Rg+	CH/RG:–1,–2,–3, –4,–5,–6,11,12	Asp	Pro	Cys	Leu	Asp	Asn	Val	Leu
C4A*1	Ch+Rg–	CH/RG:1,–2,3, –4,5,6,–11,–12	Gly	Pro	Cys	Leu	Asp	Ser	Ala	Arg
C4A*3	Ch–Rg+ WH+	CH/RG:–1,–2, –3,–4,–5,6,11,–12	Asp	Pro	Cys	Leu	Asp	Ser	Val	Leu
C4B*3	Ch+Rg–	CH/RG:1,2,3, 4,5,6,–11,–12	Gly	Leu	Ser	Ile	His	Ser	Ala	Arg
C4B*1	Ch+Rg–	CH/RG:1,2,–3, 4,5,–6,–11,–12	Gly	Leu	Ser	Ile	His	Asn	Ala	Arg
C4B*2	Ch+Rg–	CH/RG:1,–2, 3,4,–5,6,–11,–12	Asp	Leu	Ser	Ile	His	Ser	Ala	Arg
C4B*5	Ch+Rg+ WH+	CH/RG:–1,–2, –3,4,–5,6,11,–12	Asp	Leu	Ser	Ile	His	Ser	Val	Leu

Function

There are functional differences between C4A and C4B allotypes: C4A is more effective than C4B at solubilizing immune complexes and inhibiting immune precipitation. C4B binds more effectively to the RBC surface (through sialic acid), and thus is more effective at promoting hemolysis. A single amino acid substitution at position 1106 (aspartic acid for histidine) converts the functional activity of C4B to C4A[3], whereas the substitution of cysteine for serine at position 1102 affects hemolytic activity and IgG binding.

Disease association

Inherited low levels of C4 may be a predisposing factor for diseases such as insulin-dependent diabetes and autoimmune chronic active hepatitis. Specific C4 allotypes and null genes have been associated with numerous autoimmune disorders, including Graves' disease and rheumatoid arthritis (for list, see[4]). Lack of C4B (Ch–) gives increased susceptibility to bacterial meningitis in children. Rg– individuals (lack of C4A) have a much greater susceptibility for SLE.

Phenotypes (% occurrence)

Chido phenotype	Most populations	Japanese	Rodgers phenotype	Most populations	Japanese
CH/RG: 1,2,3	88.2	75	CH/RG: 11,12	95	100
CH/RG: 1,−2,3	4.9	24	CH/ RG:11,−12	3	0
CH/RG: 1,2,−3	3.1	0	CH/RG: −11,−12	2	0
CH/RG: −1,−2,−3	3.8	1			
CH/RG: −1,2,−3	Rare	0			
CH/RG:1, −2,−3	Rare	0			
Null:	C4-deficient RBCs				

Comments

Antigens of this system are stable in stored serum or plasma. Phenotypes and antibodies of this system are most accurately defined by hemagglutination inhibition tests.

RBCs coated with C4 (+C3) by use of low ionic strength 10% sucrose solution give enhanced reactivity with anti-Ch and anti-Rg.

Sialidase-treated RBCs do not adsorb C4.

References

[1] Daniels, G., 1995. Blood group antigens as markers of complement and complement regulatory molecules. In: Cartron, J.-P., Rouger, P. (Eds.), Molecular Basis of Human Blood Group Antigens. Plenum Press, New York, NY, pp. 397–419.

[2] Yu, C.Y., et al., 1988. A structural model for the location of the Rodgers and the Chido antigenic determinants and their correlation with the human complement component C4A/C4B isotypes. Immunogenetics 27, 399–405.

[3] Carroll, M.C., et al., 1990. Substitution of a single amino acid (aspartic acid for histidine) converts the functional activity of human complement C4B to C4A. Proc Natl Acad Sci USA 87, 6868–6872.

[4] Moulds, J.M., 1994. Association of blood group antigens with immunologically important proteins. In: Garratty, G. (Ed.), Immunobiology of Transfusion Medicine. Marcel Dekker, Inc., New York, NY, pp. 273–297.

Ch1 Antigen

Terminology

ISBT symbol (number)	CH/RG1 (017001 or 17.1)
Obsolete names	Ch; Ch^a; Chido
History	Named in 1967 after Mrs. Chido, who made "anti-Chido" (considered a nebulous antibody).

Occurrence

Most populations	96%
Japanese	99%

Expression

Cord RBCs	Absent or weak
Altered	Weak on GPA-deficient RBCs

Molecular basis associated with Ch1 antigen

Requires Ala1188 and Arg1191 of C4[1,2]. See System pages.

Effect of enzymes and chemicals on Ch1 antigen on intact RBCs

Ficin/Papain	Sensitive
Trypsin	Sensitive
α-Chymotrypsin	Sensitive
DTT 200 mM	Resistant

In vitro characteristics of alloanti-Ch1

Immunoglobulin class	IgG (mostly IgG2 and IgG4)
Optimal technique	IAT
Neutralization	Antigen-positive serum or plasma

Clinical significance of alloanti-Ch1

Transfusion reaction	Not hemolytic; anaphylactic from plasma products and platelets (few reports)
HDFN	No

Comments

Soluble plasma antigen in donor blood may neutralize patient's antibody. Anti-Ch1 reacts strongly with C4-coated RBCs. Antihuman globulin without anti-IgG4 will not detect anti-Ch1.

Virtually all anti-Ch contain anti-Ch1.

The Ch antigens have been divided (Ch1, Ch2, etc.), but classification is not required for clinical purposes.

References

[1] Daniels, G., 1995. Blood group antigens as markers of complement and complement regulatory molecules. In: Cartron, J.-P., Rouger, P. (Eds.), Molecular Basis of Human Blood Group Antigens. Plenum Press, New York, NY, pp. 397–419.

[2] Giles, C.M., 1988. Antigenic determinants of human C4, Rodgers and Chido. Exp Clin Immunogenet 5, 99–114.

Ch2 Antigen

Terminology

ISBT symbol (number)	CH/RG2 (017002 or 17.2)
History	Defined in 1985, when plasma inhibition studies revealed that there are at least six Chido antigens (Ch1 to Ch6) of high prevalence.

Occurrence

Most populations	Greater than 90%
Japanese	75%

Chido/Rodgers

Molecular basis associated with Ch2 antigen

Antigen expression requires presence of Ch4 and Ch5, i.e., Gly1054, Leu1101, Ser1102, Ile1105, and His1106[1,2]. See System pages.

Comments

Anti-Ch2 + anti-Ch4 was detected in a Ch:1,−2,3,−4,5,6 Rg:1, 2 person[3].
Anti-Ch2 + anti-Ch5 was detected in a Ch:1,−2,3,4,−5,6 Rg:1, 2 person[4].

References

[1] Daniels, G., 1995. Blood group antigens as markers of complement and complement regulatory molecules. In: Cartron, J.-P., Rouger, P. (Eds.), Molecular Basis of Human Blood Group Antigens. Plenum Press, New York, NY, pp. 397–419.

[2] Giles, C.M., 1988. Antigenic determinants of human C4, Rodgers and Chido. Exp Clin Immunogenet 5, 99–114.

[3] Fisher, B., et al., 1993. A new allo anti-Ch specificity in a patient with a rare Ch positive phenotype [abstract]. Transf Med 3 (Suppl. 1), 84.

[4] Giles, C.M., et al., 1987. Allo-anti-Chido in a Ch-positive patient. Vox Sang 52, 129–133.

Ch3 Antigen

Terminology

ISBT symbol (number)	CH/RG3 (017003 or 17.3)
History	See Ch2 antigen

Occurrence

Caucasians	93%
Japanese	Greater than 99%

Molecular basis associated with Ch3 antigen

Antigen expression requires presence of Ch1 and Ch6, i.e., Ser1157, Ala1188, and Arg1191[1,2]. See System pages.

References

[1] Daniels, G., 1995. Blood group antigens as markers of complement and complement regulatory molecules. In: Cartron, J.-P., Rouger, P. (Eds.), Molecular Basis of Human Blood Group Antigens. Plenum Press, New York, NY, pp. 397–419.

[2] Giles, C.M., 1988. Antigenic determinants of human C4, Rodgers and Chido. Exp Clin Immunogenet 5, 99–114.

Ch4 Antigen

Terminology

ISBT symbol (number)   CH/RG4 (017004 or 17.4)
History   See Ch2 antigen.

Occurrence

All populations   Greater than 99%

Molecular basis associated with Ch4 antigen

Antigen expression requires presence of Leu1101, Ser1102, Ile1105, and His1106[1,2]. See System pages. Detected on all C4B allotypes.

References

[1] Daniels, G., 1995. Blood group antigens as markers of complement and complement regulatory molecules. In: Cartron, J.-P., Rouger, P. (Eds.), Molecular Basis of Human Blood Group Antigens. Plenum Press, New York, NY, pp. 397–419.

[2] Giles, C.M., 1988. Antigenic determinants of human C4, Rodgers and Chido. Exp Clin Immunogenet 5, 99–114.

Ch5 Antigen

Terminology

ISBT symbol (number)   CH/RG5 (017005 or 17.5)
History   See Ch2 antigen.

Occurrence

All populations   Greater than 99%

Molecular basis associated with Ch5 antigen

Antigen expression requires Gly1054[1,2]. See System pages.

References

[1] Daniels, G., 1995. Blood group antigens as markers of complement and complement regulatory molecules. In: Cartron, J.-P., Rouger, P. (Eds.), Molecular Basis of Human Blood Group Antigens. Plenum Press, New York, NY, pp. 397–419.

[2] Giles, C.M., 1988. Antigenic determinants of human C4, Rodgers and Chido. Exp Clin Immunogenet 5, 99–114.

Ch6 Antigen

Terminology

ISBT symbol (number)   CH/RG6 (017006 or 17.6)
History   See Ch2 antigen.

Occurrence

All populations   Greater than 99%

Molecular basis associated with Ch6 antigen

Antigen expression requires Ser1157 of C4[1,2]. See System pages.

Comments

Rare specificity, two examples reported.

References

[1] Daniels, G., 1995. Blood group antigens as markers of complement and complement regulatory molecules. In: Cartron, J.-P., Rouger, P (Eds.), Molecular Basis of Human Blood Group Antigens. Plenum Press, New York, NY, pp. 397–419.
[2] Giles, C.M., 1988. Antigenic determinants of human C4, Rodgers and Chido. Exp Clin Immunogenet 5, 99–114.

WH Antigen

Terminology

ISBT symbol (number)   CH/RG7 (017007 or 17.7)
History   Named after the person who was thought to carry a hybrid of C4A and C4B.

Occurrence

Caucasians   15%

Molecular basis associated with WH antigen

Associated with Ch:6, Rg:1,−2 phenotype. Antigen expression requires Ser1157, Val1188, and Leu1191[1,2,3]. See System pages.
In one individual (WH), a single amino acid substitution encoded by the C4A*3 gene at codon 1157 gives rise to Asp in the wild type being replaced by Ser in WH type[1,3].

Comments

Rare specificity, two examples reported[4].

References

[1] Daniels, G., 1995. Blood group antigens as markers of complement and complement regulatory molecules. In: Cartron, J.-P., Rouger, P. (Eds.), Molecular Basis of Human Blood Group Antigens. Plenum Press, New York, NY, pp. 397–419.

[2] Giles, C.M., 1988. Antigenic determinants of human C4, Rodgers and Chido. Exp Clin Immunogenet 5, 99–114.

[3] Moulds, J.M., et al., 1995. Revised model for the Chido/Rogers blood group based on DNA sequencing [abstract]. Transfusion 35 (Suppl.), 53S.

[4] Giles, C.M., Jones, J.W., 1987. A new antigenic determinant for C4 of relatively low frequency. Immunogenetics 26, 392–394.

Rg1 Antigen

Terminology

ISBT symbol (number)	CH/RG11 (017011 or 17.11)
Other names	Rodgers; Rg; Rg[a]
History	"Generic" anti-Rg reported in 1976 and named after antibody maker. All anti-Rg contain anti-Rg1 (strongest component) and anti-Rg2.

Occurrence

All populations	Greater than 98%

Expression

Cord RBCs	Absent or weak

Molecular basis associated with Rg1 antigen

Antigen expression requires Val1188 and Leu1191[1,2]. See System pages.

Effect of enzymes and chemicals on Rg1 antigen on intact RBCs

Ficin/Papain	Sensitive
Trypsin	Sensitive
α-Chymotrypsin	Sensitive
DTT 200 mM	Resistant

Chido/Rodgers

In vitro characteristics of alloanti-Rg1

Immunoglobulin class	IgG
Optimal technique	IAT
Neutralization	Antigen-positive serum or plasma

Clinical significance of alloanti-Rg1

Transfusion reaction	Not hemolytic; anaphylactic from plasma products and platelets (few reports)
HDFN	No

References

[1] Daniels, G., 1995. Blood group antigens as markers of complement and complement regulatory molecules. In: Cartron, J.-P., Rouger, P (Eds.), Molecular Basis of Human Blood Group Antigens. Plenum Press, New York, NY, pp. 397–419.

[2] Giles, C.M., 1988. Antigenic determinants of human C4, Rodgers and Chido. Exp Clin Immunogenet 5, 99–114.

Rg2 Antigen

Terminology

ISBT symbol (number)	CH/RG12 (017012 or 17.12)
History	See Rg1 antigen.

Occurrence

Most populations	95%

Molecular basis associated with Rg2 antigen

Antigen expression requires Asp1157, Val1188, and Leu1191[1,2]. See System pages.

Comments

All anti-Rg contain anti-Rg1 (strongest component) and anti-Rg2.

References

[1] Daniels, G., 1995. Blood group antigens as markers of complement and complement regulatory molecules. In: Cartron, J.-P., Rouger, P. (Eds.), Molecular Basis of Human Blood Group Antigens. Plenum Press, New York, NY, pp. 397–419.

[2] Giles, C.M., 1988. Antigenic determinants of human C4, Rodgers and Chido. Exp Clin Immunogenet 5, 99–114.

H Blood Group System

Number of antigens 1

High prevalence   H

Terminology

ISBT symbol (number)	H (018)
CD number	CD173
Obsolete name	O
History	In 1948, Morgan and Watkins suggested changing the terms "anti-O" and "O substance" to "anti-H" and "H substance," as this would differentiate it as a heterogeneic, basic or primary substance common to the great majority of red cells irrespective of their ABO phenotype.

Expression

Soluble form	Saliva and all body fluids (in secretors) except CSF
Other blood cells	Lymphocytes (in secretors), platelets
Tissues	Broad tissue distribution (see ABO section)

Gene

Chromosome	19q13.33
Name	H (FUT1)
Organization	4 exons distributed over 8 kbp of gDNA
Product	2-α-fucosyltransferase (α2Fuc-T1; 2-α-L-fucosyltransferase 1; α1,2-fucosyltransferase 1) that adds α-L-fucose to precursor type 2 chains on RBCs and other cells

The homologous gene (*FUT2*; *Se*) encoding 2-α-L-fucosyltransferase 2 (α2Fuc-T2) is 35 kbp closer to the centromere also at 19q13.33. This transferase

The Blood Group Antigen (3/e). DOI: http://dx.doi.org/10.1016/B978-0-12-415849-8.00020-X

adds α-L-fucose to precursor type 1 chains in secretions (see below). It should be noted that the two enzymes prefer the precursor chains indicated, but some cross-reactivity may occur.

Database accession numbers

GenBank   NM_000148 (mRNA), M35531 (mRNA)
Entrez Gene ID   2523 (FUT1)

Molecular basis of H antigen expression on RBCs

Changes in FUT1 can give rise to H-deficient phenotypes. Inactive alleles of FUT1 fail to express an enzyme that can synthesize H epitopes on RBCs. People having these (h) alleles in the homozygous or compound heterozygous states have the H– (Bombay, O_h) or $H+^W$ phenotype (often called Para-Bombay). In Bombay people, mutated FUT1 and FUT2 both fail to encode functional 2-α-fucosyltransferases, and these individuals lack ABH antigens on RBCs and in secretions.

There are two types of $H+^W$ (Para-Bombay) people: (i) those who lack RBC ABH antigens produced on RBCs, but possess them in secretions (inactive FUT1-encoded enzyme and functional FUT2); and (ii) those who produce very few ABH antigens on RBCs, and may or may not possess them in secretions (altered but weakly active FUT1-encoded enzyme in combination with active or inactive FUT2-encoded enzyme). When H substance is expressed in secretions, it can be absorbed by the RBCs, which then can type weakly H+ ($H+^W$). If A or B are present at the ABO locus, the phenotype may be $A+^w$ and/or $B+^w$ but H– (sometimes referred to as A_h and B_h).

Another extremely rare and principally different reason for the H-deficient phenotype is homozygosity for inactivating mutations at SLC35C1 (the GDP-fucose transporter gene), see "Disease association" below.

Molecular basis of H+ phenotype in RBCs

The reference allele, FUT1*01 (Accession number M35531), encodes 2-α-fucosyltransferase that synthesizes H type 2 antigen on RBCs and other cells. Nucleotide differences, and amino acids affected, are given. H expression will be partially masked if a functional A or B allele is also inherited.

Allele encodes	Allele name	Exon	Nucleotide	Amino acid	Ethnicity (prevalence)
H+	FUT1*02	4	35C>T	Ala12Val	All, but may be more common in Asians

Molecular bases of weak H antigens

Homozygosity or compound heterozygosity leads to the H+^{W} (Para-Bombay) phenotype.

Differences from the *FUT1*01* reference allele (Accession number M35531) are given.

Allele name	Exon	Nucleotide	Amino acid	Ethnicity (prevalence)
*FUT1*01W.01*	4	293C>T	Thr98Met	Chinese (Rare)
*FUT1*01W.02*	4	328G>A	Ala110Thr	Chinese (Rare)
*FUT1*01W.03*	4	349C>T	His117Tyr	Reunion (Rare)
*FUT1*01W.04*	4	442G>T	Asp148Tyr	Japanese (Rare)
*FUT1*01W.05*	4	460T>C	Tyr154His	Taiwanese (Rare)
*FUT1*01W.06*	4	460T>C; 1042G>A	Tyr154His; Glu348Lys	Japanese (Rare)
*FUT1*01W.07*	4	491T>A	Leu164His	North American (Rare)
*FUT1*01W.08*	4	522C>A	Phe174Leu	Chinese (Rare)
*FUT1*01W.09*	4	658C>T	Arg220Cys	Taiwanese (Rare)
*FUT1*01W.10*	4	659G>A	Arg220His	Taiwanese (Rare)
*FUT1*01W.11*	4	661C>T	Arg221Cys	Australian (Rare)
*FUT1*01W.12*	4	682A>G	Met228Val	Chinese (Rare)
*FUT1*01W.13*	4	689A>C	Gln230Pro	Portugese (Rare)
*FUT1*01W.14*	4	721T>C	Tyr241His	Japanese (Rare)
*FUT1*01W.15*	4	801G>C	Trp267Cys	Caucasian (Rare)
*FUT1*01W.16*	4	801G>T	Trp267Cys	Caucasian (Rare)
*FUT1*01W.17*	4	832G>A	Asp278Asn	(Rare)
*FUT1*01W.18*	4	904_906 insAAC	His302_Thr303 insAsn	Japanese (Rare)
*FUT1*01W.19*	4	917C>T	Thr306Ile	Brazilian (Rare)
*FUT1*01W.20*	4	990delG	330fs336 Stop	Japanese (Rare)
*FUT1*01W.21*	4	235G>C	Gly79Arg	Chinese (Rare)
*FUT1*02W.01*	4	269G>T	Gly90Val	(Rare)
*FUT1*02W.02*	4	371T>G	Phe124Cys	(Rare)

H expression will be further weakened in the presence of a functional *A* or *B* allele. Also, H expression may be weakly detectable on RBCs where *FUT1*01N* homozygosity occurs, due to the adsorption of soluble H antigen synthesized by *FUT2*.

Hh

Molecular bases of silencing of *FUT1*01*

Homozygosity or compound heterozygosity leads to an H– (Bombay, O$_h$) phenotype.

Differences from the *FUT1*01* reference allele (Accession number M35531) are given.

Allele name	Exon	Nucleotide	Amino acid	Ethnicity (prevalence)
*FUT1*01N.01*	4	422G>A	Trp141Stop	Brazilian (Rare)
*FUT1*01N.02*	4	461A>G	Tyr154Cys	Caucasian (Rare)
*FUT1*01N.03*	4	462C>A	Tyr154Stop	Japanese (Rare)
*FUT1*01N.04*	4	513G>C	Trp171Cys	Caucasian (Rare)
*FUT1*01N.05*	4	538C>T	Gln180Stop	Israeli (Rare)
*FUT1*01N.06*	4	547_548delAG	182fs248Stop	Taiwanese, Chinese (Rare)
*FUT1*01N.07*	4	586C>T	Gln196Stop	Chinese (Rare)
*FUT1*01N.08*	4	695G>A	Trp232Stop	Japanese (Rare)
*FUT1*01N.09*	4	725T>G^	Leu242Arg	Indian (Several)
*FUT1*01N.10*	4	776T>A	Val259Glu	Caucasian (Rare)
*FUT1*01N.11*	4	785G>A; 786C>A	Ser262Lys	Caucasian (Rare)
*FUT1*01N.12*	4	826C>T	Gln276Stop	North American (Rare)
*FUT1*01N.13*	4	880_881delTT	294fs333Stop	Taiwanese, Chinese (Rare)
*FUT1*01N.14*	4	944C>T	Ala315Val	Caucasian (Rare)
*FUT1*01N.15*	4	948C>G	Tyr316Stop	North American (Rare)
*FUT1*01N.16*	4	980A>C	Asn327Thr	(Rare)
*FUT1*01N.17*	4	1047G>C	Trp349Cys	Caucasian (Rare)
*FUT1*01N.18*	4	684G>A	Met228Ile	Czech (Rare)
*FUT1*01N.19*	4	694T>C	Trp232Pro	Czech (Rare)

H expression may be weakly detectable on RBCs where *FUT1*01N* homozygosity occurs, due to the adsorption of soluble H antigen synthesized by *FUT2*.
^Travels with a complete deletion of *FUT2* (together, these alterations are the genetic basis of the originally-discovered Bombay phenotype).

Molecular basis of H expression in secretions

Gene name   FUT2
Number of exons   2
GenBank   NM_000511 (mRNA), U17894 (gene)
Entrez Gene ID   2524

Differences from the *FUT2*01* reference allele (Accession number U17894) are given.

The reference allele encodes a 2-α-fucosyltransferase that synthesizes H type 1 antigen present in secretions.

Allele name	Exon	Nucleotide	Amino acid	Ethnicity (prevalence)
FUT2*02	2	4G>A	Ala2Thr	Mongolian (Rare)
FUT2*03.01	2	40A>G	Ile14Val	Xhosa (Rare)
FUT2*03.02	2	40A>G; 113C>T	Ile14Val; Ala38Val	Ghanian (Rare)
FUT2*03.03	2	40A>G; 481G>A	Ile14Val; Asp161Asn	Xhosa (Rare)
FUT2*04	2	379C>T	Arg127Cys	Xhosa (Rare)
FUT2*05	2	400G>A	Val134Ile	Samoan (Rare)
FUT2*06	2	481G>A	Asp161Asn	Xhosa (Rare)
FUT2*07	2	665G>A	Arg222His	Turkish (Rare)
FUT2*08	2	685G>A	Val229Met	(Rare)
FUT2*09	2	716G>A	Arg239Gln	Chinese (Rare)
FUT2*10	2	748_750insGTG	249_250insVal	Chinese (Rare)

Hh

Molecular bases of H+^W phenotype in secretions^

Differences from *FUT2*01* reference allele (Accession number U17894) are given.

Allele name	Exon	Nucleotide	Amino acid	Ethnicity (prevalence)
FUT2*01W.01	2	278C>T	Ala93Val	Israeli (Rare)
FUT2*01W.02	2	385A>T	Ile129Phe	Asian, Polynesian, Taiwanese (Common)
FUT2*01W.03	2	853G>A	Ala285Thr	Chinese (Rare)

^associated with the Le(a+b+) phenotype.

Molecular bases of H– phenotype in secretions

Differences from the *FUT2*01* reference allele (Accession number U17894) are given.

Allele name	Exon	Nucleotide	Amino acid	Ethnicity (prevalence)
FUT2*01N.01	2	244G>A; 385A>T	Ala82Thr; Ile129Phe	Mongolian (Rare)
FUT2*01N.02	2	428G>A^	Trp143Stop	Europeans, Africans, Iranians (Common)
FUT2*01N.03	2	569G>A	Arg190His	Turkish (Rare)
FUT2*01N.04	2	571C>T	Arg191Stop	Japanese, Filipino Polynesian, Taiwanese (Rare)
FUT2*01N.05	2	628C>T	Arg210Stop	Japanese (Rare)
FUT2*01N.06	2	658C>T	Arg220Stop	Chinese, Taiwanese (Rare)
FUT2*01N.07	2	664C>T	Arg222Cys	New Guinean (Rare)
FUT2*01N.08	2	685_686delGT	230fs234 Stop	Taiwanese (Rare)
FUT2*01N.09	2	688_690delGTC	Val230del	Filipino (Rare)
FUT2*01N.10	2	400G>A; 760G>A	Val134Ile; Asp254Asn	New Guinean (Rare)
FUT2*01N.11	2	778delC	259fs275Stop	South African (Rare)
FUT2*01N.12	2	849G>A	Trp283Stop	Filipino, Taiwanese (Rare)
FUT2*01N.13	2	868G>A	Gly290Arg	New Guinean (Rare)
FUT2*01N.14	2	950C>T	Pro317Leu	Mongolian (Rare)
FUT2*0N.01		gene deletion		Bangladeshi (Rare)
FUT2*0N.02		coding region deleted^^		Indian (Several)
FUT2*0N.03		fusion gene 1 between FUT2 and Sec1		Japanese (Rare)
FUT*0N.04		fusion gene 2 between FUT2 and Sec1		Mongolian (Rare)

^The most common allele associated with nonsecretor status, especially in the Western hemisphere. It carries multiple other SNPs not given here.
^^Travels with the *FUT1*01N.09*.

H-depleted RBC phenotypes may be due to changes in GDP-fucose transporter gene[1,2]

Changes in this gene (*SLC35C1*) result in the GDP-fucose transporter being ineffective, and as no fucose can be transported there is no fucosylation despite normal 2-α- or 3/4-α-fucosyltransferase histo-blood genes (*FUT1*, *FUT2* or *FUT3*; see also Chapter on Lewis). Thus, the RBCs have the Bombay, Le(a–b–) phenotype, and WBCs lack CD15/sialyl-Le^X. These changes give rise to the very rare leukocyte adhesion deficiency (LADII or CDGII). Differences from the *SLC35C1* (*FUCT1*) reference allele (Accession number NG_009875) are given.

Nucleotide change	Amino acid change	Ethnicity (prevalence)
439C>T	Arg147Lys	Turkish
923C>G	Thr308Arg	Arab
588delG	Ser195fs Stop	Brazilian

Amino acid sequence of α-2-L-fucosyltransferase 1 (from *FUT1*)

```
MWLRSHRQLC  LAFLLVCVLS  VIFFLHIHQD  SFPHGLGLSI  LCPDRRLVTP   50
PVAIFCLPGT  AMGPNASSSC  PQHPASLSGT  WTVYPNGRFG  NQMGQYATLL  100
ALAQLNGRRA  FILPAMHAAL  APVFRITLPV  LAPEVDSRTP  WRELQLHDWM  150
SEEYADLRDP  FLKLSGFPCS  WTFFHHLREQ  IRREFTLHDH  LREEAQSVLG  200
QLRLGRTGDR  PRTFVGVHVR  RGDYLQVMPQ  RWKGVVGDSA  YLRQAMDWFR  250
ARHEAPVFVV  TSNGMEWCKE  NIDTSQGDVT  FAGDGQEATP  WKDFALLTQC  300
NHTIMTIGTF  GFWAAYLAGG  DTVYLANFTL  PDSEFLKIFK  PEAAFLPEWV  350
GINADLSPLW  TLAKP                                           365
```

Carrier molecule[3,4]

H antigen is not the primary gene product. The *FUT1* product, the α-2-L-fucosyltransferase 1, attaches an α-L-fucose to the terminal galactose on type 2 carbohydrate precursor chains attached to proteins or lipids on cells. The immunodominant fucose constitutes the defining sugar of the H antigen, which is in turn the precursor of A and B antigens (see **ABO** blood group system).

In analogy, the *FUT2* product attaches α-L-fucose to the terminal galactose on type 1 carbohydrate precursor chains in secretions.

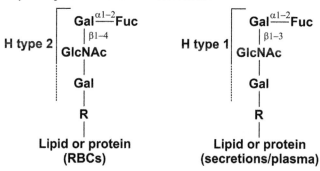

Function

Fucosylated glycans that are the products of *FUT1* and *FUT2* may serve as ligands in cell adhesion or as receptors for certain microorganisms.

Disease association

Weakened expression in acute leukemia and carcinomatous tissue cells.
Children with leukocyte adhesion deficiency (LADII; CDGII) have mental retardation and severe recurrent infections with a high white blood cell count, which is caused by an inability of leucocytes to adhere. Their RBCs are H–, and also lack A, B, Le^a, and Le^b antigens.

Phenotypes

Most RBCs have some H antigen: $O > A_2 > B > A_2B > A_1 > A_1B > H+^W$.
Null: O_h (Bombay)
Unusual: $H+^W$ (Para-Bombay)

Characteristics of phenotypes

Type	H antigen on RBCs	H antigen in secretion	Predicted genotype	Antibody
Common				
Secretor	Yes	Yes	*HH* or *Hh*; *SeSe* or *Sese*	Anti-HI
Non-secretor	Yes	No	*HH* or *Hh*; *sese*	Anti-HI
H-deficient				
Bombay	No	No	*hh*; *sese*	Anti-H
$H+^W$	Weak	No	*(H)*; *sese*	Anti-H
$H+^W$	Weak	Yes	*(H)*; *SeSe* or *Sese*	Anti-HI
H_m (dominant)^	Weak	Yes	*HH* or *Hh*; *SeSe* or *Sese*	None
LADII (CDGII)	No	No	Any genotype possible	Anti-H

^Molecular basis unknown.

For more alleles and details, see http://www.bioc.aecom.yu.edu/bgmut/index.htm.

References

[1] Hidalgo, A., et al., 2003. Insights into leukocyte adhesion deficiency type 2 from a novel mutation in the GDP-fucose transporter gene. Blood 101, 1705–1712.

[2] Lühn, K., et al., 2001. The gene defective in leukocyte adhesion deficiency II encodes a putative GDP-fucose transporter. Nat Genet 28, 69–72.

[3] Lowe, J.B., 1995. Biochemistry and biosynthesis of ABH and Lewis antigens: characterization of blood group-specific glycosyltransferases. In: Cartron, J.-P., Rouger, P. (Eds.), Molecular Basis of Human Blood Group Antigens. Plenum Press, New York, pp. 75–115.

[4] Oriol, R., 1995. ABO, Hh, Lewis, and secretion: serology, genetics, and tissue distribution In: Cartron, J.-P. Rouger, P. (Eds.), Molecular Basis of Human Blood Group Antigens, 1995. Plenum Press, New York, NY, pp. 37–73.

H Antigen

Terminology

ISBT symbol (number)	H1 (018001 or 18.1)
History	See H Blood Group System page.

Occurrence

All populations	99.9%

H-deficient people [Bombay (O_h) and $H+^W$ (Para-Bombay)]: 1 in 8,000 in Taiwan; 2 in 300,000 in Japan; 1 in 10,000 in India; 1 per million in Europe.

Expression

Adult RBCs	In decreasing order: $O>A_2>B>A_2B>A_1>A_1B>H+^W$
Cord RBCs	Weak
Altered	Weak on $H+^W$ (Para-Bombay)

Molecular basis associated with H antigen[1]

See System pages.

Effect of enzymes and chemicals on H antigen on intact RBCs

Ficin/Papain	Resistant (markedly enhanced)
Trypsin	Resistant (markedly enhanced)
α-Chymotrypsin	Resistant (markedly enhanced)
DTT 200 mM	Resistant
Acid	Resistant

In vitro characteristics of alloanti-H

Immunoglobulin class	IgM more common than IgG
Optimal technique	RT or 4°C
Neutralization	Saliva, all body fluids except CSF (secretors)
Complement binding	Some

Hh

Clinical significance of alloanti-H in Bombay (O$_h$) and H+^{W} (Para-Bombay) people

Transfusion reaction	No to severe; immediate/delayed/hemolytic; anti-H made by people with the H+^{W} phenotype is often of lower titer and less significant
HDFN	Possible in O$_h$ mothers, but no reports

Autoanti-H

Yes, usually cold reactive.

Comments

With the exception of Bombay (O$_h$) and Para-Bombay people (whose serum contains anti-A, -B and -H), anti-HI is more common than anti-H. Anti-HI is commonly found in the serum of pregnant group A$_1$ women, and can in fact be found in most people's plasma at +4°C. If reactive at higher temperatures, type-specific blood will be cross-match compatible.

For people with Bombay (O$_h$) and possibly also H+^{W} (Para-Bombay) phenotype, siblings of the patient should be tested for compatibility, and the patient urged to donate blood for cryogenic storage when his/her clinical state permits.

Reference

[1] Lowe, J.B., 1995. Biochemistry and biosynthesis of ABH and Lewis antigens: characterization of blood group-specific glycosyltransferases. In: Cartron, J.-P., Rouger, P. (Eds.), Molecular Basis of Human Blood Group Antigens. Plenum Press, New York, NY, pp. 75–115.

Kx Blood Group System

Number of antigens 1

High prevalence   Kx

Terminology

ISBT symbol (number)   XK (019)
History   Named in 1990 when the Kx antigen was assigned system status. XK was used as the ISBT symbol after the gene name.

Expression

Tissues   Fetal liver, adult skeletal muscle, brain, pancreas, heart, low levels in adult liver, kidney, spleen

Gene

Chromosome   Xp21.1
Name   *XK*
Organization   3 exons; sizes of introns have not been determined
Product   Xk protein

ATG   Stop   1 kpb

Database accession numbers

GenBank   NM_021083 (mRNA); Z32684 (gene)
Entrez Gene ID   7504

Molecular bases of silencing of *XK* (McLeod phenotype)

Homozygosity or compound heterozygosity leads to the Kx– phenotype (McLeod phenotype). Reference allele *XK*01* (Accession number

The Blood Group Antigen (3/e). DOI: http://dx.doi.oxg/10.1016/B978-0-12-415849-8.00021-1
        499

NM_021083) encodes XK1 (Kx). Nucleotide differences from reference allele, and amino acids affected, are given.

Allele name	Exon (intron)	Nucleotide change[†]	Amino acid change	Ethnicity (prevalence)
XK*01N.01	1–3	Deletion of gene	del exons 1–3	(Rare)
XK*01N.02	1	del exon 1	del exon 1	(Rare)
XK*01N.03	1	del promoter + exon 1	del exon 1	(Rare)
XK*01N.04	2	del exon 2	del 82–170	(Rare)
XK*01N.05	3	del intron 2 & exon 3	del 170–444	(Rare)
XK*01N.06	1	−272_119del	del 1–40fs	(Rare)
XK*01N.07	1	172delG	Val58fs+Leu129X	(Rare)
XK*01N.08	2	269delA	Tyr90fs	(Rare)
XK*01N.09	2	268delT	Tyr90fs	(Rare)
XK*01N.10	2	451insC	Pro150fs	(Rare)
XK*01N.11	3	686_687delTT	Phe229fs+Pro264X	(Rare)
XK*01N.12	3	771delG	Trp257fs+Ile267X	(Rare)
XK*01N.13	3	856_860delCTCTA	Leu286fs+Lys301X	(Rare)
XK*01N.14	3	938_951del	Asn313fs+Tyr336X	(Rare)
XK*01N.15	3	1013delT	Phe338fs+Ile408X	Japanese (Rare)
XK*01N.16	1	107G>A	Trp36X	(Rare)
XK*01N.17	2	397C>T	Arg133X	(Rare)
XK*01N.18	2	463C>T	Gln155X	(Rare)
XK*01N.19	3	707G>A	Trp236X	(Rare)
XK*01N.20	3	895C>T	Gln299X	(Rare)
XK*01N.21	3	941G>A	Trp314X	(Rare)
XK*01N.22	Intron 1	IVS1+1g>c	Alternative splicing	(Rare)
XK*01N.23	Intron 1	IVS1−1g>a	Alternative splicing	(Rare)
XK*01N.24	Intron 2	IVS2+1g>a	Alternative splicing	(Rare)
XK*01N.25	Intron 2	IVS2+5g>a	Alternative splicing	(Rare)
XK*01N.26	Intron 2	IVS2−1g>a	Alternative splicing	(Rare)
XK*01N.27	3	664C>G	Arg222Gly	(Rare)
XK*01N.28	3	880T>C	Cys294Arg	(Rare)
XK*01N.29	3	979G>A	Glu327Lys	(Rare)
XK*01N.30	2	452insC	151fs	Japanese (Rare)

[†]Nucleotide 1 is the first nucleotide of the translation-initiation codon, which is 82 bp downstream of the first nucleotide in earlier reports.

Amino acid sequence[1]

Note: the amino acid residues 204 and 205 listed here are the corrected data from those that are in GenBank accession number Z32684.

```
MKFPASVLAS  VFLFVAETTA  ALSLSSTYRS  GGDRMWQALT  LLFSLLPCAL   50
VQLTLLFVHR  DLSRDRPLVL  LLHLLQLGPL  FRCFEVFCIY  FQSGNNEEPY  100
VSITKKRQMP  KNGLSEEIEK  EVGQAEGKLI  THRSAFSRAS  VIQAFLGSAP  150
QLTLQLYISV  MQQDVTVGRS  LLMTISLLSI  VYGALRCNIL  AIKIKYDEYE  200
VKVKPLAYVC  IFLWRSFEIA  TRVVVLVLFT  SVLKTWVVVI  ILINFFSFFL  250
YPWILFWCSG  SPFPENIEKA  LSRVGTTIVL  CFLTLLYTGI  NMFCWSAVQL  300
KIDSPDLISK  SHNWYQLLVY  YMIRFIENAI  LLLLWYLFKT  DIYMYVCAPL  350
LVLQLLIGYC  TAILFMLVFY  QFFHPCKKLF  SSSVSEGFQR  WLRCFCWACR  400
QQKPCEPIGK  EDLQSSRDRD  ETPSSSKTSP  EPGQFLNAED  LCSA        444
```

Carrier molecule[1]

A multipass membrane protein.

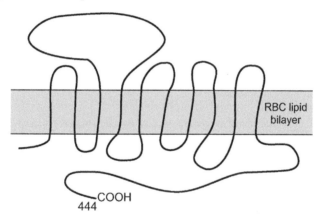

In the RBC membrane, XK protein is covalently linked at Cys72 to Cys347 of the Kell glycoprotein.

M_r (SDS-PAGE)	37,000
Glycosylation	None
Cysteine residues	16
Copies per RBC	1,000

Function

Not known, but XK has structural characteristics of a membrane transport protein, and a homolog, ced-8, is involved in regulating cell death in *C. elegans*[2]. Involved in maintenance of normal cell membrane integrity.

Disease association

Absence of XK protein is associated with acanthocytosis and the McLeod syndrome, which manifests a compensated hemolytic anemia, elevated serum creatinine kinase, and neuromuscular disorders including chorea, areflexia, skeletal muscle atrophy, and cardiomyopathy[3,4].

Some males with the McLeod phenotype have X-linked CGD.

Phenotypes

Null	McLeod (RBCs express Kell antigens weakly)
Unusual	Kx antigen has an increased expression on RBCs that lack or have a reduced expression of Kell antigens; see tables in Kell blood group system section

Comments

Very weak expression of Kx antigen (resembling a McLeod phenotype) together with extreme depression of Kell system antigens on the RBCs of a German proband were caused by the simultaneous presence of a single base change in the donor splice site of *XK*, and homozygosity for *KEL*02.03* (*Kp^a* at the *KEL* locus)[5].

References

[1] Ho, M., et al., 1994. Isolation of the gene for McLeod syndrome that encodes a novel membrane transport protein. Cell 77, 869–880.

[2] Stanfield, G.M., Horvitz, H.R., 2001. The ced-8 gene controls the timing of programmed cell deaths in *C. elegans*. Mol Cell 5, 423–433.

[3] Danek, A., et al., 2001. McLeod neuroacanthocytosis: genotype and phenotype. Ann Neurol 50, 755–764.

[4] Lee, S., et al., 2000. The Kell blood group system:Kell and XK membrane proteins. Semin Hematol 37, 113–121.

[5] Daniels, G.L., et al., 1996. A combination of the effects of rare genotypes at the *XK* and *KEL* blood group loci results in absence of Kell system antigens from the red blood cells. Blood 88, 4045–4050.

Kx Antigen

Terminology

ISBT symbol (number)	XK1 (019001 or 19.1)
Obsolete names	006015; K15
History	Named in 1975 when Kx was shown to be associated with the Kell blood group system, but controlled by a gene on the X chromosome.

Occurrence

All populations   100%

Expression

Cord RBCs   Expressed
Altered   Weak on RBCs of common Kell phenotype
                     Expression of Kx antigen is enhanced on RBCs with
                     reduced expression of Kell [K_0, K_{mod}, thiol-treated
                     RBCs, Kp(a+b–)], even though levels of XK protein
                     may be reduced[1]

Molecular basis associated with Kx antigen

For molecular basis associated with a lack of Kx antigen, see table in System
pages.

Effect of enzymes and chemicals on Kx antigen on intact RBCs

Ficin/Papain   Resistant
Trypsin   Resistant
α-Chymotrypsin   Resistant
DTT 200 mM   Resistant (enhanced)

In vitro characteristics of alloanti-Kx

Immunoglobulin class   IgG
Optimal technique   IAT

Clinical significance of alloanti-Kx

Transfusion reaction   Mild/delayed

Autoanti-Kx

One example reported in a man with common Kell phenotype.

Comments

Anti-Km is made by non-CGD McLeod males; both McLeod and K_0 blood
will be compatible.
Anti-Kx+anti-Km (sometimes called anti-KL) is made by males with the
McLeod phenotype and CGD; only McLeod blood will be compatible.
Anti-Kx has been made by one non-CGD McLeod male[2].
Anti-Kx can be prepared by adsorption of anti-Kx+anti-Km (anti-KL) onto,
and elution from, K_0 RBCs.

XK is subject to X chromosome inactivation, and female carriers of alleles that encode CGD have a mixed population of normal and acanthocytic RBCs. The range has been as much as 99%, and as few as 1% McLeod RBCs.

References

[1] Lee, S., et al., 2000. The Kell blood group system: Kell and XK membrane proteins. Semin Hematol 37, 113–121.

[2] Russo, D.C., et al., 2000. First example of anti-Kx in a person with McLeod phenotype and without chronic granulomatous disease. Transfusion 40, 1371–1375.

Gerbich Blood Group System

Number of antigens 11

Low prevalence	Wb, Ls^a, An^a, Dh^a, GEIS
High prevalence	Ge2, Ge3, Ge4, GEPL, GEAT, GETI

Terminology

ISBT symbol (number)	GE (020)
CD number	CD236
Obsolete name	ISBT Collection 201
History	Named in 1960 after one of three mothers who were found at the same time, and whose serum contained the antibody defining Gerbich; became a System in 1990.

Expression

Other blood cells	Erythroblasts
Tissues	Fetal liver, renal endothelium

Gene

Chromosome	2q14.3
Name	GE (GYPC)
Organization	4 exons distributed over 52.7 kbp of gDNA
Product	Glycophorin C (GPC) and glycophorin D (GPD)

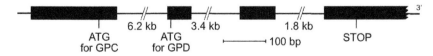

Database accession numbers

GenBank	M36284 (mRNA)
Entrez Gene ID	2995

The Blood Group Antigen (3/e). DOI: http://dx.doi.org/10.1016/B978-0-12-415849-8.00022-3

Gerbich

Molecular basis of Gerbich phenotypes

The reference allele, *GE*01* (Accession number M36284) encodes Ge2 (GE2), GE3, GE4, GE10, GE11, GE12. Nucleotide differences from this reference allele, and the amino acids affected, are given.

Allele encodes	Allele name	Exon	Nucleotide	Amino acid change	Ethnicity (prevalence)
Yus type or GE:-2,3,4	*GE*01.-02*	del Exon 2		in frame deletion→ altered GPC	Mexicans (Few) Others (Rare)
Gerbich type or GE:-2,-3,4	*GE*01.-03*	del Exon 3		in frame deletion→ altered GPC	Melanesians (up to 50%), Others (Rare)
Wb+ or GE:5	*GE*01.05*	1	23A>G	Asn8Ser in GPC	Welsh & Australians (Few), Others (Rare)
Ls(a+) or GE:6	*GE*01.06.01*	Duplicated Exon 3		in frame duplication→ altered GPC	Blacks (2%), Finns (1.6%), Others (Rare)
Ls(a+) or GE:6	*GE*01.06.02*	Triplicated Exon 3		in frame triplication→ altered GPC	Japanese (Rare)
An(a+) or GE:7	*GE*01.07*	2	67G>T	Ala23Ser in GPC Ala2Ser in GPD^	Finns (0.2%), Others (Rare)
Dh(a+) or GE:8	*GE*01.08*	1	40C>T	Leu14Phe in GPC	Scandinavians (Rare)
GEIS+ or GE:9	*GE*01.09*	2	95C>A	Thr32Asn in GPC; Thr11Asn in GPD	Japanese (Rare)
GEPL- or GE:-10	*GE*01.-10*	3	134C>T	Pro45Leu in GPC; Pro24Leu in GPD[1]	(Rare)
GEAT- or GE:-11	*GE*01.-11*	2	56A>T	Asp19Val in GPC[1]	(Rare)
GETI- or GE:-12	*GE*01.-12*	2	80C>T	Thr27Ile in GPC; Thr6Ile in GPD[1]	(Rare)
GE:2,3,4^^	*GE*01.-13*	3	173A>T	Asp58Val in GPC; Asp37Val in GPD[2]	(Rare)

^An^a^ is only expressed by GPD.
^^A woman with this allele *in trans* to an *GE*01.-03* made anti-Ge2[2].

Molecular bases of silencing of *GE*

Homozygosity and compound heterozygosity leads to the Gerbich$_{null}$ (GE:–2, –3,–4) phenotype.

Nucleotide differences from *GE*01* reference allele (Accession number M36284), and amino acids affected, are given.

Name	Allele Name	Exon	Nucleotide	Restriction Enzyme	Amino Acid	Ethnicity (Prevalence)
Leach type (PL)	*GE*01N.01*	del exons 3 & 4			Truncated protein	English (Rare)
Leach type (LN)	*GE*01N.02*	3	131G>T; 134delC	*Msp*I–	Trp44Leu; 45fs; 55Stop	North American (Rare)

Amino acid sequence[3]

Glycophorin C:

```
MWSTRSPNST  AWPLSLEPDP  GMASASTTMH  TTTIAEPDPG  MSGWPDGRME   50
TSTPTIMDIV  VIAGVIAAVA  IVLVSLLFVM  LRYMYRHKGT  YHTNEAKGTE  100
FAESADAALQ  GDPALQDAGD  SSRKEYFI                            128
```

Glycophorin D:

```
                        MASASTTMH  TTTIAEPDPG  MSGWPDGRME   29
TSTPTIMDIV  VIAGVIAAVA  IVLVSLLFVM  LRYMYRHKGT  YHTNEAKGTE   79
FAESADAALQ  GDPALQDAGD  SSRKEYFI                            107
```

The alignment of amino acids for GPD is intended to denote that the amino acid sequence for GPD is, from the second methionine residue, identical to GPC.

Carrier molecule

Single pass (type 1) membrane protein.

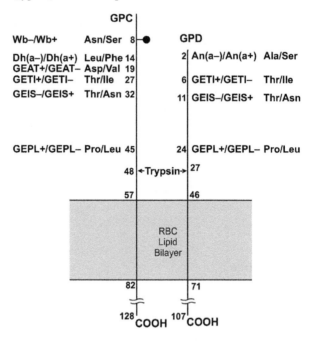

	GPC	GPD
M_r (SDS-PAGE)	40,000	30,000
CHO: N-glycan	1 site	None
CHO: O-glycan	13 sites	8 sites
Copies per RBC	135,000	50,000

Function

Maintenance of RBC membrane integrity via interaction with protein 4.1 and p55. Contributes to the negatively charged glycocalyx.

Disease association

GPC and GPD are markedly reduced in protein 4.1-deficient RBCs, and as such can be associated with hereditary elliptocytosis. RBC receptors for influenza A and influenza B and *Plasmodium falciparum*.

Phenotypes (% occurrence)

	Most populations	Melanesians
GE:2,3,4 (Ge+)	>99.9	50–90
Gerbich-negative		
GE:–2,3,4 (Yus type)	Rare	Not found
GE:–2,–3,4 (Gerbich type)	Rare	10–50
GE:–2,–3,–4 (Leach type)	Rare	Not found
Null:	Leach (PL and LN types) (GE:–2,–3,–4)	

Differentiation of Gerbich-negative phenotypes using monoclonal anti-Ge4

RBCs	Normal	Yus	Gerbich	Leach
Untreated	4+	0–2+	0–2+	0
Trypsin-treated	0	0	4+	0

Comments

The majority of RBC samples with Leach or Gerbich phenotypes have a weak expression of Kell blood group system antigens. Some anti-Vel fail to react with Ge:–2,–3,4 RBCs.

Gerbich antigens are weak on protein 4.1-deficient RBCs, due to reduced levels of GPC and GPD in these membranes.

References

[1] Poole, J., et al., 2008. Novel mutations in GYPC giving rise to lack of Ge epitopes and anti-Ge production [abstract]. Vox Sang 95 (Suppl. 1), 181.

[2] King, M-J, et al., 1997. Co-presence of a point mutation and a deletion of exon 3 in the glycophorin C gene and concomitant production of a Gerbich-related antibody. Transfusion 37, 1027–1034.

[3] Colin, Y., et al., 1986. Isolation of cDNA clones and complete amino acid sequence of human erythrocyte glycophorin C. J Biol Chem 261, 229–233.

Gerbich

Ge2 Antigen

Terminology

ISBT symbol (number)   GE2 (020002 or 20.2)
Obsolete name   Ge; 201002
History   Antigen lacking from all Gerbich-negative
                        phenotypes. Originally defined by the "Yus-
                        type" antibody found in 1961; later referred to as
                        anti-Ge1,2, and now as anti-Ge2.

Occurrence

All populations   Greater than 99.9%

Expression

Cord RBCs   Expressed
Altered   Weak on protein 4.1-deficient RBCs
                        Absent from Yus, Gerbich and Leach phenotype
                        RBCs

Molecular basis associated with Ge2 antigen[1]

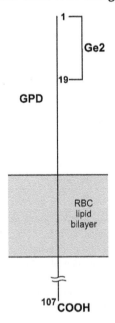

Ge2 as determined with alloanti-Ge2 is not expressed on GPC.

Effect of enzymes and chemicals on Ge2 antigen on intact RBCs

Ficin/Papain	Sensitive
Trypsin	Sensitive
α-Chymotrypsin	Weakened
Sialidase	Variable
DTT 200 mM	Variable (thus variable to WARM™ and ZZAP)
Acid	Resistant

In vitro characteristics of alloanti-Ge2

Immunoglobulin class	Usually IgG
Optimal technique	IAT
Complement binding	Yes; some hemolytic

Clinical significance of alloanti-Ge2

Transfusion reaction	No to moderate/immediate/delayed
HDFN	Positive DAT, but no clinical HDFN

Autoanti-Ge2

Yes; detects a determinant on GPC.

Comments

Alloanti-Ge2 can be made by individuals with Yus, Gerbich or Leach phenotypes, detects an antigen on GPD, and may be naturally-occurring.

The reciprocal gene to *GYPC. Yus* encodes two copies of amino acids encoded by exon 2.

Siblings of patients with anti-Ge2 should be tested for compatibility, and the patient urged to donate blood for cryogenic storage when his/her clinical state permits.

Reference

[1] Walker, P.S., Reid, M.E., 2010. The Gerbich blood group system: a review. Immunohematology 26, 60–65.

Ge3 Antigen

Terminology

ISBT symbol (number)	GE3 (020003 or 20.3)
Obsolete name	Ge; 201003
History	Antigen originally defined by the "Ge-type" serum (identified in 1960). The defining antibody was termed anti-Ge1,2,3, and later renamed to anti-Ge3.

Occurrence

Most populations	>99.9%
Melanesians	50%

Expression

Cord RBCs	Expressed
Altered	Weak on protein 4.1-deficient RBCs
	Absent from Gerbich and Leach phenotype RBCs

Molecular basis associated with Ge3 antigen[1]

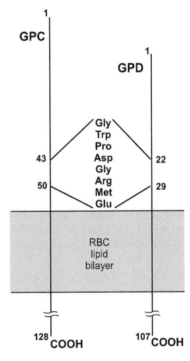

The Ge3 antigen amino acid sequence is encoded by exon 3 of *GYPC*.

Effect of enzymes and chemicals on Ge3 antigen on intact RBCs

Ficin/Papain	Resistant
Trypsin	Variable
α-Chymotrypsin	Resistant
Pronase	Sensitive
Sialidase	Resistant
DTT 200 mM	Resistant
Acid	Resistant

Gerbich

In vitro characteristics of alloanti-Ge3

Immunoglobulin class   IgG more common than IgM
Optimal technique   IAT
Complement binding   Yes; some hemolytic

Clinical significance of alloanti-Ge3

Transfusion reaction   No to moderate, immediate or delayed
HDFN   Positive DAT to severe[2]
                       (see Comments)

Autoanti-Ge3

Yes; can be clinically insignificant or cause severe *in vivo* hemolysis.

Comments

Similar to the mechanism of erythroid suppression described in HDFN caused by anti-K, anti-Ge3 has been associated with antibody-dependent hemolysis, as well as suppression of erythroid progenitor cell growth in the infant. In these cases, the affected infants may require initial treatment at delivery, followed by monitoring for signs of anemia for several weeks after birth.

Alloanti-Ge3 and autoanti-Ge3 detect the antigen on both GPC and GPD[3,4].

Alloanti-Ge3 can be made by individuals with either Gerbich or Leach phenotypes.

Siblings of patients with anti-Ge3 should be tested for compatibility, and the patient urged to donate blood for cryogenic storage when his/her clinical state permits.

References

[1] Walker, P.S., Reid, M.E., 2010. The Gerbich blood group system: a review. Immunohematology 26, 60–65.

[2] Arndt, P., et al., 2002. First example of anti-Ge associated with severe hemolytic disease of the newborn [abstract]. Transfusion 42 (Suppl.), 19S.

[3] Blackall, D.P., et al., 2008. Hemolytic disease of the fetus and newborn due to anti-Ge3: combined antibody-dependent hemolysis and erythroid precursor cell growth inhibition. Am J Perinatol 25, 541–545.

[4] Denomme, G.A., et al., 2006. Inhibition of erythroid progenitor cell growth by anti-Ge3. Br J Haematol 133, 443–444.

Ge4 Antigen

Terminology

ISBT symbol (number)   GE4 (020004 or 20.4)
Obsolete name   201004

History   Ge4 was given the next number when an antibody
                           was found that agglutinated Gerbich-positive and
                           Gerbich-negative (both Yus and Gerbich type) RBCs,
                           but not RBCs with the Leach (Ge$_{null}$) phenotype.

Occurrence

All populations   100%

Expression

Cord RBCs   Expressed
Altered   Weak on protein 4.1-deficient RBCs
                           Absent from Leach phenotype RBCs

Molecular basis associated with Ge4 antigen[1]

Effect of enzymes and chemicals on Ge4 antigen on intact RBCs

Ficin/Papain   Sensitive
Trypsin   Sensitive
α-Chymotrypsin   Resistant
Sialidase   Sensitive
DTT 200 mM   Resistant
Acid   Resistant

In vitro characteristics of alloanti-Ge4

Immunoglobulin class	IgG
Optimal technique	IAT

Clinical significance of alloanti-Ge4

No information because only one alloanti-Ge4 has been described.

Comments

Ge4 is expressed on the N-terminal domain of normal and all variants of GPC (GPC.Yus, GPC.Gerbich, GPC.Wb, GPC.Ls^a, etc.).

Siblings of patients with anti-Ge4 should be tested for compatibility, and the patient urged to donate blood for cryogenic storage when his/her clinical state permits.

Reference

[1] Walker, P.S., Reid, M.E., 2010. The Gerbich blood group system: a review. Immunohematology 26, 60–65.

Wb Antigen

Terminology

ISBT symbol (number)	GE5 (020005 or 20.5)
Obsolete names	Webb; 201005 or 700009
History	Found in 1963 and named after the donor whose group O RBCs were agglutinated by a high-titer ABO typing serum. Shown to be on a variant form of GPC in 1986.

Occurrence

Most populations	<0.01%
Wales and Australia	<0.1%

Expression

Cord RBCs	Presumed expressed

Molecular basis associated with Wb antigen[1]

Amino acid	Ser8 of GPC. This substitution results in a loss of the N-glycan, and possibly a gain of an O-glycan[2]. Thus, GPC.Wb has an M_r of approximately 2,700 less than GPC.
Nucleotide	G at bp 23 in exon 1
Wb– (wild type)	Asn8 and A at bp 23

Gerbich

Effect of enzymes and chemicals on Wb antigen on intact RBCs

Ficin/Papain	Sensitive
Trypsin	Sensitive
α-Chymotrypsin	Resistant
Sialidase	Sensitive
DTT 200 mM	Resistant
Acid	Resistant

In vitro characteristics of alloanti-Wb

Immunoglobulin class	IgM and IgG
Optimal technique	RT; IAT

Clinical significance of alloanti-Wb

Transfusion reaction	No
HDFN	No

Comments

Anti-Wb are usually naturally-occurring[3].

References

[1] Walker, P.S., Reid, M.E., 2010. The Gerbich blood group system: a review. Immunohematology 26, 60–65.

[2] Reid, M.E., et al., 1987. Structural relationships between human erythrocyte sialoglycoproteins beta and gamma and abnormal sialoglycoproteins found in certain rare human erythrocyte variants lacking the Gerbich blood-group antigen(s). Biochem J 244, 123–128.

[3] Bloomfield L., et al., The Webb (Wb) antigen in South Wales donors. Hum. Hered, 36, 352–356.

Ls^a Antigen

Terminology

ISBT symbol (number)	GE6 (020006 or 20.6)
Obsolete names	Lewis II; RI^a (Rosenlund); 700007; 700024; 201006
History	Anti-Ls^a identified in an anti-B typing serum in 1963. Originally called Lewis II after the antigen-positive donor, but later renamed Ls^a to avoid confusion with the established Lewis antigens. Associated with Gerbich in 1990.

Occurrence

Most populations	<0.01%
Blacks	2%
Finns	1.6%

Expression

Cord RBCs   Presumed expressed
Altered   Increased on RBCs with three copies of amino acids
                     encoded by exon 3

Molecular basis associated with Ls^a antigen[1,2]

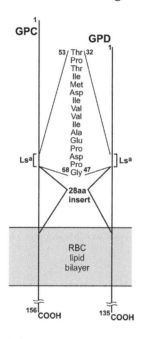

Ls^a antigen is located within an amino acid sequence encoded by nucleotides
at the junction of the replicated exon 3 to exon 3.

Effect of enzymes and chemicals on Ls^a antigen on intact RBCs

Ficin/Papain   Sensitive
Trypsin   Sensitive
α-Chymotrypsin   Resistant
Sialidase   Resistant
DTT 200 mM   Resistant

In vitro characteristics of alloanti-Ls^a

Immunoglobulin class   IgM and IgG
Optimal technique   RT; IAT

Gerbich

Clinical significance of alloanti-Ls[a]

Transfusion reaction	No data because antibody and antigen are rare
HDFN	No

Comments

Anti-Ls[a] is naturally-occurring.

References

[1] Reid, M.E., et al., 1994. Duplication of exon 3 in the glycoprotein C gene gives rise to the Ls[a] blood group antigen. Transfusion 34, 966–969.

[2] Walker, P.S., Reid, M.E., 2010. The Gerbich blood group system: a review. Immunohematology 26, 60–65.

An[a] Antigen

Terminology

ISBT symbol (number)	GE7 (020007 or 20.7)
Obsolete names	Ahonen; 700020
History	Identified in 1972 and named after the donor (Ahonen), whose RBCs were agglutinated by a patient's serum. Joined Gerbich in 1990 when the antigen was located on a variant of GPD.

Occurrence

Most populations	0.01%
Finns	0.2%

Expression

Cord RBCs	Presumed expressed

Molecular basis associated with An[a] antigen[1,2]

Amino acid	Ser2 of GPD. The altered GPC (Ser23) does not express An[a]
Nucleotide	T at bp 67 in exon 2 of *GYPC*
An(a–) (wild type)	GPC has Ala23 and GPD has Ala2, and G at bp 67

Effect of enzymes and chemicals on An[a] antigen on intact RBCs

Ficin/Papain	Sensitive
Trypsin	Sensitive
α-Chymotrypsin	Weakened
Sialidase	Sensitive
DTT 200 mM	Resistant

In vitro characteristics of alloanti-An[a]

Immunoglobulin class   IgM and IgG
Optimal technique   RT; IAT

Clinical significance of alloanti-An[a]

Transfusion reaction   No data because antibody and antigen are rare
HDFN   No

Comments

Anti-An[a] may be naturally-occurring.

References

[1] Daniels, G., et al., 1993. A point mutation in the *GYPC* gene results in the expression of the blood group An[a] antigen on glycophorin D but not on glycophorin C: further evidence that glycophorin D is a product of the *GYPC* gene. Blood 82, 3198–3203.

[2] Walker, P.S., Reid, M.E., 2010. The Gerbich blood group system: a review. Immunohematology 26, 60–65.

Dh[a] Antigen

Terminology

ISBT symbol (number)   GE8 (020008 or 20.8)
Obsolete names   Duch; 700031
History   Identified in 1968 during pretransfusion testing, and named after the antigen-positive Danish blood donor. Joined Gerbich in 1990 when Dh[a] was located on a variant of GPC.

Occurrence

All populations   <0.01%

Expression

Cord RBCs   Presumed expressed

Molecular basis associated with Dh[a] antigen[1,2]

Amino acid   Phe14 of GPC
Nucleotide   T at bp 40 in exon 1
Dh(a–) (wild type)   Leu14 and C at bp 40

Gerbich

Effect of enzymes and chemicals on Dh^a antigen on intact RBCs

Ficin/Papain	Sensitive
Trypsin	Sensitive
α-Chymotrypsin	Resistant
Sialidase	Sensitive
DTT 200 mM	Resistant

In vitro characteristics of alloanti-Dh^a

Immunoglobulin class	IgM and IgG
Optimal technique	RT and IAT

Clinical significance of alloanti-Dh^a

Transfusion reaction	No data because antibody and antigen are rare
HDFN	No

Comments

Anti-Dh^a may be naturally-occurring.

References

[1] King, M.J., et al., 1992. Point mutation in the glycophorin C gene results in the expression of the blood group antigen Dh^a. Vox Sang 63, 56–58.

[2] Walker, P.S., Reid, M.E., 2010. The Gerbich blood group system: a review. Immunohematology 26, 60–65.

GEIS Antigen

Terminology

ISBT symbol (number)	GE9 (020009 or 20.9)
History	Found and named in 2004, "GE" from Gerbich and "IS" from the name of the index case.

Occurrence

Japanese	Only three GEIS+ probands have been reported

Expression

Cord RBCs	Presumed expressed

Molecular basis associated with GEIS antigen[1]

Amino acid	Asn32 in GPC and Asn11 in GPD
Nucleotide	A at bp 95 in exon 2
GEIS− (wild type)	Thr32 in GPC and Thr11 in GPD, and C at bp 95

Gerbich

Effect of enzymes and chemicals on GEIS antigen on intact RBCs

Ficin/Papain	Sensitive
Trypsin	Resistant
α-Chymotrypsin	Sensitive
Sialidase	Sensitive
DTT 200 mM	Presumed resistant

In vitro characteristics of alloanti-GEIS

Immunoglobulin class	IgG
Optimal technique	IAT

Clinical significance of alloanti-GEIS

No data because antibody and antigen are rare

Reference

[1] Yabe, R., et al., 2004. Is a new Gerbich blood group antigen located on the GPC and GPD [abstract]? Vox Sang 87 (Suppl. 3), 79.

GEPL Antigen

Terminology

ISBT symbol (number)	GE10 (020010 or 20.10)
History	Reported in 2008 and named in 2010, "GE" from Gerbich, "P" from proline and "L" from leucine.

Occurrence

Only one GEPL– proband has been described.

Expression

Cord RBCs	Presumed expressed

Molecular basis associated with GEPL antigen[1]

Amino acid	Pro45 in GPC and Pro24 in GPD
Nucleotide	C at bp 134 in exon 3
GEPL–	Leu45 in GPC and Leu24 in GPD, and T at bp 134

Effect of enzymes and chemicals on GEPL antigen on intact RBCs

Ficin/Papain	Sensitive
Trypsin	Presumed sensitive
α-Chymotrypsin	Presumed resistant
DTT 200 mM	Presumed resistant

Gerbich

In vitro characteristics of alloanti-GEPL

Immunoglobulin class   IgG
Optimal technique   IAT

Clinical significance of alloanti-GEPL

No data because antibody is rare.

Comments

The plasma of the GEPL– proband appeared to contain anti-Ge3, whereas the patient's RBCs were GE:2,3,4 with aberrant expression of Ge3.
Siblings of patients with anti-GEPL should be tested for compatibility, and the patient urged to donate blood for cryogenic storage when his/her clinical state permits.

Reference

[1] Poole, J., et al., 2008. Novel mutations in GYPC giving rise to lack of Ge epitopes and anti-Ge production [abstract]. Vox Sang 95 (Suppl. 1), 181.

GEAT Antigen

Terminology

ISBT symbol (number)   GE11 (020011 or 20.11)
History   Reported in 2008 and named in 2010, "GE" from
                        Gerbich, and "A" and "T" from the nucleotides involved.

Occurrence

Only one GEAT– proband has been reported.

Expression

Cord RBCs   Presumed expressed

Molecular basis associated with GETI antigen[1]

Amino acid	Asp19 in GPC
Nucleotide	A at bp 56 in exon 2
GEAT–	Val19 in GPC and T at bp 36

Effect of enzymes and chemicals on GEAT antigen on intact RBCs

Ficin/Papain	Sensitive
Trypsin	Sensitive
α-Chymotrypsin	Resistant
DTT 200 mM	Resistant
Acid	Resistant

In vitro characteristics of alloanti-GEAT

Immunoglobulin class	IgG
Optimal technique	IAT

Clinical significance of alloanti-GEAT

No data because antibody is rare.

Comments

Plasma from the GEAT– proband did not react with GE:–2,–3,4 or GE:–2,–3,–4 RBC samples, and gave variable/weak reactions with GE:–2,3,4 RBC samples. The GEAT– proband had the GE:2,3,4 phenotype, but some antisera reacted weakly.

Siblings of patients with anti-GEAT should be tested for compatibility, and the patient urged to donate blood for cryogenic storage when his/her clinical state permits.

Reference

[1] Poole, J., et al., 2008. Novel mutations in GYPC giving rise to lack of Ge epitopes and anti-Ge production [abstract]. Vox Sang 95 (Suppl. 1), 181.

GETI Antigen

Terminology

ISBT symbol (number)	GE12 (020012 or 20.12)
History	Reported in 2008 and named in 2010, "GE" from Gerbich, "T" from threonine, and "I" from isoleucine.

Occurrence

Only one GETI– proband has been reported.

Expression

Cord RBCs   Presumed expressed

Molecular basis associated with GETI antigen[1]

Amino acid   Thr27 in GPC and Thr6 in GPD
Nucleotide   C at bp 80 in exon 2
GETI–   Ile27 in GPC and Ile6 in GPD, and T at bp 80

Effect of enzymes and chemicals on GETI antigen on intact RBCs

Ficin/Papain   Sensitive
Trypsin   Presumed sensitive
α-Chymotrypsin   Presumed resistant
DTT 200 mM   Presumed resistant

In vitro characteristics of alloanti-GETI

Immunoglobulin class   IgG
Optimal technique   IAT

Clinical significance of alloanti-GETI

No data because antibody is rare.

Comments

Initially the antibody made by a GETI– patient, by her GETI– brother, and by another patient appeared to be anti-Ge2. The red cells from these people were GE:–2,3,4 (except that one autoanti-Ge2 reacted), and there was marginal weakening of Ge3 and Ge4.

Siblings of patients with anti-GETI should be tested for compatibility, and the patient urged to donate blood for cryogenic storage when his/her clinical state permits.

Reference

[1] Poole, J., et al., 2008. Novel mutations in GYPC giving rise to lack of Ge epitopes and anti-Ge production [abstract]. Vox Sang 95 (Suppl. 1), 181.

Cromer Blood Group System

Number of antigens 16

Low prevalence   Tc^b, Tc^c, WES^a
High prevalence   Cr^a, Tc^a, Dr^a, Es^a, IFC, WES^b, UMC, GUTI, SERF,
        ZENA, CROV, CRAM, CROZ

Terminology

ISBT symbol (number) CROM (021)
CD Number   CD55
Obsolete name   Collection 202
History   Named after the first antigen in this system, Cr^a.

Expression

Soluble form   Low levels in plasma, serum and urine
Other blood cells Leukocytes; platelets
Tissues   Apical surfaces of trophoblasts in placenta

Gene

Chromosome   1q32.2
Name   *CROM (DAF)*
Organization   11 exons distributed over 40 kbp of gDNA
Product   Decay accelerating factor (DAF; CD55)

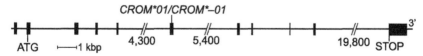

CROM*01/CROM*–01

ATG ⊢—1 kbp 4,300 5,400 19,800 STOP 3'

Database accession numbers

GenBank   NM_000573, M31516
Entrez Gene ID 1604

The Blood Group Antigen (3/e). DOI: http://dx.doi.org/10.1016/B978-0-12-415849-8.00023-5

Molecular basis of Cromer phenotypes

The reference allele is *CROM*01* or *CROM*A* (Accession number M31516); encodes Cr^a (CROM1), CROM2, CROM5, CROM6, CROM7 (IFC), CROM9, CROM10, CROM11, CROM12, CROM13, CROM14, CROM15, CROM16. Differences from this reference allele, and the amino acids affected, are given.

Allele encodes	Allele name	Exon	Nucleotide	Restriction enzyme	Amino acid^	Ethnicity (prevalence)
Cr(a–) or CROM:–1	*CROM*–01*	6	679G>C		Ala227Pro	Blacks (Many)
Tc(a–b+) or CROM:–2,3	*CROM* 01.03*	2	155G>T	RsaI–; StuI+	Arg52Leu	Blacks (Rare)
Tc(a–c+) or CROM:–2,4	*CROM* 01.04*	2	155G>C	RsaI–	Arg52Pro	Caucasians (Rare)
Dr(a–) or CROM:–5	*CROM* 01.–05*	5	596C>T†	TaqI–	Ser199Leu	Bukhara Jews (Several), Japanese (Rare)
Es(a–) or CROM:–6	*CROM* 01.–06*	2	239T>A	Sau3AI–	Ile80Asn	Mexicans, South Americans, Blacks (Rare)
WES(a+b–) or CROM:8	*CROM* 01.08*	2	245T>G	AflIII–	Leu 82Arg	Blacks, Finns (Rare)
UMC– or CROM:–10	*CROM* 01.–10*	6	749C>T		Thr250Met	Japanese (Rare)
GUTI– or CROM:–11	*CROM* 01.–11*	6	719G>A	MaeII–	Arg240His	Chileans (Rare)
SERF– or CROM:–12	*CROM* 01.–12*	5	647C>T	BstNI+	Pro216Leu	Thais (Rare)
ZENA– or CROM:–13	*CROM* 01.–13*	6	726T>G	BsrI+	His242Gln	Syrian Turks (Rare)
CROV– or CROM:–14	*CROM* 01.–14*	3	466G>A	TaqI–	Glu156Lys	Croatians (Rare)
CRAM– or CROM:–15	*CROM* 01.–15*	6	740A>G		Gln247Arg	Somali (Rare)
CROZ– or CROM:–16	*CROM* 01.–16*	3	389G>A		Arg130His	Australian (Rare)

^Change from historical counting of #1 as Asp of the mature (membrane-bound protein); thus, all amino acid numbers have increased by 34.

†This transition results in two cDNA transcripts, one encoding full length DAF with the single amino acid change. The other, more abundant, transcript uses the novel branch point, which leads to use of a downstream cryptic acceptor splice site, a 44 bp deletion, and a frame-shift in exon 5 (proband KZ)[1].

Molecular bases of silencing of *CROM*

Homozygosity or compound heterozygosity leads to Cromer$_{null}$ (CR:–7; Inab) phenotype.

Differences from *CROM*01* reference allele (Accession number M31516) are given.

Allele name	Exon	Nucleotide	Restriction enzyme	Amino acid[^]	Ethnicity (prevalence)
*CROM*01N.01*	2	261G>A	*Bc* II+	Trp87Stop	Japanese (Rare)
*CROM*01N.02*	2	263C>A	*Mbo* II–	Ser88Stop	(Rare)
*CROM*01N.03*	4	508C>T		Arg170Stop[2]	Japanese (Rare)
*CROM*01N.04*	3	367insA		Thr123fs; Glu128Stop[3]	Moroccan (Rare)

[^]Change from historical counting of #1 as Asp of the mature (membrane-bound protein); thus, all amino acid numbers have increased by 34.

Amino acid sequence[4]

```
MTVARPSVPA  ALPLLGELPR  LLLLVLLCLP  AVWGDCGLPP  DVPNAQPALE   50
GRTSFPEDTV  ITYKCEESFV  KIPGEKDSVI  CLKGSQWSDI  EEFCNRSCEV  100
PTRLNSASLK  QPYITQNYFP  VGTVVEYECR  PGYRREPSLS  PKLTCLQNLK  150
WSTAVEFCKK  KSCPNPGEIR  NGQIDVPGGI  LFGATISFSC  NTGYKLFGST  200
SSFCLISGSS  VQWSDPLPEC  REIYCPAPPQ  IDNGIIQGER  DHYGYRQSVT  250
YACNKGFTMI  GEHSIYCTVN  NDEGEWSGPP  PECRGKSLTS  KVPPTVQKPT  300
TVNVPTTEVS  PTSQKTTTKT  TTPNAQATRS  TPVSRTTKHF  HETTPNKGSG  350
TTSGTTRLLS  GHTCFTLTGL  LGTLVTMGLL  T                       381
```

A signal peptide of 34 amino acids is cleaved from the membrane-bound protein.

The 28 carboxyl terminal amino acids are cleaved prior to attachment of DAF to its GPI-linkage.

Carrier molecule

A GPI-linked glycoprotein.

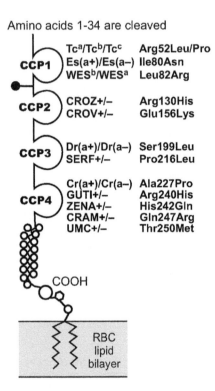

M_r (SDS-PAGE)	Reduced:	64,000–73,000
	Non-reduced:	60,000–70,000
CHO: N-glycan	1 site	
CHO: O-glycan	15 sites (32 potential)	
Cysteine residues	14	
Copies per RBC	20,000	

Function

Complement regulation: DAF inhibits assembly and accelerates decay of C3 and C5 convertases.

Disease association

Five of six known individuals with the Inab phenotype have intestinal disorders. PNH III RBCs are deficient in DAF.

Phenotypes

Null	Inab (IFC–)
Unusual	Dr(a–) RBCs weakly express inherited Cromer antigens

Comments

Antibodies in the Cromer blood group system do not cause HDFN. DAF is strongly expressed on the apical surface of placental trophoplasts[5], and will absorb antibodies in the Cromer system.

Antibodies to Cromer antigens identified early in pregnancy are often below detectable levels in late stages of pregnancy, but reappear some weeks after the birth of the baby.

References

[1] Lublin, D.M., et al., 1994. Molecular basis of reduced or absent expression of decay-accelerating factor in Cromer blood group phenotypes. Blood 84, 1276–1282.

[2] Hue-Roye, K., et al., 2005. Novel molecular basis of an Inab phenotype. Immunohematology 21, 53–55.

[3] Karamatic Crew, V., et al., 2010. Two unusual cases within the Cromer blood group system: (I) A novel high incidence antigen CROZ; and (II) A novel molecular basis of Inab phenotype. Transfus Med 20 (Suppl. 1): 12.

[4] Lublin, D.M., Atkinson, J.P., 1989. Decay-accelerating factor: biochemistry, molecular biology and function. Ann Rev Immunol 7, 35–58.

[5] Holmes, C.H., et al., 1990. Preferential expression of the complement regulatory protein decay accelerating factor at the fetomaternal interface during human pregnancy. J Immunol 144, 3099–3105.

Cr^a Antigen

Terminology

ISBT symbol (number)	CROM1 (021001 or 21.1)
Obsolete names	Go^b; 202001; 900013
History	Named in 1975 after Mrs. Cromer, a black antenatal patient who made the antibody. Cr^a was originally thought to be antithetical to Go^a.

Occurrence

Most populations	100%
Blacks	Greater than 99%
Hispanics	One Cr(a–) found

Expression

Cord RBCs	Expressed
Altered	Weak on Dr(a–) and negative on PNH III RBCs

Molecular basis associated with Cr^a antigen[1]

Amino acid	Ala227 (previously reported as 193) in CCP4
Nucleotide	G at bp 679 in exon 6
Cr(a–)	Pro227 and C at bp 679

Effect of enzymes and chemicals on Cr^a antigen on intact RBCs

Ficin/Papain	Resistant
Trypsin	Resistant
α-Chymotrypsin	Sensitive
Pronase	Sensitive
DTT 200 mM/50 mM	Weakened/resistant (thus weakened by WARM™ and ZZAP)
Acid	Resistant

In vitro characteristics of alloanti-Cr^a

Immunoglobulin class	IgG
Optimal technique	IAT
Neutralization	With concentrated plasma/serum/urine

Clinical significance of alloanti-Cr^a

Transfusion reaction	No to moderate
HDFN	No, because DAF on apical surface of trophoblasts in placenta absorbs maternal antibody

Comments

Siblings of patients with anti-Cr^a should be tested for compatibility, and the patient urged to donate blood for cryogenic storage when his/her clinical state permits.

Reference
[1] Lublin, D.M., et al., 2000. Molecular basis of Cromer blood group antigens. Transfusion 40, 208–213.

Tc^a Antigen

Terminology

ISBT symbol (number)	CROM2 (021002 or 21.2)
Obsolete names	202002; 900020
History	Named in 1980, and placed in the Cromer system when the antibody was shown to be non-reactive with Inab RBCs. The initials of the first two probands to have the antibody were GT and DLC, hence Tc^a.

Occurrence

Most populations	100%
Blacks	>99%

Antithetical antigen

Tc^b (**CROM3**); Tc^c (**CROM4**)

Expression

Cord RBCs	Expressed
Altered	Weak on Dr(a–) and negative on PNH III RBCs

Molecular basis associated with Tc^a antigen[1]

Amino acid	Arg52 (previously reported as 18) in CCP1
Nucleotide	G at bp 155 in exon 2

Effect of enzymes and chemicals on Tc^a antigen on intact RBCs

Ficin/Papain	Resistant
Trypsin	Resistant
α-Chymotrypsin	Sensitive
Pronase	Sensitive
DTT 200 mM/50 mM	Weakened/resistant (thus weakened by WARM™ and ZZAP)
Acid	Resistant

In vitro characteristics of alloanti-Tc^a

Immunoglobulin class	IgG
Optimal technique	IAT
Neutralization	With concentrated serum/plasma/urine

Clinical significance of alloanti-Tc^a

Transfusion reaction	No to severe[2]
HDFN	No, because DAF on apical surface of trophoblasts in placenta absorbs maternal antibody

Comments

Siblings of patients with anti-Tc^a should be tested for compatibility, and the patient urged to donate blood for cryogenic storage when his/her clinical state permits.

Only three examples of anti-Tc^a have been reported but a few others have been found. All Tc(a–) Blacks are Tc(b+); Tc(a–) Caucasians are Tc(c+).

References

[1] Lublin, D.M., et al., 2000. Molecular basis of Cromer blood group antigens. Transfusion 40, 208–213.

[2] Kowalski, M.A., et al., 1999. Hemolytic transfusion reaction due to anti-Tc(a). Transfusion 39, 948–950.

Tc^b Antigen

Terminology

ISBT symbol (number)	CROM3 (021003 or 21.3)
Obsolete names	202003; 700035
History	Original antibody found in a serum containing anti-Go^a; named in 1985 when it was recognized to be antithetical to Tc^a.

Occurrence

Caucasians	None found
Blacks	6%

Antithetical antigen

Tc^a (**CROM2**); Tc^c (**CROM4**)

Expression

Cord RBCs	Expressed

Molecular basis associated with Tc^b antigen[1]

Amino acid	Leu52 (previously reported as 18) in CCP1
Nucleotide	T at bp 155 in exon 2

Effect of enzymes and chemicals on Tc^b antigen on intact RBCs

Ficin/Papain	Resistant
Trypsin	Resistant
α-Chymotrypsin	Sensitive
Pronase	Sensitive
DTT 200 mM/50 mM	Weakened/resistant (thus weakened by WARM™ and ZZAP)
Acid	Resistant

In vitro characteristics of alloanti-Tc^b

Immunoglobulin class	IgG
Optimal technique	IAT

Clinical significance of alloanti-Tc^b

No data because antigen and antibody are rare.

Reference

[1] Lublin, D.M., et al., 2000. Molecular basis of Cromer blood group antigens. Transfusion 40, 208–213.

Tc^c Antigen

Terminology

ISBT symbol (number)	CROM4 (021004 or 21.4)
Obsolete names	202004; 700036
History	Described in 1982, and named when it was recognized to be antithetical to Tc^a.

Occurrence

Less than 0.01%; two Tc(a–b–c+) have only been found in two Caucasian families.

Antithetical antigen

Tc^a (**CROM2**); Tc^b (**CROM3**)

Expression

Cord RBCs	Expressed

Molecular basis associated with Tc^c antigen[1]

Amino acid	Pro52 (previously reported as 18) in CCP1
Nucleotide	C at bp 155 in exon 2

Effect of enzymes and chemicals on Tc^c antigen on intact RBCs

Ficin/Papain	Presumed resistant
Trypsin	Presumed resistant
α-Chymotrypsin	Presumed sensitive
DTT 200 mM/50 mM	Presumed weakened/resistant (thus weakened by WARM™ and ZZAP)

In vitro characteristics of alloanti-Tc^c

Immunoglobulin class	IgG
Optimal technique	IAT

Cromer

Clinical significance of alloanti-Tc^c

Transfusion reaction	No to mild
HDFN	No, because DAF on apical surface of trophoblasts in placenta absorbs maternal antibody

Comments

A female with the rare Tc(a–b–c+) phenotype made an antibody that appears to be an inseparable anti-Tc^{a}Tc^b.

Reference

[1] Lublin, D.M., et al., 2000. Molecular basis of Cromer blood group antigens. Transfusion 40, 208–213.

Dr^a Antigen

Terminology

ISBT symbol (number)	CROM5 (021005 or 21.5)
Obsolete names	202005; 900021
History	Reported in 1984, and named after the Israeli Dr(a–) proband, Mrs. Drori.

Occurrence

Dr(a–) phenotype has been reported only in Jews from Bukhara and in Japanese.

Expression

Cord RBCs	Expressed
Altered	Absent from PNH III RBCs

Molecular basis associated with Dr^a antigen[1]

Amino acid	Ser199 (previously reported as 165) in CCP3
Nucleotide	C at bp 596 in exon 5
Dr(a–)	Leu199; the T at bp 596 introduces a branch point that leads to use of a downstream cryptic acceptor splice site, deletion of 44 bp, and a frame-shift

Effect of enzymes and chemicals on Dr^a antigen on intact RBCs

Ficin/Papain	Resistant
Trypsin	Resistant
α-Chymotrypsin	Sensitive
Pronase	Sensitive
DTT 200 mM/50 mM	Weakened/resistant (thus weakened by WARM™ and ZZAP)

In vitro characteristics of alloanti-Dr^a

Immunoglobulin class	IgG
Optimal technique	IAT
Neutralization	With concentrated serum/plasma/urine

Clinical significance of alloanti-Dr^a

Transfusion reaction	No to mild
HDFN	No, because DAF on apical surface of trophoblasts in placenta absorbs maternal antibody

Comments

All inherited Cromer antigens are expressed weakly on Dr(a–) RBCs, due to a markedly reduced copy number of DAF[1].

Siblings of patients with anti-Dr^a should be tested for compatibility, and the patient urged to donate blood for cryogenic storage when his/her clinical state permits.

Dr^a is the receptor for uropathogenic *E. coli*[2].

References

[1] Lublin, D.M., et al., 1994. Molecular basis of reduced or absent expression of decay-accelerating factor in Cromer blood group phenotypes. Blood 84, 1276–1282.

[2] Hasan, R.J., et al., 2002. Structure-function analysis of decay-accelerating factor: identification of residues important for binding of the *Escherichia coli* Dr adhesin and complement regulation. Infect Immun 70, 4485–4493.

Es^a Antigen

Terminology

ISBT symbol (number)	CROM6 (021006 or 21.6)
Obsolete names	202006; 900022
History	Named in 1984 after Mrs. Escandon, whose Mexican parents were cousins.

Occurrence

Three Es(a–) probands are known: one Mexican; one South American; and one Black[1].

Expression

Cord RBCs	Expressed
Altered	Weak on Dr(a–), WES(a+b–), and negative on PNH III RBCs

Cromer

Molecular basis associated with Es[a] antigen[2]

Amino acid	Ile80 (previously reported as 46) in CCP1
Nucleotide	T at bp 239 in exon 2
Es(a–)	Asn80 and A at bp 239

Effect of enzymes and chemicals on Es[a] antigen on intact RBCs

Ficin/Papain	Resistant
Trypsin	Resistant
α-Chymotrypsin	Sensitive
Pronase	Sensitive
DTT 200 mM/50 mM	Weakened/resistant (thus weakened by WARM™ and ZZAP)

In vitro characteristics of alloanti-Es[a]

Immunoglobulin class	IgG
Optimal technique	IAT

Clinical significance of alloanti-Es[a]

Transfusion reaction	One report of a mild transfusion reaction
HDFN	No, because DAF on apical surface of trophoblasts in placenta absorbs maternal antibody

Comments

Siblings of patients with anti-Es[a] should be tested for compatibility, and the patient urged to donate blood for cryogenic storage when his/her clinical state permits.

Es(a–) RBCs have a weak expression of WES[b].

References

[1] Reid, M.E., et al., 1996. A second example of anti-Es[a], an antibody to a high-incidence Cromer antigen. Immunohematology 12, 112–114.

[2] Lublin, D.M., et al., 2000. Molecular basis of Cromer blood group antigens. Transfusion 40, 208–213.

IFC Antigen

Terminology

ISBT symbol (number)	CROM7 (021007 or 21.7)
Obsolete name	202007
History	Anti-IFC is made by people with the Inab phenotype. Named in 1986 from the names of the first three IFC– probands.

Cromer

Occurrence

Rare IFC– (Inab phenotype) people have been mostly from Japan, but Caucasians (one was Swedish; a brother and sister were Italian American), an African American, and a Moroccan have been reported.

Expression

Cord RBCs	Expressed
Altered	Weak on Dr(a–) and absent from PNH III RBCs

Molecular bases associated with IFC antigen

For molecular bases associated with an absence of IFC refer to System pages.

Effect of enzymes and chemicals on IFC antigen on intact RBCs

Ficin/Papain	Resistant
Trypsin	Resistant
α-Chymotrypsin	Sensitive
Pronase	Sensitive
DTT 200 mM/50 mM	Weakened/resistant (thus weakened by WARM™ and ZZAP)

In vitro characteristics of alloanti-IFC

Immunoglobulin class	IgG
Optimal technique	IAT
Neutralization	With concentrated serum/plasma/urine

Clinical significance of alloanti-IFC

Transfusion reaction	No to mild
HDFN	No, because DAF on apical surface of trophoblasts in placenta absorbs maternal antibody

Comments

The only phenotype that lacks IFC is the Inab phenotype, because the RBCs do not express DAF[1].

Siblings of patients with anti-IFC should be tested for compatibility, and the patient urged to donate blood for cryogenic storage when his/her clinical state permits.

A few patients with an acquired (transient) form of the Inab phenotype who made anti-IFC have been reported including: one with thalassemia who had splenic infarctions[2]; another with chronic CLL[3]; and a young child with multiple medical problems[4].

References

[1] Lublin, D.M., et al., 1994. Molecular basis of reduced or absent expression of decay-accelerating factor in Cromer blood group phenotypes. Blood 84, 1276–1282.

[2] Matthes, T., et al., 2002. Acquired and transient RBC CD55 deficiency (Inab phenotype) and anti-IFC. Transfusion 42, 1448–1457.

[3] Banks, J., et al., 2004. Transient loss of Cromer antigens and anti-IFC in a patient with chronic lymphatic leukaemia [abstract]. Vox Sang 87 (Suppl. 3), 37.

[4] Yazer, M.H., et al., 2006. Case report and literature review: transient Inab phenotype and an agglutinating anti-IFC in a patient with a gastrointestinal problem. Transfusion 46, 1537–1542.

WES[a] Antigen

Terminology

ISBT symbol (number)	CROM8 (021008 or 21.8)
Obsolete names	202008; 700042
History	Named in 1987 after the first antibody producer.

Occurrence

Most populations	<0.01%
Blacks (America)	0.48%
Blacks (N. London)	2.04%
Finns	0.56%

Antithetical antigen

WES[b] (**CROM9**)

Expression

Cord RBCs	Expressed

Molecular basis associated with WES[a] antigen[1]

Amino acid	Arg82 (previously reported as 48) in CCP1
Nucleotide	G at bp 245 in exon 2

Effect of enzymes and chemicals on WES[a] antigen on intact RBCs

Ficin/Papain	Resistant
Trypsin	Resistant
α-Chymotrypsin	Sensitive
Pronase	Sensitive
DTT 200 mM/50 mM	Weak/resistant (thus weakened by WARM™ and ZZAP)

In vitro characteristics of alloanti-WES^a

Immunoglobulin class	IgG
Optimal technique	IAT
Neutralization	With concentrated serum/plasma/urine

Clinical significance of alloanti-WES^a

Transfusion reaction	No to mild
HDFN	No, because DAF on apical surface of trophoblasts in placenta absorbs maternal antibody

Comments

WES(a+b–) RBCs have a weak expression of Es^a.

Reference

[1] Lublin, D.M., et al., 2000. Molecular basis of Cromer blood group antigens. Transfusion 40, 208–213.

WES^b Antigen

Terminology

ISBT symbol (number)	CROM9 (021009 or 21.9)
Obsolete names	202004; 900033
History	Named in 1987 when it was recognized to be antithetical to WES^a.

Occurrence

WES(a+b–) probands have been found in people of African ancestry and in Finns.

Antithetical antigen

WES^b (**CROM8**)

Expression

Cord RBCs	Expressed
Altered	Weak on Dr(a–) and Es(a–), and negative on PNH III RBCs

Molecular basis associated with WES^b antigen[1]

Amino acid	Leu80 (previously reported as 48) in CCP1
Nucleotide	T at bp 245 in exon 2

Cromer

Effect of enzymes and chemicals on WES[b] antigen on intact RBCs

Ficin/Papain	Resistant
Trypsin	Resistant
α-Chymotrypsin	Sensitive
Pronase	Sensitive
DTT 200 mM/50 mM	Weakened/resistant (thus weakened by WARM™ and ZZAP)

In vitro characteristics of alloanti-WES[b]

Immunoglobulin class	IgG
Optimal technique	IAT
Neutralization	Concentrated serum/plasma/urine

Clinical significance of alloanti-WES[b]

Transfusion reaction	No data
HDFN	Few examples of anti-WES[b], produced as a result of pregnancy, are described. The baby's RBCs had a positive DAT, but there were no clinical signs of HDFN. HDFN is unlikely because DAF on apical surface of trophoblasts in placenta absorbs maternal antibody.

Comments

WES(a+b–) RBCs have a weak expression of Es[a] antigen.

Siblings of patients with anti-WES[b] should be tested for compatibility, and the patient urged to donate blood for cryogenic storage when his/her clinical state permits.

Reference
[1] Lublin, D.M., et al., 2000. Molecular basis of Cromer blood group antigens. Transfusion 40, 208–213.

UMC Antigen

Terminology

ISBT symbol (number)	CROM10 (021010 or 21.10)
Obsolete name	202010
History	Named in 1989, from the name of the first producer of the antibody.

Occurrence

Only one UMC– proband and her UMC– brother (Japanese) have been described.

Expression

Cord RBCs   Expressed
Altered   Weak on Dr(a–) and absent from PNH III RBCs

Molecular basis associated with UMC antigen[1]

Amino acid   Thr250 (previously reported as 216) in CCP4
Nucleotide   C at bp 749 in exon 6
UMC–   Met250 and T at bp 479

Effect of enzymes and chemicals on UMC antigen on intact RBCs

Ficin/Papain   Resistant
Trypsin   Resistant
α-Chymotrypsin   Sensitive
Pronase   Sensitive
DTT 200 mM/50 mM   Weakened/resistant (thus weakened by WARM™
                      and ZZAP)

In vitro characteristics of alloanti-UMC

Immunoglobulin class IgG
Optimal technique   IAT
Neutralization   With concentrated serum/plasma/urine

Clinical significance of alloanti-UMC

Transfusion reaction No data
HDFN   The proband had three children with no signs or
                      symptoms of HDFN. HDFN is unlikely, because
                      DAF on apical surface of trophoblasts in placenta
                      absorbs maternal antibody.

Comments

Siblings of patients with anti-UMC should be tested for compatibility, and the patient urged to donate blood for cryogenic storage when his/her clinical state permits.

Reference

[1] Lublin, D.M., et al., 2000. Molecular basis of Cromer blood group antigens. Transfusion 40, 208–213.

GUTI Antigen

Terminology

ISBT symbol (number)	CROM11 (021011 or 21.11)
History	Named in 2002 after the first producer of the antibody. The immunogen was a transfusion following a motorcycle accident.

Occurrence

Only one GUTI– proband (Chilean) and his sister have been reported. 15% of Mapuche Indians are heterozygotes for *CROM*01.–11*.

Expression

Cord RBCs	Expressed
Altered	Weak on Dr(a–) and negative on PNH III RBCs

Molecular basis associated with GUTI antigen[1]

Amino acid	Arg240 (previously reported as 206) in CCP4
Nucleotide	G at bp 719 in exon 6
GUTI–	His240 and A at bp 719

Effect of enzymes and chemicals on GUTI antigen on intact RBCs

Ficin/Papain	Resistant
Trypsin	Resistant
α-Chymotrypsin	Sensitive
Pronase	Sensitive
DTT 200 mM/50 mM	Weakened/resistant (thus weakened by WARM™ and ZZAP)
Acid	Resistant

In vitro characteristics of alloanti-GUTI

Immunoglobulin class	IgG
Optimal technique	IAT

Clinical significance of alloanti-GUTI

No data because anti-GUTI has only been found in one male.
HDFN is unlikely because DAF on the apical surface of trophoblasts in placenta absorbs maternal antibody.

Comments

Siblings of patients with anti-GUTI should be tested for compatibility, and the patient urged to donate blood for cryogenic storage when his/her clinical state permits.

Reference

[1] Storry, J.R., et al., 2003. GUTI: a new antigen in the Cromer blood group system. Transfusion 43, 340–344.

SERF Antigen

Terminology

ISBT symbol (number)	CROM12 (021012 or 21.12)
History	Named in 2004 after the first producer of the antibody.

Occurrence

The only two SERF– probands reported (and the sister of one) were Thai[1].

Expression

Cord RBCs	Expressed
Altered	Weak on Dr(a–) and negative on PNH III RBCs

Molecular basis associated with SERF antigen[2]

Amino acid	Pro216 (previously reported as 182) in CCP3
Nucleotide	C at bp 647 in exon 5
SERF–	Leu216 and T at bp 647

Effect of enzymes and chemicals on SERF antigen on intact RBCs

Ficin/Papain	Resistant
Trypsin	Resistant
α-Chymotrypsin	Sensitive
DTT 200 mM/50 mM	Weakened/resistant (thus weakened by WARM™ and ZZAP)

In vitro characteristics of alloanti-SERF

Immunoglobulin class	IgG
Optimal technique	IAT

Clinical significance of alloanti-SERF

No data because only one anti-SERF has been described.
HDFN is unlikely because DAF on apical surface of trophoblasts in placenta absorbs maternal antibody.

Comments

Siblings of patients with anti-SERF should be tested for compatibility, and the patient urged to donate blood for cryogenic storage when his/her clinical state permits.

References

[1] Palacajornsuk, P., et al., 2005. Analysis of SERF in Thai blood donors. Immunohematology 21, 66–69.

[2] Banks, J., et al., 2004. SERF: a new antigen in the Cromer blood group system. Transfus Med 14, 313–318.

ZENA Antigen

Terminology

ISBT symbol (number)	CROM13 (021013 or 21.13)
History	Named in 2004; derived from the given name of the ZENA– proband.

Occurrence

The only reported ZENA– proband was a Syrian Turk.

Expression

Cord RBCs	Expressed
Altered	Weak on Dr(a–) and absent from PNH III RBCs

Molecular basis associated with ZENA antigen[1]

Amino acid	His242 (previously reported as 208) in CCP4
Nucleotide	T at bp 726 in exon 6
ZENA–	Gln242 and G at bp 726

Effect of enzymes and chemicals on ZENA antigen on intact RBCs

Ficin/Papain	Resistant
Trypsin	Resistant
α-Chymotrypsin	Sensitive
DTT 200 mM/50 mM	Weakened/resistant (thus weakened by WARM™ and ZZAP)

In vitro characteristics of alloanti-ZENA

Immunoglobulin class	IgG
Optimal technique	IAT

Clinical significance of alloanti-ZENA

No data because only one example of anti-ZENA has been described. The proband's baby was born with a normal hemoglobin level and no clinical evidence of HDFN. The baby's RBCs were negative in the direct antiglobulin test.

Comments

Siblings of patients with anti-ZENA should be tested for compatibility, and the patient urged to donate blood for cryogenic storage when his/her clinical state permits.

Reference

[1] Hue-Roye, K., et al., 2007. Three new high-prevalence antigens in the Cromer blood group system. Transfusion 47, 1621–1629.

CROV Antigen

Terminology

ISBT symbol (number)	CROM14 (021014 or 21.14)
History	Named in 2005 from "CRO" from Croatia (and the Cromer blood group system), and the first initial of the town (Vinkovci) from whence the CROV– proband hailed.

Occurrence

The only reported CROV– proband was from Croatia.

Expression

Cord RBCs	Expressed
Altered	Weak on Dr(a–) and absent from PNH III RBCs

Molecular basis associated with CROV antigen[1]

Amino acid	Glu156 (previously reported as 122) in CCP2
Nucleotide	G at bp 466 in exon 3
CROV–	Lys156 and A at bp 466

Effect of enzymes and chemicals on CROV antigen on intact RBCs

Ficin/Papain	Resistant
Trypsin	Resistant
α-Chymotrypsin	Sensitive
DTT 200 mM/50 mM	Weakened/resistant (thus weakened by WARM™ and ZZAP)

In vitro characteristics of alloanti-CROV

Immunoglobulin class	IgG
Optimal technique	IAT

Clinical significance of alloanti-CROV

No data because only one example of anti-CROV has been described.
HDFN is unlikely because DAF on apical surface of trophoblasts in placenta absorbs maternal antibody.

Comments

Siblings of patients with anti-CROV should be tested for compatibility, and the patient urged to donate blood for cryogenic storage when his/her clinical state permits.

Reference

[1] Hue-Roye, K., et al., 2007. Three new high-prevalence antigens in the Cromer blood group system. Transfusion 47, 1621–1629.

CRAM Antigen

Terminology

ISBT symbol (number)	CROM15 (021015 or 21.15)
History	Named in 2006 from "CR" for the system and "AM" from the CRAM– proband's name.

Occurrence

The only reported CRAM– proband was from Somalia.

Expression

Cord RBCs	Expressed
Altered	Weak on Dr(a–) and absent from PNH III RBCs

Molecular basis associated with CRAM antigen[1]

Amino acid	Gln247 (previously reported as 213) in CCP4
Nucleotide	A at bp 740 in exon 6
CRAM–	Arg247 and G at bp 740

Effect of enzymes and chemicals on CRAM antigen on intact RBCs

Ficin/Papain	Resistant
Trypsin	Resistant
α-Chymotrypsin	Sensitive
DTT 200 mM/50 mM	Weakened/resistant (thus weakened by WARM™ and ZZAP)

In vitro characteristics of alloanti-CRAM

Immunoglobulin class	IgG
Optimal technique	IAT

Clinical significance of alloanti-CRAM

No data because only one anti-CRAM has been described.
HDFN is unlikely because DAF on apical surface of trophoblasts in placenta absorbs maternal antibody.

Comments

Siblings of patients with anti-CRAM should be tested for compatibility, and the patient urged to donate blood for cryogenic storage when his/her clinical state permits.

Reference

[1] Hue-Roye, K., et al., 2007. Three new high-prevalence antigens in the Cromer blood group system. Transfusion 47, 1621–1629.

CROZ Antigen

Terminology

ISBT symbol (number)	CROM16 (021016 or 21.16)
History	Named in 2010 from "CR" for the system, and "OZ" for "Australia," from whence the proband hailed.

Occurrence

The only reported CROZ– proband was from Australia and may be of Italian descent.

Expression

Cord RBCs	Expressed
Altered	Weak on Dr(a–) and absent from PNH III RBCs

Molecular basis associated with CROZ antigen[1]

Amino acid	Arg130 (previously reported as 96) in CCP2
Nucleotide	G at bp 389 in exon 3
CROZ–	His130 and A at bp 389

Effect of enzymes and chemicals on CROZ antigen on intact RBCs

Ficin/Papain	Resistant
Trypsin	Resistant
α-Chymotrypsin	Sensitive
DTT 200 mM/50 mM	Weakened/resistant (thus weakened by WARM™ and ZZAP)

In vitro characteristics of alloanti-CROZ

Immunoglobulin class	IgG
Optimal technique	IAT

Clinical significance of alloanti-CROZ

No data because only one example of anti-CROZ has been reported.
HDFN is unlikely because DAF on apical surface of trophoblasts in placenta absorbs maternal antibody.

Comments

Siblings of patients with anti-CROZ should be tested for compatibility, and the patient urged to donate blood for cryogenic storage when his/her clinical state permits.

Reference

[1] Karamatic Crew, V., et al., 2010. Two unusual cases within the Cromer blood group system: (I) A novel high incidence antigen CROZ; and (II) A novel molecular basis of Inab phenotype [abstract]. Transfus Med 20 (Suppl. 1), 12.

Cromer

Knops Blood Group System

Number of antigens 9

Polymorphic	Sl^a, McC^b, Vil, KCAM in Blacks
Low prevalence	Kn^b, McC^b, Sl3, Vil in non Blacks
High prevalence	Sl^a, KCAM in non Blacks and Kn^a, McC^a, Yk^a in all populations

Terminology

ISBT symbol (number)	KN (022)
CD number	CD35
Obsolete name	ISBT Collection 205
History	Reported in 1970, and named in honor of the first patient who made anti-Kn^a. Knops was established as a system in 1992, when the antigens were found to be located on complement receptor 1 (CR1).

Expression

Soluble form	Present in low levels in plasma
Other blood cells	Granulocytes, B cells, a subset of T cells, monocytes, macrophages, neutrophils, eosinophils
Tissues	Glomerular podocytes, follicular dendritic cells in spleen and lymph nodes, peripheral nerve fibers

Gene

Chromosome	1q32.2
Name	KN (CR1)
Organization	Distributed over 130 to 160 kbp of gDNA: CR1*1 has 39 exons; CR1*2 has 47 exons; CR1*3 has 30 exons; and CR1*4 has 31 exons[1,2]
Product	Complement receptor type 1 (CR1; CD35)

The Blood Group Antigen (3/e). DOI: http://dx.doi.org/10.1016/B978-0-12-415849-8.00024-7

Knops

Database accession numbers

GenBank	NM_000573; Y00816 (mRNA)
Entrez Gene ID	1378

Molecular bases of Knops phenotypes

The reference allele is *KN*01* or *KN*A* (accession number Y00816), which encodes Kn^a (KN1), KN3, KN4, KN5, KN8, KN9. Differences from this reference allele, and the amino acids affected, are given.

Allele encodes	Allele name	Exon	Nucleotide[†]	Restriction enzyme	Amino acid	Ethnicity (prevalence)
Kn(a–b+) or KN:–1,2	*KN*02* or *KN*B*	29	4681G>A	NdeI+	Val1561Met	Caucasians (Several), Blacks (Few)
Yk(a–) or KN:–5	*KN*01.–05*	26	4223C>T		Thr1408Met	Caucasians (Several), Blacks (Few)
McC(a–b+) or KN:–3,6	*KN*01.06*	29	4768A>G		Lys1590Glu	Blacks (Common), Caucasians (Few)
Sl(a–)Vil+ or KN:–4,7	*KN*01.07*	29	4801A>G		Arg1601Gly	Blacks (Common), Caucasians (Rare)
Sl3– or KN:–8	*KN*01.–08*	29	4828T>A		Ser1610Thr^	Caucasian (Rare)
KCAM– or KN:–9	*KN*01.–09*	29	4843A>G		Ile1615Val	Blacks (Common), Caucasians (Several)

[†]Nucleotides are numbered from the initiation codon (AUG), so may differ from some earlier publications by –27 nucleotides.
^Arg1601 and Ser1610 are required for Sl3 expression.

An allele found in Brazilians (*KN*4619A>G*)[3] has not been associated with an antigen.

Amino acid sequence of the *CR1*1* product[4]

```
MGASSPRSPE  PVGPPAPGLP  FCCGGSLLAV  VVLLALPVAW  GQCNAPEWLP    50
FARPTNLTDE  FEFPIGTYLN  YECRPGYSGR  PFSIICLKNS  VWTGAKDRCR   100
RKSCRNPPDP  VNGMVHVIKG  IQFGSQIKYS  CTKGYRLIGS  SSATCIISGD   150
TVIWDNETPI  CDRIPCGLPP  TITNGDFIST  NRENFHYGSV  VTYRCNPGSG   200
GRKVFELVGE  PSIYCTSNDD  QVGIWSGPAP  QCIIPNKCTP  PNVENGILVS   250
DNRSLFSLNE  VVEFRCQPGF  VMKGPRRVKC  QALNKWEPEL  PSCSRVCQPP   300
PDVLHAERTQ  RDKDNFSPGQ  EVFYSCEPGY  DLRGAASMRC  TPQGDWSPAA   350
PTCEVKSCDD  FMGQLLNGRV  LFPVNLQLGA  KVDFVCDEGF  QLKGSSASYC   400
VLAGMESLWN  SSVPVCEQIF  CPSPPVIPNG  RHTGKPLEVF  PFGKAVNYTC   450
DPHPDRGTSF  DLIGESTIRC  TSDPQGNGVW  SSPAPRCGIL  GHCQAPDHFL   500
FAKLKTQTNA  SDFPIGTSLK  YECRPEYYGR  PFSITCLDNL  VWSSPKDVCK   550
RKSCKTPPDP  VNGMVHVITD  IQVGSRINYS  CTTGHRLIGH  SSAECILSGN   600
AAHWSTKPPI  CQRIPCGLPP  TIANGDFIST  NRENFHYGSV  VTYRCNPGSG   650
GRKVFELVGE  PSIYCTSNDD  QVGIWSGPAP  QCIIPNKCTP  PNVENGILVS   700
DNRSLFSLNE  VVEFRCQPGF  VMKGPRRVKC  QALNKWEPEL  PSCSRVCQPP   750
PDVLHAERTQ  RDKDNFSPGQ  EVFYSCEPGY  DLRGAASMRC  TPQGDWSPAA   800
PTCEVKSCDD  FMGQLLNGRV  LFPVNLQLGA  KVDFVCDEGF  QLKGSSASYC   850
VLAGMESLWN  SSVPVCEQIF  CPSPPVIPNG  RHTGKPLEVF  PFGKAVNYTC   900
DPHPDRGTSF  DLIGESTIRC  TSDPQGNGVW  SSPAPRCGIL  GHCQAPDHFL   950
FAKLKTQTNA  SDFPIGTSLK  YECRPEYYGR  PFSITCLDNL  VWSSPKDVCK  1000
RKSCKTPPDP  VNGMVHVITD  IQVGSRINYS  CTTGHRLIGH  SSAECILSGN  1050
TAHWSTKPPI  CQRIPCGLPP  TIANGDFIST  NRENFHYGSV  VTYRCNLGSR  1100
GRKVFELVGE  PSIYCTSNDD  QVGIWSGPAP  QCIIPNKCTP  PNVENGILVS  1150
DNRSLFSLNE  VVEFRCQPGF  VMKGPRRVKC  QALNKWEPEL  PSCSRVCQPP  1200
PEILHGEHTP  SHQDNFSPGQ  EVFYSCEPGY  DLRGAASLHC  TPQGDWSPEA  1250
PRCAVKSCDD  FLGQLPHGRV  LFPLNLQLGA  KVSFVCDEGF  RLKGSSVSHC  1300
VLVGMRSLWN  NSVPVCEHIF  CPNPPAILNG  RHTGTPSGDI  PYGKEISYTC  1350
DPHPDRGMTF  NLIGESTIRC  TSDPHGNGVW  SSPAPRCELS  VRAGHCKTPE  1400
QFPFASPTIP  INDFEFPVGT  SLNYECRPGY  FGKMFSISCL  ENLVWSSVED  1450
NCRRKSCGPP  PEPFNGMVHI  NTDTQFGSTV  NYSCNEGFRL  IGSPSTTCLV  1500
SGNNVTWDKK  APICEIISCE  PPPTISNGDF  YSNNRTSFHN  GTVVTYQCHT  1550
GPDGEQLFEL  VGERSIYCTS  KDDQVGVWSS  PPPRCISTNK  CTAPEVENAI  1600
RVPGNRSFFS  LTEIIRFRCQ  PGFVMVGSHT  VQCQTNGRWG  PKLPHCSRVC  1650
QPPPEILHGE  HTLSHQDNFS  PGQEVFYSCE  PSYDLRGAAS  LHCTPQGDWS  1700
PEAPRCTVKS  CDDFLGQLPH  GRVLLPLNLQ  LGAKVSFVCD  EGFRLKGRSA  1750
SHCVLAGMKA  LWNSSVPVCE  QIFCPNPPAI  LNGRHTGTPF  GDIPYGKEIS  1800
YACDTHPDRG  MTFNLIGESS  IRCTSDPQGN  GVWSSPAPRC  ELSVPAACPH  1850
PPKIQNGHYI  GGHVSLYLPG  MTISYTCDPG  YLLVGKGFIF  CTDQGIWSQL  1900
DHYCKEVNCS  FPLFMNGISK  ELEMKKVYHY  GDYVTLKCED  GYTLEGSPWS  1950
QCQADDRWDP  PLAKCTSRAH  DALIVGTLSG  TIFFILLIIF  LSWIILKHRK  2000
GNNAHENPKE  VAIHLHSQGG  SSVHPRTLQT   NEENSRVLP            2039
```

Signal peptide: 41 amino acid residues.

Carrier molecule[4]

The *CR1*1* product (hereafter called "CR1") has 30 complement control protein (CCP) repeats, each comprising about 60 amino acids with sequence homology [also called short consensus repeats (SCRs)]. Each CCP has four cysteine residues, and is maintained in a folded conformation by two disulfide bonds. Seven CCPs comprise one long homologous repeat (LHR) domain. The other allotypes have a similar structure.

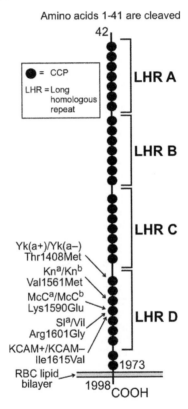

Amino acids 1-41 are cleaved

M_r (SDS-PAGE)	A allotype (*CR*1–1*) 220,000; B allotype (*CR*1–2*) 250,000; C allotype (*CR*1–3*) 190,000; D allotype (*CR*1–4*) 280,000 under non-reducing conditions
CHO: N-glycan	25 sites: probably 6–8 occupied
CHO: O-glycan	No sites
Cysteine residues	Four per CCP
Copies per RBC	20–1,500[5]

Function

CR1 binds C3b and C4b, and has an inhibitory effect on complement activation by classical and alternative pathways, protecting RBCs from autohemolysis. Erythrocyte CR1 is important in processing immune complexes by binding them for transport to the liver and spleen for removal from the circulation.
CR1 binds particles coated with C3b and C4b, thereby mediating phagocytosis by neutrophils and monocytes. The presence of CR1 on other blood cells and tissues suggests it has multiple roles in the immune response, e.g., activation of B lymphocytes.

Disease association

Knops antigens (CR1 copy number) are depressed in: SLE, CHAD, PNH, hemolytic anemia, insulin-dependent diabetes mellitus, AIDS, some malignant tumors, any condition with increased clearance of immune complexes. Low levels of CR1 on RBCs may result in deposition of immune complexes on blood vessel walls, with subsequent damage to the walls.
CR1 is a ligand for the rosetting of *Plasmodium falciparum* infected RBCs to uninfected RBCs[6].
Almost 75% of HIV-1+ patients have an *in vivo* CR1 cleavage fragment of M_r 160,000, suggesting that RBC CR1 may have a role in HIV infection. This compares with 6.5% of healthy donors, and 13.5% of patients with immune complex diseases[7].

Phenotypes (% occurrence)

Phenotype	Caucasians	Blacks
Kn(a+b−)	94.5	99.9
Kn(a−b+)	>1	0
Kn(a+b+)	3.5	0.1
McC(a+)	98	94
Sl(a+)	98	60
Yk(a+)	92	98
KCAM+	98	20

Null: Some RBCs (e.g., Helgeson) type as Kn(a−b−), McC(a−), Sl(a−), Yk(a−), and KCAM− because these RBCs have low copy numbers of CR1 (approximately 10% of normal)[5].

Knops

Comments

The CR1 copy number per RBC may be decreased in stored samples.

References

[1] Moulds, J.M., 2010. The Knops blood-group system: a review. Immunohematology 26, 2–7.

[2] Vik, D.P., Wong, W.W., 1993. Structure of the gene for the F allele of complement receptor type 1 and sequence of the coding region unique to the S allele. J Immunol 151, 6214–6224.

[3] Covas, D.T., et al., 2007. Knops blood group haplotypes among distinct Brazilian populations. Transfusion 47, 147–153.

[4] Cohen, J.H., et al., 1999. The C3b/C4b receptor (CR1, CD35) on erythrocytes: methods for study of the polymorphisms. Mol Immunol 36, 819–825.

[5] Moulds, J.M., et al., 1992. Antiglobulin testing for CR1-related (Knops/McCoy/Swain-Langley/York) blood group antigens: negative and weak reactions are caused by variable expression of CR1. Vox Sang 62, 230–235.

[6] Moulds, J.M., et al., 2001. Molecular identification of Knops blood group polymorphisms found in long homologous region D of complement receptor 1. Blood 97, 2879–2885.

[7] Moulds, J.M., et al., 1995. HIV-1 patients exhibit a novel cleavage fragment of the Knops (CR1) protein [abstract]. Transfusion 35 (Suppl.), 59S.

Kn^a Antigen

Terminology

ISBT symbol (number)	KN1 (022001 or 22.1)
Obsolete names	Knops; COST4; 205004
History	Named after the Kn(a–) patient who made anti-Kn^a. The three Kn(a–) siblings in the original paper (1970) were later shown to have the Helgeson phenotype.

Occurrence

Caucasians	94.5%
Blacks	99.9%

Antithetical antigen

Kn^b (KN2)

Expression

Cord RBCs	Weakened
Altered	Weak on dominant Lu(a–b–) RBCs and weak on RBCs from patients with diseases causing RBC CR1 deficiency, e.g., autoimmune diseases

Molecular basis associated with Knᵃ antigen[1]

Amino acid	Val1561 in CCP 24 (LHR-D)
Nucleotide	G at bp 4681 (previously reported as 4708) in exon 29

Effect of enzymes and chemicals on Knᵃ antigen on intact RBCs

Ficin/Papain	Weakened (especially ficin)
Trypsin	Sensitive
α-Chymotrypsin	Sensitive
DTT 200 mM/50 mM	Sensitive/resistant (thus sensitive to WARM™ and ZZAP)
Acid	Resistant

In vitro characteristics of alloanti-Knᵃ

Immunoglobulin class	IgG
Optimal technique	IAT

Clinical significance of alloanti-Knᵃ

Transfusion reaction	No
HDFN	No

Comments

Anti-Knᵃ is frequently found in multispecific sera.

Disease processes causing RBC CR1 deficiency can lead to "false" negative Knᵃ typing. Variable results in tests on different samples from the same patient have been described[2].

References

[1] Moulds, J.M., et al., 2001. Molecular identification of Knops blood group polymorphisms found in long homologous region D of complement receptor 1. Blood 97, 2879–2885.

[2] Rolih, S., 1990. A review: antibodies with high-titer, low avidity characteristics. Immunohematology 6, 59–67.

Knᵇ Antigen

Terminology

ISBT symbol (number)	KN2 (022002 or 22.2)
Obsolete names	COST5; 205005
History	Identified in 1980 when it was found to be antithetical to Knᵃ.

Knops

Occurrence

Caucasians	3.5%
Blacks	<0.01%

Antithetical antigen

Kn^a (**KN1**)

Expression

Cord RBCs	Weak
Altered	Weak on RBCs from patients with diseases causing RBC CR1 deficiency, e.g., autoimmune diseases

Molecular basis associated with Kn^b antigen[1]

Amino acid	Met1561 in CCP 24 (LHR-D)
Nucleotide	A at bp 4681 (previously reported as 4708) in exon 29

Effect of enzymes and chemicals on Kn^b antigen on intact RBCs

Ficin/Papain	Weakened (especially ficin)
Trypsin	Sensitive
α-Chymotrypsin	Sensitive
DTT 200 mM/50 mM	Presumed sensitive/resistant (thus presumed sensitive to WARM™ and ZZAP)

In vitro characteristics of alloanti-Kn^b

Immunoglobulin class	IgG
Optimal technique	IAT

Clinical significance of alloanti-Kn^b

No data available. Only one example of anti-Kn^b, in a serum containing anti-Kp^a, has been reported[2].

Comments

Disease processes causing RBC CR1 deficiency can lead to "false" negative Kn^b typing. Variable results in tests on different samples from the same patient have been described[3].

References

[1] Moulds, J.M., et al., 2001. Molecular identification of Knops blood group polymorphisms found in long homologous region D of complement receptor 1. Blood 97, 2879–2885.

[2] Mallan, M.T., et al., 1980. The Hall serum: detecting Kn^b, the antithetical allele to Kn^a [abstract]. Transfusion 20, 630–631.

[3] Rolih, S., 1990. A review: antibodies with high-titer, low avidity characteristics. Immunohematology 6, 59–67.

McCᵃ Antigen

Terminology

ISBT symbol (number)	KN3 (022003 or 22.3)
Obsolete names	McCoy; COST6; 205006
History	Identified in 1978 and named after the patient who made the first anti-McCᵃ. Associated with Knᵃ because 53% of McC(a–) RBCs were also Kn(a–).

Occurrence

Caucasians	98%
Blacks	94%

Antithetical antigen

McCᵇ (**KN6**)

Expression

Cord RBCs	Weak
Altered	Weak on dominant Lu(a–b–) RBCs and weak on RBCs from patients with diseases causing RBC CR1 deficiency, e.g., autoimmune diseases

Molecular basis associated with McCᵃ antigen[1]

Amino acid	Lys1590 in CCP 25 (LHR-D)
Nucleotide	A at bp 4768 (previously reported as 4795) in exon 29

Effect of enzymes and chemicals on McCᵃ antigen on intact RBCs

Ficin/Papain	Weakened (especially ficin)
Trypsin	Sensitive
α-Chymotrypsin	Sensitive
DTT 200 mM/50 mM	Sensitive/resistant (thus sensitive to WARM™ and ZZAP)
Acid	Resistant

In vitro characteristics of alloanti-McCᵃ

Immunoglobulin class	IgG
Optimal technique	IAT

Clinical significance of alloanti-McCᵃ

Transfusion reaction   No
HDFN   No

Comments

Disease processes causing RBC CR1 deficiency can lead to "false" negative typing. Variable results in tests on different samples from the same patient have been described.

Reference

[1] Moulds, J.M., et al., 2001. Molecular identification of Knops blood group polymorphisms found in long homologous region D of complement receptor 1. Blood 97, 2879–2885.

Slᵃ Antigen

Terminology

ISBT symbol (number)   KN4 (022004 or 22.4)
Obsolete names   Sl1; Swain-Langley; 205007; COST7; McCᶜ
History   Reported in 1980 and named after the first two
                         antibody producers: Swain and Langley.

Occurrence

Caucasians   98%
Blacks   50 to 60%; 30% in West Africa

Antithetical antigen

Vil (Sl2; **KN7**)

Expression

Cord RBCs   Weak
Altered   Weak on dominant Lu(a–b–) RBCs, and weak on
              RBCs from patients with diseases causing RBC CR1
              deficiency, e.g., autoimmune diseases

Molecular basis associated with Slᵃ antigen[1]

Amino acid   Arg1601 in CCP 25 (LHR-D)
Nucleotide   A at bp 4801 (previously reported as 4828) in exon 29
See Sl3 [**KN8**].

Effect of enzymes and chemicals on Sl^a antigen on intact RBCs

Ficin/Papain	Weakened (especially ficin)
Trypsin	Sensitive
α-Chymotrypsin	Sensitive
DTT 200 mM/50 mM	Sensitive/resistant (thus sensitive to WARM™ and ZZAP)
Acid	Resistant

In vitro characteristics of alloanti-Sl^a

Immunoglobulin class	IgG
Optimal technique	IAT

Clinical significance of alloanti-Sl^a

Transfusion reaction	No
HDFN	No

Comments

Sl^a has been subdivided, see Sl3 [**KN8**][2].

Anti-Sl^a is a common specificity produced by Blacks, and initially may be confused with anti-Fy3 because most Fy(a–b–) RBCs are also likely to be Sl(a–).

Disease processes causing RBC CR1 deficiency can lead to "false" negative typing. Variable results in tests on different samples from the same patient have been described.

References

[1] Moulds, J.M., et al., 2001. Molecular identification of Knops blood group polymorphisms found in long homologous region D of complement receptor 1. Blood 97, 2879–2885.

[2] Moulds, J.M., et al., 2002. Expansion of the Knops blood group system and subdivision of Sl^a. Transfusion 42, 251–256.

Yk^a Antigen

Terminology

ISBT symbol (number)	KN5 (022005 or 22.5)
Obsolete names	York; COST3; 205003
History	Briefly described in 1969, and initially thought to be anti-Cs^a because the serum was non-reactive with two Cs(a–) RBC samples. Named in 1975 after the first producer of the antibody, Mrs. York.

Knops

Occurrence

Caucasians	92%
Blacks	98%

Expression

Cord RBCs	Weak
Altered	Weak on dominant Lu(a–b–) RBCs and weak on RBCs from patients with diseases causing RBC CR1 deficiency, e.g., autoimmune diseases

Molecular basis associated with Yk^a antigen[1]

Amino acid	Thr1408 in CCP 22 (LHR-D)
Nucleotide	C at bp 4223 in exon 26
Yk(a–)	Met1408 and T at bp 4223

Effect of enzymes and chemicals on Yk^a antigen on intact RBCs

Ficin/Papain	Weakened (especially ficin)
Trypsin	Sensitive
α-Chymotrypsin	Sensitive
DTT 200 mM/50 mM	Sensitive/resistant (thus sensitive to WARM™ and ZZAP)
Acid	Resistant

In vitro characteristics of alloanti-Yk^a

Immunoglobulin class	IgG
Optimal technique	IAT

Clinical significance of alloanti-Yk^a

Transfusion reaction	No
HDFN	No

Comments

Approximately 12% of Caucasian Yk(a–) RBCs and 16% of Black Yk(a–) RBCs are Cs(a–)[2].

Disease processes causing RBC CR1 deficiency can lead to "false" negative typing. Variable results in tests on different samples from the same patient have been described.

References

[1] Veldhuisen, B., et al., 2011. Molecular analysis of the York antigen of the Knops blood group system. Transfusion 51, 1389–1396.

[2] Rolih, S., 1990. A review: antibodies with high-titer, low avidity characteristics. Immunohematology 6, 59–67.

McC^b Antigen

Terminology

ISBT symbol (number)	KN6 (022006 or 22.6)
History	Identified in 1983; antibody recognized an antigen antithetical to McC^a on RBCs of Blacks. Confirmed by molecular analysis and became a Knops system antigen in 2000.

Occurrence

Caucasians	<0.1%
Blacks	45%

Antithetical antigen

McC^a (**KN3**)

Expression

Cord RBCs	Weak
Altered	Weak on RBCs from patients with diseases causing RBC CR1 deficiency, e.g., autoimmune diseases

Molecular basis associated with McC^b antigen[1]

Amino acid	Glu1590 in CCP 25 (LHR-D)
Nucleotide	G at bp 4768 (previously reported as 4795) in exon 29

Effect of enzymes and chemicals on McC^b antigen on intact RBCs

Ficin/Papain	Variable
Trypsin	Presumed sensitive
α-Chymotrypsin	Presumed sensitive
DTT 200 mM/50 mM	Presumed sensitive/resistant (thus presumed sensitive to WARM™ and ZZAP)

Knops

In vitro characteristics of alloanti-McC^b

Immunoglobulin class   IgG
Optimal technique   IAT

Clinical significance of alloanti-McC^b

No data but unlikely to be significant.

Comments

Disease processes causing RBC CR1 deficiency can lead to "false" negative typing. Variable results in tests on different samples from the same patient have been described.

Reference
[1] Moulds, J.M., et al., 2001. Molecular identification of Knops blood group polymorphisms found in long homologous region D of complement receptor 1. Blood 97, 2879–2885.

Vil Antigen

Terminology

ISBT symbol (number)   KN7 (022007 or 22.7)
Obsolete names   Villien; McC^d
History   Reported in 1980, and named after the first patient who made the antibody before the antithetical relationship to Sl^a was established. Joined the Knops system in 2000 after molecular analysis confirmed the relationship with Sl^a.

Occurrence

Caucasians   <0.01%
Blacks   80%

Antithetical antigen

Sl^a (Sl1; **KN4**)

Expression

Cord RBCs   Weak
Altered   Weak on RBCs from patients with diseases causing RBC CR1 deficiency, e.g., autoimmune diseases

Molecular basis associated with Vil antigen[1]

Amino acid	Gly1601 in CCP 25 (LHR-D)
Nucleotide	G at bp 4801 (previously reported as 4828) in exon 29

See Sl3 [**KN8**].

Effect of enzymes and chemicals on Vil antigen on intact RBCs

Ficin/Papain	Presumed weakened
Trypsin	Presumed sensitive
α-Chymotrypsin	Presumed sensitive
DTT 200 mM/50 mM	Presumed sensitive/resistant (thus presumed sensitive to WARM™ and ZZAP)

In vitro characteristics of alloanti-Vil

Immunoglobulin class	IgG
Optimal technique	IAT

Clinical significance of alloanti-Vil

No data but unlikely to be significant.

Reference

[1] Moulds, J.M., et al., 2001. Molecular identification of Knops blood group polymorphisms found in long homologous region D of complement receptor 1. Blood 97, 2879–2885.

Sl3 Antigen

Terminology

ISBT symbol (number)	KN8 (022008 or 22.8)
Obsolete name	KMW
History	Subdivision of Sl[a] reported in 2002, when differences were noted in the reactivity of various anti-Sl[a] (used for population studies). The definitive anti-Sl[a] (anti-Sl3) was made by a Caucasian woman (KMW).

Occurrence

All populations	100%

Only one Sl:1,–2,–3 person has been reported[1].

Expression

Cord RBCs	Weak
Altered	Weak on RBCs from patients with diseases causing RBC CR1 deficiency, e.g., autoimmune diseases

Molecular basis associated with Sl3 antigen[1]

Amino acid	Arg1601 and Ser1610 in CCP 25 (LHR-D)
Nucleotide	A at bp 4801 (previously reported as 4828) and A at 4828 (previously reported as 4855) in exon 29
Sl:1,–2,–3	Thr1610 and G at bp 4855; see table below

Effect of enzymes and chemicals on Sl3 antigen on intact RBCs

Ficin/Papain	Presumed weakened
Trypsin	Presumed sensitive
α-Chymotrypsin	Presumed sensitive
DTT 200 mM/50 mM	Presumed sensitive/resistant (thus presumed sensitive to WARM™ and ZZAP)

In vitro characteristics of alloanti-Sl3

Immunoglobulin class	IgG
Optimal technique	IAT

Clinical significance of alloanti-Sl3

No data, but unlikely to be significant.

Relationship of Sl phenotypes

Phenotype	Amino acid 1601	Amino acid 1610	Ethnic association
Sl:1,–2,3	Arg	Ser	Most common in Caucasians
Sl:–1,2,–3	Gly	Ser	Common in Blacks
Sl:1,–2,–3	Arg	Thr	Found only in one Caucasian (KMW)

Reference
[1] Moulds, J.M., et al., 2002. Expansion of the Knops blood group system and subdivision of Sl[a]. Transfusion 42, 251–256.

KCAM Antigen

Terminology

ISBT symbol (number) KN9 (022009 or 22.9)

History Named "KAM" in 2005, but because an existing antigen "Kamhuber" was abbreviated "KAM", it was changed to "KCAM" in 2007 when it was placed in the Knops system. "KC" was for Kansas City, and "AM" were the initials of the proband.

Occurrence

Most populations 98%

Blacks 20%

Expression

Cord RBCs Presumed weak

Altered Weak on RBCs from patients with diseases causing RBC CR1 deficiency, e.g., autoimmune diseases

Molecular basis associated with KCAM antigen[1,2]

Amino acid Ile1615 in CCP 25 (LHR-D)

Nucleotide A at bp 4843 (previously reported as 4870) in exon 29

KCAM– Val1615 and G at bp 4843

Effect of enzymes and chemicals on KCAM antigen on intact RBCs

Ficin/Papain Weakened

Trypsin Sensitive

α-Chymotrypsin Sensitive

DTT 200 mM/50 mM Sensitive/resistant (thus sensitive to WARM™ and ZZAP)

In vitro characteristics of alloanti-KCAM

Immunoglobulin class IgG

Optimal technique IAT

Clinical significance of alloanti-KCAM

No data, but unlikely to be significant.

Knops

References

[1] Moulds, J.M., et al., 2005. KAM: a new allele in the Knops blood group system [abstract]. Transfusion 45 (Suppl.), 27A.

[2] Westhoff, C., et al., 2008. Two examples of Anti-KCAM, an antibody to an antigen in the Knops system [abstract]. Transfusion 48 (Suppl.), 189A.

Indian Blood Group System

Number of antigens 4

Low prevalence In^a
High prevalence In^b, INFI, INJA

Terminology

ISBT symbol (number)   IN (023)
CD number   CD44
Obsolete name   ISBT Collection 203
History   Named because the first In(a+) people were from India.

Expression

Other blood cells   Neutrophils, lymphocytes, monocytes
Tissues   Brain, breast, colon epithelium, gastric, heart, kidney, liver, lung, placenta, skin, spleen, thymus, fibroblasts

Gene

Chromosome   11p13
Name   *IN (CD44)*
Organization   At least 19 exons distributed over 50 kbp of gDNA (10 exons are variable). The hemopoietic isoform uses exons 1 to 5, 15 to 17, and 19
Product   CD44, Indian glycoprotein, Hermes antigen

IN*01/IN*02

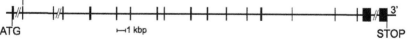

Database accession numbers

GenBank   M59040 (mRNA); NG_008937 (gene)
Entrez Gene ID   960

The Blood Group Antigen (3/e). DOI: http://dx.doi.org/10.1016/B978-0-12-415849-8.00025-9

Molecular basis of Indian phenotypes

The reference allele, *IN*02* or *IN*B* (Accession number M59040) encodes In^b (IN2), IN3, IN4. Nucleotide differences from this reference allele, and the amino acids affected, are given.

Allele encodes	Allele name	Exon	Nucleotide	Amino acid	Ethnicity (prevalence)
In(a+b−) or IN:1,−2	*IN*01* or *IN*A*	2 3 6	137G>C 326A>C 716G>A	Arg46Pro Tyr109Ser Gly239Glu	Arabs, Iranians, South Asian Indians (Several), Asians, Blacks, Caucasians (Rare)
INFI− or IN:−3	*IN*02.−03*	3	255C>G	His85Gln	Moroccan (Rare)
INJA− or IN:−4	*IN*02.−04*	5	488C>A	Thr163Lys^	Pakistani (Rare)

^= Originally reported, incorrectly, as Thr163Arg[1].

Molecular basis of IN:−1,−2 phenotype (suppression)

KLF1 encodes erythroid Krüppel-like factor (EKLF). Several nucleotide changes in this gene are responsible for the dominant Lu(a−b−) phenotype, which is also known as the In(lu) phenotype (see Lutheran blood group system). The *KLF1* gene, which is located at 19p13.1–p13.12, has 3 exons; the initiation codon is in exon 1, and the stop codon is in exon 3. GenBank accession numbers are U37106 (gene) and NM_006563 (mRNA). Differences from the *KLF1*01* reference allele (Accession number NM_006563), and amino acid affected, are given.

Allele name	Exon	Nucleotide	Amino acid	Ethnicity (Prevalence)
*KLF1*BGM10*	3	973G>A^	Glu325Lys	Rare

^= This change caused dyserythropoietic anemia and suppression of CO, IN, and LW antigens[1,2].

Amino acid sequence[2]

```
MDKFWWHAAW   GLCLVPLSLA   QIDLNITCRF   AGVFHVEKNG   RYSISRTEAA    50
DLCKAFNSTL   PTMAQMEKAL   SIGFETCRYG   FIEGHVVIPR   IHPNSICAAN   100
NTGVYILTYN   TSQYDTYCFN   ASAPPEEDCT   SVTDLPNAFD   GPITITIVNR   150
DGTRYVQKGE   YRTNPEDIYP   SNPTDDDVSS   GSSSERSSTS   GGYIFYTFST   200
VHPIPDEDSP   WITDSTDRIP   ATRDQDTFHP   SGGSHTTHES   ESDGHSHGSQ   250
EGGANTTSGP   IRTPQIPEWL   IILASLLALA   LILAVCIAVN   SRRRCGQKKK   300
LVINSGNGAV   EDRKPSGLNG   EASKSQEMVH   LVNKESSETP   DQFMTADETR   350
NLQNVDMKIG   V                                                  361
```

Signal peptide: 20 amino acid residues; which are sometimes cleaved.

Carrier molecule[2,3]

A single pass (type 1) membrane glycoprotein.

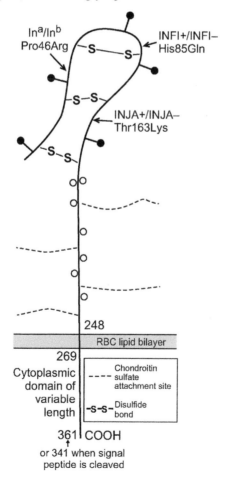

M_r (SDS-PAGE)	Reduced 80,000
CHO: N-glycan	6 sites
CHO: O-glycan	Depends on isoform
Chondroitin sulfate	4 sites
Cysteine residues	Depends on isoform
Copies per RBC	2,000 to 5,000 on mature RBCs

Two amino acid domains have been identified as being important in hyaluronate binding: residues 18 to 26, and 130 to 142.

Function

CD44 is an adhesion molecule in lymphocytes, monocytes, and other tumor cells, binds to hyaluronate and other components of the extracellular matrix, and is also involved in immune stimulation and signaling between cells. Its function in RBCs is not known.

Disease association

In(a–b–) individuals with CDA may also be Co(a–b–)[4], and have weak expression of LW antigens.

Joint fluid from patients with inflammatory synovitis has higher than normal levels of soluble CD44[3]. Serum CD44 is elevated in some patients with lymphoma.

Phenotypes (% occurrence)

Phenotype	Caucasians & Blacks	Indians (South Asians)	Iranians & Arabs
In(a+b–)	Rare	Rare	Rare
In(a–b+)	99.9	96	90
In(a+b+)	<0.1	4	10

Comments

CD44 is present in reduced (variable) amounts in dominant type Lu(a–b–) RBCs, but is expressed normally in other cells from these people.

Ser-Gly is a potential chondroitin sulfate linkage site. After Thr202, various sequences can be generated by alternative splicing of at least 10 exons. Different splicing events occur during different stages of hemopoiesis. In mature RBCs, nine exons are usually encoded. A protein of 361 amino acids is the predominant type in the RBC membrane.

References

[1] Poole, J., et al., 2007. Correction to "Two missense mutations in the *CD44* gene encode two new antigens of the Indian blood group system". Transfusion 47, 1306–1311. Transfusion 47, 1741.

[2] Spring, F.A., et al., 1988. The In[a] and In[b] blood group antigens are located on a glycoprotein of 80,000 MW (the CDw44 glycoprotein) whose expression is influenced by the *In(Lu)* gene. Immunology 64, 37–43.

[3] Moulds, J.M., 1994. Association of blood group antigens with immunologically important proteins. In: Garratty, G. (Ed.), Immunobiology of Transfusion Medicine. Marcel Dekker, Inc., New York, NY, pp. 273–297.

[4] Parsons, S.F., et al., 1994. A novel form of congenital dyserythropoietic anemia associated with deficiency of erythroid CD44 and a unique blood group phenotype [In(a−b−), Co(a−b−)]. Blood 83, 860–868.

In[a] Antigen

Terminology

ISBT symbol (number)	IN1 (023001 or 23.1)
Obsolete name	203001
History	"In" is an abbreviation of Indian, in honor of the ethnic group in which this antigen was first found.

Occurrence

Caucasians	0.1%
Asians and Blacks	0.1%
Indians (South Asians)	4%
Iranians	10.6%
Arabs	11.8%

Antithetical antigen

In[b] (**IN2**)

Expression

Cord RBCs	Weak
Altered	Weak on RBCs from pregnant women

Molecular basis associated with In[a] antigen[1]

Amino acid	Pro46
Nucleotide	C at bp 137 in exon 2

Effect of enzymes and chemicals on In[a] antigen on intact RBCs

Ficin/Papain	Sensitive
Trypsin	Sensitive

Indian

α-Chymotrypsin	Sensitive
DTT 200 mM/50 mM	Sensitive/sensitive (thus sensitive to WARM™ and ZZAP)
Acid	Resistant

In vitro characteristics of alloanti-In^a

| Immunoglobulin class | IgG; IgM |
| Optimal technique | IAT; RT |

Clinical significance of alloanti-In^a

| Transfusion reaction | Decreased cell survival |
| HDFN | Positive DAT; no clinical HDFN |

Comments

Anti-In^a can be naturally-occurring.

Reference

[1] Telen, M.J., et al., 1996. A blood group-related polymorphism of CD44 abolishes a hyaluronan-binding consensus sequence without preventing hyaluronan binding. J Biol Chem 271, 7147–7153.

In^b Antigen

Terminology

ISBT symbol (number)	IN2 (023002 or 23.2)
Obsolete names	203002; Salis
History	Named when its antithetical relationship to In^a was identified.

Occurrence

| Caucasians | 99% |
| Indians (South Asians) | 96% |

Antithetical antigen

In^a (**IN1**)

Expression

Cord RBCs	Weak
Altered	Weak on dominant Lu(a–b–) RBCs
	Weak on RBCs from pregnant women

Molecular basis associated with In[b] antigen[1]

Amino acid   Arg46
Nucleotide   G at bp 137 in exon 2

Effect of enzymes and chemicals on In[b] antigen on intact RBCs

Ficin/Papain	Sensitive
Trypsin	Sensitive
α-Chymotrypsin	Sensitive
DTT 200 mM/50 mM	Sensitive/sensitive (thus sensitive to WARM™ and ZZAP)
Acid	Resistant

In vitro characteristics of alloanti-In[b]

Immunoglobulin class	IgG
Optimal technique	IAT

Clinical significance of alloanti-In[b]

Transfusion reaction	No to severe/delayed and hemolytic[2]
HDFN	Positive DAT, but no clinical HDFN[3]

References

[1] Telen, M.J., et al., 1996. A blood group-related polymorphism of CD44 abolishes a hyaluronan-binding consensus sequence without preventing hyaluronan binding. J Biol Chem 271, 7147–7153.

[2] Joshi, S.R., 1992. Immediate haemolytic transfusion reaction due to anti-In[b]. Vox Sang 63, 232–233.

[3] Longster, G.H., et al., 1981. Four further examples of anti-In[b] detected during pregnancy. Clin Lab Haemat 3, 351–356.

INFI Antigen

Terminology

ISBT symbol (number)   IN3 (023003 or 23.3)
History   Reported and named in 2006, "IN" from the Indian system and "FI" from the INFI– proband's name.

Occurrence

The three INFI– probands (found as a result of anti-INFI in pregnancy) were from Morocco.

Expression

Cord RBCs	Weak
Altered	Weak on dominant Lu(a–b–) RBCs
	Weak on RBCs from pregnant women

Molecular basis associated with INFI antigen[1,2]

Amino acid	His85
Nucleotide	C at bp 255 in exon 3
INFI–	Gln85 and G at bp 255

Effect of enzymes and chemicals on INFI antigen on intact RBCs

Ficin/Papain	Sensitive
Trypsin	Sensitive
α-Chymotrypsin	Sensitive
DTT 200 mM/50 mM	Sensitive/sensitive (thus sensitive to WARM™ and ZZAP)
Acid	Resistant

In vitro characteristics of alloanti-INFI

Immunoglobulin class	IgG
Optimal technique	IAT

Clinical significance of alloanti-INFI

Transfusion reaction	No data because anti-INFI is rare[1,2]
HDFN	The baby of a woman with anti-INFI had mild HDFN

Comments

INFI– RBCs have a weaker than normal expression of In[b].

References

[1] Poole, J., et al., 2007. Two missense mutations in the CD44 gene encode two new antigens of the Indian blood group system. Transfusion 47, 1306–1311.

[2] Poole, J., et al., 2007. Correction to Two missense mutations in the *CD44* gene encode two new antigens of the Indian blood group system. Transfusion 47, 1306–1311. Transfusion 47, 1741.

Indian

INJA Antigen

Terminology

ISBT symbol (number)	IN4 (023004 or 23.4)
History	Reported and named in 2006, "IN" from the Indian system and "JA" from the INJA– proband's name.

Occurrence

The two INJA– probands (both found as a result of anti-INJA in pregnancy) were from Pakistan.

Expression

Cord RBCs	Weak
Altered	Weak on dominant Lu(a–b–) RBCs
	Weak on RBCs from pregnant women

Molecular basis associated with INJA antigen[1,2]

Amino acid	Thr163
Nucleotide	C at bp 488 in exon 5
INJA–	Lys163 (not Arg as originally reported) and A at bp 488

Effect of enzymes and chemicals on INJA antigen on intact RBCs

Ficin/Papain	Sensitive
Trypsin	Sensitive
α-Chymotrypsin	Sensitive
DTT 200 mM/50 mM	Sensitive/sensitive (thus sensitive to WARM™ and ZZAP)
Acid	Resistant

In vitro characteristics of alloanti-INJA

Immunoglobulin class	IgG
Optimal technique	IAT

Clinical significance of alloanti-INJA

No data because only two examples of anti-INJA have been described.

References

[1] Poole, J., et al., 2007. Two missense mutations in the CD44 gene encode two new antigens of the Indian blood group system. Transfusion 47, 1306–1311.

[2] Poole, J., et al., 2007. Correction to "Two missense mutations in the CD44 gene encode two new antigens of the Indian blood group system". Transfusion 47, 1306–1311. Transfusion 47, 1741.

Ok Blood Group System

Number of antigens 3

High prevalence Ok^a, OKGV, OKVM

Terminology

ISBT symbol (number) OK (024)
CD number CD147
History The Ok^a antigen achieved system status, becoming the OK system in 1998 when the antigen was found to be located on CD147.

Expression

Other blood cells White blood cells, platelets
Tissues Epithelium in kidney cortex and medullary, liver, acinar cells of pancreas, trachea, cervix, testes, colon, skin, smooth muscle, neural cells, forebrain, cerebellum[1–3]

Gene

Chromosome 19p13.3
Name OK (BSG, EMPRIN)
Organization 7 exons distributed over 1.8 kbp of gDNA
Product CD147 glycoprotein (OK glycoprotein; basigin, EMMPRIN[4]; M6 leukocyte activation antigen)

The Blood Group Antigen (3/e). DOI: http://dx.doi.org/10.1016/B978-0-12-415849-8.00026-0

Database accession numbers

GenBank   L10240 (mRNA); NM_001728 (mRNA); AY942196
                             (gene)

Entrez Gene ID   682

Molecular bases of Ok phenotypes

The reference allele, *OK*01* or *OK*A* (Accession number AY942196),
encodes Ok^a (OK1), OK2, and OK3. Nucleotide differences from this reference allele, and the amino acids affected, are given.

Allele encodes	Allele name	Exon	Nucleotide	Amino acid	Ethnicity (prevalence)
Ok(a–) or OK:–1	*OK*–01*	4	274G>A	Glu92Lys	Japanese (Rare)
OKGV– or OK:–2	*OK*–02*	2	176G>T	Gly59Val	(Rare)
OKVM– or OK:–3	*OK*–03*	2	178G>A	Val60Met	(Rare)

Amino acid sequence[5]

```
MAAALFVLLG  FALLGTHGAS  GAAGTVFTTV  EDLGSKILLT  CSLNDSATEV   50
TGHRWLKGGV  VLKEDALPGQ  KTEFKVDSDD  QWGEYSCVFL  PEPMGTANIQ  100
LHGPPRVKAV  KSSEHINEGE  TAMLVCKSES  VPPVTDWAWY  KITDSEDKAL  150
MNGSESRFFV  SSSQGRSELH  IENLNMEADP  GQYRCNGTSS  KGSDQAIITL  200
RVRSHLAALW  PFLGIVAEVL  VLVTIIFIYE  KRRKPEDVLD  DDDAGSAPLK  250
SSGQHQNDKG  KNVRQRNSS                                       269
```

OK encodes a leader sequence of 21 amino acids.

Carrier molecule[1,3]

Single pass type I membrane glycoprotein with two IgSF domains.

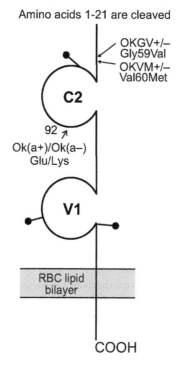

Amino acids 1-21 are cleaved

OKGV+/–
Gly59Val
OKVM+/–
Val60Met

C2

92
Ok(a+)/Ok(a–)
Glu/Lys

V1

RBC lipid bilayer

COOH

M_r (SDS-PAGE)	35,000–69,000
CHO: N-glycan	3
Cysteine residues	4
Copies per RBC	3,000

Function

The protein encoded by this gene is a plasma membrane protein that is important in spermatogenesis, embryo implantation, and neural network formation. Human CD147 (EMMPRIN – extracellular matrix metalloproteinase inducer) on tumor cells is thought to bind to fibroblasts, which stimulates collagenase and other extracellular matrix metalloproteinases, thus enhancing tumor cell invasion and metastases[4,5]. The monocarboxylate (lactate) transporters, MCT1 and MCT4, require CD147 for their correct plasma membrane expression and function[6].

OK

Disease association

Expression is increased on granulocytes in rheumatoid and reactive arthritis; may be involved in tumor metastases. Basigin/CD147 is a receptor essential for erythrocyte invasion by *Plasmodium falciparum*[7].

References

[1] Anstee, D.J., Spring, F.A., 1989. Red cell membrane glycoproteins with a broad tissue distribution. Transfusion Med Rev 3, 13–23.

[2] Spring, F.A., et al., 1997. The Ok[a] blood group antigen is a marker for the M6 leukocyte activation antigen, the human homolog of OX-47 antigen, basigin and neurothelin, an immunoglobulin superfamily molecule that is widely expressed in human cells and tissues. Eur J Immunol 27, 891–897.

[3] Williams, B.P., et al., 1988. Biochemical and genetic analysis of the OK[a] blood group antigen. Immunogenetics 27, 322–329.

[4] Biswas, C., et al., 1995. The human tumor cell-derived collagenase stimulatory factor (renamed EMMPRIN) is a member of the immunoglobulin superfamily. Cancer Res 55, 434–439.

[5] Barclay, A.N., et al., 1997.. In: Leucocyte Antigen FactsBook, second ed. Academic Press, San Diego, CA.

[6] Wilson, M.C., et al., 2002. Fluorescence resonance energy transfer studies on the interaction between the lactate transporter MCT1 and CD147 provide information on the topology and stoichiometry of the complex *in situ*. J Biol Chem 277, 3666–3672.

[7] Crosnier, et al., 2011. Basigin is a receptor essential for erythrocyte invasion by Plasmodium falciparum. Nature 480 (7378), 534–537.

Ok[a] Antigen

Terminology

ISBT symbol (number)	OK1 (024001 or 24.1)
Obsolete names	901006; 900016
History	Named in 1979 after the family name of the patient (S.Ko.G.) whose RBCs lacked the antigen and whose plasma contained the antibody.

Occurrence

All eight Ok(a–) probands are Japanese.

Expression

Cord RBCs	Expressed
Other blood cells	All tested[1,2]
Tissues	All tested[1,2]

Molecular basis associated with Ok[a] antigen[3]

Amino acid	Glu92
Nucleotide	G at bp 274 in exon 4
Ok(a–)	Lys92 and A at bp 274

Effect of enzymes and chemicals on Ok^a antigen on intact RBCs

Ficin/Papain	Resistant
Trypsin	Resistant
α-Chymotrypsin	Resistant
DTT 200 mM	Resistant
Acid	Resistant

In vitro characteristics of alloanti-Ok^a

Immunoglobulin class	IgG
Optimal technique	IAT

Clinical significance of alloanti-Ok^a

Transfusion reaction	^{51}Cr cell survival studies indicated reduced RBC survival
HDFN	No

Comments

Anti-Ok^a react variably with OKGV– RBCs.

References

[1] Anstee, D.J., Spring, F.A., 1989. Red cell membrane glycoproteins with a broad tissue distribution. Transfusion Med Rev 3, 13–23.

[2] Williams, B.P., et al., 1988. Biochemical and genetic analysis of the OK^a blood group antigen. Immunogenetics 27, 322–329.

[3] Spring, F.A., et al., 1997. The Ok^a blood group antigen is a marker for the M6 leukocyte activation antigen, the human homolog of OX-47 antigen, basigin and neurothelin, an immunoglobulin superfamily molecule that is widely expressed in human cells and tissues. Eur J Immunol 27, 891–897.

OKGV Antigen

Terminology

ISBT symbol (number)	OK2 (024002 or 24.2)
History	Described in 2003, and named in 2010 from "OK" for the blood group system and "G and V" for the glycine to valine change[1].

Occurrence

One Iranian OKGV– proband.

Molecular basis associated with OKGV antigen[2]

Amino acid	Gly59
Nucleotide	G at bp 176 in exon 2
OKGV–	Val59 and T at bp 176

Effect of enzymes and chemicals on OKGV antigen on intact RBCs

Ficin/Papain	Resistant
Trypsin	Presumed resistant
α-Chymotrypsin	Presumed resistant
DTT 200 mM	Presumed resistant

In vitro characteristics of alloanti-OKGV

Immunoglobulin class	IgG
Optimal technique	IAT

Clinical significance of alloanti-OKGV

No data because only one anti-OKGV has been reported.

Comments

OKGV– RBCs react variably with anti-Ok[a].

References

[1] Storry, J.R., et al., 2011. International society of blood transfusion working party on red cell immunogenetics and blood group terminology: Berlin report. Vox Sang 101, 77–82.
[2] Karamatic Crew, V., et al., 2003. A new variant in the Ok blood group system [abstract]. Transfus Med 13 (Suppl. 1), 32.

OKVM Antigen

Terminology

ISBT symbol (number)	OK3 (024003 or 24.3)
History	Described in 2006, and named in 2010 from "OK" for the blood group system and "V and M" for the valine to methionine change[1].

Occurrence

One Hispanic OKVM– proband.

OK

Molecular basis associated with OKVM antigen[2]

Amino acid	Val60
Nucleotide	G at bp 178 in exon 2
OKVM–	Met60 and A at bp 178

Effect of enzymes and chemicals on OKVM antigen on intact RBCs

Ficin/Papain	Resistant
Trypsin	Presumed resistant
α-Chymotrypsin	Presumed resistant
DTT 200 mM	Presumed resistant

In vitro characteristics of alloanti-OKVM

Immunoglobulin class	IgG
Optimal technique	IAT

Clinical significance of alloanti-OKVM

No data because only one anti-OKVM has been reported.

References

[1] Storry, J.R., et al., 2011. International society of blood transfusion working party on red cell immunogenetics and blood group terminology: Berlin report. Vox Sang 101, 77–82.

[2] Karamatic Crew, V., et al., 2006. A novel variant in the Ok blood group system [abstract]. Transfus Med 16 (Suppl. 1), 41.

Raph Blood Group System

Number of antigens 1

High prevalence   MER2

Terminology

ISBT symbol (number)	RAPH (025)
CD number	CD151
History	This system was established in 1998, and named RAPH after the first producer of alloanti-MER2. The antigen had been previously recognized by the monoclonal antibody, MER2, and the only antigen in the system retains this name. MER2 was shown to be expressed by tetraspanin (CD151) in 2004.

Expression

Other blood cells	CD34+ cells; there is a rapid decrease in expression during *ex vivo* erythropoiesis
Tissues	Epithelium, endothelium, muscle, renal glomeruli, and tubules, Schwann and dendritic cells, fibroblasts

Gene

Chromosome	11p15.5
Name	*CD151 (RAPH)*
Organization	8 exons over 4.3 kbp of gDNA
Product	Raph, CD151, tetraspanin, TM4SF

The Blood Group Antigen (3/e). DOI: http://dx.doi.org/10.1016/B978-0-12-415849-8.00027-2

585

Database accession numbers

GenBank	BT007397 (mRNA); D29963 (mRNA); NM_004357 (mRNA)
Entrez Gene ID	977

Molecular basis of MER− phenotype[1]

The reference allele, *RAPH*01* (Accession number BT007397) encodes MER2 (RAPH1). Nucleotide differences from this reference allele, and the amino acids affected, are given.

Allele encodes	Allele name	Exon	Nucleotide	Amino acid	Ethnicity (prevalence)
MER2− or RAPH:−1	*RAPH*−01.01*	6	511C>T^	Arg171Cys	Israeli, Pakistani, Turkish (Rare)
MER2− or RAPH:−1	*RAPH*−01.02*	6	533G>A	Arg178His	Turkish (Rare)

^May be *in cis* with 579 A>G in exon 6 (Gly193Gly).

Molecular basis of silencing *RAPH* (RAPH_null phenotype)[1]

Nucleotide difference from *RAPH*01* reference allele (Accession number BT007397), and amino acids affected, are given.

Allele name	Exon	Nucleotide	Amino acid	Ethnicity (prevalence)
*RAPH*01N.01*	5	383insG	Lys127fs → Glu140Stop	Indian Jew (Rare)

Amino acid sequence

```
MGEFNEKKTT  CGTVCLKYLL  FTYNCCFWLA  GLAVMAVGIW  TLALKSDYIS   50
LLASGTYLAT  AYILVVAGTV  VMVTGVLGCC  ATFKERRNLL  RLYFILLLII  100
FLLEIIAGIL  AYAYYQQLNT  ELKENLKDTM  TKRYHQPGHE  AVTSAVDQLQ  150
QEFHCCGSNN  SQDWRDSEWI  RSQEAGGRVV  PDSCCKTVVA  LCGQRDHASN  200
IYKVEGGCIT  KLETFIQEHL  RVIGAVGIGI  ACVQVFGMIF  TCCLYRSLKL  250
EHY                                                         253
```

Carrier molecule

A glycoprotein with four membrane passes.

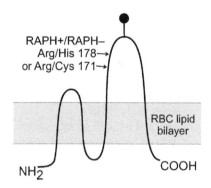

RAPH+/RAPH−
Arg/His 178→
or Arg/Cys 171→

RBC lipid
bilayer

NH$_2$

COOH

M_r (SDS-PAGE)	40,000
CHO: N-glycan	1
Cysteine residues	15

Function

CD151 interacts with α-β integrins and laminin participating in cell adhesion, proliferation, and differentiation. CD151 may be involved in the structure and development of glomerular basement membranes. In cancer cells, enhances cell motility, invasion, and metastasis.

The function of CD151 in the RBC membrane is unknown.

Disease association

An absence of CD151 has been associated with glomerulonephritis and renal failure in three people (two probands); also associated with pretibial epidermolysis bullosa and sensorineural deafness[2].

Comments

The 178His and 171Cys are spacially distinct from the ^{194}Gln-Arg-Asp^{196} motif that is likely the integrin-binding site on CD151. This is consistent with the Arg178His, and the Arg171Cys changes producing a protein lacking the MER2 epitope but retaining the integrin-binding function[3].

References

[1] Crew, V.K., et al., 2004. CD151, the first member of the tetraspanin (TM4) superfamily detected on erythrocytes, is essential for the correct assembly of human basement membranes in kidney and skin. Blood 104, 2217–2223.

RAPH

[2] Kagan, A., et al., 1988. Occurrence of hereditary nephritis, pretibial epidermolysis bullosa and beta-thalassemia minor in two siblings with end-stage renal disease. Nephron 49, 331–332.

[3] Kazarov, A.R., et al., 2002. An extracellular site on tetraspanin CD151 determines $\alpha 3$ and $\alpha 6$ integrin-dependent cellular morphology. J Cell Biol 158, 1299–1309.

MER2 Antigen

Terminology

ISBT symbol	RAPH1 (025001 or 25.1)
Obsolete names	Raph; Raf; 901011
History	The first red cell surface polymorphism to be defined by a monoclonal antibody (MER2) was described in 1987[1]. The alloantibody was described over a decade later, but the antigen retained the MER2 name.

Occurrence

All populations	92% (see Comments)

Expression

Cord RBCs	Not expressed or weakly expressed
Altered	Weakened (slightly) on RBCs with the dominant type of Lu(a–b–)

Effect of enzymes and chemicals on MER2 antigen on intact RBCs

Ficin/Papain	Resistant
Trypsin	Sensitive
α-Chymotrypsin	Sensitive
DTT 200 mM	Variable

In vitro characteristics of alloanti-MER2

Immunoglobulin class	IgG
Optimal technique	IAT
Complement binding	Yes; in three of four human anti-MER2

Clinical significance of alloanti-MER2

Transfusion reaction	No to moderate (chills and rigor)[2]
HDFN	No information

Comments

The 8% of people whose RBCs type MER2–, express MER2 on their erythroid precursors, and on their platelets. It is likely that MER2 is expressed on their RBCs, but is too weak to be detected by hemagglutination. These people have not made anti-MER2, nor have nucleotide changes been found in their *CD151*. Antigen strength varies on RBC samples from different people.

Three individuals (two probands) with anti-MER2 had renal failure requiring dialysis; two had made the antibody before receiving transfusion. All were Indian (South Asian) Jews. A fourth example of alloanti-MER2 was in a healthy Turkish blood donor (the 3rd proband), who had never been transfused but had been pregnant twice. A 4th proband was a Pakistani with two pregnancies and no transfusions, and the 5th proband was a Turk who had been pregnant and transfused[2]. The three Jews originating from India who made alloanti-MER2 had a silenced *CD151* (RAPH$_{null}$). The anti-MER2 made by the Turkish blood donor (*CD151*533G>A*, Arg178His) and the Pakistani and Turkish patients (*CD151*511C>T*, or Arg171Cys) likely have more precise specificities. However, limited cross-testing indicated all examples of alloanti-MER2 had the same specificity[2].

References

[1] Daniels, G.L., et al., 1988. Human alloantibodies detecting a red cell antigen apparently identical to MER2. Vox Sang 55, 161–164.

[2] Karamatic Crew, V., et al., 2008. Two MER2-negative individuals with the same novel CD151 mutation and evidence for clinical significance of anti-MER2. Transfusion 48, 1912–1916.

John Milton Hagen Blood Group System

Number of antigens 6

High prevalence   JMH, JMHK, JMHL, JMHG, JMHM, JMHQ

Terminology

ISBT symbol (number)	JMH (026)
CD number	CD108
Obsolete names	901007; Sema7A
History	JMH became a system in 2000 after it was shown that the JMH glycoprotein is CD108[1], and the gene encoding CD108 was cloned[2,3].

Expression

Red blood cells	Aberrant expression on RBCs can be (transiently) acquired, and study of people with JMH– or JMH-weak phenotypes suggests a post-transcriptional or post-translational mechanism may affect RBC expression[4].
Other blood cells	Weak on lymphocytes, strong on activated lymphocytes and activated macrophages[1].
Tissues	Neurons of central nervous system; respiratory epithelium, placenta, testes, spleen[5], keratinocytes and fibroblasts in the skin[6], and odontoblasts[7]; low expression in brain and thymus

The Blood Group Antigen (3/e). DOI: http://dx.doi.org/10.1016/B978-0-12-415849-8.00028-4

Gene[2,3]

Chromosome	15q24.1
Name	*JMH (SEMA7A, CD108, SEMA-L)*
Organization	14 exons over 9 kb of gDNA
Product	Semaphorin 7A; JMH; CD108; H-Sema-L

Database accession numbers

GenBank	AF069493 (mRNA); AY885237 (gene); NM_003612
Entrez Gene ID	8482

Molecular bases of John Milton Hagen phenotypes

The reference allele, *JMH*01* (Accession number AY885237) encodes JMH (JMH1), JMH2, JMH3, JMH4, JMH5, JMH6. Nucleotide differences from this reference allele, and the amino acids affected, are given.

Allele encodes	Allele name	Exon	Nucleotide	Amino acid	Ethnicity (prevalence)
JMHK– or JMH:–2	*JMH*01.–02*	6	619C>T	Arg207Trp	Japanese (Rare)
JMHL– or JMH:–3	*JMH*01.–03*	6	620G>A	Arg207Gln	Canadian, German (Rare)
JMHG– or JMH:–4	*JMH*01.–04*	11	1379G>A	Arg460His	American (Rare)
JMHM– or JMH:–5	*JMH*01.–05*	11	1381C>T	Arg461Cys	Polish (Rare)
JMHQ– or JMH:–6	*JMH*01.–06*	9	1040G>T	Arg347Leu	Native American (Few)

Amino acid sequence[2]

```
MTPPPPGRAA  PSAPRARVPG  PPARLGLPLR  LRLLLLLWAA  AASAQGHLRS   50
GPRIFAVWKG  HVGQDRVDFG  QTEPHTVLFH  EPGSSSVWVG  GRGKVYLFDF  100
PEGKNASVRT  VNIGSTKGSC  LDKRDCENYI  TLLERRSEGL  LACGTNARHP  150
SCWNLVNGTV  VPLGEMRGYA  PFSPDENSLV  LFEGDEVYST  IRKQEYNGKI  200
PRFRRIRGES  ELYTSDTVMQ  NPQFIKATIV  HQDQAYDDKI  YYFFREDNPD  250
KNPEAPLNVS  RVAQLCRGDQ  GGESSLSVSK  WNTFLKAMLV  CSDAATNKNF  300
NRLQDVFLLP  DPSGQWRDTR  VYGVFSNPWN  YSAVCVYSLG  DIDKVFRTSS  350
LKGYHSSLPN  PRPGKCLPDQ  QPIPTETFQV  ADRHPEVAQR  VEPMGPLKTP  400
LFHSKYHYQK  VAVHRMQASH  GETFHVLYLT  TDRGTIHKVV  EPGEQEHSFA  450
FNIMEIQPFR  RAAAIQTMSL  DAERRKLYVS  SQWEVSQVPL  DLCEVYGGGC  500
HGCLMSRDPY  CGWDQGRCIS  IYSSERSVLQ  SINPAEPHKE  CPNPKPDKAP  550
LQKVSLAPNS  RYYLSCPMES  RHATYSWRHK  ENVEQSCEPG  HQSPNCILFI  600
ENLTAQQYGH  YFCEAQEGSY  FREAQHWQLL  PEDGIMAEHL  LGHACALAAS  650
LWLGVLPTLT  LGLLVH                                          666
```

A signal peptide of 46 amino acids is cleaved after membrane attachment.
A GPI anchor motif of 19 amino acids is cleaved.

Carrier molecule[1]

A GPI-linked glycoprotein.

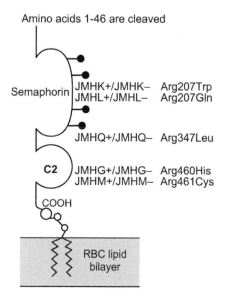

M_r (SDS-PAGE)   68,000 to 76,000
CHO: N-glycan   5 potential; 4 likely
Cysteine residues   19

Function

Secreted and membrane-bound Sema7A function as signals which guide axons in developing nervous tissue[8]. Sema7A is also involved in immune responses, particularly in the effector phase of cellular immunity[9]. It has been described to stimulate human monocytes, and be an effector molecule in T cell-mediated inflammation, and may have an important role in limiting autoimmune response[10]. CD108 contains an Arg-Gly-Asp (267-269) cell attachment motif, which is common in adhesion molecules. The main receptors for Sema7A identified so far are $\alpha_1\beta_1$ integrin[5,6], and plexin C1[11]. Function of Sema7A on RBCs is not known.

References

[1] Mudad, R., et al., 1995. Evidence of CDw108 membrane protein bears the JMH blood group antigen. Transfusion 35, 566–570.

[2] Lange, C., et al., 1998. New eukaryotic semaphorins with close homology to semaphorins of DNA viruses. Genomics 51, 340–350.

[3] Yamada, A., et al., 1999. Molecular cloning of a glycosylphosphatidylinositol-anchored molecule CDw108. J Immunol 162, 4094–4100.

[4] Seltsam, A., et al., 2007. The molecular diversity of Sema7A, the semaphorin that carries the JMH blood group antigens. Transfusion 47, 133–146.

[5] Bobolis, K.A., et al., 1992. Isolation of the JMH antigen on a novel phosphatidylinositol-linked human membrane protein. Blood 79, 1574–1581.

[6] Scott, G.A., et al., 2008. Semaphorin 7a promotes spreading and dendricity in human melanocytes through beta1-integrins. J Invest Dermatol 128, 151–161.

[7] Maurin, J.C., et al., 2005. Odontoblast expression of semaphorin 7A during innervation of human dentin. Matrix Biol 24, 232–238.

[8] Pasterkamp, R.J., et al., 2003. Semaphorin 7A promotes axon outgrowth through integrins and MAPKs. Nature 424, 398–405.

[9] Suzuki, K., et al., 2007. Semaphorin 7A initiates T-cell-mediated inflammatory responses through alpha1beta1 integrin. Nature 446, 680–684.

[10] Czopik, A.K., et al., 2006. Semaphorin 7A is a negative regulator of T cell responses. Immunity 24, 591–600.

[11] Tamagnone, L., et al., 1999. Plexins are a large family of receptors for transmembrane, secreted, and GPI-anchored semaphorins in vertebrates. Cell 99, 71–80.

JMH Antigen

Terminology

ISBT symbol (number)	JMH1 (026001 or 26.1)
Obsolete names	John Milton Hagen; "Old Boys"; 900018
History	Named after the first antibody producer, John Milton Hagen.

Occurrence

All populations   100%

Expression

Cord RBCs	Weak (some variation)
Altered	JMH variants
Absent	PNH III RBCs

Molecular basis associated with JMH antigen

The original JMH– phenotype was the acquired type.
For Molecular basis of JMH phenotypes see table in System pages.

Effect of enzymes and chemicals on JMH antigen on intact RBCs[1]

Ficin/Papain	Sensitive
Trypsin	Sensitive
α-Chymotrypsin	Sensitive
DTT 200 mM/50 mM	Sensitive/sensitive (thus sensitive to WARM™ and ZZAP)
Acid	Resistant

In vitro characteristics of alloanti-JMH

Immunoglobulin class	IgG (predominantly IgG4 in people with the acquired JMH-negative phenotype)
Optimal technique	IAT

Clinical significance of alloanti-JMH

Transfusion reaction	No
HDFN	No

Autoanti-JMH

Autoanti-JMH is often found in elderly persons with an acquired absent or weak expression of JMH; the DAT may be positive.

Comments

One family has shown dominant inheritance of the JMH-negative phenotype in three generations[2].
Alloanti-JMH, present in JMH-positive individuals are due to variant forms of CD108; see JMHK, JMHL, JMHG, JMHM, and JMHQ below.

References

[1] Mudad, R., et al., 1995. JMH variants: serologic, clinical, and biochemical analyses in two cases. Transfusion 35, 925–930.

[2] Kollmar, M., et al., 1981. Evidence of a genetic mechanism for the production of the JMH negative phenotype [abstract]. Transfusion 21, 612.

JMHK Antigen

Terminology

ISBT symbol (number)	JMH2 (026002 or 26.2)
History	Found in 1984, and named in 2006 from "JMH" for the system, and "K" from the first initial of the family name of the JMHK– proband[1].

Occurrence

Only one Japanese JMHK– proband has been reported.

Antithetical antigen

JMHL (**JMH3**)

Expression

Cord RBCs	Presumed expressed (may be some variation)
Absent	PNH III RBCs

Molecular basis associated with JMHK antigen[2]

Amino acid	Arg207
Nucleotide	C at bp 619 in exon 6
JMHK–	Trp207 and T at bp 619

Effect of enzymes and chemicals on JMHK antigen on intact RBCs[3]

Ficin/Papain	Sensitive
Trypsin	Sensitive
α-Chymotrypsin	Presumed sensitive
DTT 200 mM/50 mM	Presumed sensitive

In vitro characteristics of alloanti-JMHK

Immunoglobulin class	IgG (IgG1, IgG2, and/or IgG3)
Optimal technique	IAT

Clinical significance of alloanti-JMHK

Transfusion reaction   Decreased RBC survival[1]
HDFN   No data

Comments

Cross-testing showed that the JMHK and JMHL variants are mutually compatible[2].

References

[1] Moulds, J.J. 2011. Personal Communication.

[2] Seltsam, A., et al., 2007. The molecular diversity of Sema7A, the semaphorin that carries the JMH blood group antigens. Transfusion 47, 133–146.

[3] Mudad, R., et al., 1995. JMH variants: serologic, clinical, and biochemical analyses in two cases. Transfusion 35, 925–930.

JMHL Antigen

Terminology

ISBT symbol (number)   JMH3 (026003 or 26.3)
History   Found in 1984, and named in 2006 from "JMH" for the system and "L" from the first initial of the family name of the JMHL– proband[1].

Occurrence

Two JMHL– probands, one from Canada and one from Germany.

Antithetical antigen

JMHK (**JMH2**)

Expression

Cord RBCs   Presumed expressed (may be some variation)
Absent   PNH III RBCs

Molecular basis associated with JMHL antigen[2]

Amino acid   Arg207
Nucleotide   G at bp 620 in exon 6
JMHL–   Gln207 and A at bp 620

Effect of enzymes and chemicals on JMHL antigen on intact RBCs[3]

Ficin/Papain	Sensitive
Trypsin	Sensitive
α-Chymotrypsin	Presumed sensitive
DTT 200 mM/50 mM	Presumed sensitive

In vitro characteristics of alloanti-JMHL

Immunoglobulin class	IgG (IgG1, IgG2, and/or IgG3)
Optimal technique	IAT

Clinical significance of alloanti-JMHL

Transfusion reaction	Decreased RBC survival[1]
HDFN	No data

Comment

Cross-testing showed that the JMHL and JMHK variants are mutually compatible[2].

References

[1] Moulds J.J. 2011. Personal Communication.

[2] Seltsam, A., et al., 2007. The molecular diversity of Sema7A, the semaphorin that carries the JMH blood group antigens. Transfusion 47, 133–146.

[3] Mudad, R., et al., 1995. JMH variants: serologic, clinical, and biochemical analyses in two cases. Transfusion 35, 925–930.

JMHG Antigen

Terminology

ISBT symbol (number)	JMH4 (026004 or 26.4)
History	Found in 1981, and named in 2006 from "JMH" for the system and "G" from the first initial of the family name of the JMHG– proband[1].

Occurrence

One JMHG– proband from Florida, USA.

Expression

Cord RBCs	Presumed expressed (may be some variation)
Absent	PNH III RBCs

Molecular basis associated with JMHG antigen[2]

Amino acid	Arg460
Nucleotide	G at bp 1379 in exon 11
JMHG–	His460 and A at bp 1379

Effect of enzymes and chemicals on JMHG antigen on intact RBCs

Ficin/Papain	Sensitive
Trypsin	Sensitive
α-Chymotrypsin	Presumed sensitive
DTT 200 mM/50 mM	Presumed sensitive

In vitro characteristics of alloanti-JMHG

Immunoglobulin class	IgG (IgG1, IgG2, and/or IgG3)
Optimal technique	IAT

Clinical significance of alloanti-JMHG

No data because only one anti-JMHG has been described.

Comment

The anti-JMHG was compatible with JMHM– RBCs, but anti-JMHM reacted with JMHG– RBCs[2].

References

[1] Moulds, J.J. 2011. Personal Communication.

[2] Seltsam, A., et al., 2007. The molecular diversity of Sema7A, the semaphorin that carries the JMH blood group antigens. Transfusion 47, 133–146.

JMHM Antigen

Terminology

ISBT symbol (number)	JMH5 (026005 or 26.5)
History	Found in 1980, and named in 2006 from "JMH" for the system and "M" from the first initial of the family name of the JMHM– proband[1].

Occurrence

One JMHM– proband from Israel; of Polish descent.

Expression

Cord RBCs	Presumed expressed (may be some variation)
Absent	PNH III RBCs

Molecular basis associated with JMHM antigen[2]

Amino acid	Arg461
Nucleotide	C at bp 1381 in exon 11
JMHM–	461Cys and T at bp 1381

Effect of enzymes and chemicals on JMHM antigen on intact RBCs

Ficin/Papain	Sensitive
Trypsin	Sensitive
α-Chymotrypsin	Sensitive
DTT 200 mM/50 mM	Sensitive/sensitive (thus sensitive to WARM™ and ZZAP)
Acid	Resistant

In vitro characteristics of alloanti-JMHM

Immunoglobulin class	IgG (IgG1, IgG2, and/or IgG3)
Optimal technique	IAT

Clinical significance of alloanti-JMHM

No data because only one anti-JMHM has been described.

Comment

The anti-JMHG was compatible with JMHM− RBCs, but anti-JMHM reacted with JMHG− RBCs[2].

References

[1] Moulds, J.J. 2011. Personal Communication.

[2] Seltsam, A., et al., 2007. The molecular diversity of Sema7A, the semaphorin that carries the JMH blood group antigens. Transfusion 47, 133–146.

JMHQ Antigen

Terminology

ISBT symbol (number)	JMH6 (026006 or 26.6)
History	Found and named in 2010 from "JMH" for the system and "Q" for Quebec when four JMHQ– probands were found.

Occurrence

All populations   100%

Five JMHQ– probands were Native Americans from a reservation northwest of Quebec City.

Expression

Cord RBCs   Presumed expressed (may be some variation)
Absent   PNH III RBCs

Molecular basis associated with JMHQ antigen[1]

Amino acid   Arg347
Nucleotide   G at bp 1040 in exon 9
JMHQ–   Leu347 and T at bp 1040

Effect of enzymes and chemicals on JMHQ antigen on intact RBCs

Ficin/Papain   Sensitive
Trypsin   Sensitive
α-Chymotrypsin   Presumed sensitive
DTT 200 mM/50 mM   Presumed sensitive

In vitro characteristics of alloanti-JMHQ

Immunoglobulin class   IgG
Optimal technique   IAT

Clinical significance of alloanti-JMHQ

No data because anti-JMHQ is rare.

Comments

The alloanti-JMHQ is compatible with JMH1– RBCs. JMHQ– RBCs are JMH1+[1].

Reference

[1] Richard, M., et al., 2011. A new *SEMA7A* variant found in Native Americans with alloantibody. Vox Sang 100, 322–326.

I Blood Group System

Number of antigens 1

High prevalence I

Terminology

ISBT symbol (number)	I (27)
Obsolete names	207; Ii collection
History	The I antigen was placed in a system in 2002, when mutations of the *I* gene encoding the glycosyltransferase responsible for converting i-active straight oligosaccharide chains to I-active branched chains were identified.

Expression

Soluble form	Human milk, saliva, amniotic fluid, urine, ovarian cyst fluid (small amounts in serum/plasma)
Other blood cells	Lymphocytes, monocytes, granulocytes, platelets
Tissues	Wide tissue distribution

Gene

Chromosome	6p24.2
Name	*I (GCNT2, IGnT)*
Organization	3 exons spread over approximately 100 kbp of gDNA; three forms of exon 1 are differentially spliced to give one of three transcripts: IGnTA, IGnTB or IGnTC[1,2,3]
Product	6-β-*N*-acetylglucosaminyltransferase (β6GlcNAc-transferase, β6GlcNAc-T); the branching enzyme for I antigen expression on RBCs is encoded by *IGnTC*; expression of I antigen on lens epithelium is encoded by *IGnTB*

The Blood Group Antigen (3/e). DOI: http://dx.doi.org/10.1016/B978-0-12-415849-8.00029-6

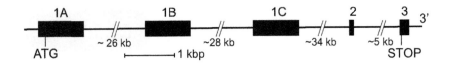

Database accession numbers

GenBank   NM_145655.3; AF458026 (mRNA)
Entrez Gene ID   2651

Molecular bases of weak I antigen

Homozygosity or compound heterozygosity for weakened expression of *GCNT2* alleles leads to the I+^{W} phenotype. The reference allele, *GCNT2*01* (Accession number NM_145655.3) encodes I (I1). Nucleotide differences from this allele, and amino acids affected, are given.

Allele encodes	Allele name	Exon	Nucleotide	Amino acid	Ethnicity (prevalence)
I+^{W}	*GCNT2*01W.01*	1C	243T>A	Asn81Lys	Caucasians, Taiwanese (Rare)
I+^{W}	*GCNT2*01W.02*	1C	505G>A	Ala169Thr	Caucasians (Rare)
I+^{W}	*GCNT2*01W.03*	1C	683G>A	Arg228Gln	Caucasians (Rare)

Molecular bases of silencing of *GCNT2*

Homozygosity or compound heterozygosity for silent *GCNT2* alleles leads to the I– (adult i) phenotype. Differences from *GCNT2*01* reference allele (accession number NM_145655.3) are given.

Allele name	Exon	Nucleotide	Amino acid	Ethnicity (prevalence)
GCNT2*01N.01	3	1049G>A	Gly350Glu	Japanese, Taiwanese (Rare)
GCNT2*01N.02	3	1154G>A	Arg385His	Japanese, Taiwanese (Rare)
GCNT2*01N.04	1C 2	816G>C; 1006G>A	Glu272Asp; Gly336Arg	(Rare)
GCNT2*01N.05	2	984G>A	Trp328Stop	Arabs (Rare)
GCNT2*01N.06	1B, 1C, 2, 3	del exons 1B, 1C, 2, 3	No protein	Taiwanese (Rare) Pakistani (Rare)
GCNT2*01N.07	1C	651delA	Val244Stop	Japanese (Rare)
GCNT2*01N.08	2	935G>A	Gly312Asp	Persian Jews (Rare)

Amino acid sequence for IGnTC β6GlcNAc-transferase[1,2,3]

```
MNFWRYCFFA  FTLLSVVIFV  RFYSSQLSPP  KSYEKLNSSS  ERYFRKTACN   50
HALEKMPVFL  WENILPSPLR  SVPCKDYLTQ  NHYITSPLSE  EEAAFPLAYV  100
MVIHKDFDTF  ERLFRAIYMP  QNVYCVHVDE  KAPAEYKESV  RQLLSCFQNA  150
FIASKTESVV  YAGISRLQAD  LNCLKDLVAS  EVPWKYVINT  CGQDFPLKTN  200
REIVQHLKGF  KGKNITPGVL  PPDHAIKRTK  YVHQEHTDKG  GFFVKNTNIL  250
KTSPPHQLTI  YFGTAYVALT  REFVDFVLRD  QRAIDLLQWS  KDTYSPDEHF  300
WVTLNRVSGV  PGSMPNASWT  GNLRAIKWSD  MEDRHGGCHG  HYVHGICIYG  350
NGDLKWLVNS  PSLFANKFEL  NTYPLTVECL  ELRHRERTLN  QSETAIQPSW  400
YF                                                         402
```

Carrier molecule

The *GCNT2* gene product adds β6GlcNAc to i-active, linear oligosaccharide chains of repeating *N*-acetyllactosamine units on glycolipids and glycoproteins on RBCs, and to glycoproteins in plasma (see figure in Section III). Present on proteins with polylactosamine-containing N-glycans (band 3, glucose transporter, etc.)[4].

A range of copy numbers per RBC has been reported[4].

Function

Not known.

Disease association

A decreased expression of I antigen and concomitant increased expression of the reciprocal i antigen are associated with leukemia, Tk polyagglutination, thalassemia, sickle cell disease, HEMPAS, Diamond Blackfan anemia, myeloblastic erythropoiesis, sideroblastic erythropoiesis, and any condition that results in stress hematopoiesis. Congenital cataracts are associated with a lack or marked reduction of I antigen on RBCs and lens[2]. Caucasians without cataracts have a markedly reduced β6GlcNAc-transferase activity[1]. Asians with cataracts have no β6GlcNAc-transferase activity[2,3]. Anti-I is associated with cold hemagglutinin disease (CHAD) and pneumonia due to *Mycoplasma pneumoniae*.

Phenotypes associated with I antigen and the reciprocal i antigen

RBCs	Antigen expression		Occurrence
	I	i	
Adult	Strong	Weak	Common
Cord	Weak	Strong	All
i Adult	Trace	Strong	Rare

Comments

I antigens occur at the branching points of A-, B-, and H-active oligosaccharide chains.

Branching is under developmental control regulated by phosphorylation of key residues in the C/EBPα transcription factor, which acts on the *GCNT2* promoter. Once the gene is activated, the level of the I antigen expression on RBCs of the newborn child begins to increase[5].

References

[1] Yu, L.-C., et al., 2001. Molecular basis of the adult i phenotype and the gene responsible for the expression of the human blood group I antigen. Blood 98, 3840–3845.

[2] Yu, L.C., et al., 2003. The molecular genetics of the human I locus and molecular background explaining the partial association of the adult i phenotype with congenital cataracts. Blood 101, 2081–2087.

[3] Inaba, N., et al., 2003. A novel I-branching β-1,6-N-acetylglucosaminyltransferase involved in human blood group I antigen expression. Blood 101, 2870–2876.

[4] Cooling, L., 2010. Polyactosamines, there's more than meets the "Ii:" a review of the I system. Immunohematology 26, 133–155.

[5] Yu, L.C., Lin, M., 2011. Molecular genetics of the blood group I system and the regulation of I antigen expression during erythropoiesis and granulopoiesis. Curr Opin Hematol 18, 421–426.

I Antigen

Terminology

ISBT symbol (number)	I1 (027001 or 27.1)
Obsolete names	900026; 207001; Individual
History	Reported in 1956; named I to emphasize the high degree of the "Individuality" of blood samples failing to react with a potent cold agglutinin. Placed in a collection with i antigen in 1990, and made a one-antigen system in 2002 when the gene encoding the branching transferase was cloned.

Occurrence

Adults	>99%

Reciprocal antigen

i [See Ii Collection (207)].

Expression

Cord RBCs	Weaker than on adult RBCs; frequently appear to be I-negative
Altered	Weakened on RBCs produced under hematopoietic stress, and on South East Asian ovalocytes (see also Disease association).

Molecular basis associated with I antigen

See System pages for molecular bases associated with I-negative (adult i) phenotype.

Effect of enzymes and chemicals on I antigen on intact RBCs

Ficin/Papain	Resistant (markedly enhanced)
Trypsin	Resistant (markedly enhanced)
α-Chymotrypsin	Resistant (markedly enhanced)
Sialidase	Resistant (enhanced)
DTT 200 mM	Resistant

Acid   Resistant

In vitro characteristics of anti-I

Immunoglobulin class	IgM (rarely IgG)
Optimal technique	RT or 4°C
Complement binding	Yes; some hemolytic

Clinical significance of anti-I

| Transfusion reaction | No (may need to infuse through an approved blood warmer). Increased destruction of I+ RBCs transfused to people with the adult i phenotype and alloanti-I |
| HDFN | No |

Autoanti-I

Most people have cold-reactive autoanti-I in their plasma.
A common specificity in CHAD and pregnancy.

Comments

So-called compound antigens have been described: IA, IB, IAB, IH, IP1, ILe^{bH}.
Alloanti-I is rare because the I− (adult i) phenotype is rare.

Globoside Blood Group System

Number of antigens 1

High prevalence   P

Terminology

ISBT symbol (number)   GLOB (028)

History   The P antigen was promoted to Globoside (GLOB) Collection and in 2002, when the molecular basis of globoside deficiency was defined, P became a blood group system (GLOB).

Expression

Other blood cells   Erythroid precursor cells, lymphocytes, monocytes

Tissues   Endothelium, placenta (trophoblasts and interstitial cells), fibroblasts, fetal liver, fetal heart, kidney, prostate, peripheral nerves

Gene

Chromosome   3q26.1

Name   *GLOB* (*B3GALNT1*); the previously widely used *B3GALT3* should not be used

Organization   At least five exons (multiple transcripts exist so exact number is still unclear), distributed over ~19 kbp

Product   UDP-N-acetylgalactosamine: globotriaosylceramide 3-β-N-acetylgalactosaminyltransferase (Gb$_4$Cer/ globoside synthase EC2.4.1.79; β3GalNAcT1; P synthase)[1]

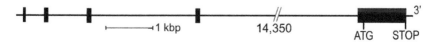

The Blood Group Antigen (3/e). DOI: http://dx.doi.org/10.1016/B978-0-12-415849-8.00030-2

Database accession numbers

GenBank   NM_033169 (mRNA); AB050855 (gene)
Entrez Gene ID   8706

Molecular bases of the P^k (P–, GLOB:–1) phenotype due to changes in *B3GALNT1*[1]

Nucleotide differences from the reference allele, *GLOB*01* (*B3GALNT1*01*; Accession number AB050855), and amino acids affected, are given. This reference allele encodes a 3-β-*N*-acetylgalactosaminyltransferase, which adds *N*-acetylgalactosamine to the lactosylceramide (P^k antigen) to form globoside (P antigen). The null phenotype caused by these alleles can be either P1+ or P1–, i.e., the RBCs have P_1^k or P_2^k phenotype.

Allele name	Exon	Nucleotide	Amino acid	Ethnicity (prevalence)
GLOB*01N.01	5	202C>T	Arg67Stop	Finnish (Few)
GLOB*01N.02	5	292_293insA	Arg97fs102Stop	Italian (Rare)
GLOB*01N.03	5	433C>T	Arg145Stop	North American (Rare)
GLOB*01N.04	5	537_538insA	Asp180fs182Stop	Arabian (Few)
GLOB*01N.05	5	648A>C	Arg216Ser	Canadian (Rare)
GLOB*01N.06	5	797A>C	Glu266Ala	French (Rare)
GLOB*01N.07	5	811G>A	Gly271Arg	European (Few)
GLOB*01N.08	5	959G>A	Trp320Stop	Swiss (Rare)
GLOB*01N.09	5	203delG	Arg68fs84Stop	Maghreb (Rare)
GLOB*01N.10	5 5	376G>A 598delT	Asp126Asn Ser200fs209Stop	French (Rare)
GLOB*01N.11	5	456T>G	Tyr152Stop	Saudi Arabian (Rare)
GLOB*01N.12	5	449A>G	Asp150Gly	Turkish (Rare)

Molecular basis of p (PP1P^k–) phenotype due to changes in *A4GALT*

See P1PK blood group system.

Amino acid sequence of 3-β-*N*-acetylgalactosaminyltransferase

```
MASALWTVLP   SRMSLRSLKW   SLLLLSLLSF   FVMWYLSLPH   YNVIERVNWM    50
YFYEYEPIYR   QDFHFTLREH   SNCSHQNPFL   VILVTSHPSD   VKARQAIRVT   100
WGEKKSWWGY   EVLTFFLLGQ   EAEKEDKMLA   LSLEDEHLLY   GDIIRQDFLD   150
TYNNLTLKTI   MAFRWVTEFC   PNAKYVMKTD   TDVFINTGNL   VKYLLNLNHS   200
EKFFTGYPLI   DNYSYRGFYQ   KTHISYQEYP   FKVFPPYCSG   LGYIMSRDLV   250
PRIYEMMGHV   KPIKFEDVYV   GICLNLLKVN   IHIPEDTNLF   FLYRIHLDVC   300
QLRRVIAAHG   FSSKEIITFW   QVMLRNTTCH   Y                        331
```

Carrier molecule

The P antigen is not a primary gene product; it is located on glycolipids.

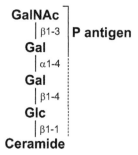

For a diagram of the biosynthetic pathway, see P1PK system.

Copies per RBC   ~15,000,000

Function

The enzyme transfers GalNAc to the terminal Gal of the P^k antigen to synthesize the P antigen.

Disease association[2]

P is a receptor for Parvovirus B19 and some P-fimbriated *E. coli.*
Anti-P is associated with paroxysmal cold hemoglobinuria (PCH).
Cytotoxic IgM and IgG3 antibodies directed against P and/or P^k antigens are associated with a higher than normal rate of spontaneous abortion in women with the rare p [Tj(a–)], P_1^k, and P_2^k phenotypes.

Phenotypes

Phenotype	Occurrence	RBC Antigens	Antibody
P_1	80%^	P, P1, P^k	None
P_2	20%^	P, P^k	Anti-P1
P_1^k	Rare	P1, P^k	Anti-P
P_2^k	Rare	P^k	Anti-P (and anti-P1)
p	Rare	None	Anti-PP1P^k (formerly anti-Tj^a)

Null: P_1^k and P_2^k phenotypes (p phenotype also lacks P but depends on nucleotide changes in *A4GALT*, see P1PK system, **003**).
^= in Caucasians; for other population groups, see P1PK system, **003**.

References

[1] Hellberg, A., et al., 2002. Molecular basis of the Globside-deficient P^k blood group phenotype. Identification of four inactivating mutations in the UDP-N-acetylgalactosamine: globotriaosyl-ceramide 3-beta-N-acetylgalactosaminyltransferase gene. J Biol Chem 277, 29455–29459.

[2] Moulds, J.M., et al., 1996. Human blood groups: incidental receptors for viruses and bacteria. Transfusion 36, 362–374.

P Antigen

Terminology

ISBT symbol (number)	GLOB1 (028001 or 28.1)
Obsolete names	Globoside; Gb$_4$Cer; Gb4; 003002; 209001
History	Anti-P was recognized in 1955 as a component in sera of people with the p phenotype, and in 1959 as the specificity made by people with the P^k phenotype. This resulted in renaming the original anti-P as anti-P$_1$ (now called anti-P1; see P1PK system, **003**).

Occurrence

All populations   >99.9%

Antigen-negative RBCs have mainly been found in Scandinavians, Israelis, Amish, Finns and Arabs.

Expression

Cord RBCs   Expressed

Molecular basis associated with P antigen

For molecular basis of P-negative phenotypes, see System pages.

Effect of enzymes and chemicals on P antigen on intact RBCs

Ficin/Papain	Resistant (markedly enhanced)
Trypsin	Resistant (markedly enhanced)
α-Chymotrypsin	Resistant (markedly enhanced)
DTT 200 mM	Resistant
Acid	Resistant

In vitro characteristics of alloanti-P

Immunoglobulin class	IgM and IgG
Optimal technique	RT; 37°C; IAT
Complement binding	Yes; some hemolytic

Clinical significance of alloanti-P

Transfusion reaction	No to severe (rare) because anti-P is rare (cross-match would be incompatible)
HDFN	No to mild (in P^k mothers with anti-P)
Spontaneous abortions	Cytotoxic IgM and IgG3 antibodies directed against P and/or P^k antigens are associated with a higher than normal rate of spontaneous abortion in women with the rare p [Tj(a–)], P_1^k, and P_2^k phenotypes.

Autoanti-P

Yes, as a biphasic autohemolysin in PCH, detected by the Donath-Landsteiner test. May occur after viral illness, particularly in children.

Comments

Anti-P is compatible with P_1^k and P_2^k phenotype RBCs. Some anti-P react weakly with untreated p phenotype RBCs, while many anti-P react positively when tested against papain-treated p RBCs. This is due to the PX2 antigen (see GLOB Collection).

Experts recommend that if transfusion is necessary and P^k phenotype RBCs are not available, p RBCs can be tried as the clinical significance of anti-PX2 is unknown. If p units cannot be obtained, the recommendation is to transfuse with P-positive washed RBCs (to remove complement) that is infused through an approved blood warmer.

Siblings of patients with anti-P should be tested for compatibility, and the patient urged to donate blood for cryogenic storage when his/her clinical state permits.

Gill Blood Group System

Number of antigens   1

High prevalence   GIL

Terminology

ISBT symbol (number)   GIL (029)

History   Named after the family name of the first antigen-negative proband. Became a system in 2002 after the antigen was shown to be located on aquaglyceroporin.

Expression

Other blood cells   Absent from platelets

Tissues   Kidney medulla and cortex, basolateral membrane of collecting duct cells, small intestine, stomach, colon, spleen, airways, skin, eye[1–3]

Gene[2,4]

Chromosome   9p13.3

Name   (*AQP3*)

Organization   Six exons spanning approximately 6 kbp of gDNA

Product   Aquaglyceroporin, AQP3; a member of the major intrinsic protein (MIP) family of water channels

Database accession numbers

GenBank   NM_004925 (mRNA)

Entrez Gene ID   360

The Blood Group Antigen (3/e). DOI: http://dx.doi.org/10.1016/B978-0-12-415849-8.00031-4

Molecular basis of silencing *GIL* (GIL_{null} phenotype)[5]

The reference allele *GIL*01* (Accession number NM_004925) encodes GIL (GIL1). The nucleotide difference from this allele, and amino acid affected, are given.

Allele encodes	Allele name	Exon	Nucleotide	Restriction enzyme	Ethnicity (prevalence)
GIL− or GIL:−1	*GIL*N.01*	Exon 5 skipped, fs, Stop	IVS5+1g>a	*PmlI+*	Americans, French, Germans (Rare)

Amino acid sequence

```
MGRQKELVSR  CGEMLHIRYR  LLRQALAECL  GTLILVMFGC  GSVAQVVLSR   50
GTHGGFLTIN  LAFGFAVTLG  ILIAGQVSGA  HLNPAVTFAM  CFLAREPWIK  100
LPIYTLAQTL  GAFLGAGIVF  GLYYDAIWHF  ADNQLFVSGP  NGTAGIFATY  150
PSGHLDMING  FFDQFIGTAS  LIVCVLAIVD  PYNNPVPRGL  EAFTVGLVVL  200
VIGTSMGFNS  GYAVNPARDF  GPRLFTALAG  WGSAVFTTGQ  HWWWVPIVSP  250
LLGSIAGVFV  YQLMIGCHLE  QPPPSNEEEN  VKLAHVKHKE  QI          292
```

Carrier molecule

A multipass membrane protein.

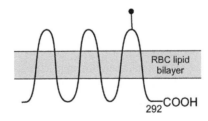

M_r (SDS-PAGE)	46,000; after N-glycosidase F treatment to 26,000
CHO: N-glycan	1
Cysteine residues	6
Copies per RBC	25,000

Function

A water channel that also transports nonionic small molecules such as urea and glycerol.

AQP3 is important in regulating epidermal structure and function[6].
RBCs from a GIL-negative proband had reduced glycerol permeability.

Disease association

No disease association has been noted.

Comments

By Western blotting, RBC membranes from different people have different
levels of expression of AQP3. AQP3 is present in the membrane as dimers,
trimers, and tetramers[7].

References

[1] Agre, P., et al., 2002. Aquaporin water channels: from atomic structure to clinical medicine.
J Physiol London 542, 3–16.

[2] Ishibashi, K., et al., 1994. Molecular cloning and expression of a member of the aquaporin family with permeability to glycerol and urea in addition to water expressed at the basolateral membrane of kidney collecting duct cells. Proc Natl Acad Sci USA 91, 6269–6273.

[3] Roudier, N., et al., 2002a. AQP3 deficiency in humans and the molecular basis of a novel blood
group system, GIL. J Biol Chem 277, 45854–45859.

[4] Inase, N., et al., 1995. Isolation of human aquaporin 3 gene. J Biol Chem 270, 17913–17916.

[5] Roudier, N., et al., 2002b. Erythroid expression and oligomeric state of the AQP3 protein. J Biol
Chem 277, 7664–7669.

[6] Qin, H., et al., 2011. Aquaporin-3 in keratinocytes and skin: its role and interaction with phos-
pholipase D2. Arch Biochem Biophys 508, 138–143.

[7] Ledvinova, J., et al., 1997. Blood group B glycosphingolipids in α-galactosidase deficiency
(Fabry disease): influence of secretor status. Biochim Biophys Acta 1345, 180–187.

GIL Antigen

Terminology

ISBT symbol (number)	GIL1 (029001 or 29.1)
Obsolete name	Gill
History	Reported in 1981; name derived from the first antigen-negative proband who made the alloantibody.

Occurrence

Five GIL– probands were American, French, and German.

Expression

Cord RBCs	Slightly weaker than on RBCs from adults

Molecular basis associated with GIL antigen

GIL-negative RBCs lack aquaglyceroporin (AQP3).

Effect of enzymes and chemicals on GIL antigen on intact RBCs

Ficin/Papain	Resistant (enhanced)
Trypsin	Resistant (enhanced)
α-Chymotrypsin	Resistant (enhanced)
DTT 200 mM	Resistant
Acid	Resistant

In vitro characteristics of alloanti-GIL

Immunoglobulin class	IgG
Optimal technique	IAT
Complement binding	Yes

Clinical significance of alloanti-GIL

Transfusion reaction	Hemolytic
HDFN	Positive DAT, but no clinical HDFN

Comments

There may be heterogeneity among the five reported anti-GIL[1].

Reference

[1] Daniels, G.L., et al., 1998. GIL: a red cell antigen of very high frequency. Immunohematology 14, 49–52.

RHAG

Rh-Associated Glycoprotein Blood Group System

Number of antigens   4

Low prevalence   Olᵃ, RHAG4
High prevalence   Duclos, DSLK (Duclos-like)

Terminology

ISBT Symbol (number) RHAG (030)
CD number   CD241
History   RhAG was elevated to a blood group system in 2008
                      when the molecular basis of Duclos, Olᵃ, and DSLK
                      were determined to be due to nucleotide changes in
                      RHAG.

Expression

Cord RBCs   Expressed
Tissues   RhAG homologs

Gene

Chromosome   6p21.3
Name   *RHAG*
Organization   10 exons distributed over 32 kbp of DNA
Product   Rh-associated glycoprotein (RhAG)

RHAG*01/RHAG*–01

The Blood Group Antigen (3/e). DOI: http://dx.doi.org/10.1016/B978-0-12-415849-8.00032-6

Database accession numbers

GenBank	X64594 (mRNA); NG_011704 (gene); NM_000324 (mRNA)
Entrez Gene ID	6005

Molecular basis of Rh-associated glycoprotein antigens[1]

Reference allele, *RHAG*01* (Accession number X64594) encodes Duclos (RHAG1), RHAG3. Differences from this allele are given.

Allele encodes	Allele name	Exon	Nucleotide	Amino acid	Ethnicity (prevalence)
Duclos− or RHAG:−1	*RHAG*−01*	2	316C>G	Gln106Glu	(Rare)
Ol(a+) or RHAG:2	*RHAG*01.02*	5	680C>T^	Ser227Leu	Norwegian, Japanese (Several)
DSLK− or RHAG:−3	*RHAG*01.−03*	3	490A>C	Lys164Gln	(Rare)
RHAG:4	*RHAG*01.04*	6	808G>A	Val270Ile	(Rare)

^= Homozygosity for *RHAG*01.02* encodes an Rh_{mod} phenotype.

Molecular bases of silencing of *RHAG*

Homozygosity and compound heterozygosity leads to regulator Rh_{null} phenotype.

Assignment of null (*N*) alleles has been made (by the ISBT working party) according to the lack of phenotypic expression of RhD and RhCE antigens or lack of reactivity with monoclonal anti-RhAG (e.g., 2D10, L18.18). Differences from *RHAG*01* reference allele (Accession number X64594) are given.

Allele name	Exon (intron)	Nucleotide	Restriction enzyme	Amino acid	Ethnicity (prevalence)
RHAG*01N.01	2	154–157CCTC>GA	MnlI–	Tyr51fs, Ile107Stop	South African (Rare)
RHAG*01N.02	8	1086delA	PvuII–	Ala362fs, Val374Stop	(Rare)
RHAG*01N.03	Intron 1	IVS1+1g>a		Alternative splicing	White American (Rare)
RHAG*01N.04	Intron 6	IVS6+1g>a		Alternative splicing[2]	(Rare)
RHAG*01N.05	Intron 6	IVS6–1g>a		Alternative splicing	Spanish (Rare)
RHAG*01N.06	Intron 6	IVS6–1g>t		Alternative splicing	Japanese (Rare)
RHAG*01N.07	Intron 7	IVS7+1g>a	PmlI–	Alternative splicing	Japanese (Rare)
RHAG*01N.08	6	808G>A^ 838G>A		Val270Ile Gly280Arg	Japanese (Rare)
RHAG*01N.09	6	836G>A	MnlI–	Gly279Glu	Australian (English/French, Irish/Scottish) (Rare)
RHAG*01N.10	8	1094T>G		Leu365Arg[2]	(Rare)
RHAG*01N.11	9	1139G>T		Gly380Val + alternative splicing	Japanese (Rare)
RHAG*01N.12	5	762C>A		Ser224Arg[3]	Chinese (Rare)

See Rh blood group system for molecular bases of amorph Rh_{null} phenotype.
^ = This single change encodes RHAG:4.

Molecular bases of weak Rh-Associated Glycoprotein antigens

Homozygosity, compound heterozygosity, or heterozygosity for *RHAG*01M.08* or *RHAG*01M.09* leads to Rh$_{mod}$ phenotype.

Assignment of mod (*M*) alleles has been made (by the ISBT working party) according to the weak phenotypic expression of RhD and RhCE antigens or depressed RhAG.

Differences from *RHAG*01* reference allele (Accession number X64594) are given.

Allele name	Exon	Nucleotide	Amino acid	Ethnicity (prevalence)
RHAG*01M.01	9	1183delA	Asn395fs + 52 additional amino acids	Japanese (Rare)
RHAG*01M.02	1	3G>T	Met1Ile	Jewish Russian (Rare)
RHAG*01M.03	2	236G>A	Ser79Asn	Caucasian (Rare)
RHAG*01M.04	2	269G>T	Gly90Val[4]	(Rare)
RHAG*01M.05	3	398T>C	Leu133Pro[5#]	(Rare)
RHAG*01M.06	4	560G>A	Gly187Asp[4]	(Rare)
RHAG*01M.07	9	1195G>T	Asp399Tyr	French (Rare)
RHAG*01M.08	2	182T>G	Ile61Arg	(Rare)
RHAG*01M.09	2	194T>C	Phe65Ser	(Few)
*RHAG*01.02^*	5	680C>T^	Ser227Leu	Japanese

^ = Homozygosity causes a Rh$_{mod}$ phenotype. This allele encodes the Ol^a antigen.
\# = In the abstract 133 is, incorrectly, given as Arg.

Amino acid sequence

```
MRFTFPLMAI  VLEIAMIVLF  GLFVEYETDQ  TVLEQLNITK  PTDMGIFFEL   50
YPLFQDVHVM  IFVGFGFLMT  FLKKYGFSSV  GINLLVAALG  LQWGTIVQGI  100
LQSQGQKFNI  GIKNMINADF  SAATVLISFG  AVLGKTSPTQ  MLIMTILEIV  150
FFAHNEYLVS  EIFKASDIGA  SMTIHAFGAY  FGLAVAGILY  RSGLRKGHEN  200
EESAYYSDLF  AMIGTLFLWM  FWPSFNSAIA  EPGDKQCRAI  VDTYFSLAAC  250
VLTAFAFSSL  VEHRGKLNMV  HIQNATLAGG  VAVGTCADMA  IHPFGSMIIG  300
SIAGMVSVLG  YKFLTPLFTT  KLRIHDTCGV  HNLHGLPGVV  GGLAGIVAVA  350
MGASNTSMAM  QAAALGSSIG  TAVVGGLMTG  LILKLPLWGQ  PSDQNCYDDS  400
VYWKVPKTR                                                   409
```

Carrier protein

A multipass membrane glycoprotein.

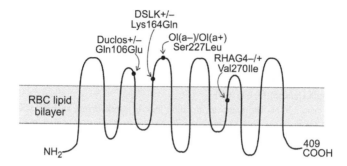

M_r (SDS-PAGE)	45,000 to 100,000 with a predominant band of 50,000
CHO: N-glycan	1 (plus 1 potential site)
Cysteine residues	5
Copies per RBC	100,000–200,000

Function

RhAG forms a core complex with RhD and/or RhCE. This complex is also associated with GPB, LW, and CD47, which in turn is associated with the band 3/GPA complex that links to the RBC membrane skeleton via ankyrin and protein 4.2. This complex maintains erythrocyte membrane integrity, as demonstrated by the abnormal morphology of stomatocytic Rh_{null} RBCs. Involved in NH_4^+/NH_3 or CO_2/O_2, and cation transport across the membrane[6].

Disease association

Compensated hemolytic anemia occurs in some individuals with regulator type Rh_{null} or Rh_{mod} RBCs. Over-hydrated stomatocytosis (OHSt) is associated with dominant mutations in RhAG: Phe65Ser (six families) and Ile61Arg (one family), and cation leak[7].

Comments

RhAG is essential for expression of Rh antigens[8].

References

[1] Tilley, L., et al., 2010. A new blood group system, RHAG: three antigens resulting from amino acid substitutions in the Rh-associated glycoprotein. Vox Sang 98, 151–159.

RHAG

[2] Tsuneyama, H., et al., 2005. Identification of two new mutations in the RhAG gene of Japanese with Rh_{null} phenotype [abstract]. Transfusion 45 (Suppl.), 130A.

[3] Tian, L., et al., 2011. A family study of the Chinese Rh(null) individual of the regulator type: a novel single missense mutation identified in RHAG gene. Transfusion 51, 2686–2689.

[4] Scharberg, A., et al., 2006. RHMOD phenotype caused by double heterozygosity for two new alleles of the RHAG gene [abstract]. Vox Sang 91 (Suppl. 3), 129.

[5] Tsuneyama, H., et al., 2008. Identification of two new mutations in the RhAG gene of Japanese with Rh_{mod} phenotype [abstract]. Transfusion 48 (Suppl.), 185A–186A.

[6] Burton, N.M., Anstee, D.J., 2008. Structure, function and significance of Rh proteins in red cells. Curr Opin Hematol 15, 625–630.

[7] Bruce, L.J., et al., 2009. The monovalent cation leak in overhydrated stomatocytic red blood cells results from amino acid substitutions in the Rh-associated glycoprotein. Blood 113, 1350–1357.

[8] Huang, C.-H., et al., 2000. Molecular biology and genetics of the Rh blood group system. Semin Hematol 37, 150–165.

Duclos Antigen

Terminology

ISBT symbol (number)	RHAG1 (030001 or 30.1)
History	Identified in 1978. Mrs Duclos made an antibody to a high-prevalence antigen that was non-reactive with Rh_{null} and U– RBCs.

Occurrence

Only one Duclos– proband has been reported.

Expression

Cord RBCs	Expressed

Molecular basis associated with Duclos antigen[1]

Amino acid	Gln106
Nucleotide	C at bp 316 in exon 2
Duclos–	Glu106 and G at bp 316

Effect of enzymes and chemicals on Duclos antigen on intact RBCs

Ficin/Papain	Resistant (enhanced)
Trypsin	Resistant (enhanced)
α-Chymotrypsin	Weakened
DTT 200mM/50mM	Sensitive/resistant (thus sensitive to WARM™ and ZZAP)

In vitro characteristics of alloanti-Duclos

Immunoglobulin class	IgG
Optimal technique	IAT

Clinical significance of alloanti-Duclos

No information is available because only one example of anti-Duclos has been described.

Comments

Duclos is absent from RBCs with either the U– Rh_{null} or U– Rh_{mod} phenotype.

Reference

[1] Tilley, L., et al., 2010. A new blood group system, RHAG: three antigens resulting from amino acid substitutions in the Rh-associated glycoprotein. Vox Sang 98, 151–159.

Ol^a Antigen

Terminology

ISBT symbol (number)	RHAG2 (030002 or 30.2)
Obsolete name	Oldeide
History	Identified in 1986 by screening random donors with the multispecific serum Kirk. Named from the initials of the Ol(a+) index case whose C+c+ RBCs had a weak expression of C.

Occurrence

All populations	<0.01%

Expression

Cord RBCs	Presumed expressed

Molecular basis associated with Ol^a antigen[1]

Amino acid	Leu227
Nucleotide	T at bp 680 in exon 5
Ol(a–) (wild type)	Ser227 and C at bp 680

Effect of enzymes and chemicals on Ol^a antigen on intact RBCs

Ficin/Papain	Resistant (enhanced)
Trypsin	Resistant

RHAG

In vitro characteristics of alloanti-Ol[a]

Immunoglobulin class	IgG
Optimal technique	IAT

Clinical significance of alloanti-Ol[a]

No data because antigen is rare.

Comments

Homozygosity for *RHAG*01.02* encodes an Rh$_{mod}$ phenotype.
Ol(a+) RBCs have a weakened expression of Rh antigens (C and E) when *in trans* haplotype is informative.

Reference

[1] Tilley, L., et al., 2010. A new blood group system, RHAG: three antigens resulting from amino acid substitutions in the Rh-associated glycoprotein. Vox Sang 98, 151–159.

DSLK Antigen

Terminology

ISBT symbol (number)	RHAG3 (030003 or 30.3)
History	Identified in 1996 and named in 2010: "DS" from "Duclos," and "LK" from "like" because the proband's RBCs were not agglutinated by the Duclos plasma.

Occurrence

Only one DSLK– proband has been reported.

Expression

Cord RBCs	Presumed expressed

Molecular basis associated with DSLK antigen[1]

Amino acid	Lys164
Nucleotide	A at bp 490 in exon 5
DSLK–	Gln164 and C at bp 490

Effect of enzymes and chemicals on DSLK antigen on intact RBCs

Ficin/Papain	Resistant (enhanced)
Trypsin	Resistant

RHAG

In vitro characteristics of alloanti-DSLK

Immunoglobulin class	IgG
Optimal technique	IAT

Clinical significance of alloanti-DSLK

No data because only one example of anti-DSLK has been described.

Comments

DSLK is absent from RBCs with either the U– Rh_{null} or U– Rh_{mod} phenotype.

Reference

[1] Tilley, L., et al., 2010. A new blood group system, RHAG: three antigens resulting from amino acid substitutions in the Rh-associated glycoprotein. Vox Sang 98, 151–159.

RHAG4 Antigen

Terminology

ISBT symbol (number)	RHAG4 (030004 or 30.4)
History	Identified in 2011, when the antibody caused severe HDFN.

Occurrence

The only antigen-positive proband was of African ancestry.

Expression

Cord RBCs	Expressed

Molecular basis associated with RHAG4 antigen[1]

Amino acid	Ile270
Nucleotide	A at bp 808 in exon 6
RHAG– (wild type)	Val270 and G at bp 808

Effect of enzymes and chemicals on RHAG4 antigen on intact RBCs

Ficin/Papain	Resistant (enhanced)
Trypsin	Resistant
α-Chymotrypsin	Presumed weakened
DTT 200 mM/50 mM	Presumed sensitive/resistant (thus sensitive to WARM™ and ZZAP)

In vitro characteristics of alloanti-RHAG4

Immunoglobulin class   IgG
Optimal technique   IAT

Clinical significance of alloanti-RHAG4

Transfusion reaction   No data because antigen is rare
HDFN   Severe[1]

Reference

[1] Poole, J., et al., 2011. A novel RHAG blood group antigen associated with severe HDFN [abstract]. Vox Sang 101 (Suppl. 1), 70.

FORS Blood Group System

Number of antigens 1

Low prevalence   FORS1

Terminology

ISBT symbol (number)   FORS (031)

History A_{pae} was reported in 1987 as a subgroup of A in three English families[1]. In 2011, it was shown to be independent of ABO, and was indeed the Forssman antigen[2]. Thus, A_{pae} was renamed in honor of John Forssman who first discovered this antigen that bears his name. At the time of printing, Forssman had been provisionally assigned the ISBT System number 031 and the name "FORS".

Expression

Other blood cells   Not normally expressed on blood cells

Tissues   Reports about Forssman glycolipid expression in normal human gastric and colonic mucosa, lung, and kidney

Gene

Chromosome   9q34.2

Name   *FORS (GBGT1, A3GALNT)*

Organization   7 exons spread over approximately 11 kbp of gDNA

Product   Globoside 3-α-N-acetylgalactosaminyltransferase 1 (Forssman glycolipid synthetase)

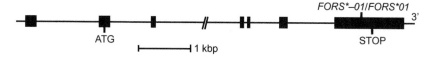

The Blood Group Antigen (3/e). DOI: http://dx.doi.org/10.1016/B978-0-12-415849-8.00033-8

Database accession numbers

GenBank   NM_021996 (mRNA); NC_000009.11
EMBL   HE583597
Entrez Gene ID   26301

Molecular bases of the FORS1+ RBC phenotype[2]

*GBGT1*01.01* (EMBL accession number HE583597) encodes FORS1 on RBCs. The nucleotide difference from the reference allele (NM_021996, *GBGT1*01N.01* below), and amino acid affected, are given.

Allele encodes	Allele name^	Exon	Nucleotide	Amino acid	Ethnicity (prevalence)
FORS:1 (FORS1+)	*GBGT1*01.01*[a]	7	887G>A	Arg296Gln	British (Rare)
FORS:1 (FORS1+)	*GBGT1*01.02*[b]	2	58C>T	Leu20Phe	British (Rare)
		7	887G>A	Arg296Gln	

^The sequences encoding these alleles have been deposited in the EMBL database under the following accession numbers: HE583597[a] and HE583598[b].

Molecular bases of the FORS1– RBC phenotype[2,4]

*GBGT1*01.01* (EMBL accession number HE583597) encodes FORS1 on RBCs. The reference allele (NM_021996, *GBGT1*01N.01*) is a null allele and is compared with other null alleles in the table below. Nucleotides of importance and amino acids affected are given.

Allele encodes	Allele name^	Exon	Nucleotide	Amino acid	Ethnicity (prevalence)
FORS:–1 (FORS1–)	*GBGT1*01N.01*[a]	7	887G	Arg296	(Common)
FORS:–1 (FORS1–)	*GBGT1*01N.02*[b]	2	58C>T	Leu20Phe	(Common)
		7	887G	Arg296	
FORS:–1 (FORS1–)	*GBGT1*01N.03*[c]	7	363C>A	Tyr121Stop	(Several)

^The sequences encoding these alleles have been deposited in the EMBL database under the following accession numbers: HE583599[a] (consensus), HE583600[b], HE583596[c].

Amino acid sequence for globoside 3-α-*N*-acetylgalactosaminyltransferase 1

```
MHRRRLALGL  GFCLLAGTSL  SVLWVYLENW  LPVSYVPYYL  PCPEIFNMKL   50
HYKREKPLQP  VVWSQYPQPK  LLEHRPTQLL  TLTPWLAPIV  SEGTFNPELL  100
QHIYQPLNLT  IGVTVFAVGK  YTHFIQSFLE  SAEEFFMRGY  RVHYYIFTDN  150
PAAVPGVPLG  PHRLLSSIPI  QGHSHWEETS  MRRMETISQH  IAKRAHREVD  200
YLFCLDVDMV  FRNPWGPETL  GDLVAAIHPS  YYAVPRQQFP  YERRRVSTAF  250
VADSEGDFYY  GGAVFGGQVA  RVYEFTRGCH  MAILADKANG  IMAAWREESH  300
LNRHFISNKP  SKVLSPEYLW  DDRKPQPPSL  KLIRFSTLDK  DISCLRS     347
```

Carrier molecule

The *GBGT1* gene product adds α1-3GalNAc to globoside (the P antigen).

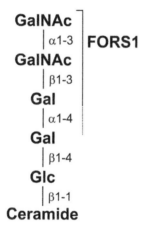

Function

The *GBGT1*-encoded glycosyltransferase catalyzes the formation of Forssman glycolipids via the addition of *N*-acetylgalactosamine (GalNAc) in α-1,3-linkage to its acceptor substrate globoside, the P antigen.

Disease association

Glycolipids serve as involuntary receptors for the adherence of selected pathogens. P-fimbriated strains (expressing the PrsG adhesin that binds to terminal α3GalNAc) of *E. coli* attach to non-primate mammal cells expressing FORS1 antigen[3]. A_{pae} RBCs bind nephritogenic PrsG+ *E. coli* strains *in vitro*[4]. It is possible that FORS1 expression on human cells may increase the susceptibility for infections with *E. coli* that normally prefer non-primate mammal hosts such as dogs and sheep. Cells expressing Forssman glycolipids are less susceptible to the effects of Shiga toxin[5].

Several studies have shown appearance of Forssman glycolipid in human cancer cells, such as lung, colon, and stomach malignancies.

Comments

Forssman glycolipid is widely considered an animal structure with unequal distribution (for instance present in mouse, sheep, dog, cat, and horse, but not in rat, rabbit, and primates). The amino acid sequence of Forssman synthetase in humans differs from that of the canine enzyme by substitution of 58 residues, one of which is amino acid 296 that is altered to the canine version in A_{pae} individuals resulting in FORS1+ RBCs.

References

[1] Stamps, R., et al., 1987. A new variant of blood group A. Apae. Transfusion 27, 315–318.

[2] Hult, A.K., et al., 2011. Forssman expression on human red cells: biochemical and genetic basis of a novel histo-blood group system candidate [abstract]. Transfusion 51 (Suppl. 3), 1A.

[3] Xu, H., et al., 1999. Characterization of the human Forssman synthetase gene. an evolving association between glycolipid synthesis and host–microbial interactions. J Biol Chem 274, 29390–29398.

[4] Hult, A.K., et al., 2011. Genetic basis of Forssman antigen expression on human red cell blood cells [abstract]. Vox Sang 101 (Suppl. 2), 33.

[5] Elliott, S.P., et al., 2003. Forssman synthetase expression results in diminished shiga toxin susceptibility: a role for glycolipids in determining host-microbe interactions. Infect Immun 71, 6543–6552.

FORS1 Antigen

Terminology

ISBT symbol (number)	FORS1 (031001 or 31.1)
Obsolete names	A_{pae}
History	Forssman antigen has been known since 1911, following Prof. Forssman's experiments in which extracts of guinea pig kidney were injected into rabbits[1]. The resulting immune sera hemolyzed sheep erythrocytes.
	A century later, the supposed ABO subgroup A_{pae} was shown to be independent of the ABO system, but dependent on expression of Forssman glycolipids on RBCs, and the phenotype was renamed FORS1+.

Occurrence

Caucasians	<0.1%

Molecular basis associated with expression of FORS1 antigen on RBCs[2]

Amino acid	Gln296
Nucleotide	A at bp 887 in exon 7
FORS– (wild type)	Arg296 and G at bp 887

Effect of enzymes and chemicals on FORS1 antigen on intact RBCs

Ficin/Papain	Resistant (enhanced)
Trypsin	Resistant (enhanced)
α-Chymotrypsin	Resistant (enhanced)
DTT 200mM	Resistant

In vitro characteristics of anti-FORS1

Immunoglobulin class	IgM (some IgG)
Optimal technique	RT or 4°C; enzymes

Clinical significance of anti-FORS1

Not known.

Comments

Group O RBCs expressing FORS1 antigen are agglutinated strongly by *Helix pomatia*, but not by *Dolichos biflorus*, and weakly by some polyclonal anti-A and anti-A,B reagents, but not by monoclonal anti-A. The terminal α3GalNAc attached to the H carbohydrate structure confers the A antigen, but when attached to the P carbohydrate structure confers the FORS1 antigen. This provides an explanation of the cross-reactivity with some anti-A.

FORS1+ donor RBCs may result in a weakly or strongly positive cross-match reaction due to naturally-occurring anti-FORS1 in the plasma of ABO-compatible FORS1– individuals.

References

[1] Forssman, J., 1911. Die Herstellung hochwertiger spezifisher Schafhämolysine ohne Verwendung von Schafblut: Ein Beitrag Zur Lehre von heterologer Antikörperbildung. Biochemische Zeitung 37, 78–115.

[2] Hult, A.K., et al., 2011. Forssman expression on human red cells: biochemical and genetic basis of a novel histo-blood group system candidate [abstract]. Transfusion 51 (Suppl. 3), 1A.

FORS

JR Blood Group System

Number of antigens 1

High prevalence Jr^a

Terminology

ISBT symbol (number)   JR (032)
CD number   CD338
History   The Jr^a antigen was promoted from the 901 Series of High-Incidence antigens to a System in 2012, when it was shown that *ABCG2* null alleles define the Jr(a–) phenotype[1].

Expression

Tissues   Highly expressed in placenta (syncytiotrophoblasts). Low expression in epithelial cells of small and large intestines, liver ducts, colon, lobules of the breast, endothelial cells of veins and capillaries, and brain microvessel endothelium, stem cells, lung, and in the apical membrane of proximal tubules of the kidney. It is unregulated in breast and brain tumors.

Gene

Chromosome   4q22.1
Name   *JR (ABCG2)*
Organization   16 exons spread over approximately 68.6 kbp of gDNA
Product   Jr glycoprotein (ATP-binding cassette, sub-family G, member 2 [ABCG2]; breast cancer resistance protein [BCRP])

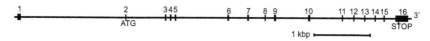

The Blood Group Antigen (3/e). DOI: http://dx.doi.org/10.1016/B978-0-12-415849-8.00034-X

Database accession numbers

GenBank   NM_004827.2 (DNA)
Entrez Gene ID   9429

Molecular bases of silencing *JR* [JR$_{null}$ (Jr(a–), JR:–1)] phenotype[1-3]

The reference allele, *ABCG2* (Accession number NM_004827.2) encodes Jr^a (JR1). Nucleotide differences from this reference allele, and amino acids affected, are given.

Allele name	Exon	Nucleotide	Restriction enzyme	Amino acid	Ethnicity (prevalence)
*ABCG2*01N.01* or *JR*01N.01*	4	376C>T	*Rsa*I–	Gln126Stop	Asian (Many)
*ABCG2*01N.02* or *JR*01N.02*	7	706C>T	*Bsm*FI–	Arg236Stop	Europeans (Few)
*ABCG2*01N.03* or *JR*01N.03*	7	736C>T	*Taq*I–	Arg246Stop	(Few)
*ABCG2*01N.04* or *JR*01N.04*	4	337C>T	*Tsp*45I+	Arg113Stop	(Rare)
*ABCG2*01N.05* or *JR*01N.05*	7	784G>T	*Nla*III+	Gly262Stop	(Few)
*ABCG2*01N.06* or *JR*01N.06*	13	1591C>T	*Hpy*188I–	Gln531Stop	(Rare)
*ABCG2*01N.07* or *JR*01N.07*	2	187_197del ATATTATCGAA		Ile63TyrfsStop	(Rare)
*ABCG2*01N.08* or *JR*01N.08*	6	542_543insA		Phe182ValfsStop fsStop	(Rare)
*ABCG2*01N.09* or *JR*01N.09*	7	730C>T		Gln244Stop	(Rare)
*ABCG2*01N.10* or *JR*01N.10*	7	791_792delTT		Leu264His fsStop	(Few)
*ABCG2*01N.11* or *JR*01N.11*	8	875_878dupACTT		Phe293Leu fsStop	(Rare)

Amino acid sequence

```
MSSSNVEVFI  PVSQGNTNGF  PATASNDLKA  FTEGAVLSFH  NICYRVKLKS   50
GFLPCRKPVE  KEILSNINGI  MKPGLNAILG  PTGGGKSSLL  DVLAARKDPS  100
GLSGDVLING  APRPANFKCN  SGYVVQDDVV  MGTLTVRENL  QFSAALRLAT  150
TMTNHEKNER  INRVIQELGL  DKVADSKVGT  QFIRGVSGGE  RKRTSIGMEL  200
ITDPSILFLD  EPTTGLDSST  ANAVLLLLKR  MSKQGRTIIF  SIHQPRYSIF  250
KLFDSLTLLA  SGRLMFHGPA  QEALGYFESA  GYHCEAYNNP  ADFFLDIING  300
DSTAVALNRE  EDFKATEIIE  PSKQDKPLIE  KLAEIYVNSS  FYKETKAELH  350
QLSGGEKKKK  ITVFKEISYT  TSFCHQLRWV  SKRSFKNLLG  NPQASIAQII  400
VTVVLGLVIG  AIYFGLKNDS  TGIQNRAGVL  FFLTTNQCFS  SVSAVELFVV  450
EKKLFIHEYI  SGYYRVSSYF  LGKLLSDLLP  MRMLPSIIFT  CIVYFMLGLK  500
PKADAFFVMM  FTLMMVAYSA  SSMALAIAAG  QSVVSVATLL  MTICFVFMMI  550
FSGLLVNLTT  IASWLSWLQY  FSIPRYGFTA  LQHNEFLGQN  FCPGLNATGN  600
NPCNYATCTG  EEYLVKQGID  LSPWGLWKNH  VALACMIVIF  LTIAYLKLLF  650
LKKYS                                                      655
```

Carrier molecule

A multipass membrane glycoprotein with one nucleotide-binding domain (NBD; residues 1 to ~396), followed by one membrane-spanning domain (MSD; residues ~397 to 655). The functional molecule is likely a homodimer.

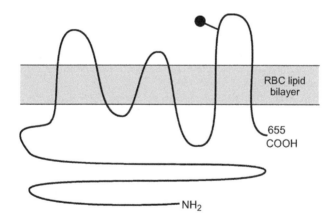

M_r (SDS-PAGE)	72,000 reduced; 180,000 non-reduced
CHO: N-glycan	Three potential; one likely
Cysteine residues	12

Function

ABCG2 is an ATP-dependent transport protein that has broad substrate specificity (including for urate). It is involved in multidrug resistance in tumor cells, particularly in breast cancer, and may function in the defense of normal cells against toxic agents, and have a role in folate homeostasis. Transport of PPIX suggests that ABCG2 may be important for homeostasis of endogenous por-phyrins. The abcg2 expression in mice conferred a survival advantage during hypoxia.

Xenobiotic transporter that may play an important role in the exclusion of xenobiotics from the brain; may be involved in brain-to-blood efflux; appears to play a major role in the multidrug resistance phenotype of several can-cer cell lines. When overexpressed, the transfected cells become resistant to mitoxantrone, daunorubicin, and doxorubicin, display diminished intracellular accumulation of daunorubicin, and manifest an ATP-dependent increase in the efflux of rhodamine.

Significant expression of this protein has been observed in the placenta, which may suggest a potential role for this molecule in placenta tissue[4,5].

Disease association

The Gln126Stop and Gln141Lys variants of ABCG2 are associated with an increased risk for gout[5].

References

[1] Zelinski, T., et al., 2012. ABCG2 null alleles define the Jr(a–) blood group phenotype. Nat Genet 44, 131–132.

[2] Reid, M.E., et al. 2012. The JR Blood Group System (ISBT 032): Molecular Characterization of Three New Null Alleles. Transfusion, submitted.

[3] Saison, C., et al., 2012. Null alleles of *ABCG2* encoding the breast cancer resistance protein define the new blood goup system Junior. Nat Genet 44, 174–177.

[4] Doyle, L.A., et al., 1998. A multidrug resistance transporter from human MCF-7 breast cancer cells. Proc Natl Acad Sci USA 95, 15665–15670.

[5] Woodward, O.M., et al., 2011. ABCG transporters and disease. FEBS Journal 278, 3215–3225.

Jr^a Antigen

Terminology

ISBT symbol (Number)	JR1 (032001 or 32.1)
Obsolete names	Junior; 900012; 901005
History	The first five examples of anti-Jr^a were reported in 1970. Named for the first maker of anti-Jr^a, Rose Jacobs, and not for "Junior" as some believed.

Occurrence

All populations	>99%

The Jr(a–) phenotype has been found mostly in Japanese and other Asians, but also in persons of northern European extraction, Bedouin Arabs, and in one Mexican.

Expression

Cord RBCs   Expressed

Molecular basis associated with Jr^a antigen

See "Molecular basis of JR$_{null}$ [Jr(a–), JR:–1] phenotype" in the System pages.

Effect of enzymes and chemicals on Jr^a antigen on intact RBCs

Ficin/Papain   Resistant (enhanced)
Trypsin   Resistant
α-Chymotrypsin   Resistant
DTT 200mM   Resistant
Acid   Resistant

In vitro characteristics of alloanti-Jr^a

Immunoglobulin class   IgG more common than IgM
Optimal technique   IAT
Complement binding   Some

Clinical significance of alloanti-Jr^a

Transfusion reaction ^{51}Cr cell survival studies indicated reduced RBC
                        survival; a patient with anti-Jr^a developed rigor after
                        transfusion of 150mL of cross-match incompatible
                        blood
HDFN   Positive DAT but usually no HDFN; however, one
                        fatal case of HDFN[1]

Comments

Siblings of patients with anti-Jr^a should be tested for compatibility, and the patient urged to donate blood for cryogenic storage when his/her clinical state permits.

Reference

[1] Peyrard, T., et al., 2008. Fatal hemolytic disease of the fetus and newborn associated with anti-Jr^a 48, 1906–1911.

Lan Blood Group System

Number of antigens 1

High prevalence   Lan

Terminology

ISBT symbol (number)   LAN (033)

History   Lan, which stems from the name Langereis, was promoted from the 901 Series of High-Incidence antigens to a System in 2012 when it was shown that homozygosity for *ABCB6* null alleles define the Lan– phenotype.

Expression

Tissues   Widely expressed; high expression in heart, skeletal muscles, and fetal liver; also in mitochondrial membrane, eye, and Golgi apparatus

Gene

Chromosome   2q36

Name   *LAN (ABCB6)*

Organization   19 exons spread over approximately 9.2 kbp of gDNA

Product   Lan glycoprotein (ATP-binding cassette, sub-family B [MDR/TAP], member 6 [ABCB6])

Database accession numbers

GenBank   NM_005689.1 (DNA)

Entrez Gene ID   10058

The Blood Group Antigen (3/e). DOI: http://dx.doi.org/10.1016/B978-0-12-415849-8.00035-1

Molecular bases of LAN$_{null}$ (Lan–, LAN:–1) phenotype[1]

The reference allele, *ABCB6* (Accession number NM_005689.1) encodes Lan (LAN1). Nucleotide differences from this reference allele, and amino acids affected, are given.

Allele name	Exon(intron)	Nucleotide	Amino acid	Ethnicity (prevalence)
*ABCB6*01N.01* or *LAN*01N.01*	1	197_198insG	Ala66Gly fs	Caucasian (Rare)
*ABCB6*01N.02* or *LAN*01N.02*	3	717G>A	Gln239Stop	Caucasian (Rare)
*ABCB6*01N.03* or *LAN*01N.03*	4	953_956delGTGG	Gly318Ala fs	Caucasian (Rare)
*ABCB6*01N.04* or *LAN*01N.04*	9	1533_1543 dupCGGCTCCCTGC	Leu515Pro fs	Caucasian (Rare)
*ABCB6*01N.05* or *LAN*01N.05*	11	1709_1710delAG	Glu570Gly fs	Caucasian (Rare)
*ABCB6*01N.06* or *LAN*01N.06*	11	1690_1691delAT	Met564Val fs	Caucasian (Rare)
*ABCB6*01N.07* or *LAN*01N.07*	14	1867 delinsAACAGGTGA	Gly623Asn fs	Caucasian (Few)
*ABCB6*01N.08* or *LAN*01N.08*	14	1942C>T	Arg648Stop	Caucasian (Few)
*ABCB6*01N.09* or *LAN*01N.09*	15	1985_1986delTC	Leu662Pro fs	Caucasian (Rare)
*ABCB6*01N.10* or *LAN*01N.10*	(16)	2256+2t>g	Splicing defect	Japanese (Rare)

Amino acid sequence

```
MVTVGNYCEA  EGPVGPAWMQ  DGLSPCFFFT  LVPSTRMALG  TLALVLALPC   50
RRRERPAGAD  SLSWGAGPRI  SPYVLQLLLA  TLQAALPLAG  LAGRVGTARG  100
APLPSYLLLA  SVLESLAGAC  GLWLLVVERS  QARQRLAMGI  WIKFRHSPGL  150
LLLWTVAFAA  ENLALVSWNS  PQWWWARADL  GQQVQFSLWV  LRYVVSGGLF  200
VLGLWAPGLR  PQSYTLQVHE  EDQDVERSQV  RSAAQQSTWR  DFGRKLRLLS  250
GYLWPRGSPA  LQLVVLICLG  LMGLERALNV  LVPIFYRNIV  NLLTEKAPWN  300
SLAWTVTSYV  FLKFLQGGGT  GSTGFVSNLR  TFLWIRVQQF  TSRRVELLIF  350
SHLHELSLRW  HLGRRTGEVL  RIADRGTSSV  TGLLSYLVFN  VIPTLADIII  400
GIIYFSMFFN  AWFGLIVFLC  MSLYLTLTIV  VTEWRTKFRR  AMNTQENATR  450
ARAVDSLLNF  ETVKYYNAES  YEVERYREAI  IKYQGLEWKS  SASLVLLNQT  500
QNLVIGLGLL  AGSLLCAYFV  TEQKLQVGDY  VLFGTYIIQL  YMPLNWFGTY  550
YRMIQTNFID  MENMFDLLKE  ETEVKDLPGA  GPLRFQKGRI  EFENVHFSYA  600
DGRETLQDVS  FTVMPGQTLA  LVGPSGAGKS  TILRLLFRFY  DISSGCIRID  650
GQDISQVTQA  SLRSHIGVVP  QDTVLFNDTI  ADNIRYGRVT  AGNDEVEAAA  700
QAAGIHDAIM  AFPEGYRTQV  GERGLKLSGG  EKQRVAIART  ILKAPGIILL  750
DEATSALDTS  NERAIQASLA  KVCANRTTIV  VAHRLSTVVN  ADQILVIKDG  800
CIVERGRHEA  LLSRGGVYAD  MWQLQQGQEE  TSEDTKPQTM  ER          842
```

Carrier molecule

In the RBC, ABCB6 is presumed to be a multipass membrane protein, with one nucleotide binding domain (NBD) oriented to the cytoplasm. In the mitochondria, ABCB6 passes through the membrane 11 times, with the Walker A, Walker B, and Signature motifs on the outer surface, i.e., oriented to the cytoplasm.

M_r (SDS-PAGE)	80,000
CHO: N-glycan	Four potential
Cysteine residues	10

Function

Binds heme and porphyrins, and functions in their ATP-dependent uptake into the mitochondria. Plays a crucial role in heme synthesis[2,3], although expression of ABCB6 does not appear to be required for normal erythropoiesis[1].

Disease association

The eye developmental defect coloboma is associated with changes in *ABCB6*, but Lan– individuals appear healthy[4].

References

[1] Helias, V., et al., 2012. ABCB6 is dispensable for erythropoiesis and specifies the new blood group system Langereis. Nat Genet 44, 170–173.

[2] Krishnamurthy, P.C., et al., 2006. Identification of a mammalian mitochondrial porphyrin transporter. Nature 443, 586–589.

[3] Mitsuhashi, N., et al., 2000. MTABC3, a novel mitochondrial ATP-binding cassette protein involved in iron homeostasis. J Biol Chem 275, 17536–17540.

[4] Wang, L., et al., 2012. ABCB6 mutations cause ocular coloboma. Am J Hum Genet 90, 40–48.

Lan Antigen

Terminology

ISBT symbol (number)	Lan (033001 or 33.1)
Obsolete names	Gn^a; Gonsowski; So; 900003; 901002
History	Reported in 1961; named after the first antigen-negative proband (Langereis) to make anti-Lan.

Occurrence

All populations	>99%

The Lan– phenotype occurs in about 1 in 20,000 people; found in Blacks[1,2], Caucasians, and Japanese.

Expression

Cord RBCs	Expressed
Altered	A weak form of Lan has been reported[3]

Effect of enzymes and chemicals on Lan antigen on intact RBCs

Ficin/Papain	Resistant
Trypsin	Resistant
α-Chymotrypsin	Resistant
DTT 200 mM	Resistant
Acid	Resistant

In vitro characteristics of alloanti-Lan

Immunoglobulin class	IgG
Optimal technique	IAT
Complement binding	Some

Clinical significance of alloanti-Lan

Transfusion reaction	No to severe/hemolytic
HDFN	No to mild

Autoanti-Lan

One example in a patient with depressed Lan antigens.

References

[1] Ferraro, M.L., et al., 2000. The rare red cell phenotype, Lan–, in an African-American [abstract]. Transfusion 40 (Suppl.), 121S.

[2] Sturgeon, J.K., et al., 2000. Report of an anti-Lan in an African American [abstract]. Transfusion 40 (Suppl.), 115S.

[3] Storry, J.R., Øyen, R., 1999. Variation in Lan expression. Transfusion 39, 109–110.

Lan

Blood Group Collections

Antigens in each Collection have serological, biochemical, and/or classic genetic connection. The i antigen remained the sole antigen in the Ii Collection when the gene controlling the I antigen was identified, and I was promoted to the I System.

Collection			Antigen number					
Symbol	Name	Number	001	002	003	004	005	006
COST	Cost	205	Cs^a	Cs^b				
I	Ii	207	...	i				
ER	Er	208	Er^a	Er^b	Er3			
GLOB	Globoside	209	...	...	LKE	PX2		
	Unnamed	210	Le^c	Le^d				
VEL	Vel	212	Vel	ABTI				
MN CHO^		213	Hu	M$_1$	Tm	Can	Sext	Sj

^= M and N antigens associated with different sialic acid-carrying oligosaccharides on GPA. Obsolete Collections: 201 (GE), 202 (CROM), 203 (IN), 204 (AU), 206 (GY), and 211 (WR). For obsolete antigens see ISBT Working Party publications or the ISBT website: www.isbt-web.org.

Cost Blood Group Collection

Number of antigens 2

Polymorphic	Cs^b
High prevalence	Cs^a

Terminology

ISBT symbol (number)	COST (205)
Obsolete name	Cost-Sterling

The Blood Group Antigen (3/e). DOI: http://dx.doi.org/10.1016/B978-0-12-415849-8.00036-3

History	This collection of phenotypically-associated antigens was established in 1988, and named after the two original patients who made anti-Cs^a (Copeland and Sterling). Five of the original antigens from this collection are now in the Knops system because they are carried on CR1.

Phenotypes

Null	Some RBCs type as Kn(a–b–), McC(a–), Sl(a–), Yk(a–), Cs(a–b–), and have very low copy numbers of CR1; however, Cs^a and Cs^b do not appear to be carried on CR1[1]

Reference

[1] Moulds, J.M., et al., 1992. Antiglobulin testing for CR1-related (Knops/McCoy/Swain- Langley/ York) blood group antigens: negative and weak reactions are caused by variable expression of CR1. Vox Sang 62, 230–235.

Cs^a Antigen

Terminology

ISBT symbol (number)	COST1 (205001 or 205.1)
Obsolete names	Cost-Sterling; 900004
History	Named in 1965 after two of the original patients (Mrs. Copeland and Mrs. Sterling) who made anti-Cs^a.

Occurrence

Most populations	>98%
Blacks	95%

Antithetical antigen

Cs^b (**COST2**)

Expression

Cord RBCs	Expressed; may be slightly weaker

Effect of enzymes and chemicals on Csᵃ antigen on intact RBCs

Ficin/Papain	Resistant
Trypsin	Resistant
α-Chymotrypsin	Resistant
DTT 200 mM/50 mM	Variable
Acid	Resistant

In vitro characteristics of alloanti-Csᵃ

Immunoglobulin class	IgG
Optimal technique	IAT

Clinical significance of alloanti-Csᵃ

Transfusion reaction	No
HDFN	No

Comments

Csᵃ has variable expression on RBCs from different people. RBCs of approximately 12% of Caucasians, and 15% of Blacks with the Yk(a–) phenotype are also Cs(a–)[1].

Reference

[1] Rolih, S., 1990. A review: antibodies with high-titer, low avidity characteristics. Immunohematology 6, 59–67.

Csᵇ Antigen

Terminology

ISBT symbol (number)	COST2 (205002 or 205.2)
History	Identified in 1987, and named when the antigen it was shown to be antithetical to Csᵃ.

Occurrence

Most populations	34%

Antithetical antigen

Csᵃ (**COST1**)

Expression

Cord RBCs	Presumed expressed

Effect of enzymes and chemicals on Cs^b antigen on intact RBCs

Ficin/Papain	Resistant
Trypsin	Resistant
α-Chymotrypsin	Resistant
DTT 200 mM/50 mM	Variable

In vitro characteristics of alloanti-Cs^b

Immunoglobulin class	IgG
Optimal technique	IAT

Clinical significance of alloanti-Cs^b

No data because only one example of antibody published.

Ii Blood Group Collection

Number of antigens   1

High prevalence   i

Terminology

ISBT symbol (number)   I (207)

History   I and i antigens were placed in the Ii Blood Group
                       Collection in 1990. In 2002 the I antigen was
                       promoted to a blood group system, leaving i alone in
                       Blood Group Collection 207.

Gene

The genetic basis of i expression is unknown.

Carrier molecule

The i antigen is on unbranched carbohydrate chains of repeating *N*-acetyl-lactosamine units on glycolipids and glycoproteins on RBCs, and on proteins in plasma. With the action of the branching enzyme, β6GlcNAc-transferase, these i antigen-carrying chains become the I antigen (see I Blood Group System [027])[1], i.e. i is the precursor structure for I.

Disease association

Enhanced expression of i antigens is associated with leukemia, Tk polyagglutination, thalassemia, sickle cell disease, HEMPAS, Diamond Blackfan anemia, myeloblastic erythropoiesis, sideroblastic erythropoiesis, and any condition that results in stress hemopoiesis.

Anti-i is associated with infectious mononucleosis and other lymphoproliferative disorders (e.g., Hodgkins disease), and occasionally with CHAD.

The Blood Group Antigen (3/e). DOI: http://dx.doi.org/10.1016/B978-0-12-415849-8.00037-5

651

Phenotypes associated with i antigen and the reciprocal I antigen

See I Blood Group System [027].

Molecular basis of adult i phenotype

See I Blood Group System [027].

Comments

The i antigen occurs on unbranched A-, B-, and H-active oligosaccharide chains.

Reference

[1] Cooling, L., 2010. Polyactosamines, there's more than meets the "Ii": a review of the I system. Immunohematology 26, 133–155.

i Antigen

Terminology

ISBT symbol (number)	I2 (207002 or 207.2)
Obsolete name	900027
History	Named in 1960 because of its reciprocal, but not classical antithetical, association with the I antigen.

Occurrence

All RBCs of adults have at least trace amounts of i antigen. The adult i phenotype is rare.

Reciprocal antigen

I (See I Blood Group System [027]).

Expression

Cord RBCs	Strong
Altered	Enhanced on CDA II RBCs and RBCs produced under hemopoietic stress

Molecular basis associated with i antigen[1]

Linear type 2 chains	Galβ1-4(GlcNAcβ1-3Galβ1-4)$_n$-Glc-Cer, but also as part of unbranched glycans on glycoproteins

See I Blood Group System (027) pages for molecular basis associated with adult i phenotype.

Effect of enzymes and chemicals on i antigen on intact RBCs

Ficin/Papain	Resistant (markedly enhanced)
Trypsin	Resistant (markedly enhanced)
α-Chymotrypsin	Resistant (markedly enhanced)
DTT 200 mM	Resistant
Acid	Resistant

In vitro characteristics of anti-i

Immunoglobulin class	IgM (rarely IgG)
Optimal technique	RT or 4°C
Complement binding	Yes; some hemolytic

Clinical significance of anti-i

Transfusion reaction	No
HDFN	Rare

Autoanti-i

Anti-i are considered to be autoantibodies. Transient autoanti-i can occur in infectious mononucleosis and some lymphoproliferative disorders.

Comments

So-called compound antigens have been described: iH, iP1, iHLe^b.
RBCs with the dominant Lu(a–b–) phenotype have a depressed expression of i antigen, whereas RBCs with the X-linked form of the Lu(a–b–) phenotype have enhanced expression of i antigen. The i antigen expression is often enhanced on RBCs from patients with hemopoietic stress, due to the rapid transit through ER and Golgi. Horse RBCs have a strong expression of i antigen, and can be used as a diagnostic tool for infectious mononucleosis.

Reference

[1] Roelcke, D., 1995. Serology, biochemistry, and pathology of antigens defined by cold agglutinins. In: Cartron, J-P, Rouger, P (Eds.), Molecular Basis of Human Blood Group Antigens. Plenum Press, New York, NY, pp. 117–152.

Er Blood Group Collection

Number of antigens 3

High prevalence   Er^a, Er3
Low prevalence   Er^b

Terminology

ISBT symbol (number)   ER (208)
History   Became a blood group collection in 1990.

Phenotypes

Null   Er(a–b–)

Er^a Antigen

Terminology

ISBT symbol (number)   ER1 (208001 or 208.1)
Obsolete names   Rosebush; Ros; Min; Rod
History   Reported in 1982; named after the first proband to
                       make the antibody.

Occurrence

Caucasians   1 in 100,000
                       With the exception of one Japanese woman (see below)
                       all Er(a–) probands have been of European ancestry

Antithetical antigen

Er^b (**ER2**)

Expression

Cord RBCs   Expressed
Altered   RBCs from a Japanese woman and two of her
                       siblings reacted with three of eight anti-Er^a

The Blood Group Antigen (3/e). DOI: http://dx.doi.oxg/10.1016/B978-0-12-415849-8.00038-7

Effect of enzymes and chemicals on Er^a antigen on intact RBCs

Ficin/Papain	Resistant
Trypsin	Resistant
α-Chymotrypsin	Resistant
DTT 200 mM	Resistant
Acid	Sensitive

In vitro characteristics of alloanti-Er^a

Immunoglobulin class	IgG
Optimal technique	IAT

Clinical significance of alloanti-Er^a

Transfusion reaction	No to reduced RBC survival[1]
HDFN	Positive DAT but no clinical HDFN

Comments

The mode of inheritance of Er^a is unclear: one Er(a–) proband has two siblings, a mother, two aunts and an uncle, all of whom were Er(a–) suggesting the presence of a third allele.

The anti-Er^a made by the Japanese proband gave slightly weakened reactions with trypsin-treated RBCs.

Reference

[1] Thompson, H.W., et al., 1985. Survival of Er(a+) red cells in a patient with allo-anti-Er^a. Transfusion 25, 140–141.

Er^b Antigen

Terminology

ISBT symbol (number)	ER2 (208002 or 208.2)
History	Reported in 1988, when the antibody was shown to recognize the antithetical low prevalence antigen to Er^a.

Occurrence

Most populations	<0.01%

Antithetical antigen

Er^a (**ER1**)

Expression

Cord RBCs   Expressed

Effect of enzymes and chemicals on Er^b antigen on intact RBCs

Ficin/Papain   Resistant
Trypsin   Resistant
α-Chymotrypsin   Resistant
AET   Resistant

In vitro characteristics of alloanti-Er^b

Immunoglobulin class   IgG
Optimal technique   IAT

Clinical significance of alloanti-Er^b

Limited data because only two examples of anti-Er^b have been reported.
HDFN   DAT+, but no clinical HDFN[1]

Reference

[1] Poole, J., et al., 2010. The second example of anti-Er^b and its clinical significance in pregnancy [abstract]. Vox Sang 99 (Suppl. 1), 340.

Er3 Antigen

Terminology

ISBT symbol (number)   ER3 (208003 or 208.3)
History   An antibody made by a person with ER:−1,−2 RBCs with characteristics of antibodies in the ER collection was identified in 2000 and reported in detail in 2003[1]. The antigen recognized by this antibody was named Er3 in 2004.

Occurrence

Most populations   >99.9%

Expression

Cord RBCs   Expressed

Er

Effect of enzymes and chemicals on Er3 antigen on intact RBCs

Ficin/Papain	Presumed resistant
Trypsin	Presumed resistant
α-Chymotrypsin	Presumed resistant
DTT 200 mM	Presumed resistant
Acid	Sensitive

In vitro characteristics of alloanti-Er3

Immunoglobulin class	IgG
Optimal technique	IAT

Clinical significance of alloanti-Er3

Transfusion reaction	Mild in the only reported patient with anti-Er3
HDFN	No data because the only example of anti-Er3 was made by a male

Comments

Anti-Er3 reacted with RBCs from the only other Er(a−b−) person, whereas the antibody made by that person was compatible with RBCs from the maker of anti-Er3.

Reference

[1] Arriaga, F., et al., 2003. A new antigen of the Er collection. Vox Sang 84, 137–139.

Globoside Blood Group Collection

Number of antigens 2

High prevalence   LKE, PX2

Terminology

ISBT symbol (number)   GLOB (209)
Obsolete name   GLOBO
History   P, P^k, and LKE were removed from the P system
                                    (see **003**) because these antigens belong to the
                                    globoseries of glycolipids, while the P1 antigen is
                                    part of the neolactoseries. They were gathered into
                                    an unnamed Collection in 1990 because of their
                                    serological and biochemical relationship (all three
                                    antigens are based on lactosylceramide); in 1991, the
                                    name Globoside (GLOB) Collection was assigned.
                                    In 2002, P was upgraded to its own system (**GLOB
                                    028**), and in 2010, P^k was moved to the P system,
                                    which was then renamed the P1PK system. In 2010,
                                    the remaining antigen in the GLOB Collection,
                                    LKE, was joined by PX2.

Carrier molecule

The sequential action of multiple gene products is required for expression of
these antigens.
See P1PK (**003**) and GLOB (**028**) Blood Group Systems, and Section III.

Disease association

LKE and disialo-LKE are associated with metastasis in renal cell carcinoma.

The Blood Group Antigen (3/e). DOI: http://dx.doi.org/10.1016/B978-0-12-415849-8.00039-9

Null phenotypes

For the LKE antigen the null phenotypes are p, P_1^k, and P_2^k, and for the PX2 antigen the null phenotypes are P_1^k and P_2^k.

LKE Antigen

Terminology

ISBT symbol (number)	GLOB3 (209003 or 209.3)
Obsolete names	Luke; SSEA-4; MSGG (monosialo-galactosyl-globoside)
History	In 1986, the name LKE was proposed for the antigen detected by the Luke serum, which was reported in 1965. LKE joined the GLOB collection in 1990.

Occurrence

All populations   98%

Expression

Cord RBCs   Expressed

Molecular basis associated with LKE antigen[1]

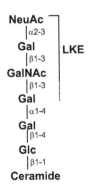

Effect of enzymes and chemicals on LKE antigen on intact RBCs

Ficin/Papain	Resistant (markedly enhanced)
Trypsin	Resistant (markedly enhanced)
α-Chymotrypsin	Resistant
Sialidase	Sensitive
DTT 200 mM	Resistant
Acid	Resistant

In vitro characteristics of alloanti-LKE

Immunoglobulin class	IgM
Optimal technique	RT or lower
Complement binding	Some

Clinical significance of alloanti-LKE

Transfusion reaction	None reported
HDFN	No

Comments

Anti-LKE in humans is a rare specificity.

The expression of LKE and P^k antigens is inversely related: LKE-negative RBCs express almost twice the P^k expressed by LKE+ (strong) RBCs[2].

Terminal NeuNAc is crucial for the LKE determinant; standard methods for sialidase treatment of RBCs do not affect reaction of RBCs with monoclonal anti-LKE.

The presence of *Se* decreases LKE expression; secretors have a 3-4 fold decreased risk of *E. coli* infections.

There are three LKE phenotypes[3]:

LKE+^{S}	80% to 90%
LKE+^{W}	10% to 20% (correlated to alterations in *B3GALT5*[3])
LKE–	1% to 2%

References

[1] Bailly, P., Bouhours, J.F., 1995. P blood group and related antigens. In: Cartron, J.-P., Rouger, P. (Eds.), Molecular Basis of Human Blood Group Antigens. Plenum Press, New York, NY, pp. 300–329.

[2] Cooling, L.L., Kelly, K., 2001. Inverse expression of P^k and Luke blood group antigens on human RBCs. Transfusion 41, 898–907.

[3] Cooling, L., 2002. A missense mutation in β3GalT5, the glycosyltransferase responsible for galactosylgloboside and Lewis c synthesis, may be associated with the LKE-weak phenotype in African Americans [abstract]. Transfusion 42 (Suppl.): 9S.

PX2 Antigen

Terminology

ISBT symbol (number)	GLOB4 (209004 or 209.4)
History	In 2010, PX2 was added to the GLOB Collection after antibodies in plasma from P^k people were shown to agglutinate RBCs with the p phenotype (PP1P^K–).

Occurrence

All populations   >99.9%

Expression

Cord RBCs   Expressed

Molecular basis associated with PX2 antigen[1,2,3]

A terminal β3-*N*-acetylgalactosamine (β3GalNAc) on paragloboside

```
        GalNAc ⎤
         | β1-3
         Gal   | PX2
         | β1-4 |
        GlcNAc  |
         | β1-3 |
         Gal    ⎦
         | β1-4
         Glc
         | β1-1
        Ceramide
```

Effect of enzymes and chemicals on PX2 antigen on intact RBCs

Ficin/Papain   Resistant (enhanced)

In vitro characteristics of alloanti-PX2

Immunoglobulin class   IgG (possibly mixture of IgM and IgG)
Optimal technique   IAT with papain-treated RBCs

Clinical significance of alloanti-PX2

No data available

Comments

Weak/variable reactivity in cross matching tests of plasma from P^k individuals with RBCs of the p phenotype detects the PX2 antigen.
Anti-PX2 appears to be naturally-occurring.
PX2 is expressed more strongly on RBCs with the p phenotype than on RBCs with other phenotypes[2,3].

References

[1] Kannagi, R., et al., 1982. A new glycolipid antigen isolated from human erythrocyte membranes reacting with antibodies directed to globo-N-tetraosylceramide (globoside). J Biol Chem 257, 4438–4442.

[2] Olsson, M.L., et al., 2011. PX2: a new blood group antigen with implications for transfusion recommendations in P1K and P2K individuals [abstract]. Vox Sang 101 (Suppl. 1), 53.

[3] Thorn, J.J., et al., 1992. Structural characterization of x2 glycosphingolipid, its extended form, and its sialosyl derivatives: accumulation associated with the rare blood group p phenotype. Biochemistry (Mosc) 31, 6509–6517.

Globoside

Unnamed Blood Group Collection

Number of antigens 2

Low prevalence	Le^c
Polymorphic	Le^d

Terminology

ISBT number   210

Carrier molecule

Glycosphingolipid adsorbed onto RBCs.

Le^c Antigen

Terminology

ISBT number   210001 or 210.1

Occurrence

Most populations   1%

Molecular basis associated with Le^c antigen[1]

Gal
| β1-3
GlcNAc
| β1-3
Gal
| β1-4
Glc
| β1-1
Ceramide

The Blood Group Antigen (3/e). DOI: http://dx.doi.org/10.1016/B978-0-12-415849-8.00040-5

Unnamed

Comments

Anti-Le^c agglutinates Le(a–b–) RBCs from non-secretors.
Anti-Le^c has been made by humans, and in goats.

Reference

[1] Mollison, P.L., et al., 1997. Blood Transfusion in Clinical Medicine, tenth ed. Blackwell Science, Oxford, UK.

Le^d Antigen

Terminology

ISBT number   210002 or 210.2

Occurrence

Most populations   6%

Molecular basis associated with Le^d antigen[1]

$$\text{Fuc} \xrightarrow{\alpha 1\text{-}2} \text{Gal}$$
$$| \beta 1\text{-}3$$
$$\text{GlcNAc}$$
$$| \beta 1\text{-}3$$
$$\text{Gal}$$
$$| \beta 1\text{-}4$$
$$\text{Glc}$$
$$| \beta 1\text{-}1$$
$$\text{Ceramide}$$

H type 1

Comments

Anti-Le^d agglutinates Le(a–b–) RBCs from non-secretors.
Anti-Le^d has been made in goats.

Reference

[1] Mollison, P.L., et al., 1997. Blood Transfusion in Clinical Medicine, tenth ed. Blackwell Science, Oxford, UK.

Vel Blood Group Collection

Number of antigens 2

High prevalence   Vel, ABTI

Terminology

ISBT symbol (number)   VEL (212)
History   Became a Collection in 2008 when it was
                       recognized that ABTI– RBCs are Vel+^{W}.

Carrier molecule

Possibly a small glycoprotein (M_r SDS-PAGE ~35 kDa under non-reducing
conditions, and ~20 kDa under reducing conditions)[1].

Reference

[1] Storry, J.R., et al., 2010. Investigation into the carrier molecule of the Vel blood group system
[abstract]. Transfusion 50 (Suppl.), 28A.

Vel Antigen

Terminology

ISBT symbol (Number)   Vel (212001 or 212.1)
Other names   Ve^a; 900001
History   Reported in 1952, and named after the first antigen-
                       negative proband who made anti-Vel.

Occurrence

Vel– RBCs have been found in 1 in ~4,000 people and 1 in ~1,700 in
Norwegians and Swedes.

Expression

Cord RBCs   Weak

The Blood Group Antigen (3/e). DOI: http://dx.doi.org/10.1016/B978-0-12-415849-8.00041-7

Adult RBCs	Expression is variable; RBCs with a weak expression of the Vel antigen may be mistyped as Vel–[1]

Effect of enzymes and chemicals on Vel antigen on intact RBCs

Ficin/Papain	Resistant (markedly enhanced)
Trypsin	Resistant (markedly enhanced)
α-Chymotrypsin	Resistant (markedly enhanced)
DTT 200 mM	Sensitive or resistant[2]
Acid	Resistant

In vitro characteristics of alloanti-Vel

Immunoglobulin class	IgM and IgG (usually as a mixture)
Optimal technique	IAT; enzyme IAT
Complement binding	Yes; some hemolytic

Clinical significance of alloanti-Vel

Transfusion reaction	No to severe/hemolytic
HDFN	Positive DAT to severe[3]

Autoanti-Vel

Yes

Comments

Three of 14 anti-Vel did not react with 4 Ge:–2,–3,4 samples[4].

A disproportional number of Vel– samples have the P_2 phenotype[5].

Six of eight Vel– RBC samples were weakly reactive, and one was non-reactive with anti-ABTI[6].

References

[1] Issitt, P.D., Anstee, D.J., 1998. Applied Blood Group Serology, fourth ed. Montgomery Scientific Publications, Durham, NC.

[2] Rainer, T., et al., 2004. The effects of dithiothreitol-tested red blood cells with anti-Vel [abstract]. Transfusion 44 (Suppl.), 122A.

[3] Le Masne, A., et al., 1992. [Severe form of neonatal hemolytic disease by anti-Vel allo-immunization]. Arch Fr Pediatr 49, 899–901.

[4] Issitt, P., et al., 1994. Phenotypic association between Ge and Vel [abstract]. Transfusion 34 (Suppl.), 60S.

[5] Cedergren, B., et al., 1976. The Vel blood group in northern Sweden. Vox Sang 31, 344–355.

[6] Banks, J., et al., 2004. Two new cases of anti-ABTI showing an association between ABTI and Vel [abstract]. Vox Sang 87 (Suppl. 3), 38.

ABTI Antigen

Terminology

ISBT symbol (Number)   ABTI (212002 or 212.2)

History   Anti-ABTI reported in 1996 in three multiparous women, members of an inbred Israeli Arab family. Named after this family[1].

Occurrence

ABTI-negative phenotype found in one Israeli Arab family[1], one Bavarian, and one German[2].

Expression

Cord RBCs   Presumed expressed

Altered   Vel-negative RBCs are ABTI+^{W} (1 was ABTI–)[2]

Effect of enzymes and chemicals on ABTI antigen on intact RBCs

Ficin/Papain   Resistant

Trypsin   Resistant

α-Chymotrypsin   Resistant

DTT 200 mM   Resistant

In vitro characteristics of alloanti-ABTI

Immunoglobulin class   IgG (IgG1 plus IgG3)

Optimal technique   IAT

Clinical significance of alloanti-ABTI

Transfusion reaction   No data

HDFN   No

Comments

ABTI– RBCs have a weak expression of Vel.

References

[1] Schechter, Y., et al., 1996. ABTI (901015), a new red cell of high frequency [abstract]. Transfusion 36 (Suppl.), 25S.

[2] Banks, J., et al., 2004. Two new cases of anti-ABTI showing an association between ABTI and Vel [abstract]. Vox Sang 87 (Suppl. 3), 38.

MN CHO Blood Group Collection

Number of antigens 6

Polymorphic   Hu, M_1, Tm, Can, Sext, Sj

Terminology

ISBT symbol (number)   MN CHO (213)

History   Became a Collection in 2010. The antigens are
                             expressed on GPA with altered NeuNAc or GlcNAc
                             levels, and are associated with M or N antigens
                             (MNS [002] system).

Carrier molecules

The N-terminal domain of GPA^M or GPA^N when O-glycans contain altered
sialic acid (NeuNAc) or GlcNAc[1].

Disease association

It is possible that GlcNAc-containing O-glycans confer a selective advantage
against invasion by *Plasmodium falciparum* merozoites.

Reference

[1] Dahr, W., et al., 1991. Studies on the structures of the Tm, Sj, M_1, Can, Sext and Hu blood group
antigens. Biol Chem Hoppe-Seyler 372, 573–584.

Hu Antigen

Terminology

ISBT symbol (number)   MN CHO1 (213001 or 213.1)
Obsolete name   Hunter
History   Reported in 1934, and named after Charles Hunter,
                             the name of the donor of the RBCs used for
                             immunizing rabbits[1].

The Blood Group Antigen (3/e). DOI: http://dx.doi.oxg/10.1016/B978-0-12-415849-8.00042-9
671

Occurrence

Caucasians	1%
African Americans	7%
West Africans	22%

Molecular basis associated with Hu antigen

Predominantly GPA^N when O-glycans contain sialic acid (NeuNAc) or GlcNAc[2].

Effect of enzymes and chemicals on Hu antigen on intact RBCs

Ficin/Papain	Sensitive
Trypsin	Sensitive
α-chymotrypsin	Resistant

In vitro characteristics of alloanti-Hu

Immunoglobulin class	Antibody only found in rabbit plasma immunized with Hu+ RBCs
Optimal technique	RT

Comments

Shows strong association with Sext.

References

[1] Landsteiner, K., et al., 1934. An agglutination reaction observed with some human bloods, chiefly among negroes. J Immunol 27, 459–472.

[2] Dahr, W., et al., 1991. Studies on the structures of the Tm, Sj, M_1, Can, Sext and Hu blood group antigens. Biol Chem Hoppe-Seyler 372, 573–584.

M_1 Antigen

Terminology

ISBT symbol (number)	MN CHO2 (213002 or 213.2)
History	Reported in 1960, and used to describe a strong reaction with anti-M^1.

Occurrence

Caucasians	0.5%
Blacks	16.5%

Molecular basis associated with M_1 antigen

Predominantly GPA^M when O-glycans contain sialic acid (NeuNAc) or GlcNAc[2].

Effect of enzymes and chemicals on M_1 antigen on intact RBCs

Ficin/Papain	Sensitive
Trypsin	Sensitive
α-chymotrypsin	Resistant

In vitro characteristics of alloanti-M_1

Optimal technique	RT

Comments

Some anti-M react only with M_1+ RBCS; however "anti-M_1" and anti-M are not separable by differential absorption. Most anti-M_1 have been found in the plasma of M− people, but a small number were made by people with M+N+ RBCs.
M_1 is expressed only on M+ RBCs but is not simply an enhanced form of M antigen.

References

[1] Jack, J.A., et al., 1960. M_1, a subdivision of the human blood-group antigen M. Nature 186, 642.
[2] Dahr, W., et al., 1991. Studies on the structures of the Tm, Sj, M_1, Can, Sext and Hu blood group antigens. Biol Chem Hoppe-Seyler 372, 573–584.

Tm Antigen

Terminology

ISBT symbol (number)	MN CHO3 (213003 or 213.3)
Obsolete name	Sheerin
History	Reported in 1965, and named "T" because it was next in sequence to S and s, but because "T" had already been used for polyagglutination, "m" was used to denote the association with MN[1].

Occurrence

Caucasians	25%
Blacks	31%

Molecular basis associated with Tm antigen

Predominantly GPA^N when O-glycans contain sialic acid (NeuNAc) or $GlcNAc$[1]

Effect of enzymes and chemicals on Tm antigen on intact RBCs

Ficin/Papain	Sensitive
Trypsin	Sensitive
α-chymotrypsin	Resistant

In vitro characteristics of alloanti-Tm

Optimal technique   RT

Comments

Many M+N+ Tm+ samples are also M_1+.
When tested against RBCs that have been treated with neuraminidase, anti-Tm appears to have anti-N specificity.

Reference

[1] Issitt, PD., et al., 1965. Anti-Tm, an antibody defining a new antigenic determinant within the MN blood-group system. Vox Sang 10, 742–743.

Can Antigen

Terminology

ISBT symbol (number)	MN CHO4 (213004 or 213.4)
Obsolete name	Canner
History	Reported in 1979, and named after the first antigen-positive proband[1].

Occurrence

Anti-Can reacted with a higher proportion of RBCs from African Americans (60%) than from Caucasians (27%). 87% of the Can+ RBCs were M+.

Molecular basis associated with Can antigen

Predominantly GPA^M when O-glycans contain sialic acid (NeuNAc) or $GlcNAc^2$

Effect of enzymes and chemicals on Can antigen on intact RBCs

Ficin/Papain	Sensitive
Trypsin	Sensitive
α-chymotrypsin	Resistant

In vitro characteristics of alloanti-Can

Optimal technique   RT

References

[1] Judd, W.J., et al., 1979. The Can serum: demonstrating further polymorphism of M and N blood group antigens. Transfusion 19, 7–11.

[2] Dahr, W., et al., 1991. Studies on the structures of the Tm, Sj, M_1, Can, Sext and Hu blood group antigens. Biol Chem Hoppe-Seyler 372, 573–584.

Sext Antigen

Terminology

ISBT symbol (number)	MN CHO5 (213005 or 213.5)
History	Reported in 1974, and named after the antibody producer[1]

Occurrence

Caucasians	Not found
Blacks	24% of N+

Molecular basis associated with Sext antigen

Predominantly GPA^N when O-glycans contain sialic acid (NeuNAc) or GlcNAc[2].

Effect of enzymes and chemicals on Sext antigen on intact RBCs

Ficin/Papain	Sensitive
Trypsin	Sensitive
Sialidase	Sensitive

In vitro characteristics of alloanti-Sext

Optimal technique	RT

Comments

Shows strong association with Hu.

References

[1] Giles, C.M., Howell, P., 1974. An antibody in the serum of an MN patient which reacts with the M1 antigen. Vox Sang 27, 43–51.

[2] Dahr, W., et al., 1991. Studies on the structures of the Tm, Sj, M₁, Can, Sext and Hu blood group antigens. Biol Chem Hoppe-Seyler 372, 573–584.

Sj Antigen

Terminology

ISBT symbol (number)	MN CHO6 (213006 or 213.6)
History	Reported in 1968 when the Sheerin serum (anti-Tm) was shown to have a second antibody, which was named "Sj" after two employees at the New York Blood Center, Stenbar and James, who had strongly reactive RBCs[1].

MN CHO

Occurrence

| Caucasians | 2% |
| Blacks | 4% |

Molecular basis associated with Sj antigen

Predominantly GPA^N when O-glycans contain sialic acid (NeuNAc) or $GlcNAc^2$.

Effect of enzymes and chemicals on Sj antigen on intact RBCs

| Ficin/Papain | Sensitive |

In vitro characteristics of alloanti-Sj

| Immunoglobulin class | IgM |
| Optimal technique | RT |

References

[1] Issitt, P.D., et al., 1968. Sj, a new antigen in the MN system, and further studies on Tm. Vox Sang 15, 1–14.

[2] Dahr, W., et al., 1991. Studies on the structures of the Tm, Sj, M_1, Can, Sext and Hu blood group antigens. Biol Chem Hoppe-Seyler 372, 573–584.

700 Series

The 700 Series of Low-Incidence Antigens

700 Series of Low-Incidence Antigens

Antigens in this series occur in less than 1% in most populations; as the responsible genes are not known, they cannot be placed in a Blood Group System, and do not meet the criteria for being placed in a Collection.

The Blood Group Antigen (3/e). DOI: http://dx.doi.oxg/10.1016/B978-0-12-415849-8.00043-0

Number	Symbol	Name	# of probands	# of antibodies	Found as				Caused HDFN	Enzymes Chemicals	Comments
					Immune	Stimulus not known	Multi-specific	Mono-specific			
700.002	By	Batty	Few	Many	✓	✓	✓		+DAT	Papain/ficin/α-chymotrypsin resistant; AET resistant	Original anti-By stimulated by pregnancy; Antibody found in AIHA.
700.003	Chr^a	Christiansen	Two	Few		✓		✓	No	Trypsin resistant	Found in Danes
700.005	Bi	Biles	Few	Few	✓	✓			Probably	Trypsin sensitive	Original anti-Bi stimulated by pregnancy
700.006	Bx^a	Box	Few	Few		✓	✓			Papain resistant	Antibody found in AIHA.
700.017	To^a	Torkildsen	Few	Many		✓	✓	✓		Ficin resistant (lytic)	Some cold reactive, IgM bind complement, others are IgG; Scandinavian
700.018	Pt^a	Peters	Few	Many		✓	✓			Papain/ficin/α-chymotrypsin sensitive; Trypsin resistant; AET resistant	M_r approx. 31,600
700.019	Re^a	Reid	Few	Few	✓				Mild +DAT	Papain/ficin/trypsin/α-chymotrypsin resistant	
700.021	Je^a	Jensen	Few	Few		✓			No	Papain/ficin sensitive	Danish; IgM

No.	Symbol	Name								HDFN	Enzymes/chemicals	Comments
700.028	Li^a	Livesey	Few	Few	✓	Few		✓		Mild	Papain/ficin resistant	Maybe part of Lutheran system
700.039		Milne	1	Many	✓	✓		✓			Papain/ficin resistant (lytic)	IgM
700.040	RASM	Rasmussen	1	1	✓	✓			✓	+DAT	Papain/ficin/trypsin resistant; DTT/AET sensitive	IgG
700.044	JFV		2	Few	✓	✓			✓	+DAT to moderate	Papain resistant (↑); DTT/AET resistant	German/Dutch probands
700.045	Kg	Katagiri	1	1	✓	✓				Severe	Papain/trypsin resistant	Japanese
700.047	JONES	Jones	2	Few	✓	✓			✓	Moderate	Papain/ficin/trypsin/α-chymotrypsin resistant (↑); AET resistant	May be an Rh antigen
700.049	HJK		1	1	✓	✓			✓	Severe	Papain/ficin/trypsin/α-chymotrypsin resistant	
700.050	HOFM		1	1	✓	✓			✓	Mild	Papain resistant (↑); DTT/AET resistant	May be an Rh antigen; associated with weak C antigen; Dutch
700.052	SARA	SARAH	Few	Few	✓	✓	✓	✓	✓	Severe (Towns et al., 2011)	Papain/trypsin/α-chymotrypsin resistant; DTT/AET resistant	Australian; Canadian
700.054	REIT		Few	1						Severe	AET resistant	

Few = 1–5 examples; Several = 6–12 examples; Many = 13 or more examples; Not known = apparently naturally-occurring.
(↑) = enhanced reactivity.
For obsolete numbers see www.isbt.org.
Towns D, et al. (2011). Hemolytic disease of the fetus and newborn caused by an antibody to a low-prevalence antigen, anti-SARA. *Transfusion*, 51: 1977–1979.
Note, not all the antibody/antigen combinations have been tested with all enzymes and chemicals. We have given those that have been reported.

The 901 Series of High-Incidence Antigens

901 Series of High-Incidence Antigens

Antigens in this series occur in more than 90% of people; as the responsible genes are not known, they cannot be placed in a Blood Group System, and do not meet the criteria for being placed in a Collection.

Originally, high-incidence antigens were in the 900 series. At the 1988 meeting of the ISBT Working Party on Terminology[1], many of the 900 series antigens were transferred to Blood Group Systems or the newly-established Blood Group Collections, thereby generating many obsolete 900 numbers. Consequently, the 900 Series was replaced by the 901 Series.

Number	Symbol	Name
901003 or 901.3	At^a	August
901008 or 901.8	Emm	
901009 or 901.9	AnWj	Anton
901012 or 901.12	Sd^a	Sid
901014 or 901.14	PEL	
901016 or 901.16	MAM	

Obsolete: 901001 Vel, 901002 Lan, 901004 Jo^a, 901005 Jr^a, 901006 Ok^a, 901007 JMH, 901010 Wr^b, 901011 MER2, 901013 Duclos, and 901015 ABTI.

Reference

[1] Lewis, M., et al., 1990. Blood group terminology 1990. ISBT working party on terminology for red cell surface antigens. Vox Sang 58, 152–169.

The Blood Group Antigen (3/e). DOI: http://dx.doi.oxg/10.1016/B978-0-12-415849-8.00044-2

At^a Antigen

Terminology

ISBT symbol (number)	At^a (901003 or 901.3)
Obsolete names	El; Elridge; 900006
History	Reported in 1967; named after the first antigen-negative proband (Mrs Augustine) to make anti-At^a.

Occurrence

Most populations	100%
Blacks	>99%

Expression

Cord RBCs	Expressed

Effect of enzymes and chemicals on At^a antigen on intact RBCs

Ficin/Papain	Resistant
Trypsin	Resistant
α-Chymotrypsin	Resistant
DTT 200 mM	Resistant
Acid	Resistant

In vitro characteristics of alloanti-At^a

Immunoglobulin class	IgG
Optimal technique	IAT

Clinical significance of alloanti-At^a

Transfusion reaction	No to severe[1,2,3]
HDFN	Most At(a+) babies born to At(a−) mothers were not affected; only one mild case

Comments

Many At(a−) probands originate from the Caribbean or Southern USA.
In three patients, the anti-At^a was concomitant with autoimmune disease[2].

References

[1] Cash, K.L., et al., 1999. Severe delayed hemolytic transfusion reaction secondary to anti-At(a). Transfusion 39, 834–837.

[2] Ramsey, G., et al., 1995. Clinical significance of anti-At^a. Vox Sang 69, 135–137.

[3] Sweeney, J.D., et al., 1995. At(a−) phenotype: Description of a family and reduced survival of At(a+) red cells in the proposita with anti-At^a. Transfusion 35, 63–67.

Emm Antigen

Terminology

ISBT symbol (number)	Emm (901008 or 901.8)
Obsolete names	Emma; 900028
History	Reported in 1987, and named after the first antigen-negative proband to make anti-Emm.

Occurrence

Six probands with the Emm– phenotype have been found: three Americans, a French Madagascan, a Pakistani, and a French Canadian.

Expression

Cord RBCs	Expressed

Effect of enzymes and chemicals on Emm antigen on intact RBCs

Ficin/Papain	Resistant
Trypsin	Resistant
α-Chymotrypsin	Resistant
DTT 200 mM	Resistant
Acid	Resistant

In vitro characteristics of alloanti-Emm

Immunoglobulin class	IgG more common than IgM (4 of 5)
Optimal technique	IAT; 4°C (the original anti-Emm)
Complement binding	Some (2 of 5)

Clinical significance of alloanti-Emm

No data are available. Six of the seven examples of anti-Emm were in non-transfused males[1].

Comments

Emm is carried on a GPI-linked protein in the RBC membrane[2], thus, absent from PNHIII RBCs.

.

References

[1] Daniels, G.L., et al., 1987. Emm. A red cell antigen of very high frequency. Transfusion 27, 319–321.

[2] Telen, M.J., et al., 1990. Evidence that several high-frequency human blood group antigens reside on phosphatidylinositol-linked erythrocyte membrane proteins. Blood 75, 1404–1407.

AnWj Antigen

Terminology

ISBT symbol (number)	AnWj (901009 or 901.9)
Obsolete names	Wj; 005015; Lu15
History	Reported in 1982 as an alloantibody to an antigen called Anton, and in 1983 as an autoantibody to an antigen called Wj. In 1985, it was shown that both antibodies detected the same antigen, and the name AnWj was applied.

Occurrence

Genetic form of AnWj-negative found in two Israeli women and one Arab-Israeli family.

Expression

Cord RBCs	Not expressed
Altered	Weak on dominant Lu(a–b–) RBCs
	Expression varies from person to person

Molecular basis associated with AnWj antigen

Carried on CD44 proteoglycan[1,2]. The AnWj epitope is likely to reside in the glycosylated region encoded by exons 5 and 15. If this is true, all information regarding CD44 (Indian blood group system, [**IN**]) would apply.

Effect of enzymes and chemicals on AnWj antigen on intact RBCs

Ficin/Papain	Resistant
Trypsin	Resistant
α-Chymotrypsin	Resistant
DTT 200 mM	Variable
Acid	Resistant

In vitro characteristics of alloanti-AnWj

Immunoglobulin class	IgG
Optimal technique	IAT
Complement binding	Rare

Clinical significance of alloanti-AnWj

Transfusion reaction	Severe in one case[3]
HDFN	No

Autoanti-AnWj

Yes, may appear to be an alloantibody because of transient suppression of AnWj antigen. Such patients frequently tolerate transfusion of AnWj+ RBCs.

Comments

Only two examples of alloanti-AnWj (both in Israeli women) have been described. The AnWj– phenotype is usually the result of transient (often long-term) suppression of AnWj.
AnWj antigen is the receptor for *Haemophilus influenzae*[4].

References

[1] Telen, M.J., et al., 1993. The ANWJ blood group antigen/hemophilus influenzae receptor resides on a high-molecular-weight protein expressed by CD44 transfectants [abstract]. Clin Res 41, 161A.

[2] Udani, M., et al., 1995. Erythroid progenitors express a CD44 variant with reduced hyaluronic acid binding ability and reduced expression of the epitope associated with *Hemophilus influenzae* hemagglutination [abstract]. Blood 86 (Suppl. 1), 472a.

[3] de Man, A.J., et al., 1992. An example of anti-AnWj causing haemolytic transfusion reaction. Vox Sang 63 238–238.

[4] van Alphen, L., et al., 1986. The Anton blood group antigen is the erythrocyte receptor for *Haemophilis influenzae*. FEMS Microbiol Lett 37, 69–71.

Sd[a] Antigen

Terminology

ISBT symbol (number)	Sd[a] (901012 or 901.12)
Other names	Sid
History	Reported in 1967 after many years of investigation. Named for Sidney Smith, head of the maintenance department at the Lister Institute in London which housed the laboratory of Race and Sanger. For many years, his RBCs were frequently used as a Sd(a++) control.

Occurrence

All populations	91% of RBC samples express Sd[a]; however, 96% of urine samples have Sd[a] substance; 4% of people are truly Sd(a–)

Expression

Soluble form	Urine (Tamm-Horsfall glycoprotein)
Newborns	Not expressed on RBCs; expressed in saliva, urine, and meconium

Adult RBCs	Strength of expression varies greatly; the strongest expression is on Cad phenotype RBCs
Altered	Marked reduction of Sd^a expression occurs in pregnancy
Other tissues	Stomach, colon, kidney, lymph nodes

Molecular basis associated with antigen[1]

Sd^a-active Pentasaccharide from GPA

$$\text{GalNAc} \xrightarrow{\beta 1-4} \text{Gal} \xrightarrow{\beta 1-3} \text{GalNAc} \text{——— Ser/Thr}$$
$$\qquad\qquad\quad |\alpha 2-3 \qquad\quad |\alpha 2-6$$
$$\qquad\qquad\quad \text{NeuAc} \qquad\quad \text{NeuAc}$$

Sd^a-active ganglioside

$$\text{GalNAc} \xrightarrow{\beta 1-4} \text{Gal} \xrightarrow{\beta 1-4} \text{GlcNAc} \xrightarrow{\beta 1-3} \text{Gal} \xrightarrow{\beta 1-4} \text{Glc-Cer}$$
$$\qquad\qquad\quad |\alpha 2-3$$
$$\qquad\qquad\quad \text{NeuAc}$$

Sd^a-active Tamm-Horsfall glycoprotein

$$\text{GalNAc} \xrightarrow{\beta 1-4} \text{Gal} \xrightarrow{\beta 1-4} \text{GlcNAc} \xrightarrow{\beta 1-3} \text{Gal}$$
$$\qquad\qquad\quad |\alpha 2-3$$
$$\qquad\qquad\quad \text{NeuAc}$$

Effect of enzymes and chemicals on Sd^a antigen on intact RBCs

Ficin/Papain	Resistant (enhanced)
Trypsin	Resistant (enhanced)
α-Chymotrypsin	Resistant (enhanced)
Sialidase	Usually resistant
DTT 200 mM	Resistant
Acid	Resistant

In vitro characteristics of alloanti-Sd^a

Immunoglobulin class	IgM more common than IgG
Optimal technique	RT; IAT
Neutralization	Urine (guinea pig and human)
Complement binding	Yes, some

Clinical significance of alloanti-Sd[a]

Transfusion reaction	Two cases reported associated with transfusion of Sd(a++) RBCs
HDFN	No

Comments

Agglutinates are typically small and refractile in a sea of free RBCs. Anti-Rx (formerly anti-Sd[x]) can be confused with anti-Sd[a] because of similar type of agglutination[2] .

Tamm-Horsfall protein in urine binds specifically to type 1[3] fimbriated *E. coli*, thereby preventing adherence of pathogenic *E. coli* to urothelial receptors.

Hemagglutination inhibition tests with urine are the most reliable way of determining Sd[a] status.

Urine inhibition tests (particularly using guinea pig urine) are useful for the identification of anti-Sd[a].

In 2003, two groups[4,5] independently reported the characterization of a genetic locus (*B4GALNT2*, also known as *GALGT2*) located at 17q21.32 that appeared to encode a glycosyltransferase with the enzyme activity (β4GalNAcT) required for synthesis of Sd[a]-active epitopes by addition of GalNAc in a β1,4 linkage to relevant precursor structures (Gal substituted with an α-2,3-NeuNAc). However, there is still no published evidence that mutations in this gene give rise to the Sd(a+[weak]) and/or Sd(a–) phenotypes. It is also unclear if the acceptor profile of this enzyme is compatible with its identity as the Sd[a] histo-blood group synthetase. Notably, two transcript forms code for a short and a long form of the enzyme by using different exon 1 sequences (exon 1S or 1L, respectively), but the same exons 2–11. It is tempting to speculate that these two forms, which are driven by two different promoters, may be differentially expressed in erythroid and other tissues in a way similar to the transcript forms of the *GCNT2* locus responsible for I antigen expression. In summary, however, more work is needed before the Sd[a] antigen can be promoted to a blood group system.

References

[1] Watkins, W.M., 1995. Sd[a] and Cad antigens. In: Cartron, J.-P., Rouger, P. (Eds.), Molecular Basis of Human Blood Group Antigens. Plenum Press, New York, NY, pp. 51–375.

[2] Issitt, P.D., 1991. The antigens Sd[a] and Cad. In: Moulds, J.M., Woods, L.L. (Eds.), *Blood groups: P, I, Sd[a] and Pr*. American Association of Blood Banks, Arlington, VA, pp. 53–71.

[3] Pak, J., et al., 2001. Tamm-Horsfall protein binds to type 1 fimbriated *Escherichia coli* and prevents *E. coli* from binding to uroplakin Ia and Ib receptors. J Biol Chem 276, 9924–9930.

[4] Lo, P.L., et al., 2003. Molecular cloning of the human beta1,4 N-acetylgalactosaminyltransferase responsible for the biosynthesis of the Sd[a] histo-blood group antigen: the sequence predicts a very long cytoplasmic domain. *J Biochem* (Tokyo) 134, 675–682.

[5] Montiel, M.D., et al., 2003. Molecular cloning, gene organization and expression of the human UDP-GalNAc:Neu5Acα2-3Galβ-R β1,4-*N*-acetylgalactosaminyltransferase responsible for the biosynthesis of the blood group Sd[a]/Cad antigen: evidence for an unusual extended cytoplasmic domain. Biochem J 373, 369–379.

PEL Antigen

Terminology

ISBT symbol (number)	PEL (901014 or 901.14)
Obsolete name	Pelletier
History	Identified in 1980, and reported in 1996 when the antigen was named after the first antigen-negative proband who made anti-PEL.

Occurrence

PEL– phenotype has been found in two French Canadian families.

Expression

Cord RBCs	Expressed
Altered	Weak expression (shown by absorption studies) in two French Canadian families[1]

Effect of enzymes and chemicals on PEL antigen on intact RBCs

Ficin/Papain	Resistant
Trypsin	Resistant
α-Chymotrypsin	Resistant
DTT 200 mM	Resistant

In vitro characteristics of alloanti-PEL

Immunoglobulin class	Presumed IgG
Optimal technique	IAT

Clinical significance of alloanti-PEL

Transfusion reaction	Reduced survival of ^{51}Cr-labelled RBCs
HDFN	No

Comments

An antibody (provisionally named anti-MTP) with similar specificity to anti-PEL was made by the probands from the two French Canadian families with suppressed PEL expression[1].

Reference

[1] Daniels, G.L., et al., 1996. PEL, a "new" high-frequency red cell surface antigen. Vox Sang 70, 31–33.

MAM Antigen

Terminology

ISBT symbol (number)	MAM (901016 or 901.16)
History	Reported in 1993; assigned to the 901 Series in 1999; name is derived from the initials of the first antigen-negative proband to make anti-MAM[1].

Occurrence

Four MAM– probands have been reported: one with Irish and Cherokee descent; two Arabs; and a Jordanian.

Expression

Cord RBCs	Expressed
Other blood cells	Lymphocytes, granulocytes, monocytes, and probably on platelets

Molecular basis associated with MAM antigen[2]

M_r assessment based on SDS-PAGE revealed a diffuse banding pattern of approximately 23,000–80,000, but with a discrete band also at 18,000. Possibly on N-glycosylated proteins[3].

Effect of enzymes and chemicals on MAM antigen on intact RBCs

Ficin/Papain	Resistant
Trypsin	Resistant
α-Chymotrypsin	Resistant
DTT 200 mM	Resistant

In vitro characteristics of alloanti-MAM

Immunoglobulin class	IgG
Optimal technique	IAT

Clinical significance of alloanti-MAM

Transfusion reaction	Monocyte monolayer assay suggests anti-MAM is potentially clinically significant
HDFN	No to severe

Comments

Anti-MAM may also cause neonatal thrombocytopenia; however, two MAM+ babies born to two different women with anti-MAM had no thrombocytopenia or anemia[4].

References

[1] Anderson, G., et al., 1993. An antibody to a high frequency antigen found on red cells, platelets, lymphocytes, and monocytes [abstract]. Transfusion 33 (Suppl.), 23S.

[2] Montgomery Jr., W.M., et al., 2000. MAM: a "new" high-incidence antigen found on multiple cell lines. Transfusion 40, 1132–1139.

[3] Li, W., Denomme, G.A., 2002. MAM is an N-glycan linked carbohydrate antigen expressed on all blood cells [abstract]. Transfusion 42 (Suppl.), 10S.

[4] Denomme, G.A., et al., 2000. First example of maternal-fetal incompatibility due to anti-MAM with an absence of thrombocytopenia [abstract]. Transfusion 40 (Suppl.), 28S.

Other Useful Facts

Usefulness of the effect of enzymes and DTT on antigens in antibody identification[1,2]

The following table shows *general* patterns of reactions; for more detail, see individual antigen sheets. It is important to control for "anti-enzyme" reactivity, i.e., by testing autologous RBCs treated in parallel or by testing an eluate. The patterns given are a useful guide in antibody identification, but remember that not all antibodies read the FactsBook!

Ficin/ Papain	Trypsin	α-Chymo-trypsin	200 mM DTT/AET	Possible specificity
Negative	Negative	Negative	Positive	Bp^a; Ch/Rg; XG
Negative	Negative	Negative	Negative	IN; JMH
Negative	Negative	Positive	Positive	M, N, En^{a}TS; Ge2, Ge4
Negative	Positive	Negative	Positive	'N'; Fy^a, Fy^b
Variable	Positive	Negative	Positive	S, s
Variable	Positive	Negative	Weak or negative	YT
Negative	Positive	Positive	Positive	En^{a}FS
Positive	Negative	Negative	Weak or negative	LU, MER2
Positive – Papain Weak or negative – Ficin	Negative	Negative	Negative	KN
Positive	Negative	Weak	Negative	DO
Positive	Positive	Negative	Weak	CROM
Positive	Positive	Negative	Positive	Some DI (3rd loop)
Positive	Positive	Positive/weak	Negative	LW
Positive	Positive/weak	Positive/weak	Positive	SC
Positive	Positive^	Positive^	Negative	KEL^ (except KALT, which is trypsin sensitive)
Positive	Positive	Positive	Positive	ABO; En^{a}FR, U; P1PK; RH; LE; Fy3; JK; most DI; CO; H; Ge3; OK; I/i; P, FORS; JR; LAN, Cs^a; ER; LKE, PX2; Vel,† ABTI; At^a; Emm; AnWj; Sd^a; PEL; MAM
Positive	Positive	Positive	Enhanced	Kx

^Kell blood group system antigens are sensitive to treatment with a mixture of trypsin and α-chymotrypsin.
†DTT may be variable.

Effect of acid on antigen expression

EDTA/glycine/acid-treated RBCs do not express antigens in the KEL blood group system, the ER collection or Bg antigens, and antigens of the JK blood group system may be weakened.

Effect of chloroquine diphosphate on antigen expression

A modified technique of treating RBCs with chloroquine for 30 mins at 37°C weakens Mt^a, Lu^b, Fy^b, Yt^a, Bg^a, and antigens of the RH, DO, KN, and JMH blood group systems.

Substrate specificity of selected enzymes for peptide and CHO bonds

Classification	Enzyme (source)	Substrate specificity (in order of preference)
Thiol endoprotease has an essential cysteine in the active site and may require a sulfhydryl compound to activate it	Bromelin (Pineapple)	Hydrolyzes C-terminal peptide bond of Lys, Ala, Tyr, Gly
	Ficin (Fig tree latex)	Hydrolyzes C-terminal peptide bond of Lys, Ala, Tyr, Gly, Asp, Leu, Val
	Papain (Papaya)	Hydrolyzes C-terminal peptide bond of Arg, Lys, and bond next but one to Phe
Metallo endoprotease requires a specific metal ion in the active site	Pronase (*Streptomyces griseus*)	Hydrolyzes C-terminal peptide bond of any hydrophobic amino acid
Serine endoprotease requires serine and histidine residues at the enzyme site for enzymatic activity	α-chymotrypsin (Bovine pancreas)	Hydrolyzes C-terminal peptide bond of Phe, Trp, Tyr, Leu
	Proteinase K (*Tritirachium album*)	Hydrolyzes C-terminal peptide bond of aromatic or hydrophobic amino acids
	Trypsin (Bovine or Porcine pancreas)	Hydrolyzes C-terminal peptide bond of Arg, Lys
	V8 protease (*Staphylococcus aureus* strain V8)	Hydrolyzes C-terminal peptide bond of Glu, Asp

(Continued)

(Continued)

Classification	Enzyme (source)	Substrate specificity (in order of preference)
Carboxyl endoprotease has an essential COOH in the active site	Pepsin (Porcine stomach mucosa)	Hydrolyzes C-terminal peptide bond of Phe, Leu, Trp, Tyr, Asp, Glu
Exoglycosidase	Sialidase/Neuraminidase (*Vibrio cholerae*)	Hydrolyzes glycosidic bond between terminal NeuAc in any linkage to any sugar
	α-Galactosidase (GH^27/36 from e.g., coffee bean)	Hydrolyzes glycosidic bond of α-linked terminal Gal in B, P^k, and P1 antigens
	A-zyme (GH^109, bacterial α3-*N*-acetylgalactosaminidase)	Hydrolyzes glycosidic bond of α3-linked terminal GalNAc in A antigen, leaving H antigen
	B-zyme (GH^110, bacterial α3-galactosidase)	Hydrolyzes glycosidic bond of α3-linked terminal Gal in B antigen, leaving H antigen
Endoglycosidase	Endo F (*Flavobacterium meningosepticum*)	Hydrolyzes the glycosidic bond between the two core GlcNAc residues in biantennary N-glycans, and leaves one GlcNAc attached to the Asn residue of the protein

Endo = Internal substrate bonds. Exo = Terminal substrate bonds.
Note: Bacterial deacetylases may modify the side-chains of sugars, e.g., the acquired B phenomenon results from deacetylation of *N*-acetylgalactosamine to galactosamine. Organisms such as *E. coli*, *Clostridium tertium*, and *Proteus mirabilis* have been implicated in this phenomenon.
^GH = glycoside hydrolase family, see www.cazy.org.

Cord RBCs are

Negative for Le^a, Le^b (sometimes); Ch, Rg; AnWj; Sd^a
Weak for   A, B; H; I; P1; Lu^a, Lu^b; Yt^a; JMH; sometimes Xg^a; Vel; Bg; KN; and DO antigens; Fy3 as detected by anti-Fy3 made by Blacks
Strong for   LW system antigens; i antigen

Antigens with variable expression on different RBCs in the same sample and on RBCs from different donors (presumed to be due to different antigen copy number)

Carbohydrate antigens A, B; FORS1; H; I; Le^a, Le^b; P1, P^k; P; Sd^a

Protein antigens Lu^a, Lu^b; Xg^a; KN; MER2; JMH; Jr^a; Vel; Lan; AnWj; Ch/Rg

Plasma adsorbed antigens Le^a, Le^b; Ch, Rg

Mixed Field Agglutination may be observed in:

Transfused patients

Maternal-fetal hemorrhage or fetal-maternal hemorrhage

Stem cell transplant recipients

Chimera (genetic)

Genetic variants of antigen, e.g., A_3, A_{finn}, A_{mos}, B_{mos}

Chromosomal abnormalities resulting in two populations of RBCs, e.g., ABO and RH in leukemia

Low density of antigen sites, e.g., Xg^a, Sd^a, Lu^a

Polyagglutination, e.g., Tn

Modification by bacterial enzymes, e.g., deacetylation in acquired B

X-inactivation, Kx in female carriers.

Blood group antigens absent (altered) on selected RBC phenotypes

Phenotype	Absent or altered, usually reduced (in parentheses) antigens
O_h (Bombay)	ABO and H systems; rarely LE
A_h/B_h (para-Bombay)	(Very weak A and/or B antigens)
En(a–)Fin	M, N, GPA-associated; Wr^a, Wr^b
U–	S, s, U, He, and GPB-associated
M^kM^k	MNS; Wr^a, Wr^b (some antigens, not in the MNS system, may appear to be enhanced due to reduced sialic acid)
p [previously Tj(a–)]	P; P1, P^k, NOR; LKE (PX2 elevated)

(Continued)

(Continued)

Phenotype	Absent or altered, usually reduced (in parentheses) antigens
P_1^k	P; LKE, PX2
P_2^k	P; P1; LKE, PX2
Rh_{null}	RH; LW; RHAG; Fy5 (S, s, U may be weak)
Rh_{mod}	(Weak RH; LW; RHAG; S, s, U; Fy5)
Recessive Lu(a–b–)	LU system
Dominant Lu(a–b–) [*In(Lu)*]	(Weak LU; KN; IN; P1; MER2; AnWj)
X-Linked Lu(a–b–)	(Weak LU system; I; strong i)
K_0	KEL system (Kx increased)
K_{mod}	(Weak KEL system; Kx increased)
Kp(a + b–)	(Weak KEL system; Kx slightly increased)
Fy^x	(Weak Fy^b often requires adsorption/elution for detection, Fy3, Fy6 weak in homozygotes)
Recessive Jk(a–b–)	JK system
Dominant Jk(a–b–) [*In(Jk)*]	(Very weak JK antigens)
Gy(a–)	DO system
Hy–	Hy, Jo^a (Weak Gy^a, Do^b, DOYA, DOMR, and sometimes Jo^a)
McLeod	Kx (Weak KEL system)
Leach (Ge:–2,–3,–4)	GE system (Weak KEL system)
Gerbich (Ge:–2,–3,4)	Ge2, Ge3 [Weak KEL system (some)]
Yus (Ge:–2,3,4)	Ge2
Inab	CROM system
Dr(a–)	Dr^a (dramatically weak CROM system)
Helgeson	KN system, Cs^a
Vel–	ABTI (can be weak)
ABTI–	(Vel)

Biosynthetic pathways

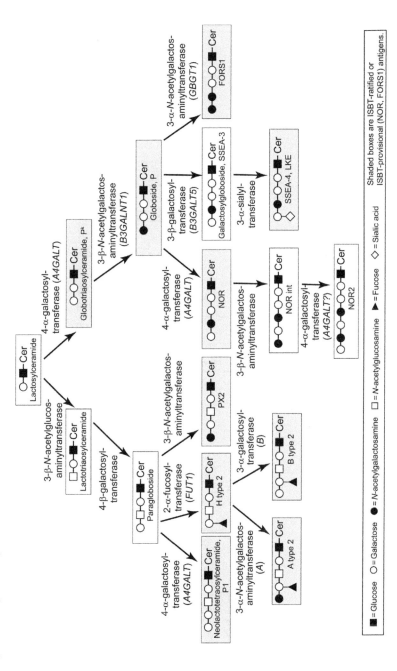

Antigens with lactosylceramide as a precursor

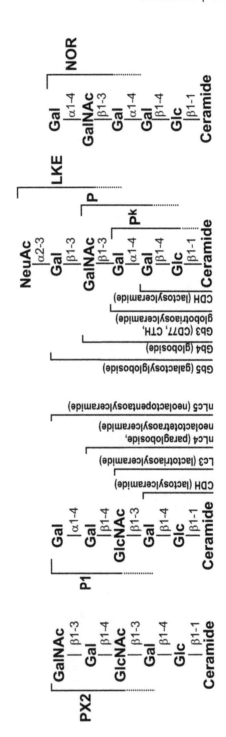

High prevalence antigens absent (and selected phenotypes) in certain ethnic populations

Phenotype	Population (Any = may be found in any population; >=more prevalent than)
AnWj–	Transient in any >>Israeli Arabs (inherited type)
At(a–)	Blacks
Cr(a–)	Blacks
Di(b–)	South Americans > Native Americans > Japanese
DISK–	Dutch > Europeans > Any
Dr(a–)	Jews from Bukhara > Japanese
En(a–)	Finns > Canadians > English > Japanese
Es(a–)	Mexicans, South Americans, Blacks
Fy(a–b–)	Blacks >> Arabs/Jews > Mediterraneans >> Caucasians
Ge:–2,–3 (Gerbich phenotype)	Papua New Guineans >> Melanesians >> Caucasians > Any
Ge:–2,3 (Yus phenotype)	Mexicans > Israelis > Mediterraneans > Any
Ge:–2,–3,–4 (Leach phenotype)	Any
GUTI–	Chileans
Gy(a–)	Eastern Europeans (Romany) > Japanese
hr^B–	Blacks
hr^S–	Blacks
Hy–	Blacks
IFC (Cr$_{null}$, Inab)	Japanese > Any
In(b–)	Indians > Iranians > Arabs
Jk(a–b–)	Polynesians >> Finns > Japanese > Any
Jo(a–)	Blacks

(Continued)

(Continued)

Phenotype	Population (Any = may be found in any population; >=more prevalent than)
Jr(a−)	Japanese > Asians > Europeans > Bedouin Arabs > Any
Js(b−)	Blacks
k−	Caucasians >> Any
K_0 (K_{null})	Reunion Islanders > Finns > Japanese > Any
K12 −	Caucasians
K14−	French-Cajuns
K22−	Israelis
KCAM−	Blacks >>> Any
Kn(a−)	Caucasians > Blacks > Any
Kp(b−)	Caucasians > Japanese
KUCI−	Native Americans
Lan−	Caucasians > Japanese > Blacks > Any
Lu(a−b−)	Any
Lu20−	Israelis
Lu21−	Israelis
LW(a−b−)	Transient in any >> inherited type in Canadians
LW(a−)	Balts
MAM−	Arabs > Any
MAR−	Finns > Any
McC(a−)	Blacks > Caucasians > Any
MER2−	Indian Jews, Turks, Portugese
M^kM^k	Swiss > Japanese
O_h (Bombay)	Indians > Japanese > Any
Ok(a−)	Japanese
P−	Japanese > Finns > Israelis > Any
Para-Bombay	Reunion Islanders > Indians > Any

(Continued)

(Continued)

Phenotype	Population (Any = may be found in any population; >=more prevalent than)
PEL–	French-Canadians
PP1P^k–	Swedes > Amish > Israelis > Japanese > Any
Sl(a–)	Blacks >> Caucasians > Any
Tc(a–b+c–)	Blacks
Tc(a–b–c+)	Caucasians
SERF–	Thais
U– and S–s–U+^{var}	Blacks
UMC–	Japanese
Vel–	Swedes > Any
WES(b–)	Finns > Blacks > Any
Yk(a–)	Caucasians > Blacks > Any
Yt(a–)	Arabs > Jews > Any

Low-prevalence antigens present in certain ethnic populations

Phenotype	Population (Any = may be found in any population; >= more prevalent than)
An(a+)	Finns
Be(a+)	Germans > Poles
Bp(a+)	English, Italians
Cl(a+)	Scottish, Irish
Crawford+	Blacks > Hispanic
C^W+	Latvians > Finns > Caucasians
C^X+	Finns > Caucasians > Somalis
DAK+	Blacks >>> Caucasians
DANE+	Danes > Europeans

(Continued)

(Continued)

Phenotype	Population (Any = may be found in any population; >= more prevalent than)
Dantu+	Blacks
Di(a+)	South American Indians > Japanese > Native Americans > Chinese, Poles
Evans+	Celts >>> Any
E^W+	Germans > Any
FPTT+	Any
FORS1+	English
Fr(a+)	Mennonites
Go(a+)	Blacks
HAG+	Israelis
He+	Xhosas > Blacks
Hg(a+)	Welsh > Australians
Hil+	Chinese > Any
Hop+	Thais > Any
In(a+)	Arabs > Iranians > Indians > Any
JAL+	English, French-speaking Swiss, Brazilians, Blacks
Jn(a+)	Poles, Slovaks
Js(a+)	Blacks
K+	Arabs > Iranian Jews > Caucasians > Any
K24+	French-Cajuns
Kn(b+)	Caucasians > Blacks
Kp(a+)	Caucasians > Any
Kp(c+)	Japanese
KREP+	Poles
Ls(a+)	Blacks > Finns > Any
Lu14+	English > Danes > Any
LW(b+)	Estonians > Finns > Balts > Europeans
MARS+	Choctaw tribe of Native Americans
M^c+	Europeans
M^g+	Swiss > Sicilians > Any
Mi(a+)	Thais > Taiwanese > Chinese > Any
MINY+	Thais > Taiwanese > Chinese > Any
Mit+	Western Europeans

(Continued)

(Continued)

Phenotype	Population (Any = may be found in any population; >= more prevalent than)
Mo(a+)	Belgians, Norwegians
Mt(a+)	Thais > Swiss > Caucasians > Blacks
Mur+	Thais > Taiwanese > Chinese > Any
MUT+	Chinese > Any
NFLD+	French Canadians, Japanese
NOR+	Poles
Ny(a+)	Norwegians > Swiss > Any
Or+	Japanese, Australians, Blacks, Jamaicans
Os(a+)	Japanese
Rd+	Danes > Canadians > Jews > Blacks > Any
Rh32+	Blacks > Caucasians > Japanese
Rh33+	Germans > Caucasians
Rh35+	Danes
Rh42+	Blacks
SAT+	Japanese
Sc2+	Mennonites > Northern Europeans
St(a+)	Japanese > Asian > Caucasians >>> Any
STEM+	Blacks
Tc(b+)	Blacks
Tc(c+)	Caucasians
TSEN+	Thais > Any
Ul(a+)	Finns > Japanese
V+	Blacks
Vr+	Dutch
VS+	Bantus > Blacks
Vw+	Swiss > Caucasians
WARR+	Native Americans
Wb+	Welsh > Australians
Wd(a+)	Hutterites
WES(a+)	Finns > Blacks
Wu+	Scandinavians > Dutch > Blacks
Yt(b+)	Arabs > Jews > Europeans > Any

Clinical significance of some alloantibodies to blood group antigens [3,4]

Usually clinically significant	Sometimes clinically significant	Clinically insignificant if not reactive at 37°C	Generally clinically insignificant
A and B	AnWj	A1	Chido/Rodgers
Diego	At^a	H	Cost
Duffy	Colton	Le^a	JMH
H in O$_h$	Cromer	Lutheran	HLA/Bg
Kell	Dombrock	M, N†	Knops
Kidd	Gerbich	P1	Le^b
P	Indian	Sd^a	Xg^a
PP1P^k	Jr^a		
Rh	Kx		
S, s, U	Lan		
Vel	Landsteiner-Wiener		
	Scianna		
	Yt^a		

†Rule out that 37°C reactivity is not due to carry-over agglutination.

Characteristics of some blood group alloantibodies

Antibody specificity	IgM (direct)	IgG (indirect)	Clinical transfusion reaction	HDFN
ABO	Most	Some	Immediate Mild to severe	Common Mild to moderate
Rh	Some	Most	Immediate/delayed Mild to severe	Common Mild to severe
Kell	Some	Most	Immediate/delayed	Sometimes

(Continued)

(Continued)

Antibody specificity	IgM (direct)	IgG (indirect)	Clinical transfusion reaction	HDFN
			Mild to severe	Mild to severe
Kidd	Few	Most	Immediate/delayed Mild to severe	Rare; mild
Duffy	Rare	Most	Immediate/delayed Mild to severe	Rare; mild
M	Some	Most	Delayed (rare)	Rare; usually mild
N	Most	Rare	None	None
S	Some	Most	Delayed/mild	Rare; mild to severe
s	Rare	Most	Delayed/mild	Rare; mild to severe
U	Rare	Most	Immediate/delayed Mild to severe	Rare; mild to severe
P1	Most	Rare	None (rare)	None
Lutheran	Some	Most	Delayed	Rare; mild
Le^a	Most	Few	Immediate (rare)	None
Le^b	Most	Few	None	None
Diego	Some	Most	Delayed; None to severe	Mild to severe
Colton	Rare	Most	Delayed; mild	Rare; mild to severe
Dombrock	Rare	Most	Immediate/delayed Mild to severe	Rare; mild
LW	Rare	Most	Delayed; none to mild	Rare; mild
Yt^a	Rare	Most	Delayed (rare); mild	None
I	Most	Rare	None	None
Ch/Rg	Rare	Most	Anaphylactic (3)	None
JMH	Rare	Most	Delayed (rare)	None
Knops	Rare	Most	None	None
Xg^a	Rare	Most	None	None

Antigen-negative prevalence for some polymorphic antigens

System	Antigen	Prevalence of antigen-negativity	
		Caucasian	Black
Rh	D	0.15	0.08
	C	0.32	0.73
	E	0.71	0.78
	c	0.20	0.04
	e	0.02	0.02
	f	0.35	0.08
	C^W	0.98	0.99
	V	>0.99	0.70
	VS	>0.99	0.73
MNS	M	0.22	0.26
	N	0.30	0.25
	S	0.48	0.69
	s	0.11	0.06
	M–S–	0.15	0.19
	M–s–	0.01	0.02
	N–S–	0.10	0.16
	N–s–	0.06	0.02
P1PK	P1	0.21	0.06
Lewis	Le^a	0.78	0.77
	Le^b	0.28	0.45
Lutheran	Lu^a	0.92	0.95
	Lu^b	<0.01	<0.01
Kell	K	0.91	0.98
	k	0.002	<0.001
	Kp^a	0.98	>0.99
	Kp^b	<0.01	<0.01
	Js^a	>0.99	0.80
	Js^b	<0.001	0.01
Duffy	Fy^a	0.34	0.90
	Fy^b	0.17	0.77
Kidd	Jk^a	0.23	0.08
	Jk^b	0.26	0.51
Dombrock	Do^a	0.33	0.45
	Do^b	0.18	0.11
Colton	Co^a	<0.001	<0.001
	Co^b	0.90	0.90

To determine the average number of blood donor samples to screen when searching for antigen-negative units, multiply the antigen-negative prevalence for each antigen, and divide the resultant percentage into 100.

For example, to screen for blood for a patient with anti-Fy^a + anti-K + anti-S:

$(0.34) \times (0.91) \times (0.48)$	=	0.148521
0.148521×100	=	14.851
$100 \div 14.851$	=	6.7 rounds up to 7

Thus, approximately 1 in 7 (or 14 in 100) donor samples will lack Fy^a, K, and S antigens.

Potentially useful information for problem-solving in immunohematology

Available information	Considerations
Patient demographics	Diagnosis, age, sex, ethnicity, transfusion, and/ or pregnancy history, drugs, IV fluids (Ringer's lactate, IV-IgG, Rh-immune globulin, other plasma-containing products, anti-lymphocyte globulin (ALG), anti-thymocyte globulin (ATG), infections, malignancies, hemoglobinopathies, stem cell transplantation
Initial serological results	ABO, Rh, DAT, phenotype, antibody detection results, autologous control, cross-match results
Hematology/chemistry values	Hemoglobin, hematocrit, bilirubin, LDH, reticulocyte count, haptoglobin, hemoglobinuria, albumin:globulin ratio, RBC morphology
Sample characteristics	Site and technique of collection, age of sample, anticoagulant, hemolysis, lipemic, color of serum/ plasma, agglutinates/aggregates in the sample
Other	Check records in current and previous institutions for previously identified antibodies
Antibody identification	Auto control, phase of reactivity, potentiator (saline, albumin, LISS, PEG), reaction strength, effect of chemicals on antigen (proteases, thiol reagents), pattern of reactivity (single antibody or mixture of antibodies), characteristics of reactivity (mixed field, rouleaux), hemolysis, preservatives/antibiotics in reagents

Alloantibodies that may have *in vitro* hemolytic properties

Anti-A, -B, -A,B, -H (in O_h people), -I, -i, -Le^a, -Le^b, -PP1P^k, -P, -Jk^a, -Jk^b, -Jk3, -Ge3, -Vel, and rare examples of anti-Sc1, -Lan, -Jr^a, -Co3, -Emm, and -Milne.

Conditions associated with suppression (sometimes total) or with alteration of antigen expression

Condition	Antigens affected
Pregnancy	ABO; H; I; LE; LW; P1; JMH; Sd^a; some Jk^a; Gy^a; AnWj
Carcinoma	ABO; H; I; P1; KN
Leukemia	ABO; H; I; RH; Yt^a; CO (chromosome 7 rearrangements)
Infection	ABO; A with appearance of Tn, A with appearance of acquired B; T activation; H; I; K
Hodgkin's lymphoma	ABO; H; LW
LADII (CDG-II)	ABO; H; LE
PNH	CROM; YT; DO; MER2; JMH; Emm
CDA	CO; LW; IN
AIHA	En^a; U; RH; KEL; JK; DI; LW; SC; GE; Vel; AnWj
SLE	CH/RG; KN; Yt^a
Hemopoietic stress	ABO; H; I (concomitant increased expression of i)
Diseases with increased clearance of immune complexes (e.g. AIDS)	KN
Old age	ABO; H; JMH
South-East Asian ovalocytes	En^a, S, s, U; Di^b, Wr^b; D, C, e; Kp^b; Jk^a, Jk^b; Xg^a; LW; Sc1^5

Causes of apparent *in vivo* hemolysis

Immune

ABO incompatibility
Clinically significant alloantibody
Anamnestic alloantibody response
Autoimmune hemolytic anemia
Cold agglutinin disease
HDFN
Drug-induced hemolytic anemia

Polyagglutination (sepsis T-active plasma)
Paroxysmal cold hemoglobinuria
TTP/HUS*-microangiopathic process

*Thrombotic thrombocytopenic purpura/hemolytic uremic syndrome.

Non-immune

Mechanical
Poor sample collection
Small-bore needle used for transfusion
Excessive pressure during transfusion
Malfunctioning blood warmer
Donor blood exposed to excessive heat or cold
Urinary catheter
Crush trauma
Prosthetic heart valves
Aortic stenosis
March hemoglobinuria

Microbial
Sepsis
Malaria
Contamination of donor blood

Chemical
Inappropriate solutions infused
Drugs infused
Serum phosphorus <0.2 mg/dL
Water irrigation of bladder
Azulfidine
Dimethyl sulfoxide
Venom (snake, bee, Brown Recluse spider)[6]
Certain herbal preparations, teas, enemas

Inherent RBC abnormalities
Paroxysmal nocturnal hemoglobinuria
Sickle cell anemia
Spherocytosis
Hemoglobin H
G6PD deficiency (in recipient or donor)

Warm autoantibodies to the following blood group antigens have been described[7,8] (listed alphabetically)

A, B	JMH
AnWj	K, k, Kp^b, Js^b, K13, Kell protein
Co3	Kx
Di^b, Wr^b	LW^a, LW^{ab}
En^a, U, M, N, S, Pr	Rh, in particular e, Rh17
Fy^b	Rx

Ge2, Ge3	Sc1
H	Vel
I^T	Yt^a
Jk^a, Jk^b, Jk3	Xg^a

Target antigen suppression

In some autoimmune cases, the target antigen may be weakened to the extent that the patient's RBCs are negative in the DAT. The following antigens have been implicated[9] (listed alphabetically):

AnWj	Ge3	JMH	Rh	U
Co3	Jk^a	Kp^b	Sc1	Vel
En^a	Jk^b	LW	Sc3	

Drugs associated with immune hemolytic anemia and/or positive DAT in which drug-dependent antibodies were detected[8,10–13]

Drugs can cause the production of antibodies that may be against the drug itself, RBC membrane components or an antigen formed by the drug and the RBC membrane. Such antibodies may cause a positive DAT, immune hemolytic anemia or both. A drug may also cause a positive DAT through non-immunologic protein adsorption onto the RBC. The mechanisms involved eliciting an immune response to drugs are not well-understood, and various theories have been proposed.

In the table[10], when an antibody is indicated to react by two methods, it does not necessarily mean that all examples of antibodies to that drug were detected by both methods.

Drug (alternative name)	Therapeutic category	HA	Positive DAT	Method of detecting serum antibody		Reactive without drug added *in vitro*
				Drug-coated RBCs	Serum+ drug + RBCs	
Aceclofenac	NSAID	✓	✓	–	✓	–
Acetaminophen (Paracetamol)	NSAID	✓	✓	–	✓	–
Acyclovir	Anti-viral	✓	✓	✓	–	–
Aminopyrine (Piramidone)	NSAID	✓	–	✓	–	–
Amoxicillin	Anti-microbial	✓	✓	✓	–	–

(Continued)

(Continued)

Drug (alternative name)	Therapeutic category	HA	Positive DAT	Method of detecting serum antibody		Reactive without drug added *in vitro*
				Drug-coated RBCs	Serum+ drug + RBCs	
Amphotericin B	Anti-microbial	✓	✓	–	✓†	–
Ampicillin	Anti-microbial	✓	✓	✓	✓	–
Antazoline	Anti-histamine	✓	✓	–	✓	–
Aspirin	Analgesic, antipyretic, anti-inflammatory	✓	–	–	✓	–
Azapropazone (Apazone)	Anti-inflammatory, analgesic	✓	✓	✓	–	✓
Buthiazide (Butizide)	Diuretic, anti-hypertensive	✓	✓	–	✓†	–
Carbimazole	Anti-thyroid	✓	✓	✓	✓	✓
Carboplatin	Anti-neoplastic	✓	✓	✓	✓	✓
Carbromal	Sedative, hypnotic	–	✓	✓	–	–
Catechin [(+)-Cyanidanol-3] (Cianidanol)	Anti-diarrheal	✓	✓	✓	✓†	✓
Cefamandole	Anti-microbial	✓	✓	✓	–	–
Cefazolin	Anti-microbial	✓	✓	✓	–	–
Cefixime	Anti-microbial	✓	–	✓	✓	–
Cefotaxime	Anti-microbial	✓	✓	✓	✓	✓^
Cefotetan	Anti-microbial	✓	✓	✓¶	✓	✓
Cefoxitin	Anti-microbial	✓	✓	✓	✓	✓
Cefpirome	Anti-bacterial	–	✓	–	✓	–
Ceftazidime	Anti-microbial	✓	✓	✓	✓	✓
Ceftizoxime	Anti-microbial	✓	✓	✓	✓	✓^
Ceftriaxone	Anti-microbial	✓	✓	–	✓†	✓^
Cefuroxime	Anti-bacterial	✓	✓	✓	–	–

(Continued)

Drug (alternative name)	Therapeutic category	HA	Positive DAT	Method of detecting serum antibody		Reactive without drug added *in vitro*
				Drug-coated RBCs	Serum+ drug + RBCs	
Cephalexin	Anti-microbial	✓	✓	✓[¶]	–	–
Cephalothin	Anti-microbial	✓	✓	✓[¶]	✓	–
Chloramphenicol	Anti-bacterial	✓	✓	✓	–	✓
Chlorinated hydrocarbons	Insecticides	✓	✓	✓	✓	✓
Chlorpromazine	Anti-emetic, anti-psychotic	✓	✓	✓	–	✓
Chlorpropamide	Anti-diabetic	✓	✓	–	✓	✓[^]
Cimetidine[14]	Anti-ulcerative	✓	✓	✓	✓	–
Ciprofloxacin	Anti-bacterial	✓	✓	–	✓	✓
Cisplatin (Cisdiaminodi-chloroplatinum)	Anti-neoplastic	✓	✓	✓[¶]	✓	–
Cloxacillin	Anti-bacterial	–	✓	–	–	✓
Cyclofenil	Gonad-stimulating principle	✓	✓	–	✓	✓
Cyclosporin (Cyclosporine)	Immuno-suppressant	✓	✓	✓	–	✓
Dexchlor-pheniramine maleate (Chlorpheniramine)	Anti-histaminic	✓	✓	–	✓	–
Diclofenac	NSAID	✓	✓	✓	✓[†]	✓[^]
Diethylstilbestrol (Stilboestrol)	Estrogen	✓	✓	–	✓	–
Dipyrone	NSAID	✓	✓	✓	✓	–
Erythromycin	Anti-microbial	✓	✓	✓	–	–
Etodolac	NSAID	✓	✓	–	✓[†]	–
Ethambutol	Anti-bacterial	✓	✓	✓	✓	–
Fenoprofen	NSAID	✓	✓	–	✓	✓[^]

(Continued)

(Continued)

Drug (alternative name)	Therapeutic category	HA	Positive DAT	Method of detecting serum antibody		Reactive without drug added *in vitro*
				Drug-coated RBCs	Serum + drug + RBCs	
Fluconazole	Anti-fungal	✓	✓	✓	✓	–
Fluorescein	Injectable dye	✓	✓	✓	✓	✓∧
Fluorouracil	Anti-neoplastic	✓	✓	–	✓	–
Furosemide	Diuretic	–	✓	–	✓	–
Glafenine (Glaphenine)	Analgesic	✓	✓	–	–	✓
Hydralazine	Anti-hypertensive	✓	✓	✓	–	–
Hydro chlorothiazide	Diuretic	✓	✓	✓	✓	✓∧
Hydrocortisone[15]	Glucocorticoid	✓	✓	✓	✓	–
9-Hydroxy-methyl-ellipticinium (Elliptinium acetate)	Anti-neoplastic	✓	✓	–	✓	–
Ibuprofen	NSAID	✓	✓	–	✓	✓
Imatinib mesylate	Anti-neoplastic	✓	✓	✓	–	–
Insulin	Anti-diabetic	✓	✓	✓	–	–
Isoniazid	Anti-microbial	✓	✓	✓	✓	–
Latamoxef (Moxalactam)	Anti-microbial	✓	✓	–	–	✓
Levofloxacin (Ofloxacin)	Anti-bacterial	✓	✓	✓	✓	✓
Mefloquine	Anti-microbial	✓	✓	✓	✓	✓∧
Melphalan	Anti-neoplastic	✓	–	–	✓	–
6-Mercaptopurine	Anti-neoplastic	✓	✓	✓	–	–
Methadone	Analgesic	–	✓	✓	–	–
Methotrexate	Anti-neoplastic, anti-rheumatic	✓	✓	✓	✓	✓

(Continued)

(Continued)

Drug (alternative name)	Therapeutic category	HA	Positive DAT	Method of detecting serum antibody		Reactive without drug added *in vitro*
				Drug-coated RBCs	Serum+ drug + RBCs	
Metrizoate-based radiographic contrast media		✓	✓	✓	✓	✓
Minocycline	Anti-bacterial	✓	✓	–	✓	–
Nabumetone analgesic	Anti-inflammatory	✓	✓	–	✓[†]	✓
Nafcillin	Anti-microbial	✓	✓	✓	✓[–]	–
Naproxen	Anti-inflammatory, analgesic, anti-pyretic	✓	✓	–	✓	–
Nifedipine[16]	Anti-hypertensive	✓	✓	–	✓	–
Nitrofurantoin	Anti-bacterial	✓	–	–	✓	–
Nomifensine[§]	Anti-depressant	✓	✓	–	✓[†]	✓^
Norfloxacin	Anti-microbial	–	✓	✓	–	–
Oxaliplatin	Anti-neoplastic	✓	✓	✓[¶]	✓	✓^
p-Aminosalicylic acid (PAS) (para-aminosalicylsaure)	Anti-microbial	✓	✓	–	✓	–
Penicillin G	Anti-microbial	✓	✓	✓	✓	–
Phenacetin (Acetophenetidin)	NSAID	✓	✓	–	✓	✓
Phenytoin (Fenitoine)	Anti-convulsant, anti-arrhythmic	✓	✓	✓	–	–
Piperacillin	Anti-microbial	✓	✓	✓	✓	✓^
Probenecid	Uricosuric	✓	✓	–	✓	✓^
Propyphenazone	NSAID	✓	✓	–	✓	–
Pyrazinamide	Anti-bacterial	✓	✓	✓	✓	–
Pyrimethamine (Pirimetamine)	Anti-microbial	✓	✓	✓	–	–
Quinidine	Anti-arrhythmic, anti-microbial	✓	✓	✓	✓	✓^

(Continued)

(Continued)

Drug (alternative name)	Therapeutic category	HA	Positive DAT	Method of detecting serum antibody		Reactive without drug added *in vitro*
				Drug-coated RBCs	Serum+ drug + RBCs	
Quinine	Anti-microbial	✓	–	–	✓	✓
Ranitidine	Anti-ulcerative	✓	✓	✓	✓	–
Rifabutin	Anti-bacterial	✓	✓	–	✓	–
Rifampin (Rifampicin)	Anti-bacterial	✓	✓	✓	✓	✓^
Stibophen	Anti-microbial	✓	✓	–	✓	–
Streptokinase	Thrombolytic	✓	✓	✓	–	✓
Streptomycin	Anti-microbial	✓	✓	✓	✓	✓
Sulfasalazine	Anti-inflammatory	✓	✓	–	✓	–
Sulfisoxazole	Anti-bacterial	✓	✓	✓	✓	–
Sulindac	Anti-inflammatory	✓	✓	✓	✓	✓^
Suprofen	NSAID	✓	✓	–	✓	✓^
Tartrazine	Colorant	✓	✓	✓	✓	–
Teicoplanin	Anti-microbial	✓	✓	–	✓	✓
Temafloxacin§	Anti-microbial	✓	✓	–	✓	–
Teniposide	Anti-neoplastic	✓	✓	–	✓	✓
Tetracycline	Anti-microbial	✓	✓	✓	–	–
Thiopental sodium	Anesthetic	✓	–	–	✓	–
Ticarcillin	Anti-microbial	✓	✓	✓	–	✓
Tolbutamide	Anti-diabetic	✓	✓	✓	–	–
Tolmetin	NSAID	✓	✓	–	✓	✓^
Triamterene	Diuretic	✓	✓	✓	✓	–
Trimellitic anhydride	Used in preparation of resins, dyes, adhesives, etc.	✓	–	✓	–	–
Trimethoprim and sulfamethoxazole	Anti-bacterial	✓	✓	✓	✓	✓

(Continued)

(Continued)

| Drug (alternative name) | Therapeutic category | HA | Positive DAT | Method of detecting serum antibody | | Reactive without drug added *in vitro* |
				Drug-coated RBCs	Serum+ drug + RBCs	
Vancomycin	Anti-bacterial	✓	✓	–	✓	–
Zomepirac	NSAID	✓	✓	–	✓	✓

HA = Hemolytic anemia; NSAID = Nonsteroidal anti-inflammatory drug.
†Positive or gives the strongest reactions when the drug metabolite is present.
§No longer manufactured.
¶Associated with nonimmunologic protein adsorption.
^Positive, possibly due to the presence of circulating drug or drug–antibody immune complexes.

Drugs associated with cases of immune hemolytic anemia and/or positive DAT caused by drug-independent antibodies (autoantibodies)

Drug (alternative name)	Therapeutic category	HA	Positive DAT	More evidence needed
Captopril	Anti-hypertensive	✓	✓	✓
Chaparral	Herbal	–	✓	✓
Cladribine (2-chlorode-oxyadenosine)	Anti-neoplastic	✓	✓	–
Fenfluramine	Anorexic	✓	✓	✓
Fludarabine	Anti-neoplastic	✓	✓	–
Interferon	Anti-neoplastic, anti-viral	✓	✓	✓
Interleukin-2	Anti-neoplastic	✓	✓	✓
Ketoconazole	Anti-fungal	✓	✓	✓

(Continued)

(Continued)

Drug (alternative name)	Therapeutic category	HA	Positive DAT	More evidence needed
Lenalidomide	Immunomodulatory	✓	✓	✓
Levodopa (L-dopa)	Anti-parkinsonian	✓	✓	–
Mefenamic acid	NSAID	✓	✓	–
Mesantoin (Mephenytoin)	Anti-convulsant	✓	✓	✓
Methyldopa	Anti-hypertensive	✓	✓	–
Nalidixic acid	Anti-bacterial	✓	✓	✓
Procainamide	Anti-arrhythmic	✓	✓	–
Tacrolimus	Immunosuppressant	✓	✓	✓

HA = Hemolytic anemia.

Drugs associated with the detection of non-immunologic protein adsorption onto RBCs

Drug (alternative name)	Therapeutic category	HA	DAT	Drug-dependent antibody(ies) also detected
Cefotetan	Anti-microbial	✓	✓	✓
Cephalothin	Anti-microbial	✓	✓	✓
Cisplatin	Anti-neoplastic	✓	✓	✓
Clavulanate potassium (Clavulanic acid)	β-Lactamase inhibitor	–	✓	–
Diglycoaldehyde (INOX)	Anti-neoplastic	–	✓	–
Oxaliplatin	Anti-neoplastic	✓	✓	✓
Sulbactam	β-Lactamase inhibitor	✓	✓	–
Suramin	Anti-helminthic, anti-protozoal	–	–	–
Tazobactam	β-Lactamase inhibitor	✓	✓	–

HA = Hemolytic anemia.

Blood group systems and their gene products

Carbohydrate based blood group systems

System	Gene product
ABO	N-acetylgalactosaminyltransferase (A glycosyltransferase) or galactosyltransferase (B glycosyltransferase)
P1PK	Galactosyltransferase
LE	Fucosyltransferase
H	Fucosyltransferase
I	N-acetylglucosaminyltransferase
GLOB	N-acetylgalactosaminyltransferase
FORS	N-acetylgalactosaminyltransferase

Blood group systems located on single pass membrane proteins

System	Gene product	Number of amino acids	N-terminus	Function in RBCs
MNS	Glycophorin A Glycophorin B	131 72	Exofacial Exofacial	Carrier of sialic acid, which contributes to the negatively charged barrier. Complement regulation. Facilitates membrane assembly of band 3.
GE	Glycophorin C Glycophorin D	128 107	Exofacial Exofacial	Carrier of sialic acid (see MNS). Interacts with band 4.1 and p55 in RBC membrane to maintain RBC shape.
KEL	Kell glycoprotein	732	Cytoplasmic	Zinc endopeptidase that cleaves big endothelin.
LU	Lutheran glycoprotein	597	Exofacial	Binds laminin.
XG	Xgª glycoprotein	180	Exofacial	Unknown.
	CD99	163	Exofacial	Adhesion molecule.
LW	LW glycoprotein	241	Exofacial	Ligand for integrins.
IN	CD44	341	Exofacial	Adhesion molecule that binds to hyaluronic acid.
KN	CD35 (CR1)	1998	Exofacial	Complement regulation.
Okª	CD147	248	Exofacial	Possible cell–cell adhesion.
SC	ERMAP	446	Exofacial	Possible adhesion.

Blood group systems located on multipass membrane proteins

System	Gene product	Number of amino acids	Predicted number of spans	Function in RBCs
RH	RhD protein	417	12	The Rh/RhAG/band 3/ complex contributes to the RBC membrane structure and transports gases.
	RhCE protein	417	12	
RHAG	RhAG	409	12	
FY	Fy glycoprotein (DARC)	336 (major product)	7^	Cytokine receptor for pro-inflammatory cytokines.
DI	Band 3 (AE1)	911	14	Anion transport (HCO_3^-/ Cl^-) essential for respiration).
CO	AQP1 (CHIP-1)	269	6	Water/CO_2 transport.
JK	Urea transporter	389	10	Urea transport.
XK	Kx glycoprotein	444	10	Possible neurotransmitter. Possible amino acid transporter.
RAPH	Tetraspanin	253	4	Cell adhesion, proliferation, differentiation.
GIL	AQP3	342	6	Glycerol/water/urea transport.
JR	ABCG2; breast cancer resistance protein	655	6	ATP-dependent transporter of a diverse range of substrates.
LAN	ABCB6	842	11	Mitochondrial transporter essential for heme biosythesis.

^=N-terminus oriented to exofacial surface, C-terminus to cytoplamic surface. All others are predicted to be oriented with both their N- and C-termini to the cytoplasmic aspect of the RBC membrane.

Blood group systems carried on glycosylphosphatidylinositol-linked proteins

System	Gene product	Number of amino acids	Function in RBCs
YT	Acetylcholinesterase	557	Enzymatic
CROM	CD55 (DAF)	347	Complement regulation
DO	Do glycoprotein	314	Possibly enzymatic
JMH	CD108	646	Adhesion molecule involved in cell migration

Blood group systems located on proteins adsorbed from the plasma

System name	Component	Antigen location	Function in RBCs
CH/RG	C' component 4 (C4)	C4d fragment	Complement regulation

Proteins altered on Rh$_{null}$ RBCs

Protein	Gene location	M_r	Copies per RBC	Comments
RhD/RhCE	1p36.11	30,000–32,000	100,000–200,000 for RhD/RhCE combined	Absent
RhAG	6p21.3	45,000–100,000	100,000–200,000	Absent
CD47	3q13	47,000–52,000	10,000–50,000	Reduced (~25% of normal)
LW	19p13.2	37,000–43,000	3,000–5,000	Absent
GPB	4q31.22	20,000–25,000	200,000	Reduced (30% of normal)
Duffy (Fy5)	1q23.2	35,000–45,000	6,000–13,000	Fy5 antigen absent

Blood Group Proteins, M_r, abundance, and selected reactive monoclonal antibodies (MAbs)

Blood group system	M_r (SDS-PAGE)	Approximate copy number/RBC	Conditions for immunoblotting	MAbs active by immunoblotting	Serology/flow cytometry MAbs
MNS (GPA)	43,000	800,000	R or NR	Sigma E3, many clones	BRIC256
MNS (GPB)	20–25,000	200,000	R or NR	Sigma E3, R1.3, Anti-N	
RH	30–32,000	100,000– 200,000, RhD and RhCE combined	NR Use 8 M Urea Do not boil	LOR15C9 (anti-D)	
LU	85,000	1,500–4,000	NR	BRIC224 (D1), BRIC221 (D4)	BRIC224 (D1), BRIC221 (D4)
KEL	93,000	3,500–18,000	NR	C-10 (R or NR), 195031 (NR only)	BRICs 18, 68, and 203, 4B10
FY	35–45,000	6,000–13,000	R or NR	MIMA107, MIMA29	Polyclonal antibodies available
JK	43,000	14,000	R or NR	MIMA128	
DI	95–105,000	1,000,000	R or NR	N-terminal: BIII-136, 2D5, BRIC170 C-terminal: BRIC155	BRIC6, BRAC18
YT	160,000	7,000–10,000		None available	Many clones
XG (Xgª)	22–29,000	9,000	R or NR	NBL-1	
XG (CD99)	32,500	200–2,000	R or NR	12E7, MEM-131, BANRS1, MSG-B1	Many clones
SC	60–68,000	Not determined	NR	IgSF: 6F8, YS-6, C8. Intracellular: 10C132	C8
DO	47–58,000	Not determined	NR	MIMA52	MIMA52
CO	28,000 Glycosylated 40–60,000	120,000–160,000	R	Loop E: 7D11. C-terminal: 1A5F6, MIMA136	7D11

(Continued)

(Continued)

Blood group system	M_r (SDS-PAGE)	Approximate copy number/RBC	Conditions for immunoblotting	MAbs active by immunoblotting	Serology/flow cytometry MAbs
LW	37–43,000	D+ 4,400 D- 2,800	NR	BS56, BS86	BS46, BS56, BS87
XK	37,000	1,000		None available	None available
GE (GPC)	40,000	135,000	R	BRIC4, BRIC10, E5, 1H3, 3H2007, BGRL-100, MIMA81	Many clones
GE (GPD)	30,000	50,000	R	BRAC11, (Human anti-Ge2)	BRAC11
CROM (CD55)	60–70,000	20,000	NR	CCP1: BRICs 128, 220, 230. CCP2: BRIC110. CCP3: BRIC216. CCP4: MEM-118. CCP2/3: MIMA28, MIMA69	CCP1: BRICs 128, 220, 230. CCP2: BRIC110. CCP3: BRIC216. CCP4: MEM-118.
KN (CR1)	220,000	20–1,500		LHR A-C: To5	Many clones
IN (CD44)	80,000	2,000–5,000	NR or R	NR: all clones R: only KZ-1, Hermes-3	Many clones
OK (CD147)	35–69,000	3,000	NR	D1: MEM-M6/1, HIM6 D2: MEM6/6, MIMA144	Many clones
RAPH (CD151)	40,000	Not determined	NR	IIG5a	IIG5a, TS151
JMH (CD108)	68–76,000	Not determined	NR	Sema7A: MEM-150 IgSF: 3D3, 1G1 Others: 9L98, 9G441.	MEM-150, KS-2, 310829
RHAG	45–100,000	100,000–200,000	NR Use 8 M Urea Do not boil	LA18.18, 2D10, MIMA77	

R = Reducing conditions; NR = Non-reducing conditions; D# = Domain number.

Changes in numbering of nucleotides and amino acids

For all alleles, the numbering for nucleotides and amino acids follows the ISBT system, i.e., nucleotides are counted as #1 being the "A" of the initiating "AUG," and amino acids are counted as #1 from the initiating methionine. This ISBT consistency policy means that the numbers for some nucleotides and amino acids may differ from those published.

System	Nucleotide change	System	Amino acid change
KEL	−120	MNS	+19
KN	−27	FY	+2 from minor product
Kx	−82	LW	+30
		CROM	+34

Some causes of pseudo-discrepancies between genotype and phenotype

The gene is present, but the expected product is not detectable in the RBC membrane.

Event	Mechanism	Blood group phenotype
Transcription	Nt change in GATA box	Fy(b−) or Fy(a−)
Alternative splicing due to nt change in splice site	Partial/complete skipping of exon	S–s–; Gy(a–)
	Deletion of nts	Dr(a–)
Premature stop codon	Deletion of nt(s)→frame-shift	Fy(a–b–); D–; Rh$_{null}$
	Insertion of nt(s)→frame-shift	Ge: −2,−3,−4; Gy(a–)
	Nt change	D–; Co(a–b–); Fy(a–b–); r′; Gy(a–); K$_0$; McLeod
Amino acid change	Missense mutation	D–; Rh$_{null}$; K$_0$; McLeod
Reduced amount of protein	Missense mutation	Fy^X, Co(a–b–)
Hybrid genes	Cross-over	GP.Vw; GP.Hil; GP.TSEN
	Gene conversion	GP.Mur; GP.Hop; D– –; R$_0$^{Har}
Interacting protein	Absence of RhAG	Rh$_{null}$
	Absence of Kx	Kell antigens are weak
	Absence of aas 59 to 76 of GPA	Wr(b–)
	Absence of protein 4.1	Ge antigens are weak
Modifying gene	EKLF [In(Lu)]	Lu(a–b–)
	In(Jk)	Jk(a–b–)

Chromosomal location of genes encoding or influencing the expression of blood groups

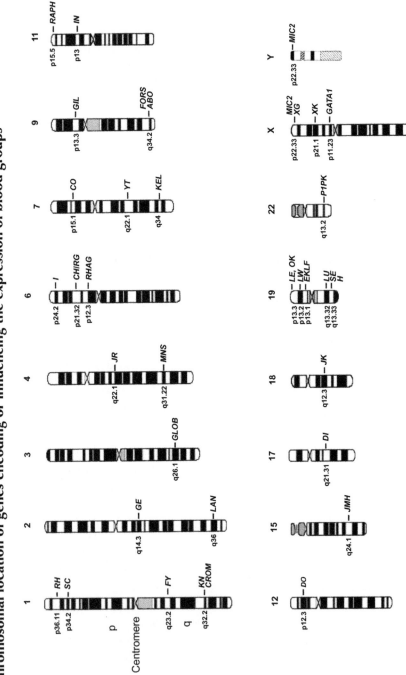

Useful definitions

Absorbed	From; away
Adsorbed	Onto
	Thus, an antibody is *absorbed* from serum, but *adsorbed* onto RBCs. Another definition is that *absorbed* is a nonspecific term (as in "absorbed" by a sponge), while *adsorbed* is a specific reaction.
Allele	Alternative form(s) of a *gene* at a given locus (antigens cannot be allelic).
Antithetical	Refers to *antigens* produced by alleles (alleles cannot be antithetical).
Haplotype	A set of alleles of a group of closely linked genes, which are usually inherited together. People have haplotypes, RBCs do not.
Propositus	Singular male or index case (singular) regardless of sex.
Propositi	Plural male or index cases (plural) regardless of sex.
Proposita	Singular female.
Propositae	Plural female.
Proband	Index case regardless of sex.
Probands	Plural for index cases regardless of sex.
Transition	Change of purine (A, G) to purine or pyrimidine (C, T) to pyrimidine.
Transversion	Change between purine and pyrimidine (A or G to C or T).
Missense mutation	Nucleotide change leading to a change of amino acid (nonsynonymous).
Nonsense mutation	Nucleotide change leading to a stop codon.
Silent mutation	Nucleotide change that, due to redundancy in the genetic code, does not change the amino acid (synonymous).
Frameshift mutation	A change in DNA that occurs when the number of nucleotides inserted or deleted is not a multiple of three, so that every codon beyond the point of insertion or deletion is shifted during translation. This results in a novel sequence of amino acids and sooner or later a stop codon.
Northern blot	Analysis of RNA.
Southern blot	Analysis of DNA.
Western blot	Analysis of proteins.

Some lectins and their simple specificities[17,18]

Lectin	Common name	Carbohydrate-binding specificity
Arachis hypogaea	Peanut	D-Galβ(1–3)GalNAc > α-D-Gal
Dolichos biflorus	Horsegram	α-D-GalNAc >> α-D-Gal
Glycine max	Soybean	α-D-GalNAc > β-D-GalNAc > α-D-Gal
Griffonia simplicifolia I^	GS1	α-D-Gal > α-D-GalNAc
Griffonia simplicifolia II^	GS2	α-D-GlcNAc = β-D-GlcNAc
Helix pomatia	Edible snail	α-D-GalNAc > α-D-GlcNAc > α-D-Gal
Leonurus cardiaca	Motherwort	α/β-D-GalNAc
Phaseolus lunatus	Lima bean	α-D-GalNAc
Salvia horminum	Clary^^	α-D-GalNAc > β-D-GalNAc
Salvia sclarea	Clary^^	α-D-GalNAc
Ulex europaeus I	Gorse/furze	α-L-Fuc
Vicia cretica		D-Gal

^Previously known as *Bandeiraea simplicifolia* (BS) lectins.
^^Both *Salvia horminum* and *Salvia sclarea* are commonly known as clary, but they are botanically different.

Polyagglutination types, and the expected reactions with group O^ RBCs and lectins[18,19]

Lectin	Acquired B	Type of polyagglutination										
		T	Tk	Th	Tx	Tn	Tr	Cad	NOR	VA	HEMPAS	HbM^^
Arachis hypogaea	0	+	+	+	+	0	↓	0	0	0	0	↓↓
Dolichos biflorus	↓	0	0	0	0	+	0	+	0	0	0	0
Glycine max	0	+	0	0	0	+	+	0	0	0	0	+
GSI	+	0	0	0	0	+	+	0	0	0	0	0
GSII	0	0	+	0	0	0	+	0	0	0	0	+
Helix pomatia	+	+	0	NT	NT	+	↓	+	0	+	+	+
Leonurus cardiaca	0	0	0	0	0	0	NT	+	0	0	0	0
Phaseolus lunatus	+	0	0	0	0	0	0	0	0	0	0	NT
Salvia horminum	0	0	0	0	0	+	↓	+	0	0	0	↓↓
Salvia sclarea	0	0	0	0	0	+	↓	0	0	0	0	0
Ulex europaeus	+	↑↑	↓	+	+	↑	+	↓	+	↓	↓	↑↑
Vicia cretica	0	+	0	+	0	0	NT	0	0	0	0	↓↓

↓ (↓↓) Weaker than normal RBCs; ↑ (↑↑) Stronger than normal RBCs.
+ Agglutination; 0 No agglutination.
^ Except acquired B, which occurs on group A; ^^ HbM-Hyde Park.

Types of Polyagglutination

T
Neuraminidase made by some organisms (*Vibrio cholerae*, *Clostridium perfringens*, pneumococci, influenza virus) cleaves sialic acid (NeuAc) from RBC-bound disialylated alkali-labile tetrasaccharides, leaving Galβ1-3GalNAc-Ser/Thr, which is recognized by anti-T.

Tk
β-Galactosidases produced by some organisms (*Bacteroides fragilis*, *Aspergillus niger*, *Serratia marcescens*, *Candida albicans*) cleaves Gal from GlcNAc in complex carbohydrate structures including A, B, H, P1, I, active carbohydrate chains.

Diagrammatic representation of Cad, Tn, T, Tk, and acquired B

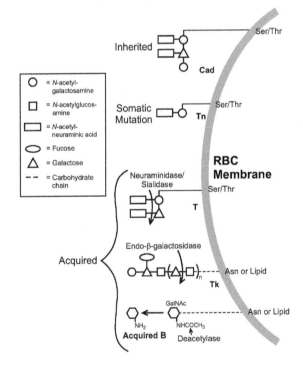

References

[1] Committee ARCNRLMM, 1993. In: Immunohematology Methods. American Red Cross National Reference Laboratory, Rockville, MD.
[2] Judd, W.J., 1994. In: Methods in Immunohematology, second ed. Montgomery Scientific Publications, Durham, NC.

[3] Daniels, G., et al., 2002. The clinical significance of blood group antibodies. Transfusion Med 12, 287–295.

[4] Reid, M.E., et al., 2000. Summary of the clinical significance of blood group alloantibodies. Semin Hematol 37, 197–216.

[5] Booth, P.B., et al., 1977. Selective depression of blood group antigens associated with hereditary ovalocytosis among Melanesians. Vox Sang 32, 99–110.

[6] Williams, S.T., et al., 1995. Severe intravascular hemolysis associated with brown recluse spider envenomation. A report of two cases and review of the literature. Am J Clin Pathol 104, 463–467.

[7] Garratty, G., 1999. Specificity of autoantibodies reacting optimally at 37°C. Immunohematology 15, 24–40.

[8] Petz, L.D., Garratty, G., 2004. In: Immune Hemolytic Anemias, second ed. Churchill Livingstone, Philadelphia.

[9] Arndt, P.A., et al., 1999. Serology of antibodies to second- and third-generation cephalosporins associated with immune hemolytic anemia and/or positive direct antiglobulin tests. Transfusion 39, 1239–1246.

[10] Garratty, G., Arndt, P.A., 2007. An update on drug-induced immune hemolytic anemia. Immunohematology 23, 105–119.

[11] Campbell, S., et al., 1992. Drug-induced positive direct antiglobulin tests and hemolytic anemia. CSTM Bulletin 4, 40–44.

[12] Johnson, S.T., et al., 2007. One center's experience: the serology and drugs associated with drug-induced immune hemolytic anemia – a new paradigm. Transfusion 47, 697–702.

[13] Roback, J.D., et al., 2011. In: Technical Manual, seventeenth ed. American Association of Blood Banks, Bethesda, MD.

[14] Arndt, P.A., et al., 2010. Immune hemolytic anemia due to cimetidine: the first example of a cimetidine antibody. Transfusion 50, 302–307.

[15] Martinengo, M., et al., 2008. The first case of drug-induced immune hemolytic anemia due to hydrocortisone. Transfusion 48, 1925–1929.

[16] Osafune, K., et al., 2000. Nifedipine and haemolytic anaemia. J Intern Med 247, 299–300.

[17] EY Laboratories I, 1992. In: *Lectins and Lectin Conjugates*. EY Laboratories, Inc, San Mateo, CA.

[18] Judd, W.J., 1992. Review: polyagglutination. Immunohematology 8, 58–69.

[19] Mollison, P.L., et al., 1997. In: Blood Transfusion in Clinical Medicine, tenth ed. Blackwell Science, Oxford, UK.

Y

Z

Printed and bound by CPI Group (UK) Ltd, Croydon, CR0 4YY

03/10/2024

01040420-0012